Crop Ferality *and* Volunteerism

Crop Ferality *and* Volunteerism

Edited by
Jonathan Gressel

OECD
ORGANISATION FOR ECONOMIC CO-OPERATION AND DEVELOPMENT

Taylor & Francis
Taylor & Francis Group
Boca Raton London New York Singapore

A CRC title, part of the Taylor & Francis imprint, a member of the
Taylor & Francis Group, the academic division of T&F Informa plc.

Published in 2005 by
CRC Press
Taylor & Francis Group
6000 Broken Sound Parkway NW, Suite 300
Boca Raton, FL 33487-2742

Printed in the United States of America on acid-free paper
10 9 8 7 6 5 4 3 2 1

International Standard Book Number-10: 0-8493-2895-0 (Hardcover)
International Standard Book Number-13: 978-0-8493-2895-4 (Hardcover)
Library of Congress Card Number 2004066419

Library of Congress Cataloging-in-Publication Data

Crop ferality and volunteerism / edited by Jonathan B. Gressel.
p. cm.
Includes bibliographical references (p.).
ISBN 0-8493-2895-0 (alk. paper)
1. Transgenic plants. 2. Crops--Genetic engineering. 3. Pollination. I. Gressel. Jonathan.

SB123.57C76 2005
631.5′23--dc22 2004066419

T&F informa
Taylor & Francis Group
is the Academic Division of T&F Informa plc.

Visit the Taylor & Francis Web site at
http://www.taylorandfrancis.com

and the CRC Press Web site at
http://www.crcpress.com

Foreword

This book evolved from the workshop on "Crop Ferality and Volunteerism: a Threat to Food Security in the Transgenic Era?" sponsored by the OECD Cooperative Research Program. The workshop was hosted by the Rockefeller Foundation Conference Center in Bellagio, Italy on May 24–28, 2004, with the cosupport of the Rockefeller Foundation as part of their Food Security Program, and with subsidiary support by BASF, Kumiai, and Monsanto, to whom all the participants are grateful.

The OECD cooperative research program for "Biological Resource Management" has existed since 1979. It focuses on work in four specific areas, one of which, "New Agricultural Products for Sustainable Farming and Industry," is supporting this workshop. This includes new approaches and possibilities for production of valuable materials and substances within agricultural systems, which could have large-scale effects at the farm level and for farming communities as a whole. The activities promoted by this program are postdoctoral fellowships (announced annually) and the organization of expert workshops, such as this.

The workshop on "Crop Ferality and Volunteerism" in relation to the increasing territory of commercial cultivation of transgenic plants was motivated by the growing concerns in the public on environmental issues. Transgenic plants were grown on more than 60 million ha all over the world last year and are expected to further increase in the future. Volunteerism is a well-known phenomenon in the daily practice of agriculture, and ferality is usually neglected. Only plant breeders are well aware of ferality due to their experience based on their long-term selection efforts. When plant breeders leave breeding lines on the same plot without any intervention for years, they will not be recognized due to evolution of ferality. The phenomena of volunteerism and ferality have to be considered in the transgenic era especially in relation to gene flow. As no one is directly performing research related to ferality *per se,* the major concern was to convene those world experts who are most actively working in related fields in the expectation that on the basis of their recent findings, we will be able to find out whether transgenics are different in this context from normally cultivated crops. Any knowledge about possible impacts on the natural or agroecosystems will be of use to the decision makers and legislators responsible for plant releases.

It was clear at the outset that the syntheses and information from a workshop such as this must not just remain in the memory of the 23 participants, as this information is otherwise unavailable in a concentrated form to the scientific community. Thus, it was decided to collate the information in a book, despite the drawbacks of multiauthored volumes. The chapters in this book were peer-reviewed prior to the workshop. The questions and answers in the long discussions following every presentation were recorded. Each author was given the prerogative of either including the issues and answers from the discussions directly as part of the revised text or including them at the end of the chapter in a separate section. Two chapters (11 and 18) could not be subjected to this second round of intensive peer review due to the last minute inability of the authors to get to Bellagio.

During the discussions it became apparent that there were other cases of ferality, or where ferality may become an issue, and sections for a special chapter (Chapter 15) were added by participants or were commissioned after the workshop. This helps round out the demonstration of the vast array of tricks plants have used to evolve ferality and the problems in trying to generalize. The participants hope that they have conveyed to the readers the excitement and stimulation that they felt while discussing this hitherto hardly portrayed problem, which in a few cases, could be exacerbated by the introduction of some transgenic crops in some areas.

Bellagio, Italy

Ervin Balázs
OECD Coordinator – Theme 1 - Biological Resource Management

Jonathan Gressel
Convenor and editor

Contributors

Hani Al-Ahmad
Plant Sciences
Weizmann Institute of Science
Rehovot, Israel

Pia Rufener Al Mazyad
Botanical Garden
University of Bern
Bern, Switzerland

Klaus Ammann
Botanical Garden
University of Bern
Bern, Switzerland

Sharon Ayal
Plant Sciences
Weizmann Institute of Science
Rehovot, Israel

Ervin Balázs
Agricultural Biotechnology Center
Institute for Environmental Biosafety Research
Gödöllö, Hungary

Detlef Bartsch
Bundesamt für Verbraucherschutz und Lebensmittelsicherheit (BVL)
Berlin, Germany

André Bervillé
INRA-UMR-DGPC
Montpellier, France

Catherine Breton
IMEP CNRS
Aix-en-Provence, France

Jutta C. Burger
Department of Botany and Plant Sciences
University of California — Riverside
Riverside, California, USA

Lesley G. Campbell
Department of Evolution, Ecology, and Organismal Biology
Ohio State University
Columbus, Ohio, USA

Ken Cunliffe
La Trobe University
Bundoora, Australia

Henri Darmency
Laboratory of Biology and Management of Weeds
INRA
Dijon, France

Sarah Driessen
Lehrstuhl für Biologie V der Rheinisch-Westfälischen Technischen Hochschule
Aachen, Germany

Gebisa Ejeta
Department of Agronomy
Purdue University
West Lafayette, Indiana, USA

Norman C. Ellstrand
Department of Botany and Plant Sciences
Center for Conservation Biology
Biotechnology Impacts Center
University of California — Riverside
Riverside, California, USA

Ana Mercedes Espinoza
CIBCM
Facultad de Agronomía
Escuela de Ciencias Agroalimentarias
University of Costa Rica
San José, Costa Rica

Aldo Ferrero
Dipartimento di Agronomia
Selvicoltura e Gestione del Territorio
Università degli Studi di Torino
Grugliasco, Italy

David R. Gealy
USDA — ARS
Dale Bumpers National Rice Research Center
Stuttgart, Arkansas, USA

Allen G. Good
Department of Biological Sciences
University of Alberta
Edmonton, Alberta, Canada

Cécile Grenier
Department of Agronomy
Purdue University
Lafayette, Indiana, USA

Jonathan Gressel
Plant Sciences
Weizmann Institute of Science
Rehovot, Israel

Robert H. Gulden
Department of Plant Agriculture
University of Guelph
Guelph, Ontario, Canada

Linda Hall
Alberta Agriculture
Food and Rural Development
University of Alberta
Edmonton, Alberta, Canada

Yolande Jacot
Botanical Garden
University of Bern
Bern, Switzerland

Akito Kaga
National Institute of Agrobiological Sciences
Tsukuba, Japan

Ingo Kowarik
Institute of Ecology
Technische Universität Berlin
Berlin, Germany

Zaida Lentini
Biotechnology Research Unit/ Rice Genetics
CIAT (International Center for Tropical Agriculture)
Cali, Colombia

Avraham A. Levy
Plant Sciences
Weizmann Institute of Sciences
Rehovot, Israel

Bao-Rong Lu
Institute of Biodiversity Science
School of Life Sciences
Fudan University
Shanghai, China

Buddhi Marambe
Department of Crop Science
Faculty of Agriculture
University of Peradeniya
Peradeniya, Sri Lanka

Frédéric Médail
IMEP CNRS
Aix-en-Provence, France

Marc A. McPherson
Department of Agriculture, Food and Nutritional Science
University of Alberta
Edmonton, Alberta, Canada

Marie-Hélène Muller
UMR-DGPC
INRA
Mauguio, France

Micheal D.K. Owen
Agronomy Department
Iowa State University
Ames, Iowa, USA

Christian Pinatel
AFIDOL, Maison des Agriculteurs
Aix-en-Provence, France

Matthias Pohl
TÜV Hannover/Sachsen-Anhalt e.V.
Division Energy and Systems Technology, Biotechnology
Hannover, Germany

Bernard Poinso
UMR-DGPC
INRA
Montpellier, France

Stephen Powles
School of Plant Biology
University of Western Australia
Crawley, Australia

M. Habibur Rahman
Department of Agriculture, Food and Nutritional Science
University of Alberta
Edmonton, Alberta, Canada

Alan Raybould
Syngenta — Ecological Sciences
Jealott's Hill International Research Centre
Bracknell, Berkshire, United Kingdom

Richard Roush
Director, Statewide IPM Program
University of California — Davis
Davis, California, USA

Paulino L. Sanchez
National Institute of Agrobiological Sciences
Tsukuba, Japan

Hervé Serieys
UMR-DGPC
INRA
Mauguio, France

Allison A. Snow
Evolution, Ecology, and Organismal Biology
Ohio State University
Columbus, Ohio, USA

C. Neal Stewart, Jr.
Department of Plant Sciences
University of Tennessee
Knoxville, Tennessee, USA

Ulrich Sukopp
Federal Agency for Nature Conservation
Division I 1.3 — Monitoring
Bonn, Germany

A. Gordon Thomas
Agriculture and Agri-Food Canada
Saskatoon, Saskatchewan, Canada

Norihiko Tomooka
National Institute of Agrobiological Sciences
Tsukuba, Japan

Jun Ushiki
National Agricultural Research Organization
Tsukuba, Japan

Bernal E. Valverde
Tropical Agriculture Research and Development
Alajuela, Costa Rica, and
Weed Science
The Royal Veterinary and Agricultural University
Fredericksberg, Denmark

Duncan A. Vaughan
National Institute of Agrobiological Sciences
Tsukuba, Japan

Francesco Vidotto
Dipartimento di Agronomia
Selvicoltura e Gestione del Territorio
Università degli Studi di Torino
Grugliasco, Italy

Suzanne I. Warwick
Environmental Health
National Program — Biodiversity
Agriculture and Agri-Food Canada
Ottawa, Ontario, Canada

Table of Contents

Chapter 1 Introduction — The Challenges of Ferality 1

Jonathan Gressel

1.1 Domestication and Ferality 1
1.1.1 Definitions 1
1.1.2 What Is Known about Feral Plants, *per se*? 2
1.2 The Need for a Synthesis of Information on Plant Ferality 2
1.2.1 Outcomes of the Syntheses 3
1.2.1.1 The Good Seed 3
1.2.1.2 Volunteers — The First Step to Ferality? 3
1.2.1.3 Will Transgenics Hasten the Evolution of Feral Forms? 4
1.2.1.4 Clear Cases of Ferality 4
1.2.1.5 Endoferality and Exoferality 5
1.3 The Biodiversity of Feral Forms and Their Evolution 5
1.4 Ferality and Scientific Terminology — A Caution 6
Literature Cited 7

Chapter 2 Crops Come from Wild Plants — How Domestication, Transgenes, and Linkage Together Shape Ferality 9

Suzanne I. Warwick and C. Neal Stewart, Jr.

2.1 Introduction 9
2.2 Domesticated Crops, Agricultural Weeds, and Ferality 9
2.2.1 The Domestication Process 9
2.2.2 Agricultural Weeds 11
2.2.3 Crop-Weed-Wild Complex 11
2.2.4 Genetics of Domestication and Weediness Traits 12
2.2.4.1 Domestication Traits 13
2.2.4.2 Weediness Traits 14
2.2.4.3 Dedomestication Process 14
2.2.5 Herbicide-Resistant Weedy Biotypes 15
2.2.6 Crop Ferality 15
2.3 Degree of Crop Domestication 16
2.3.1 The Case of Rice Weeds 22
2.3.2 The Case of Weedy Brassicas 22
2.4 The Effects of Transgenes and Genetic Linkage 22
2.4.1 Herbicide Resistance 23
2.4.2 Other Transgenes 23
2.4.3 Ameliorating Ferality 24
2.5 Conclusions 24
Acknowledgments 25
Literature Cited 25

Chapter 3 The Ecology and Detection of Plant Ferality in the Historic Records 31

Klaus Ammann, Yolande Jacot, and Pia Rufener Al Mazyad

3.1 General Introduction 31
3.2 Reversion of Crops to Wild Types 31
3.3 Historical Accounts of Feral Crops 32
3.3.1 Methods of Detection 32
3.3.1.1 Archaeobotanical Methods 32
3.3.1.2 Pollen Analysis 34
3.3.1.3 Phytolith Analysis 34
3.3.1.4 Analysis of Herbarium Specimens 35
3.3.2 Archaeobotanical Studies — The Example of Hulled Wheat 38
3.3.2.1 Einkorn 39
3.3.2.2 Emmer 40
Literature Cited 41

Chapter 4 Feral Beets — With Help from the Maritime Wild? 45

Ulrich Sukopp, Matthias Pohl, Sarah Driessen, and Detlef Bartsch

4.1 History of Beet Domestication 45
4.2 Hybridization and Gene Flow in Beet 46
4.3 Ferality in Beet Connected to the Bolting Gene "B" 48
4.4 Potential Impact of Transgenes on Ferality 50
4.5 Conclusions and Outlook 52
Acknowledgments 53
Literature Cited 54

Chapter 5 Volunteer Oilseed Rape — Will Herbicide-Resistance Traits Assist Ferality? 59

Linda M. Hall, M. Habibur Rahman, Robert H. Gulden, and A. Gordon Thomas

5.1 Introduction 59
5.2 *Brassica rapa* and *B. napus*, Origins and Biology 60
5.2.1 Introgression between Crop and Wild Brassicaceae 61
5.2.2 Crop Improvement Objectives, Domestication, and Ferality 61
5.3 Biological Characteristics Influencing Weediness 63
5.4 Presence and Persistence of Volunteer *B. rapa* and *B. napus* 66
5.4.1 Influence of Herbicide-Resistance Traits on Persistence and Ferality 68
5.4.2 Anticipated and Unintended Consequences 70
5.4.3 A Simple Scenario for Population Demographics in Western Canada 71
5.5 Conclusions 73
Literature Cited 73

Chapter 6 Incestuous Relations of Foxtail Millet (*Setaria italica*) with Its Parents and Cousins 81

Henri Darmency

6.1 Introduction 81
6.2 Domestication of Foxtail Millet 82
6.2.1 Domesticated Traits 83
6.2.1.1 Seed Shedding 83
6.2.1.2 Flowering Duration 83

6.2.1.3 Uniform Germination on Sowing 84
6.2.1.4 Other Characteristics 84
6.2.2 The Genetic Bases of Domestication and Dedomestication 85
6.2.2.1 Seed Shedding 85
6.2.2.2 Tiller Development Patterns and Flowering Duration 85
6.2.2.3 Uniform Germination and Grain Size 86
6.3 Volunteers or Weedy Hybrid Derivatives? 86
6.3.1 Why No Volunteers? 86
6.3.2 *Setaria viridis* ssp. *pycnocoma* 87
6.3.3 Gene Flow 89
6.4 Polyploid Species of the Foxtail Millet Gene Pool 90
6.4.1 *Setaria verticillata* 90
6.4.2 *Setaria faberii* 91
6.4.3 *Setaria pumila* 92
6.5 Conclusions 92
Literature Cited 93

Chapter 7 Urban Ornamentals Escaped from Cultivation 97

Ingo Kowarik

7.1 Introduction 97
7.1.1 Aims 98
7.2 Urban Ornamentals — A Heterogeneous Species Pool 98
7.3 Invasions by Ornamentals 100
7.3.1 How Many Species Will Spread? 100
7.3.2 Naturalization of Deliberately Introduced Species 101
7.3.3 Evoking Negative Effects — Problematic Plant Invasions 102
7.3.3.1 Introduced Ornamentals as Vectors of Pests 103
7.3.4 Temporal Patterns — Lag Time between Cultivation and Escape 103
7.3.5 Spatial Pattern — From Cultivation Sites to Natural Ecosystems 104
7.3.5.1 Confinement to Sites of Cultivation of Ornamentals as Historical Indicators 105
7.3.5.2 Changes in Urban Flora and Urban–Rural Gradients 105
7.4 Underlying Processes 107
7.4.1 Ecological Role of Cultivation 107
7.4.2 Spread as a Response to Environmental Changes 108
7.4.3 Dedomestication Processes 109
7.4.3.1 Endoferality — Some Examples 109
7.4.3.2 Exoferality 110
7.5 Can We Predict the Spread of Introduced Ornamentals? 112
7.6 Conclusions 113
Acknowledgments 114
Literature Cited 114

Chapter 8 Sorghum and Its Weedy Hybrids 123

Gebisa Ejeta and Cécile Grenier

8.1 Introduction 123
8.2 The Sorghum Taxa 123
8.2.1 Cultivated Sorghums 125
8.2.2 Wild and Weedy Sorghums 126

8.3 Weedy Sorghums in Agroecosystems 127
8.4 Gene Flow among Sorghums 128
8.4.1 Weed-to-Crop Gene Flow 128
8.4.2 Crop-to-Weed Gene Flow 129
8.4.3 Consequences of Recurrent Gene Flow 130
8.4.3.1 In Conventional Agriculture 130
8.4.3.2 Introgression Disrupted by Genetic Erosion 131
8.4.3.3 *In situ* Introgression in a Natural Habitat 131
8.4.3.4 Introgression Enhanced by Modern Farming Practices 132
8.4.4 Gene Flow in the Transgenic Era 132
Literature Cited 133

Chapter 9 Multidirectional Gene Flow among Wild, Weedy, and Cultivated Soybeans 137

Bao-Rong Lu

9.1 Introduction 137
9.2 Soybean and Its Weedy and Wild Relatives 138
9.2.1 Distribution and Relationships of Soybean and Its Wild Relatives 140
9.2.2 The Potential of Transgene Flow in Soybean Species 142
9.3 Possible Consequences of Gene Flow from Transgenic Soybean 143
Acknowledgments 145
Literature Cited 145
Questions and Answers 146

Chapter 10 Maize and Soybeans — Controllable Volunteerism without Ferality? 149

Micheal D.K. Owen

10.1 Introduction 149
10.1.1 Maize and Soybean 150
10.2 Current Maize and Soybean Production 150
10.2.1 North and South America 152
10.2.2 China 152
10.2.3 Status of Biotechnologically Derived Traits in Maize and Soybean 152
10.3 Origin of Maize and Soybeans 153
10.3.1 Potential for Introgression with Compatible Plant Species in the Agroecosystem 153
10.3.1.1 Maize 153
10.3.1.2 Soybean 154
10.4 Extent of Volunteerism in Maize and Soybeans 154
10.4.1 Weedy Characteristics of Volunteer Maize 155
10.4.1.1 Factors Involved with Volunteerism in Maize 155
10.4.2 Weedy Characteristics of Soybeans 156
10.4.3 Implications for Ferality and Volunteerism 157
10.5 Volunteers Need Not Be in Fields — Identity Preservation and Derived Traits 157
10.5.1 Implications of Volunteerism on Labeling Requirements and Regulations 158
10.5.1.1 Pollen Drift 159
10.5.1.2 Grain Marketing Issues 160
10.6 Conclusions 162
Literature Cited 163

Chapter 11 Wheat Domestication and Dedomestication — What Are the Odds? 167

Sharon Ayal and Avraham A. Levy

11.1 Introduction....167
11.2 Wheat Domestication 167
11.3 Genomics of Wheat Domestication 168
11.3.1 The Approach and the Materials....168
11.3.2 Design and Analysis of cDNA Microarrays 169
11.3.3 Differential Expression of Genes in Young Spikes of Wild vs. Domesticated Wheat....169
11.4 Assessing the Odds of Dedomestication....171
11.5 Semiwild Wheat from Tibet — A Case of Dedomestication? 171
11.6 Conclusions....172
Acknowledgments....172
Literature Cited 173

Chapter 12 Feral Rye — Evolutionary Origins of a Weed 175

Jutta C. Burger and Norman C. Ellstrand

12.1 Introduction....175
12.1.1 The Many Faces of Weedy Rye....176
12.1.2 Taxonomy of Rye and Its Relatives....176
12.2 History of Domestication 179
12.2.1 Distribution of Cultivated Rye in Eurasia 180
12.2.2 Wild and Weedy Rye in Eurasia....180
12.2.3 Distribution of Rye and Weedy Rye in North America 181
12.2.3.1 Rye in North America....181
12.2.3.2 Weedy Rye in North America....184
12.2.4 Weedy Rye in the Western U.S....184
12.2.5 Introductions of Mountain Rye and Hybrid Michels Grass....185
12.3 The Case of Naturalized Feral Rye in the Western U.S. 186
12.3.1 Genetic Analysis....186
12.3.2 Ecological Comparison 188
12.4 Summary and Discussion....188
Acknowledgments....189
Literature Cited 189
Questions and Answers....191

Chapter 13 Can Feral Radishes Become Weeds?....193

Allison A. Snow and Lesley G. Campbell

13.1 Introduction....193
13.2 Early Domestication 193
13.3 Modern Radish Varieties with Edible Roots....195
13.4 Dedomestication and Weed Evolution in Radishes 197
13.4.1 Characteristics of Weedy *Raphanus raphanistrum*....197
13.4.2 Weedy Traits and Endoferality....198
13.4.2.1 General Considerations 198
13.4.2.2 Earlier Flowering and a Less Swollen Root 199
13.4.2.3 Early Abscission of Mature Fruits....199

13.4.2.4 Thicker and Woodier Fruits 199
13.4.2.5 Segmented Fruit Capsules that Break into Single-Seeded Sections 199
13.4.2.6 Staggered Seed Germination 200
13.4.2.7 Resistance to Insect Herbivores and Pathogens 200
13.4.2.8 Greater Genetic Diversity 200
13.5 Exoferality via Crop-Weed Hybridization 200
13.5.1 General Considerations 200
13.5.2 Herbicide-Resistant Feral *Raphanus sativus* in Southern Brazil 201
13.5.3 Crop-Weed Hybridization in California 202
13.5.4 Crop-Weed Hybridization in Experimental Populations in Michigan 202
13.5.4.1 Overview of Experimental Populations 202
13.5.4.2 Fitness of F_1 Weed-Crop Hybrids and Persistence of Crop Alleles 203
13.5.4.3 Persistence of Feral Populations 204
13.6 Conclusions 204
Acknowledgments 205
Literature Cited 205

Chapter 14 Ferality — Risks of Gene Flow between Sunflower and Other *Helianthus* Species 209

André Bervillé, Marie-Hélène Muller, Bernard Poinso, and Hervé Serieys

14.1 Introduction 209
14.1.1 Botany and Economic Importance of *Helianthus* Species 209
14.1.2 Domestication and Breeding Sunflower and Jerusalem Artichoke 209
14.1.3 Where Did *Helianthus* Establish in the Wild Outside Its Native Area? 210
14.1.3.1 Europe 210
14.1.3.2 Other Continents 210
14.1.4 Why Do We Care about These Volunteer or Feral Populations? 211
14.1.4.1 Gene Flow from Volunteers or Feral May Modify Sunflower Oil Composition 211
14.1.4.2 Impact of Crop Alleles in the Wild 211
14.1.4.3 How Did These Introduced *Helianthus* Species Evolve from the Native Forms? 211
14.1.4.4 Feral and Volunteers May Constitute Gene Reservoirs for Crop Alleles and Become Invasive 212
14.2 Studies on *Helianthus annuus* 212
14.2.1 Escaped and Permanent Wild Sunflower Populations in Italy and Spain 212
14.2.1.1 The Problem 212
14.2.1.2 Hypotheses on Their Origins 213
14.2.1.3 Planned Work 213
14.2.2 A Case Study — Model of Wild Sunflower Establishment Near Montpellier 213
14.2.2.1 Unwanted Sunflowers Become an Experimental Plot 213
14.2.2.2 Planned Work 216
14.2.3 Existing Volunteer Populations — Record, Localization, Fate, and Establishment 216
14.3 Studies on Jerusalem Artichoke 218
14.3.1 Hybridization Rate Efficiency with Sunflower 218
14.3.2 Feral Jerusalem Artichoke Populations 219
14.3.3 Studies on Jerusalem Artichoke Populations 220
14.4 Modeling the Impact of Gene Flow and Fate of Wild Relatives 221

14.4.1 Contact 221
14.4.2 Modeling 221
14.4.3 Sunflower 222
14.4.4 Transgene Containment 223
14.5 Weediness, Ferality, and Invasiveness in *Helianthus* 223
14.5.1 Major Risks of Crop-Allele Gene Flow in Europe 223
14.5.1.1 Risk of Crop-Allele Escape 223
14.5.1.2 The Risk of Seed Volunteers Turning into Weedy Populations 224
14.5.1.3 The Risk of Seed Volunteers Turning into Feral Populations 224
14.5.1.4 The Risk of Gene Flow Back from Feral Populations to Crop 224
14.5.1.5 Evolution of Weedy and Feral Populations toward Wild Sunflower 224
14.5.2 Major Risk of *Helianthus* Gene Flow at Center of Origin 225
14.5.3 Adaptation to Drought and Poor Soils 225
14.6 Conclusion 225
14.7 Synopsis of Materials and Methods 226
14.7.1 Interspecific Hybridization 226
14.7.2 Molecular Tools Available for Sunflower 226
Acknowledgments 227
Literature Cited 227
Questions and Answers 229

Chapter 15 Issues of Ferality or Potential for Ferality in Oats, Olives, the *Vigna* Group, Ryegrass Species, Safflower, and Sugarcane 231

André Bervillé, Catherine Breton, Ken Cunliffe, Henri Darmency, Allen G. Good, Jonathan Gressel, Linda M. Hall, Marc A. McPherson, Frédéric Médail, Christian Pinatel, Duncan A. Vaughan, and Suzanne I. Warwick

15.1 Introduction 231
15.2 Oats 231
15.2.1 Cytogenetic Interrelationships and Crossability 231
15.2.2 Origin of Fatuoids 232
15.3 Olives 233
15.4 *Vigna* in Asia 235
15.4.1 Cowpea 235
15.4.2 Rice Bean 235
15.4.3 Azuki Bean 235
15.4.4 Considerations for Transgenic Biosafety for *Vigna* Crops 236
15.5 The Ryegrass Complex 236
15.5.1 Taxonomy of the *Lolium–Festuca* Complex 237
15.5.1.1 *Lolium–Festuca* Complex — A Taxonomic Problem 237
15.5.2 Ecogeographical Adaptation — Spatial Separation of the Species 237
15.5.2.1 Phylogeography within the Genus *Lolium* 237
15.5.2.2 Phylogeography of *Lolium* vs. *Festuca* 238
15.5.3 Interfertility 238
15.5.3.1 Self-Incompatibility and Self-Fertility 238
15.5.3.2 Karyotype Fertility Barrier 239
15.5.3.3 Interfertility between the Outcrossing *Lolium* Species 239
15.5.3.4 Interfertility between *Lolium perenne* and the Inbreeding *Lolium* Species 239
15.5.3.5 *Lolium–Festuca* — Interfertility beyond the Genus *Lolium* 240

15.5.4 Pollen Flow 240
15.5.4.1 Amount and Timing 240
15.5.4.2 Distance of Pollen Flow 241
15.5.4.3 Distribution of Pollen and Gene Flow over Distance 241
15.5.4.4 Pollen Viability and the Outer Limits of Gene Flow 241
15.5.5 The Control of Gene Flow Using Nonpaternally Inherited Transformation or Male Sterility 242
15.5.6 Concluding Remarks 242
15.6 Safflower — Ferality in a Plant-Made Pharmaceutical Platform 242
15.6.1 Platform for Plant-Made Pharmaceuticals 243
15.6.2 Safflower, Systematics, Biology, Biogeography 243
15.6.2.1 Systematics and Hybridization Potential between Cultivated Safflower and Its Wild Relatives 243
15.6.2.2 Carthamus — Biology and Biogeography 245
15.6.3 Domestication Traits in Safflower 245
15.6.4 Ferality of Safflower 246
15.6.5 Information Required to Improve Our Prediction for PMP Safflower Ferality 247
15.6.6 Conclusions 247
15.7 Sugarcane 247
Literature Cited 249

Chapter 16 Asian Rice and Weedy Rice — Evolutionary Perspectives 257

Duncan A. Vaughan, Paulino L. Sanchez, Jun Ushiki, Akito Kaga, and Norihiko Tomooka

16.1 Introduction 257
16.1.1 The Genus *Oryza* in Relation to Rice 257
16.1.1.1 The Antiquity of *Oryza* 257
16.1.1.2 *Oryza* Genome and Species Diversity as It Relates to Rice 258
16.1.1.3 The Perennial–Annual Axis of Diversity in the Primary Gene Pool of Rice AA Genome (Axis 1) 258
16.1.1.4 Distribution of the Wild AA Genome *Oryza* Species 258
16.2 Rice Domestication 260
16.2.1 Domestication Processes 260
16.2.1.1 The Wild Rice–Domesticated Rice Evolutionary Axis (Axis 2) 260
16.2.1.2 Indica–Japonica Evolutionary Axis (Axis 3) 260
16.2.2 Places of Domestication 262
16.2.3 Genetics of Domestication 263
16.2.3.1 Neutral Alleles and Domestication 263
16.2.3.2 Domestication Related Traits and Domestication 263
16.2.3.3 Gene Clusters (Closely Linked Genes) 263
16.2.3.4 Gene Association (Unlinked Genes or Cryptic Linkage) 263
16.3 Diversification of Rice 264
16.3.1 Evolution of Weedy Rice 265
16.4 Case Studies 267
16.4.1 Gene Flow in the Rice Crop Complex 267
16.4.1.1 Introgression 267
16.4.1.2 Gene Dispersal and Hybrid Swarms 267
16.4.2 Evolution of Weedy Rice in Rice Fields 270
16.4.2.1 Weedy Rice in Japan 270
16.4.2.2 Weedy Rice in Malaysia 271
16.5 Concluding Comments 271

Acknowledgments 273
Literature Cited 273
Questions and Answers 276

Chapter 17 The Damage by Weedy Rice — Can Feral Rice Remain Undetected? 279

Bernal E. Valverde

17.1 Introduction 279
17.2 Distribution and Diversity of Weedy Rice 279
17.2.1 Weedy Rice Species and Their Distribution 279
17.2.2 Genetic Relationship and Diversity of Weedy and Cultivated Rice 280
17.3 Agronomic and Market Impact of Weedy Rice 280
17.3.1 Direct Competitive Effects of Weedy Rice 280
17.3.2 Indirect and Market Effects of Weedy Rice 281
17.4 Field Management of Weedy Rice 281
17.4.1 Agronomic Practices and Weedy Rice Management 281
17.4.2 Chemical Control and Herbicide-Resistant Rice 283
17.5 The Spread of Weedy Rice 284
17.5.1 Domestication and Weediness 284
17.5.2 Contaminated Seed 285
17.5.3 Farmer Attitudes about Weedy Rice and Its Dissemination 286
17.6 Going Undetected 287
Acknowledgments 289
Literature Cited 289

Chapter 18 Properties of Rice Growing in Abandoned Paddies in Sri Lanka 295

Buddhi Marambe

18.1 Introduction 295
18.1.1 Land Use in Rice Cultivation in Sri Lanka 295
18.1.2 Wild and Weedy Rices in Sri Lanka 296
18.2 Methodology 297
18.3 Results and Discussion 298
18.3.1 Field Observations of Morphological Characteristics 298
18.3.2 Germination and Seed Viability 299
18.3.3 Plant Growth Characteristics under Greenhouse Conditions 299
18.4 Concluding Remarks 301
Literature Cited 302

Chapter 19 Coexistence of Weedy Rice and Rice in Tropical America — Gene Flow and Genetic Diversity 305

Zaida Lentini and Ana Mercedes Espinoza

19.1 Introduction and Dissemination of Rice in the Americas 305
19.2 *Oryza* Species in Tropical America 306
19.2.1 Overview of Species Composition and Distribution 306
19.2.2 Distributions and Genetic Diversity of Wild *Oryza* Species in Costa Rica 308
19.2.3 *Oryza glumaepatula*, the AA Genome Wild Relative of Rice in Tropical America 309
19.2.4 Is Gene Flow between Native Wild *Oryza* Species and Rice (*O. sativa*) Possible in the Field? 310

19.3 Coexistence of Weedy Rice with Domestic Rice in Fields 311
19.3.1 Costa Rican Weedy Rice 312
19.3.2 Colombian Weedy Rice 313
19.3.3 Weedy Rice Resemblance across Two Countries 315
19.4 Rice–Weedy Rice Gene Flow in Tropical America 316
19.5 Conclusions 317
Acknowledgments 319
Literature Cited 319

Chapter 20 Gene Movement between Rice (*Oryza sativa*) and Weedy Rice (*Oryza sativa*) — a U.S. Temperate Rice Perspective 323

David R. Gealy

20.1 Introduction to U.S. Temperate Rice Production 323
20.1.1 Localization and Production Practices 323
20.1.2 History of Cultivar Development 323
20.2 Weed Problems — The Red Rice Dilemma 324
20.2.1 Introduction and Distribution in the U.S. 324
20.2.2 Economic and Agronomic Impacts of Red Rice 325
20.2.3 Phenotypic and Genetic Characterization of Red Rice Populations 326
20.2.4 Differential Herbicide Resistance in Red Rice Populations 327
20.3 Herbicide-Resistant Cultivars 328
20.3.1 Background 328
20.3.2 IMI-Resistant Rice 328
20.3.3 Glufosinate-Resistant Rice 330
20.3.4 Special Management Considerations 330
20.4 Outcrossing Causes, Rates, and Consequences 330
20.4.1 Biological Basis for Outcrossing in *Oryza sativa* 330
20.4.2 Historic Estimates and Limits of Rice-Red Rice Outcrossing Rates 331
20.4.3 Biotic and Abiotic Factors Affecting Rice-Rice and Rice-Weedy Rice Outcrossing Rates 331
20.4.4 Directionality of Outcrossing 340
20.4.5 Average Outcrossing Estimates for Rice, Herbicide-Resistant Rice, and Weedy Rice Combinations 341
20.4.6 Rice-Wild Rice Outcrossing Rates Are Relatively High 342
20.4.7 Introgression Rate Considerations 342
20.5 Phenotypic Traits of Rice/Red Rice Hybrids in the U.S. 343
20.5.1 Key Traits in F_1 Hybrids 343
20.5.2 Differentiating between F_1 and F_2 Hybrids 344
20.6 Backcrossing Considerations 344
20.7 Dormancy, Shattering, and Other Keys to Domestication/Dedomestication 345
20.7.1 Dormancy and Shattering 345
20.7.2 Reproductive and Other Traits 346
20.8 Anecdotal Evidence of Gene Flow between Red Rice and Rice in the U.S.? 346
20.8.1 Aroma Chemicals Detected in Red Rice Accessions at Low Frequencies 347
20.8.2 Blast Resistance Detected in Red Rice Accessions at Low Frequencies 347
20.8.3 Short Stature, Short Awn Red Rice Plants Detected at Low Frequencies 347
20.9 Prospects for Volunteerism 348
20.10 Concluding Remarks 348
Acknowledgments 350

Literature Cited 350
Questions and Answers 354

Chapter 21 Modeling Population Dynamics to Overcome Feral Rice in Rice 355

Francesco Vidotto and Aldo Ferrero

21.1 Spread and Importance of Weedy Rice in Europe 355
21.2 Weedy Rice Biology in Relation to Population Dynamics 356
21.2.1 Seed 357
21.2.2 Seedling Emergence 357
21.2.3 Seed Production and Dispersal 358
21.3 Modeling Weedy Rice Infestation Dynamics 358
21.4 Running the Model 361
21.4.1 Validation 361
21.4.2 Sensitivity Analysis — Singling Out Crucial Factors 361
21.4.3 Simulation of Scenarios 363
21.5 Conclusions 366
Literature Cited 367

Chapter 22 Molecular Containment and Mitigation of Genes within Crops — Prevention of Gene Establishment in Volunteer Offspring and Feral Strains 371

Jonathan Gressel and Hani Al-Ahmad

22.1 Introduction — Needs for Preventing Gene Flow and Overcoming Ferality 371
22.1.1 Molecular Tools to Prevent or Overcome Ferality in Traditionally Bred Crops 371
22.1.2 Molecular Tools Are Needed to Prevent or Overcome Ferality with Transgenic Crops 372
22.2 Methods for Precluding Feral Traits from Becoming Predominant in Populations 372
22.2.1 Containing Gene Flow 373
22.2.1.1 Containment by Targeting Genes to a Cytoplasmic Genome 373
22.2.1.2 Male Sterility Coupled with Transplastomic Traits 373
22.2.1.3 Genetic Use Restriction Technologies and Recoverable Block of Function 374
22.2.1.4 Repressible Seed Lethal Technologies 374
22.2.2 Preventing Volunteer Establishment by Transgenic Mitigation 374
22.2.2.1 Demonstration of Transgenic Mitigation in Tobacco and Oilseed Rape 376
22.2.2.2 Risk that Introgression of Transgenic Mitigation Traits Will Affect Wild Relatives of the Crop 378
22.2.2.3 Following Transgene Flow to Volunteers and Feral Forms 380
22.3 Special Cases Where Transgenic Mitigation Is Needed — Special Genes 380
22.3.1 Transgenic Mitigation Genes for Crop-Produced Pharmaceuticals and Industrial Products 380
22.3.2 Mitigation for Biennial and Other Root Crops 381
22.3.3 Mitigation of Ferality in Species Used for Phytoremediation 381
22.3.4 Mitigating Endo- and Exoferality by Rendering Crops Obligatively Vegetatively Propagated 382
22.3.5 Tac-Tics for Eliminating Feral Forms of Pasture Grasses 382
22.3.5.1 The Use of Tac-Tics for Insect Control 383
22.3.5.2 Modification of Tac-Tics for Preventing Ferality of Pasture Grasses 383
22.3.5.3 TM Genes for Use in Transposons for Pasture Grasses 384

22.3.5.4 Herbicide-Mimic, Lethal *Kev* Genes 384
22.3.5.5 Chemically Induced Promoters for *Kev* Genes 385
22.3.5.6 Biosafety of Transposing Pasture Grasses with the DTs 385
22.4 Concluding Remarks 385
Acknowledgments 385
Literature Cited 386

Chapter 23 Assessing the Environmental Risks of Transgenic Volunteer Weeds 389

Alan Raybould

23.1 Introduction 389
23.2 What Is a Risk Assessment? 389
23.3 Tiered Testing and Risk Assessment 390
23.4 General Requirements for Assessing Risks from Volunteer Trangenic Crops 391
23.5 Assessment Endpoints 392
23.5.1 Regulatory Obligations 392
23.5.2 Harmful Effects of Volunteer Crops 393
23.5.3 Do Volunteer Transgenic Crops Pose New Hazards? 394
23.5.4 Making Assessment Endpoints Operational 394
23.6 Hazards, Exposure, and Risks of Volunteer Transgenic Crops 395
23.7 Hazard Assessments 396
23.7.1 Abundance 396
23.7.2 Composition 397
23.8 Exposure Assessments 398
23.9 Monitoring 399
23.10 Conclusions — Acceptable Risks and Trigger Values 399
Literature Cited 400

Chapter 24 Regulation Should Be Based on Data, Not Just Models 403

Richard Roush

24.1 Introduction 403
24.2 Trends in Regulation — Estimating Risk 404
24.3 Types of Models 405
24.3.1 Verbal Models 405
24.3.2 Statistical Models 406
24.3.3 Simulation Models 406
24.3.4 Analytical Models 407
24.4 Conclusions — What Models Can Do for Regulation and Research on Ferality 407
Literature Cited 407

Chapter 25 Epilogue 411

Ervin Balázs

25.1 Good Agricultural Practice 411
25.2 Volunteerism 411
25.3 Ferality 412

Index 413

1 Introduction — The Challenges of Ferality

Jonathan Gressel

1.1 DOMESTICATION AND FERALITY

Domestication of crops and animals has been a long and arduous, never ending process, typically requiring centuries or millennia of selection. Domestication was performed more rationally and fine-tuned in the last century by an understanding of genetics and scientific breeding. More recently, the process has been further accelerated by deoxyribonucleic acid (DNA) technology: both by marker-assisted breeding and by the transgenic introduction of novel traits not found in their own genomes or in those of interbreeding relatives. These traits came from other species, genera, families, or kingdoms, oftentimes artificially modified in the laboratory.

The return to the wild by dedomestication is much quicker, at least in animals. Cute pink English pigs escaped to the wild in the U.S. and New Zealand, and in not many generations, rough, nasty European wild boars abounded, even though they had never been introduced into those countries. It took much longer to domesticate pigs from wild boars. It took quite a few generations to select and breed foxes to be docile pets, but in a few generations without continued selection, they become feral (5). Much (but not all) of domestication, it seems, is a function of selecting for rare recessive traits requiring finding them in their ultra-low homozygous frequency. The return of a recessively domesticated species to a feral form by *dedomestication* (evolution of ferality) by back mutation to the dominant heterozygous feral traits can be rapid. This vast difference in rates is because the mutation frequency of homozygous domestic traits is exponentially lower than the frequency of heterozygous dominant *feral* (wild-type) traits.

The process of becoming feral can be hastened when and if the domesticated species can hybridize with its wild progenitor or a wild relative, where they exist nearby, which is not always the case. Thus, although dogs can evolve ferality by themselves, it is far quicker when they mate with wolves (now considered to be the same species). There are recently evolved feral dogs where there have been no wolves for centuries.

Many crops are ultimately selected for *polyploidy* (or amphiploidy resulting from a combination of genomes of different related species). Domestication had to occur before polyploidization or interspecific genome combinations, as it would take much longer to select for recessive domestication traits in multiple genomes. The reversion to feral forms using dominant traits is under no such constraints in diploids or polyploids.

1.1.1 DEFINITIONS

The feral pig looks like a wild pig, but are they the same? Similarly, a crop gone feral may look like the wild progenitor — but this is not a foregone conclusion. The context of feral used by the zoologists is "left domestication and able to reproduce outside of domestication." In the context of this book, we can define feral plants as plants derived fully or in part from crop plants that have become partially or fully dedomesticated. Feral plants (in contrast to crops) can reproduce on their own, without being dependent on managed cultivation. As will be demonstrated in many chapters, not all of the domestication process must be reversed for a plant to be feral. Thus, feral types may not necessarily resemble wild or weedy types of the same or related species. This will be seen best with rice — its conspecific feral or weedy and its wild forms.

Feral plants are more likely to evolve from *volunteer weeds* (plants that germinate in season after a crop had been cultivated) that derive from seed that the crop has shattered before or during harvest. The feral plant may be *weedy* (exist in human cultivated or disturbed, ruderal ecosystems) as emphasized in this book, or it may enter the ecosystem of wild species, as has been widely discussed.

DNA evidence has countered the speciation presumptions of splitter taxonomists. There are now many cases where wild and cultivated forms that had been given different Latin binomials have been found to be unspeciated, justifying the lumper taxonomists. Clearly, the intermediate feral form will be included in the same species — ferality is a graded continuum, a process, and thus those who prefer more definitive, stop-the-evolutionary-clock definitions may dislike the term. The term ferality, describing this continuous process, is not found in contemporary dictionaries; therefore, it may be a preferred term, due to having fewer preconceptions.

The differences between weeds and wild species are well defined in Chapter 2. Most differential definitions that the reader comes up with or prefers will be adequate in the context of this chapter.

1.1.2 What Is Known about Feral Plants, *per se*?

Far more is known about the evolution of feral animals from their domestic progenitors, even if only observational, than about the dedomestication of plants. To obtain an estimate about what is known about ferality in plants, a Thomson's ISI Web of Knowledge database search during the last 2 decades was performed. No document had "ferality" or "dedomestication," the best descriptors for the evolution of domestic to feral types, in the title, abstract, or keywords. The database turned up 2113 documents with "feral" in the title, abstract, or keywords, of which 918 had "feral" in the title. A quick perusal showed that most of the thousands of documents referred to feral animals, so the search was limited to "feral and plants" under the assumption that somewhere in an abstract about a particular feral plant species, the word "plant" would appear. Of the 115 documents, 92 referred to feral mammals, birds, or insects that devour or pollinate plants, not to feral plants. The 23 remaining abstracts appearing at an average rate of 1 per year described various situations with feral plant species. The use of "feral and plants" could provide an underestimate of the information available, as the particular crop name may be used. Thus, a separate search was made for each crop described in a chapter below, plus "feral." Five more citations came forth.

The paucity of articles on crops going feral is not quite as indicative of the totality of papers on the subject, as some authors use "weedy" or other adjectives for feral forms of crops (e.g., weedy wheat, red rice, wild rice). Feral and wild are not synonyms in the context of this book, even though some dictionaries define feral as "untamed" or "undomesticated." Herein, feral is defined as "domesticated evolving to be untamed or undomesticated." Crops gone feral in an agroecosystem will not necessarily be identical to the wild type of the species — the feral plants are more likely to have weedy characters that allow it to compete with crops.

1.2 THE NEED FOR A SYNTHESIS OF INFORMATION ON PLANT FERALITY

The dispersed nature of the scant literature, with little effort to discuss the evolutionary processes and find common threads was reason enough to gather experts dealing with various crops to attempt to understand the evolutionary processes that are common and those that are species-specific in a crop becoming feral.

The participants and authors were expected to gather and synthesize the data available, give a clear picture of the situation, and predict whether or when the evolution of problematic feral forms can be exacerbated by cultivating transgenic crops. The analyses are mainly limited to the evolution of ferality within the agroecosystem and its implications to agriculture. From the following chapters it is clear that much is known or can be deduced about the evolution of ferality in agroecosystems that could not be accessed using the keywords that should have provided information.

1.2.1 OUTCOMES OF THE SYNTHESES

It became apparent that the evolution of feral forms may be facilitated by gene flow from adjacent *ruderal* (human disturbed) or wild ecosystems. The effects of transgene flow to ecosystems outside of agriculture has been covered in many recent studies and monographs (1–4) and will not be addressed here. The actual F_1 hybrids between crops and weedy or wild species may be the same, irrespective of ecosystem where they occur, but here we deal with the consequences to the agricultural or urban ecosystems and not to effects on the wild. In genetic terms, where others have studied the F_1 backcrossing to the wild species, here we are interested in the cases where the F_1 backcrosses with the crop, reimbuing it with feral traits, or where the crop back-mutates to a feral form.

1.2.1.1 The Good Seed

There is less chance of ferality evolving when certified seed is cultivated. The provider of certified seed culls off-types in production fields just as the breeder of hogs or dogs culls off-types from a litter — the back mutations are quickly removed from the population. Crops do not have the mobility of hogs, so how can feral forms evolve?

1.2.1.2 Volunteers — The First Step to Ferality?

We envisage the first step to be via volunteer weeds — offspring of crop seed that *shattered* (prematurely dropped their seeds) prior to harvest in previous seasons (Figure 1.1). If these volunteers can remain continuously in the same fields for extended periods, there are many chances for selection of back mutations of various feral traits and then for them to recombine.

Not many genes have to mutate for a volunteer weed to become a hard-to-control feral weed. Indeed, there is the possibility that some of the initial volunteer weeds have a higher frequency of one such trait — the ability to shatter. This fills the soil seed bank for the next growing season, and only part of the seeds are harvested. There will then be further selection for shattering as well as selection, in the volunteer population, for those individual mutants that have enhanced *secondary dormancy:* the ability to germinate non-uniformly during the following season and over a few years, filling the soil seed bank for many seasons to come. This will allow an incipiently feral form of

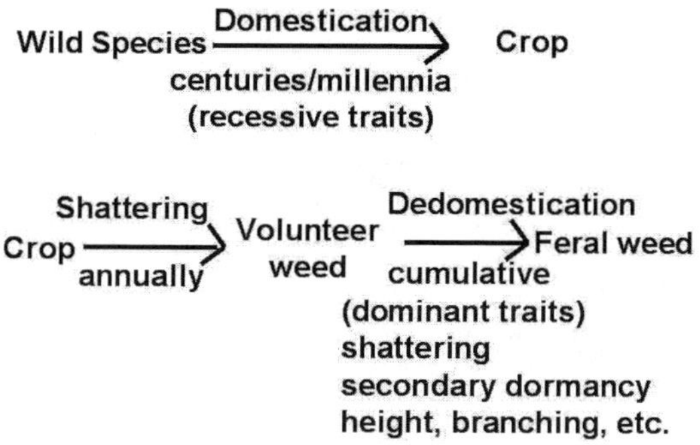

FIGURE 1.1 The processes of domestication of crops from wild progenitors and dedomestication to feral forms.

the crop species to bypass preemergence and early season herbicide treatment and cultivations, and for some, to remain dormant under rotational crops, only to come up and interbreed with the crop itself.

Many crops have been dwarfed to enhance the ratio of seed to straw (*harvest index*). Dedomesticating volunteer weeds that are taller and branch more will provide a competitive advantage over the crops. Other adaptive minor traits will soon appear as the feralizing volunteer continues to evolve. Thus, it may be important to ensure that volunteer weeds be controlled in the following season to prevent such an evolutionary scenario from having a chance. It is the hard-to-control volunteer weeds, in monoculture crops, that force farmers to use rational rotations.

Those volunteer individuals mutating to feral weedy traits will have a competitive selective advantage in successive generations and can cross with others bearing other feral traits. Then there is a more intractable problem than volunteer weeds.

1.2.1.3 Will Transgenics Hasten the Evolution of Feral Forms?

Why the worry that ferality will be exacerbated by transgenics? So many people state that transgenics are no different from traditionally bred crops. This may not be the case *vis-à-vis* ferality for two interrelated reasons:

1. Many of the traits being transformed into plants do not exist in the genome of the particular crop species, and the trait might provide a selective advantage over previous volunteers of the crop or against competing weeds.
2. The use of transgenics may change agronomic practices in ways that will enhance the rate of dedomestication.

Consider the following scenario: a valuable cash crop (such as oilseed rape) is presently grown in rotation with a less valuable crop (such as wheat). Rotation is used because the valuable crop suffers from disease, is attacked by insects, and is heavily competed by weeds. Rotation eliminates or suppresses the pathogens, arthropods, and weeds. More importantly, rotation allows control of volunteer weeds emanating from the valuable crop during the seasons when the less valuable crop is cultivated. Thus, volunteers cannot continue the evolution to ferality. Now we must consider what might occur if the valuable crop is transgenically rendered resistant to pathogens, to arthropods, and to excellent herbicides. Many growers may prefer monoculture of the valuable crop, and the volunteers could remain. The implications of whether less rotation will hasten ferality must be assessed. This can be only done after we have more basic information on the evolution of feral forms of crops.

1.2.1.4 Clear Cases of Ferality

Prior to this workshop, we have few clear indications of crops that dedomesticated. In many cases, there can be a valid argument that the feral forms one sees are the wild progenitors of the crop that is mixed with the crop or feral hybrids between the crop and the progenitor. A few cases can clearly be envisaged as evolution of ferality from the domesticated form, based on mainly on epidemiology. One is feral hexaploid wheat in Tibet (Chapter 11), which must have evolved in Tibet. There are no known wild progenitor forms of hexaploid of wheat (for the reasons described above) and Tibet is rather geographically removed from the Middle East, the center of wheat domestication and amphiploidization of wheat; thus the feral forms most probably came from dedomestication (Chapter 11). There are no known weedy relatives of wheat in Tibet that could have contributed the feral genes.

Another case is the forests of olives that are supplanting native species in Australia. These are feral forms, by definition of where they are growing, caused by bird droppings of olive pits picked up from olive groves. There is no evidence that there are major genetic differences between these and the cultivated olives, but there are minor discernable differences (at the DNA level) between feral olives and their wild oleaster progenitor. Yet if cultivated olive trees are abandoned and not pruned, they rapidly take on the oleaster phenotype (Chapter 15).

The opposite was found in a well-documented case concerning feral sugar beets. The feral weedy forms of beet evolved through crossing with wild beets in seed-production areas (2) or did they? See Chapter 4. Throughout the book there are many other cases where the authors have good reason to suspect that feral forms evolved with the assistance of genes from wild or weedy members of their own or closely related species.

1.2.1.5 Endoferality and Exoferality

We see from this that there are two ways an organism can evolve ferality, and we will coin new terms to quickly describe them:

1. Endoferality — where the species dedomesticates on its own, using back mutations to achieve the feral forms. This includes the wild pigs of the U.S. or New Zealand, alley cats throughout the world, probably the feral wheat in Tibet, and many or most weedy beet strains.
2. Exoferality — where wild relatives contribute feral genes and hasten the evolutionary process. This includes cases where couplings with wolves facilitated to the ferality of wild dogs and many feral weeds.

This author wonders if there is not a cultural bias, even among scientists, that tends to explain phenomena as exoferal that could be endoferal. It is deemed cultural as present culture tends to blame others and not take responsibility. The dog or beet must have become feral because it was mated with a wolf or wild beet; it could not conceivably have done so by itself. The present scientific culture calls what had been know as naturalized weeds as alien weeds, in a gradual cultural evolutionary process where in the middle the same species had been termed foreign.

The new DNA technologies can contribute to our determination of provenance, whether evolution of endoferality or exoferality has occurred. Interspecific gene flow between distantly related species in the same genus or family is rare in human time, but far from uncommon in evolutionary time. It is not hard to determine the origin of feral genes, thanks to polymorphisms and gene rearrangements, as discussed in later chapters.

1.3 THE BIODIVERSITY OF FERAL FORMS AND THEIR EVOLUTION

The symposium leading to this book began with a basic discussion of domestication and then of an elucidation of ferality in the historic records.

Beets, oilseed rape, and birdseed millet are crops that are not fully domesticated as they often have a high frequency of feral traits: bolting and polyspermy (multiply fused seed) in the case of beets and considerable seed shattering in birdseed millet and oilseed rape, as well as undesirable compounds in the oilseed rape. The propensity to evolve ferality of each is discussed separately. Other crops that are more domesticated are then discussed. Transgenic sorghums have yet to be released for fear of introgression with its weedy relatives and weedy hybrid forms growing in or near agricultural fields. The problems are discussed at length. The evolution of the hexaploid weedy wheat in Tibet is then described, followed by that of domestic soybeans growing in its center of origin near wild soybeans. The likelihood of soybeans and maize to dedomesticate outside their centers of origin is then evaluated. Crops such as rye, as well as *Cannabis* for hemp production, are occasionally abandoned and become feral and introduced ornamentals can escape — and some were hardly domesticated to begin with. Rye and ornamentals will be described here. Radishes and sunflowers are cases where exoferality is clearly possible due to weedy relatives in proximity; the evidence is presented.

Rice is the major staple food crop in the world, and conspecific red feral rice is the major weed problem where hand-transplantation has been abandoned (as being a cruel and inhuman procedure, mainly inflicted on women, who prefer other employment) in favor of direct seeding. Wild related species live nearby. Rice is usually cultivated in monoculture, so havoc from feral forms can be expected. Five chapters discuss five interrelated aspects of the situation with rice and the potential for further problems.

It is clear that ferality is a possibility and there are cases where it may be exacerbated by introduced transgenic traits. Thus, a description of ways is presented to contain transgenes or to mitigate their effects so that they do not become established.

A discussion of how parallel areas of evolution (pesticide resistance) has been modeled and how models are limited follows. A conceptual framework was then proposed to allow predictions of how a given transgene or crop situation may affect the transition to ferality, as it might pertain to transgene regulatory considerations.

Could one have understood ferality with fewer case studies? The evidence in the following chapters clearly demonstrates the nuances and differences in the domestication of crops and their dedomestication to feral forms. After reading reasons why one author considers the possibility of evolution of ferality in maize to be extremely remote because of the complexity of the domestication process, one discovers feral or weedy wheat, with an even more complex evolutionary background. One cultivated species needs no mutations to become feral, another many, and maize may never become feral. Another crop can hybridize with its weedy progenitor giving rise to an even more pernicious weedy form. Surprises abound in every chapter. If the reader expects to find a unified theory of ferality, there is good reason to read on to see how the diversity of the plant kingdom *vis-à-vis* mechanisms of evolution of ferality. There is no unified theory, just unified suggestions: appreciate nature, understand the intricacies, use history to predict the future.

1.4 FERALITY AND SCIENTIFIC TERMINOLOGY — A CAUTION

The topics in this book are often contentious, especially those relating to evolution, gene flow, and transgenics. Both advocacy groups as well as (unfortunately) scientists stoop to the use of politically or emotionally loaded terminology to gain support for their case or cause and to marginalize the opposition. Yellow journalism now appears in the editorial as well as research columns of esteemed scientific journals. Emotive terminology has been eschewed throughout this workshop and book with an effort to use scientifically descriptive and value neutral terms. Thus, what some will call (using fascist or racist terminology) "genetic purity," "gene pollution," or "gene contamination" will be correctly called "pollen movement," "gene flow," "hybridization," "gene introgression," or "enhanced biodiversity," depending on the issue and the data. One person's "genetic purity" is another's "inbreeding depression," one's "bastard" or "miscegenation" is another's "heterosis" or "hybrid vigor." Introduced or naturalized species are not be given the ultra-nationalistic/xenophobic "alien" or "foreign" labels. Scientifically meaningless terms, such as genetically modified, genetically modified organism (GMO), living modified organism (LMO), that were coined to mask realities will not be used to hide transgenic, genetically engineered, etc. There was no reason to use "superweeds" when "feral" is fully descriptive. The idea of the authors was not to sell books via emotive hyperbole, but to describe our present state of knowledge (and lack thereof) about ferality and its implications in a rapidly changing agricultural environment.

LITERATURE CITED

1. Abbott RJ, James JK, Milne RI, Gillies ACM. 2003. Plant introductions, hybridization and gene flow. *Phil. Trans. R. Soc. London B* 358:1123–1132.
2. Ellstrand NC. 2003. *Dangerous liaisons — when cultivated plants mate with their wild relatives.* Baltimore, MD: Johns Hopkins University Press. 244 pp.
3. Snow AI, Ed. 2002. *Ecological and agronomic consequences of gene flow from transgenic crops to wild relatives*, http://www.biosci.ohio-state.edu/~asnowlab/Proceedings.pdf, Columbus, OH: Ohio State University.
4. Stewart CN, Halfhill MD, Warwick SI. 2003. Transgene introgression from genetically modified crops to their wild relatives. *Nature Rev. Genet.* 4:806–817.
5. Trut LN. 1999. Early canid domestication; the fox farm experiment. *Am. Sci.* 87:160–169.

2 Crops Come from Wild Plants — How Domestication, Transgenes, and Linkage Together Shape Ferality

Suzanne I. Warwick and C. Neal Stewart, Jr.

2.1 INTRODUCTION

> "Weeds evolved, and are still evolving, within the man-made habitat in three principal ways: (a) from colonizers through selection towards adaptation to continuous habitat disturbance, (b) as derivatives of hybridization between wild and cultivated races of domestic species and (c) through selection towards re-establishing natural seed dispersal mechanisms in abandoned domesticates. Domesticates differ from weeds primarily in degree of dependency on man for survival. They evolved from wild food plants, which were brought into cultivation."
>
> **De Wet and Harlan (25)**

In this chapter, we outline differences between three kinds of plants — domesticated crops, agricultural weeds, and feral crop plants. Both domesticated crops and agricultural weeds have arisen from wild plants, but domesticated crops are different, as they are mostly human inventions selected for certain traits through thousands of years. We will first describe the crop domestication process and the evolution of agricultural weeds in general, contrasting "domestication" with "weediness" traits acquired during the dedomestication or ferality process, including a review of their genetic basis. Then for a select group of crop plants we will examine the degree of *inferred domestication* (ratio of domestic acquired to retained weediness traits). We will also discuss the evidence from the "world-weed" literature to analyze the extent that agronomic problems or losses involve weed species that are either crops or weedy plants that are coadapted and related to crops. As other chapters deal with specific crops, we focus on an examination of genetic factors (e.g., linkage disequilibrium) that might affect crop ferality in general and then briefly describe how biotechnological factors (e.g., transgenes) might affect ferality. However, before we do this, we must briefly examine the concepts of plant domestication, weediness, and ferality.

2.2 DOMESTICATED CROPS, AGRICULTURAL WEEDS, AND FERALITY

2.2.1 The Domestication Process

In the first phase of domestication, plants selected for cultivation were indistinguishable from their wild relatives. Once agriculture became established, humans selected traits that were useful (e.g., self-fertility, non-shattering seeds or seedpods, first tall growth habit to outgrow weeds, and then

TABLE 2.1
Components of Domestication, Weediness, and Wild Syndromes

Domestication Traits	Weedy or Invasive Traits	Wild Non-weedy Traits
Retention of the seed or fruit on the plant at maturity	Propagules that are adapted to long-distance dispersal and easily distributed	Propagules that are not adapted to long-distance dispersal
Loss of germination inhibitors	Seed dormancy	Seed dormancy
Synchrony in germination (loss of secondary dormancy)	Discontinuous germination (secondary dormancy)	Discontinuous germination (secondary dormancy)
Narrow germination requirements	Broad germination requirements	Special germination requirements
—	Ability to germinate in a wide range of conditions	Reduced ability to germinate in a wide range of conditions
Short-lived seeds (no seed bank)	Long-lived seeds (seed bank)	Long-lived seeds (seed bank)
Synchrony of flowering and fruit development	Rapid growth to flowering, annual	Slow growth to flowering, perennial
More determinate growth	Continuous seed production for as long as growing conditions permit	More determinate growth
Smaller numbers of larger fruits or inflorescences	High seed output	Lower seed output
—	Seed produced in a wide range of environmental conditions	Seed produced in a narrower range of environmental conditions
Increase in seed or fruit size	—	—
Reduction in seed dispersal (shattering)	Propagule (seed) shattering	Propagule (seed) shattering
—	Special adaptations for seed dispersal over both short and long distances	No special adaptations for seed dispersal over both short and long distances
Increase in vegetative vigor	Vigorous vegetative reproduction, if perennial	Not vigorous vegetative reproduction, if perennial
Increase in apical dominance	Plasticity of growth	Reduced plasticity of growth
Reduced competitive ability	Strong competitive ability	Reduced competitive ability
Selfing or self-incompatible	Selfing or self-incompatible, but not obligate selfer or apomictic	Selfing or self-incompatible, can be obligate selfer or apomictic
Unspecialized pollinators	Unspecialized pollinators	Specialized pollinators
Adaptation to disturbed habitats	Adaptation to disturbed habitats	Not adapted to disturbed habitats
Polyploidy frequent	Polyploidy frequent	Diploid or polyploid
Increase in starch, sugar, or oil and decrease in protein content of the seed or fruit	—	—
Loss of bitter substances in the seed or fruit	Presence of bitter substances in the seed or fruit (increased pest resistance)	Presence of bitter substances in the seed or fruit (increased pest resistance)

Source: Compilation adapted from Baker (5,6), Doebley (27), and Harlan (48).

with the advent of selective herbicides, a low growth habit to maximize harvest index), with cultivated plants diverging from their wild relatives. The domestication of plants from their wild progenitors over the last 10,000 years has led to the production of a wide variety of monocot and dicot crops that share a number of traits (27,48) (Table 2.1). This similar suite of adaptations to human cultivation is collectively known as the domestication syndrome (48). Syndrome is an appropriate term as some crops may only have a few of the symptoms of domestication, whereas other crops may have all or most of the features listed in Table 2.1. In some cases, domestication traits may be closely, genetically linked with each other, as they would have been subjected to coselection. Human-aided selection of domesticated crops continued with the introduction of foreign genes from geographically or genetically distinct forms and other species to create elite cultivars, and dissemination of these plants throughout the world. Many crops are polyploids or complex aneupolyploidal hybrids between several forms or species from wide-crosses. In some crops, a complete range of intermediates provides a link to the wild, either as weeds or as wild plants in their original range, whereas in other crops such links are no longer evident. Domestication generally results in a loss of genetic diversity compared to the wild progenitor gene pool (69).

2.2.2 Agricultural Weeds

A plant is defined as a *weed* or *weedy* if, in any specified geographical area, its populations grow entirely or predominantly in situations markedly disturbed by man (5). Weed species or even populations within a weed species are often divided into two categories — those on agricultural lands (*agrestals*) and those that occur on waste places as well as along roadsides (*ruderals*) (5). Species or populations that do not occupy such disturbed habitats are considered *wild* or *non-weedy*.

As indicated in the introductory quote, De Wet and Harlan (25) indicated three modes of weed evolution. As with domesticated crops, most agricultural weeds, too, come from wild plants. They are, however, not simply wild plants that interfere with the growth of domesticated crops. They thrive in highly disturbed agroecosystems and have been selected for agriculture (i.e., indirect and inadvertent human selection for coadaptation for high fitness in farmers' fields), including selection for staggered germination, rapid early season growth, and shattering, among other traits. A distinct suite of adaptive plant characteristics is associated with weediness equals feraliness syndrome (5,6,8) (Table 2.1 and Table 2.2). Most of the worst weeds have adapted to be global in distribution, often in a coevolutionary adaptation, as the weeds spread around the world simultaneously with the crops they infest (e.g., *Avena* spp. with wheat, *Amaranthus* spp. with maize).

Many of the major crops are closely related to weeds of agriculture, and, as with crops, they have migrated far beyond their place of origin with unintended human assistance, often evolving seed or seedling mimicry, making it hard for the farmer to distinguish between crop and weed. For example, compatible pairs of non-native North American crops and weeds include wheat and *Aegilops cylindrica* (goatgrass), sorghum and *Sorghum halepense* (johnsongrass), and cultivated and weedy rice.

2.2.3 Crop-Weed-Wild Complex

Some domesticated plants occur as part of a crop-weed-wild complex (25,60,70,104,116). Examples include crops in the genera *Avena*, *Beta*, *Capsicum*, *Chenopodium*, *Daucus*, *Helianthus*, *Hordeum*, *Lactuca*, *Medicago*, *Phaseolus*, *Raphanus*, *Saccharum*, *Sorghum*, and *Zea* (116). The crop-weed-wild complex has been defined as "a compound of crops, accompanying weeds, and wild related species, mutually influencing each other by means of introgression" (Figure 2.1). Wild plants have a genetic architecture that enables them to thrive independently of humans for propagation and independently of human-disturbed habitats, whereas both crops and weeds have a dependency on human-disturbed habitats to grow. Crops and weeds differ from each other insofar as crops are highly dependent on humans for propagation, but most weeds are independent of direct human intervention for propagation. Indeed, weeds are heavily independent of human attempts to prevent their propagation. Although domestication is the end result in areas where crops exist in a crop-weed-wild complex, all intermediary steps toward domestication may be expected (49). The domestication process often involves intentional human-mediated selection of true breeding populations. In such cases, the process of domestication is likely to involve an evolutionary shift from obligate cross-pollination to some degree of self-fertility or selection for isolating barriers that prevent the domesticate from being the female parent of hybrids. With this in mind, Ladizinsky (69,70)

TABLE 2.2
Weediness Characteristics Prevalent in Most Weeds

Weediness Characteristic	Description of the Characteristic
Rapid seedling growth	Allows for maximum capture of growth-limiting factors, such as light, water, and nutrients.
Short vegetative phase	Under those situations where a plant germinates later in the growing season, a plant can complete its life cycle and produce viable seed.
Indeterminant	Flowering throughout the season, while growing vegetatively over an extended period of time.
Self-compatible	Allows for genetic divergence from previous generations, but the plants do not require special pollinators such as bees or other insects to produce viable seed.
High seed output	Plants produce many seeds that are dispersed both spatially and temporally into as many favorable locations as possible for subsequent growth. *Conyza canadensis* (Canada fleabane) plants produce more than 100,000 seeds per plant.
Long-distance seed dispersal	Given that terrestrial plants are not mobile, the only chance for spread is via seed or other propagules. Seeds that float in water or are carried by the wind are more widely distributed. *Taraxacum officinale* (dandelion) seeds can be moved considerable distances in the wind.
Competitive with crop plants	Weeds can compete for light, nutrients, water and substantially reduce crop yield. An average of 54 *Setaria faberi* (giant foxtail) plants per 30 cm of row can reduce grain yields of soybean and corn to 24 and 28%, respectively [68].
Deep root system	Deep root systems allow weeds to thrive in drought conditions. *Convolvulus arvensis* (field bindweed) roots can penetrate up to 3 m deep into the soil.
Discontinuous dormancy	Weed seeds may be dormant in unfavorable environment, ensuring germination when environmental conditions are favorable. *Chenopodium album* (common lamb's-quarters) is a weed that displays extensive dormancy mechanisms.
Environmentally plastic	A plant can change its growth form in response to environmental factors or other control strategies. *Eleusine indica* (goose grass) has an upright growth form under field conditions, but it develops a prostrate shape when it is mowed (such as on a golf course). Both types of plants can produce viable seed.
Dual mode of reproduction	Weeds can reproduce both sexually and asexually. Although most of them reproduce by seeds, many weeds can reproduce asexually. *Cynodon dactylon* (common bermuda grass) can reproduce by stolons and also by seeds.
Allelopathy	Plants produce chemicals that discourage the growth of other plants.

Source: Adapted from Baker (5,6) and Basu et al. (8). Used with permission.

suggested that gene flow between crops and their wild relatives, if it exists, is apparently more effective in the direction from the cultivated to the wild populations. However, in some crops, barriers preventing gene flow from the wild species to the crop may not have been selected, and such gene flow is considered beneficial and contributing to genetic renewal and adaptation of landraces to local conditions (e.g., foxtail millet (*Setaria italica*), Chapter 6). Many genetic factors can affect reciprocal hybridization rates and successful introgression, including differences in outcrossing rates for the crop vs. the weed, ploidy levels of crop vs. the weed, etc. It is, therefore, not possible to generalize about the ease of transfer of genes from wild or weedy species into the crop vs. crop into the wild or weedy species.

2.2.4 Genetics of Domestication and Weediness Traits

Genetic analyses to date have revealed that both domestication and weediness traits are often under relatively simple genetic control. Quantitative trait locus (QTL) mapping can be used to estimate

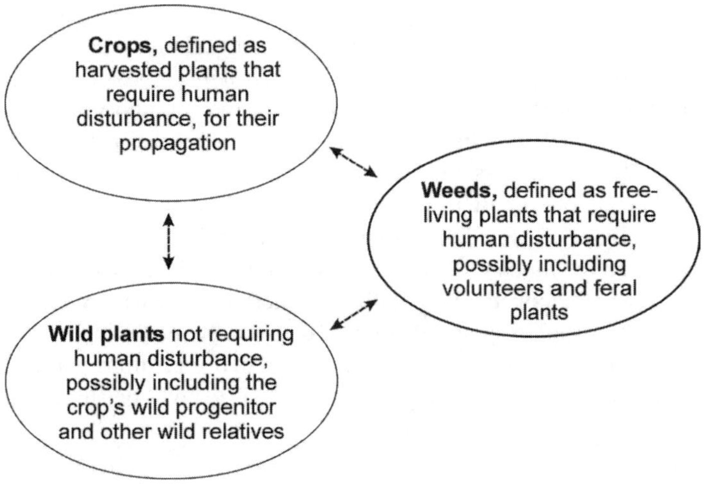

FIGURE 2.1 Crop-weed-wild complex. Arrows signify sporadic hybridization, providing opportunities for introgression. The extent of hybridization and introgression is idiosyncratic, varying with the specific populations involved. (Adapted from Ellstrand (32) with permission.)

the locations, numbers, magnitude of phenotypic effects, and modes of gene action of individual determinants that contribute to the inheritance of quantitative traits. Although one must be careful in equating QTLs with genes, in general the oxymoronic single QTLs with a major effect represent a single gene (often referred to as simple inheritance), and larger numbers of QTLs of more or less equal effects for quantitative traits represent control by larger numbers of structural genes or controlling elements. QTL mapping is an important molecular tool that has been used to better understand the properties of plant domestication (88) and is beginning to have a similar role in studies of weed evolution (8).

2.2.4.1 Domestication Traits

Genetic analyses of domestication traits via QTL mapping in maize revealed that 6 of 10 domestication traits studied, including the striking morphological differences between maize and teosinte, were explained by one major QTL each, even though the QTLs explained less than half of each variance (28,29). Koinange et al. (68) identified major QTLs for 8 of 10 traits involved in the domestication of the common bean. In contrast, genetic analysis of sunflower domestication (15) revealed few major genes (i.e., major QTLs); QTL numbers ranged from 2 to 10 QTLs per trait and only 3 of the 18 domestication-related quantitative traits studied were controlled by major genes. "The occurrence of numerous wild alleles with cultivarlike effects, combined with the lack of major QTLs, suggests that sunflower was readily domesticated" (15).

The question arises: Are there some plant species that lack the genes for domestication? For example, *Dioscorea deltoidea,* the major source of steroid precursors for estrogens and related compounds has not been domesticated despite extensive efforts, although it can be cultivated as cell cultures (113), yet is still commercially collected from the wild.

The majority of the traits that distinguish cultivated plants from their wild relatives are determined by recessive alleles at individual loci with major effects, sometimes modified by a few extra loci of minor effect (39). A recent study of domestication traits in sunflower (15) found that additive traits were the most common class and that dominant and recessive traits were about equal in frequency. In contrast, transgenic traits in plants are almost universally more or less dominant because they invariably represent a "gain of function," even when the function is to suppress a pathway. That contrast is arguably the biggest difference between transgene expression and the expression of other genes that separate crops from their wild relatives (32).

Hybrids forming with transgenic crops are necessarily *hemizygous* (i.e., a situation where the allele in the wild relative is simply an empty location on the chromosome corresponding to the point of insertion in the transgenic plant). As a result of the dominant nature of the allele, hybrids with transgenics will always express the trait, and therefore, be instantly subject to selection. However, the transgene, which in certain circumstances is fully dominant, and in others, partially dominant, would generally be expected to be expressed only half as much as in homozygous plants (43). Natural selection can act much more effectively on dominant alleles. Recent studies in foxtail millet (*Setaria italica*) have demonstrated that different herbicide resistances inherited as recessive, and to a greater extent, chloroplast gene inheritance delayed the evolution of resistant phenotypes compared to a dominant resistance (120). As a result, the majority of crop alleles may be masked in a hybrid from natural selection, whereas transgenes in hybrids will be subject to natural selection (32).

2.2.4.2 Weediness Traits

Approximately half of the weediness traits listed by Baker (5,6) are determined by single genes in species, where known. For many, substitution of a single dominant allele will produce a weedy phenotype in a plant that previously lacked it, suggesting that evolution of a weedy phenotype need not be difficult (64). A single gene difference, or at most a closely linked gene complex, in cultivated oats (*Avena sativa*) is responsible for fatuoid characters in *A. fatua* (wild oats) (121). However, many weedy traits such as rapid plant growth, dormancy, and competitiveness are quantitative traits, but not all are needed to begin the dedomestication transition. The genetic basis of seed dormancy in *Avena fatua* (wild oat) is controlled by two loci associated with promoting early germination and one locus associated with late germination (35). In contrast, multiple loci and epistatic control of seed dormancy was indicated for weedy rice, *Oryza sativa* (42). Three QTLs associated with vegetative dispersal by rhizomes of *Sorghum halepense* (johnsongrass), were identified in crosses of cultivated and wild species of *Sorghum* (89). A single QTL associated with competitiveness against the weed *Lolium rigidum* (annual ryegrass) was identified in wheat (20). Important morphological traits associated with positive competition, such as leaf width and time to anthesis, were used to identify the QTL that mapped to wheat chromosomes 2B and 2D (20).

2.2.4.3 Dedomestication Process

Domestication studies based on QTLs (88) have altered earlier hypotheses (48) that the domestication process was a "relatively slow and gradual progression from the wild state to a state of incipient domestication." Several lines of evidence now support the idea that domestication events may have been rapid, including the relatively small number of genes that appear to control many domestication-related traits, the identification of tightly linked clusters of domestication genes, the levels and patterns of deoxyribonucleic acid (DNA) variation in contemporary gene pools of crops and their ancestors, and the high degree of correspondence among gene locations (*synteny*) related to independent domestication events of diverse cereals crops on different continents. Thus, if domestication events can happen quickly, Paterson (88) suggested that the reversal of such events (i.e., dedomestication process) could also occur with similar speed. However, if domestication traits are generally recessive as discussed above, one would expect the dedomestication process to be quicker than domestication. The genetic basis for key domestication traits will be reviewed for each of the major crop species in subsequent chapters. It is clear, however, that there is a paucity of genetic information on dedomestication or weediness traits (e.g., mode of inheritance, numbers of genes) and that this lack of information currently limits our ability to generalize about the likelihood and speed of the dedomestication process.

2.2.5 Herbicide-Resistant Weedy Biotypes

One of the key adaptive traits that has evolved in agricultural weeds is herbicide resistance in response to herbicide use. Indeed, one of the main concerns about the release of herbicide-resistant transgenic crops is the increased pressure for the selection of such herbicide-resistant weed biotypes. Worldwide, the number of herbicide-resistant weeds has expanded exponentially during the past 3 decades from the first documented occurrence of triazine resistance in 1968 to 287 herbicide-resistant weed biotypes (172 species) in 59 countries by 2004 (53). Herbicide-resistant weed biotypes occur most often to acetolactate synthase (ALS)-inhibiting herbicides (83 species) and triazines (65 species). In contrast, only 6 weed species are reported to date to have evolved resistance to glyphosate, the most extensive herbicide associated with herbicide-resistant crops currently grown. Mutation rates for resistance clearly vary with herbicide type and are estimated at 10^{-6} for ALS-inhibiting herbicides and 10^{-12} to 10^{-20} for triazine herbicides, which are semidominant or recessive depending on rate (41). Mutation rates conferring herbicide resistance to these two herbicide categories may be higher or lower than the likelihood of a wild relative evolving the herbicide resistance trait through interspecific hybridization, depending on the wild species involved.

2.2.6 Crop Ferality

Unlike wild plants, that collectively occupy a variety of ecological niches, agricultural weeds are narrowly selected for inhabiting cropping systems under specific management practices. Crops and agricultural weeds occupy the same habitat and ecological niche, yet have many different selected traits. Feral crop-turned-weeds (*weed mimics*) will form a subset of agricultural weeds in the agroecosystem and this section will focus on understanding and predicting ferality. For example, when does or could a crop volunteer become or lead to a feral population? Volunteer is defined here as crop plants that grow in the same field in subsequent crops or years from a seed bank formed from seed that either shattered from the crop prior to or as a result of harvesting operations or from originally sown seed that did not germinate immediately after sowing. Volunteers rarely persist for more than a season or two, unless they evolve feral traits.

At first glance, it would appear that crops going feral should be widespread; many crops have the capacity to volunteer in subsequent crops. These volunteers may be of agronomic concern when they compete with subsequent crops and result in yield losses, contaminate seed lots and lower seed quality, or serve as alternate hosts for crop pests and diseases (Chapter 23). Such volunteers are likely to have a selective advantage over other agricultural weeds in the agroecosystem as a result of their innate "crop mimicry," as observed for agricultural weeds that mimic crops in their growth pattern (7). This should especially be the case where monoculture is practiced, and it is harder to rogue the volunteers. Similarly, crop-weed or weed-crop hybrid-derived feral forms would also have a selective advantage as the associated negative linkage effects of crop traits introgressing to the weed would not be evident in these plants (45,110). During introgression, one would expect several genes moving from weed to crop or from crop to weed to be linked together on chromosomes that are inherited, at least for a few generations. This effect is often referred to as *linkage disequilibrium,* which is a measure of the non-independence (clustering) of genes in multiple locus gametic types. Such gene clustering might be expected for domestication traits that have been subject to coselection. Therefore, during introgression of a weedy trait into a crop, that trait might be surrounded by a multitude of genes that favor weediness and not domestication. Linkage effects may be much greater for hybridization events involving different species (e.g., *Brassica napus* *B. rapa*) than for the same species (e.g., *Helianthus annuus* (sunflower) *H. annuus* (weedy sunflower); *Oryza sativa* (rice) *O. sativa* (weedy rice)). Linkage effects are also correlated with the mating system of the plant, with increased linkage disequilibrium expected in predominantly selfing and apomictic species compared to predominantly outcrossing species (110). Linkage effects may also be affected by the commonality of shared genomes, the degree of similarity in chromosomal arrangements of genes, or cytological or ploidal differences between the crop and the weed (110). Nevertheless, interspecific hybridization, and the possible heterosis associated with it, is frequently suggested as a stimulus for increased invasive or weedy capacity of new plant introductions (1,33, Chapter 7) or crop-to-weed hybrids (32,34).

In spite of the apparent advantages conferred to volunteers, these modes of weed evolution — crop-turned-weed on its own (*endoferality*) and crop-mediated ferality with the help of related weeds or wild species (*exoferality*) — are rare in nature. This may well be due to the propensity of crop rotation, where the volunteer or its hybrid are no longer competing with the parent crop, but must compete with a different crop and agronomic procedures. Reduced genetic diversity in the crop species would also be reflected in reduced genetic diversity in the volunteer populations and, similar to founder effects in introduced species (99), may hamper the ability of the crop to adapt to new situations or habitats. Although endoferality and exoferality are listed by De Wet and Harlan (25) as possibilities of weed evolution, they seem to be more hypothetical then actual.

But perhaps transgenes will change all that. In a worst-case scenario, we imagine crops (transgenic or not) leading to a specialized weediness syndrome that is the antithesis of croplike domestication. Such a viewpoint is epitomized by Ellstrand (32), who argues that most crops and weeds can interbreed, leading to more problematic weeds. We shall therefore briefly examine the degree of crop domestication, the most important weeds that are related to important crops, the role transgenes might play in dedomestication, and then finally how linkage effects might be expected to ameliorate ferality in a crop or crop-weed hybrids.

2.3 DEGREE OF CROP DOMESTICATION

Crops vary in their degree of *domestication*, that is, proportion of domestication traits vs. wild traits. We will briefly examine 7 crops that demonstrate the range in domestication: cranberry (essentially an almost completely undomesticated crop plant), maize (a highly domesticated crop), and rice, oilseed rape, sugarcane, oat, and sorghum (5 crops of intermediate domestication). The last 5 were selected as they were the only 5, among the 25 most important food crops in the world (32, Table 2.3), known to have an introgressing weedy relative in the list of 180 weeds that are estimated to cause 90% of the weed damage worldwide (55,57). This fact would suggest that crop or crop-weed ferality has not yet played a major role in the development of major weed problems worldwide.

Several food plants are essentially wild and hardly domesticated. None of these plants are responsible for providing many calories for humans. For instance, cranberry (*Vaccinium macrocarpon*) is a New World small fruit plant indigenous to peatlands in the northeastern U.S. and maritime provinces of Canada (31). Most commercial cranberry varieties are essentially wild clones chosen by incisive growers in the 1800s. About 80% of the commercial cranberry crop is composed of four clones that are vegetatively propagated (31). Cranberry, by definition, (while being cultivated in a domestic situation) is little changed from wild cranberry. There has been little breeding for cranberry, even though there does exist some genetic variation in natural populations (109).

Compared with cranberry, virtually all important crops are far more domesticated and have had far more selection for traits that lead to predictable food or fiber production. Crops such as maize and bread wheat never existed in nature. Rice and oilseed rape have undergone significant breeding leading to domestication and are not found in their so-called natural state.

Maize (*Zea mays* ssp. *mays*) putatively arose once from a single teosinte strain (*Zea mays* ssp. *parviglumis*) in southern Mexico (79). It appears that people living in the proximity of the Mexican Balsas River valley selected a proto-maize about 9000 years ago and began to cultivate it for food. Unlike its progenitor, maize carries its seeds on an ear that allows ease of harvest, a trait that also prevents it from escaping cultivation and becoming feral.

Domestication generally results in a loss of genetic diversity compared to the wild progenitor gene pool (69). When neutral alleles or non-domestication-related traits are considered for most

TABLE 2.3
World's 25 Most Important Food Crops, Related Sexually Compatible Weeds, Weed Importance*, and Weed Geographical Distribution

Rank	Crop	Scientific Name	World Area Planted[a] (M Ha)	World Yield[a] (MT)	Related Weeds: Compatible (or not) with Crop [b]	Weed Importance[c]	Citation: Compatible with Crop	Geographical Distribution (55–57)
1	Wheat	*Triticum aestivum*	208	557	***T. aestivum***			Nepal (56)
		T. turgidum durum			***Aegilops cylindrica***		100, 125	Turkey and U.S. (56)
					A. tauschii		117	Mediterranean; Iran (56)
					A. triuncialis		117	Mediterranean; Morocco and Turkey (56)
					A. ventricosa		117	Mediterranean; Morocco (56)
2	Rice	*Oryza sativa*	151	585	***O. sativa***	Group C	17, 72	Worldwide; >50 countries (57)
		O. glaberrima			*O. glaberrima*		17, 72	West Africa
					O. barthii	Group C	17, 72	Subsaharan Africa; Nigeria (56)
					O. longistaminata*		17, 72	Subsaharan Africa
					O. rufipogon	Group C	17, 72	Continental and insular Asia to New Guinea and north Australia; Latin America; Bangladesh (56)
					O. punctata	Group C	17, 72	Nigeria and Swaziland (56)
					***O. officinalis**: not**	Group C		
3	Maize	*Zea mays*	141	636	*Z. mays* ssp. *mexicana*		66	Mexico (66)
4	Soybean	*Glycine max*	84	190	***G. soya***		103	Northeast Asia: Korea, Taiwan, Japan, northeast China, Russia (Siberia); Japan (56)
5	Barley	*Hordeum vulgare*	55	139	***H. vulgare***			Argentina (56)
					H. spontaneum		50	East Mediterranean to Iran and west central Asia; Iran and Jordan (56)
					***H. murinum*: not**	Group C		Worldwide (57); 28 countries (56)
6	Sorghum	*Sorghum bicolor*	44	59	***S. bicolor***			Africa; U.S. (56)
					S. almum			Argentina, Australia, South Africa, and U.S. (56)
					S. halepense	Group A6	4	Worldwide (49); native southwest Asia and adjacent Africa; 51 countries (56)
					S. propinquum			Southeast Asia; Philippines (56)

TABLE 2.3 (continued)
World's 25 Most Important Food Crops, Related Sexually Compatible Weeds, Weed Importance*, and Weed Geographical Distribution

Rank	Crop	Scientific Name	World Area Planted[a] (M Ha)	World Yield[a] (MT)	Related Weeds: Compatible (or not) with Crop [b]	Weed Importance[c]	Citation: Compatible with Crop	Geographical Distribution (55–57)
7	Millet	*Eleusine coracana*	35	29	*E. coracana* ssp. *africana*		23	West Africa
					***E. indica*: not**	Group A5	not 23	Worldwide (55); 74 countries (56)
		Pennisetum glaucum			***P. sieberanum****		24, 78	West Africa, north Namibia
					***P. purpureum*: not**	Group B	not 24	Africa; Central America; Australia (55); 21 countries (56)
8	Cottonseed	*Gossypium hirsutum*	32	57	*G. hirsutum, feral*			Mesoamerica and Caribbean
		G. barbadense			***G. tomentosum*: compatible?**			U.S. (56)
9	Beans, dry, green, and snap	*Phaseolus vulgaris*	28	26	***P. vulgaris****: **weed-crop-wild complex**		10	Peru, Columbia
10	Groundnut (peanut)	*Arachis hypogaea*	26	37	***A. hypogaea***		n/a	Taiwan (56)
11	Rapeseed (canola)	*Brassica napus*, *B. rapa*	24	36	***B. napus***			Europe, Argentina, Australia, Canada, U.S.; 7 countries (56)
					B. juncea		36	Australia, Argentina, Canada, Fiji, Mexico, and U.S. (56)
					B. rapa (B. campestris)	Group C	46, 62, 123	Worldwide (temperate climate); >50 countries (57)
					Hirschfeldia incana (B. adpressa)		74	Europe, Australia, southern Africa, Argentina, USA; Argentina (57)
					Raphanus raphanistrum	Group C	18, 95, 123	Worldwide (temperate climate); 65 countries (57)
					Sinapis arvensis (B. kaber)	Group C	73, 82	Worldwide (temperate climate); 52 countries (57)

12	Sunflower seed	*Helianthus annuus*	21	26	***H. annuus***		76	Mexico, South America, U.S., 11 countries (56)
					H. petiolaris		96, 97	U.S. (56)
13	Sugarcane	*Saccharum officinarum*	20	1350	***S. officinarum***		59, 98	Taiwan (56)
					S. spontaneum	Group C	59, 98	Asia, Africa, Middle East, Mesoamerica; 33 countries (56)
14	Potato	*Solanum tuberosum*	19	311	***S. dulcamara*: not**		not 81	Belize, Canada, New Zealand, Turkey, and U.S. (56)
					***S. nigrum*: not**	Group B	not 81	Worldwide (55); 68 countries (56)
15	Cassava	*Manihot esculenta*	17	188	***M. esculenta****			
					Manihot spp.: all		All 61, 84	
					M . reptans		85	Southwest U.S. south to Argentina
16	Oats	*Avena sativa*	13	26	***A. fatua***	Group A13	114	Worldwide (55); native to Europe, North America, Middle East, and Central Asia; 56 countries (56)
					A. sterilis	Group A13	114	Europe, North America, Middle East, and Central Asia (55); 18 countries (56)
17	Oil palm fruit	*Elaeis guineensis*	11	139	None			
18	Coffee	*Coffea arabica* *C. canephora*	11	7	None			
19	Coconut	*Cocos nucifera*	11	50	***C. nucifera**; feral populations**		51	
20	Chickpea	*Cicer arietinum*	10	7	None			
21	Sweet potato	*Ipomoea batatas*	10	137	***I. trifida***		26	Central and South America; Honduras, and Mexico (56)
					***I. aquatica*: not**	Group C	Not 26	Africa, Asia; 32 countries (56)
					***I. triloba*: not**	Group C	Not 26	Asia, Caribbean, Mesoamerica, and South America, U.S.; 22 countries (56)
22	Cowpea	*Vigna unguiculata*	9	4	***V. unguiculata****		92	Niger, Nigeria (roadside weed)
23	Olive	*Olea europaea*	9	17	***O. europea****		127	Mediterranean basin
24	Rye	*Secale cereale*	8	16	***S. cereale***			Argentina, Finland, Iran, Turkey, U.S. (56)
					S. montanum		112, 127	Mediterranean basin east through Turkey to Iraq, Iran; Turkey (56)

TABLE 2.3 (continued)
World's 25 Most Important Food Crops, Related Sexually Compatible Weeds, Weed Importance*, and Weed Geographical Distribution

Rank	Crop	Scientific Name	World Area Planted[a] (M Ha)	World Yield[a] (MT)	Related Weeds: Compatible (or not) with Crop [b]	Weed Importance[c]	Citation: Compatible with Crop	Geographical Distribution (55–57)
25	Grape	*Vitis vinifera*	7	62	*Vitis* spp.		All 86	
					V. aestivalis			U.S. (56)
					V. candicans			U.S. (56)
					V. hastata			Malaysia (56)
					V. rotundifolia			U.S. (56)
					V. rupestris			U.S. (56)
					V. tiliaefolia			Honduras (56)
					V. trifolia			India (56)
					V. vulpina			U.S. (56)

* Weed importance status according to Holm et al. (55,57).

[a] Area of production (million ha) and world yield (million metric tons) for 2003 from the FAOSTAT Web site; http://faostat.fao.org/default.jsp.

[b] Species name in **bold:** listed as a weed in Holm et al. (56); ***Bold:** listed as a weed in Global Compendium of Weeds at Web site http://www.hear.org/gcw/index.html.

[c] Holm Classification: Group A: ranked 1 to 18 by Holm et al. (55); Group B: ranked 19 to 76 by Holm et al. (55); Group C: ranked as one of 104 worst additional weeds by Holm et al. (57). Unclassified weeds are not in the worst 180 weeds.

cereals, the loss of diversity is about 30% (13, Chapter 11). For domestication-related traits, however, the loss of diversity is much greater. Recent studies on maize have provided insight into the reduced genetic diversity that accompanies the domestication process and whether strong selection for domestication traits, *a priori*, may or may not affect diversity of linked genes. The limits of selection of and the diversity of the *tb1* gene and linked genes were examined for maize (19,119). The *tb1* gene controls apical dominance and branching and during domestication it was subject to strong positive selection for the dominant single-stalk type (unlike the recessive many branched teosinte). The chromosomal region surrounding *tb1* retains extensive diversity and has *low linkage disequilibrium*: that is, few genes segregated with *tb1* in evolutionary time (19). Indeed, it has been reported that linkage disequilibrium declines rapidly in maize with nucleotide distance ($R^2 < 0.1$ within 1500 base pairs) (93). Maize, then, is highly domesticated and does not form feral populations (only rarely volunteers), although gene flow is primarily unidirectional: from teosinte to maize (30). Many crops, such as maize, do not have any weedy derivatives.

Cultivated Asian rice (*Oryza sativa*) provides more calories to people worldwide than any other plant. Its domestication is described in Chapter 16. Both rice and its perennial progenitor *O. rufipogon* have the AA genome and are diploids. *Oryza* contains about 23 species and 9 genome designations (AA through HH and JJ) (67,118). We consider rice to be of intermediate domestication, because of its relative similarity to wild types of rice, such as *O. rufipogon* that can still be found in its site of origin, as well as red rice, a weedy biotype of *O. sativa*.

Oilseed rape (*Brassica napus*) is a crop that regularly forms volunteer populations due to extensive seed shattering and secondary dormancy traits and has several weedy relatives with which mating is possible. *Brassica napus* is an allotetraploid (AACC) that putatively arose through a hybridization event between *B. rapa* (AA) and *B. oleracea* (CC) (115, Chapter 5). Unlike *B. napus*, the progenitor species are composed of crop-type plants as well as either weedy populations (*B. rapa*) (57) or wild-type plants (*B. oleracea*).

Sugarcane (*Saccharum officinarum*) is the most important sucrose-producing crop in the tropics and subtropics and originates from New Guinea (98). Its congeneric major weedy counterpart is *S. spontaneum*, a serious weed of tropical Asia, which also occurs as a sporadic weed in Africa, Europe, and the tropical New World (57). *S. spontaneum* is cross-compatible with the crop, and both crop and weed species are polyploid. There is evidence for ample introgression and convolution of complex *Saccharaum* genomes. *S. officinarum* has chromosome numbers of $2n = 80$, and *S. spontaneum* ranges from $2n = 40$ to 128 (59,98), as discussed further in Chapter 15.

Cultivated oats (*Avena sativa*) is a temperate food and feed crop of decreasing worldwide importance. However, wild oats, mainly *A. fatua* and *A. sterilis*, are important worldwide weeds (55). All three species are allohexaploids ($2n = 42$, AACCDD) (52,124). Cultivated oats are considered to be a secondary domesticate, and the progenitor is likely one of the hexaploid wild oats species (Chapter 15, 75,126). In fact, at least one argument has been made that these three hexaploid oats species should be considered one biological species (71). Interspecific hybrids and introgressed hybrids among the three species have been produced in breeding programs (83) and introgressed genes from *A. sterilis* to *A. sativa* have been especially useful in crop improvement (37). Crop to weed introgression might be hampered by meiotic irregularities (80), and there is no molecular evidence to support crop-to-wild introgression in nature.

Sorghum (*Sorghum bicolor*) is a diploid crop with African origins that is grown for its grain and forage. It is an important crop in temperate, subtropical, and tropical climates. Its chromosome number is $2n = 20$, but the genomic compositions of the crop and its most important weedy relative are unclear (54). The crop will hybridize and introgress with wild populations of the noxious weed *Sorghum halepense* (johnsongrass) ($2n = 40$), which is of Mediterranean origin, as well as among other closely related *Sorghum* species that are not as weedy (89,94), as discussed further in Chapter 8.

In summary, for non-domesticated crops such as cranberry, there is little room for dedomestication, and for highly domesticated crops such as maize, the switch from domestication to feral may be too steep, which may explain why we do not observe many crops gone feral. What about the crops in the middle, such as rice and oilseed rape? Are these two crops typical of many other crops in having prospects for ferality or anomalies?

2.3.1 The Case of Rice Weeds

Red or weedy rice (*Oryza sativa*), which sometimes accompanies temperate rice cultivation, is so-called because it often (but not always) has a red pericarp. It has evolved as a conspecific crop mimic that gets harvested and resown with the rice crop. Red rice has earlier tillering, shattering, and seed dormancy and is also taller than most contemporary green revolution rice cultivars. Bres-Patry et al. (12) suggested that red rice is simply feral cultivated rice that has been dedomesticated, because it has appeared where no wild relative has been observed. Whether this is always the case is discussed in Chapter 16 through Chapter 21. Rice domestication and dedomestication appear to be relatively simple, because many of the major QTLs for the above-listed traits are linked, and only four chromosomal regions are involved with the major domestication traits (12,16). Both red (weedy) rice and cultivated rice are predominantly self-pollinated. When grown together, hybridization rates between red and crop rice are less than 0.5%, dropping to close to 0 at 5 m separation (38). Weed-to-crop gene flow is greater than crop-to-weed (38). The perennial progenitor *Oryza rufipogon* can also hybridize and introgress with crop rice. Hybridization rates are in the neighborhood of 1 to 2%, and there is evidence of introgression at low rates from crop-to-wild (2,63,105,106,111). There are even greater possibilities for cross-pollination in hybrid rice where the inbred lines are selected for non-cleistogamy — protruding anthers and more pollen production.

2.3.2 The Case of Weedy Brassicas

Unlike red rice, there is no such entity as conspecific weedy *B. napus* nor are wild populations of *B. napus* known. However, both volunteer and feral populations of *B. napus* can exist, and introgressive hybridization can occur between *B. napus* and weedy *B. rapa* (43,44,46,123). Volunteer oilseed rape can persist for a minimum of 4 to 5 years after production in Canadian agricultural fields and up to 10 years in Europe (122). Free-living *B. napus* populations have only a 3 to 4 year life span along roadsides in the U.K. (21), whereas another European study reported that roadside populations could be sustained for at least 8 to 9 years (90). *Brassica rapa*, not *B. napus* is listed as one of the world's worst weeds (57), but temporary feral *B. napus* populations might act as reservoir populations for hybridization and introgression with *B. rapa*. To date, there are little to no data available describing the biological changes in volunteer populations that have existed for various durations. One might expect, for example, more shattering and more secondary dormancy in the populations that are volunteers for longer periods. Such comparative data are needed.

Even though endoferality is implicated for only two important crops, it seems to be of significant risk for us to examine potential transgene effects in introgression and exacerbation of ferality. Both rice and oilseed rape have been genetically modified with recombinant DNA techniques. Herbicide-resistant transgenic oilseed rape, first released in 1995, is now commercially grown in large areas (121,122). We therefore have the ideal opportunity with oilseed rape to examine whether transgenes have contributed to increased weediness or ferality since their commercial release. We will also examine linkage effects as they relate to transgenics and non-transgenics and what linkage can tell us about mediating introgression and ferality.

2.4 THE EFFECTS OF TRANSGENES AND GENETIC LINKAGE

It is conceivable that transgenes could play a role in increasing feral tendencies of certain crops (e.g., red rice) or wild-to-crop hybrids (e.g., wild beet to sugar beet (11,22, Chapter 4): wild radish to cultivated radish (87, Chapter 13). Certain categories of transgenic crops pose special risks, particularly those that are hardy, perennial, competitive, open-pollinating, prolific, have a wide range of relatives with which they hybridize, and have an ability to colonize a range of natural and seminatural habitats (121). Examples of such plants are grasses, range, and pasture species.

If so, which transgenes are likely to play a role in evolving ferality and will crop plants that are already able to establish feral volunteer populations more readily evolve into pests if they contain multiple transgenes conferring resistances to insects, environmental stress, disease, and herbicides? As indicated earlier, crops selected by humans over a long period have had their adaptive potential modified, but generally, domesticated crops have been deprived of some of their natural resistance to environmental conditions and tend not to be competitive in the wild. Because tolerance to local stresses determines whether or not any plant, including those that have been genetically transformed can survive, breeding for stress-resistance characteristics that increase survival under domestication has the potential of increasing survival in the wild and therefore of turning cultivated plants into feral forms.

Studies on transgene flow have focused on movement from the transgenic crop to the weedy or wild relative rather than from the weed to the transgenic crop. Indeed transgenes, such as those conferring herbicide resistances, have allowed easy large-scale screening and have served as excellent markers to track interspecific gene flow to weedy relatives growing in commercially transgenic fields of oilseed rape, *Brassica napus* (122,123). Studies have also shown that the movement of genes from the weed to the crop can occur, even when gene flow in the opposite direction was not detected, such as from wild radish to oilseed rape in Australian studies (95). Seefeldt et al. (100) also detected hybrids between herbicide-resistant wheat (pollen recipient) and the weedy species *Aegilops cylindrica*.

2.4.1 Herbicide Resistance

We will first consider herbicide resistance — the most widely introduced transgenic trait (58). As we would expect, given the introduction of different herbicide-resistant oilseed rape types in Canada, the stacking of such genes in weedy crop volunteers is frequently observed (122). A single herbicide resistance trait or gene stacking of such traits does not appear to have changed plant productivity (101,102) and will only contribute to increased ferality of the crop volunteers if the herbicides play a key role in control measures (9). However, without fail-safe mechanisms to contain or mitigate gene flow, one must ponder the effect of introducing herbicide-resistant crop rice, given that *O. sativa* seems to turn feral spontaneously by back mutation and introgressed herbicide resistance has shown up quickly in red rice (Chapter 20). Even though rates of introgression are low, herbicide resistance as a trait (transgenic- or mutagenesis-derived) is likely to lead to further in-field weed problems (110). Studies have shown that introgression of the ALS resistance trait into red rice was 10 times faster than the mutation rate for ALS resistance in red rice (38).

2.4.2 Other Transgenes

With respect to other transgenes, field experiments have shown that an insect resistance transgene (*Bacillus thuringiensis cry1Ac*) can increase the fitness of oilseed rape *B. napus* (108). It seems possible that transgenes will be perpetuated in a reservoir in non-agronomic populations of feral oilseed rape, which might either enable oilseed rape to increase its ferality or be introgressed into weedy *B. rapa*. If the latter were to occur, then it is conceivable that new and novel weeds could be produced (121). In fact, the first transgenic herbicide-resistant *B. rapa* *B. napus* hybrids have been recently reported in Canada (123). The early evidence seems to indicate that growing transgenic varieties could lead to increased weediness in *B. rapa*.

In the case of traits other than herbicide resistance, the role of transgene introgression in increasing ferality is less clear. There are no published data to suggest that introgressed weedy relatives of transgenic crops would possess weedier tendencies, but perhaps more time is needed for this to occur or be seen. In fact, the opposite is the case in sunflowers (14) and oilseed rape

(45). In the case of oilseed rape weedy *B. rapa*, a Bt *cry1Ac* gene that confers lepidopteran insect resistance was transferred to a *B. rapa*-like BC_2F_2 population that was then competed against wheat. In two separate field trials, this transgenic population performed worse or equivalent to the non-transgenic *B. rapa*. In addition, hydroponic growth experiments demonstrated that the transgenic population had growth characteristics more like F_1 hybrids or *B. napus*; it did not grow like non-transgenic *B. rapa*, even though it had *B. rapa* chromosome numbers ($2n = 20$ in contrast to *B. napus*: $2n = 38$) (45). One explanation for the lower competitiveness with wheat of the transgenic weed compared to the wild type was the retention of crop-specific genes, measured in this case by the presence of 15 to 29% of *B. napus* (crop)-specific amplified fragment length polymorphism (AFLP) markers (43). Therefore, the transgenic hybrid population was more croplike genetically than non-transgenic *B. rapa*. Many of the traits, such as disease and insect resistance that are being added transgenically, replace traits lost in crops during domestication, but are still found in related weeds. Thus, there is little potential advantage associated with these kinds of transgenes at this level of backcrossing. What advantage additional backcrossing with the weed might give under selection pressure is an open question.

2.4.3 Ameliorating Ferality

Conceivably, there are several ways that one could prevent a transgenic crop from becoming feral (i.e., prevent it from being a volunteer or reduce the extent of its volunteerism). The first involves utilizing natural variation in the crop species and preselecting lines for transformation that have low dormancy levels or reduced shattering. Good stewardship, including good crop rotations and herbicide rotations are by far the best way to eliminate volunteers. It is also possible for ferality to be prevented, or at least ameliorated by technological means, where, for example, seed viability in the crop is reduced preventing formation of a seed bank or the crop made non-receptive to foreign weed pollen (110). Fitness of volunteers may also be altered by directly engineering fitness-reducing genes onto the same plasmid as the transgene of interest, called *tandem mitigation* (40,41, Chapter 22). As a demonstration of the principle, tobacco was engineered with an herbicide resistance gene coupled with one conferring dwarfism (3). If one were to combine tandem mitigation with random transgene integration into non-transmitting genomic regions, then there would be a lower probability of introgression.

2.5 CONCLUSIONS

It is the exceptional crop that goes feral; most remain tame. In contrast to the contentions of Ellstrand dealing with uncommon, rare cases (32), it seems that most crops generally stay domesticated. It might be, as with maize, that many crops have a fixed suite of domestication genes and that they are now maladapted to survive without human intervention. (Perhaps they may need us as much as we need them (91)). Maize is much like highly domesticated farm animals, such as chickens, that have not gone feral because they are too domesticated to go back. Linkage effects and competition might also prevent crops from evolving back into weeds because they cannot compete with dominant wild-type weeds (45). The trend of ever improving weed control in agriculture during the past 50 years will probably not change the crop-turned-feral situation any time soon. However, caveats to this statement are that fewer new herbicides are likely to be developed in the future and some effective herbicides are losing favor because of environmental impacts. In addition, integrated weed management practices for weed control, such as reduced tillage practices have resulted in weed shifts and may favor volunteers and increased crop ferality.

Has weed-to-crop mating resulted in increased crop ferality and subsequent increased weedy invasions? Among the 180 of the world's most damaging weeds (55,57), only 5 groups (*Avena* weed spp., Brassicaceae weed spp. — *Brassica rapa*, *Raphanus raphanistrum*, *Sinapis arvensis*), *Oryza* spp. (weedy rices), *Saccharum spontaneum*, and *Sorghum halepense*) are sexually compatible with important crop plants (Table 2.3). Although such related weeds could contribute weediness traits to crop-weed-derived feral forms, it would seem that weeds generally do not arise by crop weed hybridization, but by other means. As with most biological phenomena, there are exceptions; see, for example, Chapter 6. According to De Wet and Harlan (25), the most important mode of weed evolution is "from colonizers through selection towards adaptation to continuous habitat disturbance." Studies of gene flow from crop to weeds, to date, have found that weed crop hybrids are either less adaptive or equivalent to both the weedy and crop parent, most likely as a result of linkage effects of associated crop traits. However, such reduced fitness may not be evident in crop weed derived feral forms as crop mimetic-types are likely to be of selective advantage.

With respect to transgenes, herbicide resistance traits in volunteer feral crops would obviously increase their weedy potential and adaptiveness in agricultural settings, but only when those herbicides are used in control measures. Thus, there is a need for diversity in herbicide use in crops, a greater diversity in available herbicide resistance traits, and less reliance on fewer herbicides, and more dependency on crop and herbicide rotations. Transgenes conferring stress-tolerance in the crop (i.e., designed to expand the range of environmental habitats in which the crop can be grown), do need to be given more scrutiny, as these would have obvious affects on invasiveness of the volunteers. Many transgenes will have no effect on weedy traits.

Dedomestication of crops and associated ferality would appear to be restricted to only a few crop groups. They are only of minor importance globally with regard to invasive weed problems especially compared to other plant groups. In North America, the feral plants that cause much of the economic damage are imported horticultural plants (107). For example, Japanese privet (*Ligustrum japonicum*), Japanese honeysuckle (*Lonicera japonica*), and kudzu (*Pueraria montana* var. *lobata)* are all plants that were imported to North America as landscape plants from climatically similar eastern Asia. These vines and shrubs have become naturalized and are now more accurately characterized as problem weeds rather than beautifying ornamentals. Similarly, of the 463 exotic pasture grass and legume species imported and grown in agricultural trials between 1947 and 1985 in Australia, 13% have become weeds (77); and in northern Australia all but two of the worst weeds were introduced as pasture fodder or ornamental plants (65). Unlike annual row crops, these horticultural or forage plants are mostly perennials that have extensive sexual and asexual reproduction. It would seem that these exotic pasture and horticultural crops, turned invasive, would appear to be the feral pigs of the plant world and not the food crops.

ACKNOWLEDGMENTS

The authors wish to thank Drs. M. Halfhill and J. Gressel for helpful discussions and Susan Flood for preparing the figure.

LITERATURE CITED

1. Abbott RJ, James JK, Milne RI, Gillies ACM. 2003. Plant introductions, hybridization and gene flow. *Phil. Trans. R. Soc. Lond. B.* 358:1123–1132.
2. Akimoto M, Shimamoto Y, Morishima H. 1999. The extinction of genetic resources of Asian wild rice, *Oryza rufipogon* Griff.: a case study in Thailand. *Genetic Resources Crop Evol.* 46:419–425.
3. Al-Ahmad H, Galili S, Gressel J. 2004. Tandem constructs to mitigate transgene persistence: tobacco as a model. *Mol. Ecol.* 13:697–710.
4. Arriola PE, Ellstrand NC. 1996. Crop-to-weed gene flow in the genus *Sorghum* (Poaceae): spontaneous interspecific hybridization between johnsongrass, *Sorghum halepense*, and crop sorghum, *S. bicolor. Am. J. Bot.* 83:1153–1160.
5. Baker HG. 1965. Characteristics and modes of origin of weeds. In *The genetics of colonizing species*, Baker HG, Stebbins GL, Eds., 1st ed., pp. 147–168. New York: Academic Press.

6. Baker HG. 1974. The evolution of weeds. *Annu. Rev. Ecol. Syst.* 5:1–24.
7. Barrett SCH. 1983. Crop mimicry in weeds. *Econ. Bot.* 37:255–282.
8. Basu C, Halfhill MD, Mueller TC, Stewart CN Jr. 2004. Weed genomics: new weapons to defeat an old enemy. *Trends Plant Sci.* 9:391–398.
9. Beckie HJ, Séguin-Swartz G, Warwick SI, Johnson E. 2004. Multiple herbicide resistant canola (*Brassica napus*) can be controlled by alternative herbicides. *Weed Sci.* 52:152–157.
10. Beebe S, Toro ChO, González AV, Chacón MI, Debouck DG. 1997. Wild-weed-crop complexes of common bean (*Phaseolus vulgaris* L., Fabaceae) in the Andes of Peru and Colombia, and their implications for conservation and breeding. *Genet. Resources Crop Evol.* 44:73–91.
11. Boudry P, Mörchen M, Saumitou-Laprade P, Vernet P, Van Dijk H. 1993. The origin and evolution of weed beets: consequences for the breeding and release of herbicide-resistant transgenic sugar beets. *Theor. Appl. Genet.* 87:471–478.
12. Bres-Patry C, Lorieux M, Clément G, Bangratz M, Ghesquière A. 2001. Heredity and genetic mapping of domestication-related traits in a temperate *japonica* weedy rice. *Theor. Appl. Genet.* 102:118–126.
13. Buckler IV ES, Thornsberry JM, Kresovich S. 2001. Molecular diversity and domestication of grasses. *Genet. Res.* 77:213–218.
14. Burke JM, Rieseberg LH. 2003. Fitness effects of transgenic disease resistance in sunflowers. *Science* 300:1250.
15. Burke JM, Tang S, Knapp SJ, Rieseberg LH. 2002. Genetic analysis of sunflower domestication. *Genetics* 161:1257–1267.
16. Cai HW, Morishima H. 2002. QTL clusters reflect character associations in wild and cultivated rice. *Theor. Appl. Genet.* 104:1217–1228.
17. Chang TT. 1995. Rice. In *Evolution of crop plants*, Smartt J, Simmonds NW, Eds., 2nd ed., pp. 147–155. Harlow: Longman Group UK.
18. Chèvre AM, Eber F, Darmency H, Fleury A, Picault H, Letanneur JC, Renard M. 2000. Assessment of interspecific hybridization between transgenic oilseed rape and wild radish under agronomic conditions. *Theor. Appl. Genet.* 100:1233–1239.
19. Clark RM, Linton E, Messing J, Doebley JF. 2004. Pattern of diversity in the genomic region near the maize domestication gene *tb1*. *Proc. Natl. Acad. Sci. USA* 101:700–707.
20. Coleman RK, Gill GS, Rebetzke GJ. 2001. Identification of quantitative trait loci for traits conferring weed competitiveness in wheat (*Triticum aestivum* L.) *Aust. J. Agric. Res.* 52:1235–1246.
21. Crawley MJ, Brown SL. 1995. Seed limitation and the dynamics of feral oilseed rape on the M25 motorway. *Proc. Roy. Soc. Lond. Ser. B Biol. Sci.* 259:49–54.
22. Desplanque B, Boudry P, Broomberg K, Saumitou-Laprade P, Cuguen J, Van Dijk H. 1999. Genetic diversity and gene flow between wild, cultivated and weedy forms of *Beta vulgaris* L. (Chenopodiaceae), assessed by RFLP and microsatellite markers. *Theor. Appl. Genet.* 98:1194–1201.
23. De Wet JMJ. 1995. Finger millet. In *Evolution of crop plants*, Smartt J, Simmonds NW, Eds., 2nd ed., pp. 137–140. Harlow: Longman Group UK.
24. De Wet JMJ. 1995. Pearl millet. In *Evolution of crop plants*, Smartt J, Simmonds NW, Eds., 2nd ed., pp. 156–159. Harlow: Longman Group UK.
25. De Wet JMJ, Harlan JR. 1975. Weeds and domesticates: evolution in the man-made habitat. *Econ. Bot.* 29:99–107.
26. Diaz J, Schmiediche, P, Austin DF. 1996. Polygon of crossability between eleven species of *Ipomoea*: section *Batatas* (Convolvulaceae). *Euphytica* 88:189–200.
27. Doebley J. 1992. Molecular systematics and crop evolution. In *Molecular systematics of plants*, Soltis PS, Soltis DE, Doyle JJ, Eds., pp. 202–222. New York: Chapman and Hall.
28. Doebley J, Stec A. 1991. Genetic analysis of the morphological differences between maize and teosinte. *Genetics* 129:285–295.
29. Doebley J, Stec A. 1993. Inheritance of the morphological differences between maize and teosinte: comparison of results for two F_2 populations. *Genetics* 134:559–570.
30. Doebley J, Goodman MM, Stuber CW. 1987. Patterns of isozyme variation between maize and Mexican annual teosinte. *Econ. Bot.* 41:234–246.
31. Eck P. 1990. *The American cranberry.* New Brunswick: Rutgers University Press. 420 pp.
32. Ellstrand NC. 2003. *Dangerous liaisons? When cultivated plants mate with their wild relatives.* Baltimore, MD: Johns Hopkins University Press. 244 pp.

33. Ellstrand NC, Schierenbeck KA. 2000. Hybridization as a stimulus for the evolution of invasiveness in plants? *Proc. Natl. Acad. Sci. USA* 97:7043–7050.
34. Ellstrand NC, Prentice HC, Hancock JF. 1999. Gene flow and introgression from domesticated plants into their wild relatives. *Ann. Rev. Ecol. Syst.* 30:539–563.
35. Fennimore SA, Nyquist WE, Shaner GE, Doerge RW, Foley ME. 1999. A genetic model and molecular markers for wild oat (*Avena fatua* L.) seed dormancy. *Theor. Appl. Genet.* 99: 711–718.
36. Frello S, Hansen KR, Jensen J, Jørgensen RB. 1995. Inheritance of rapeseed (*Brassica napus*)-specific RAPD markers and a transgene in the cross *B. juncea* x (*B. juncea* x *B. napus*). *Theor. Appl. Genet.* 91:236–241.
37. Frey KJ. 1994. Remaking a crop gene pool: the case history of *Avena*. *Proc. Natl. Sci. Council Rep. China Part B Life Sci.* 18:85–93.
38. Gealy DR, Mitten DH, Rutger JN. 2003. Gene flow between red rice (*Oryza sativa*) and herbicide-resistant rice (*O. sativa*): implications for weed management. *Weed Technol.* 17:627–645.
39. Gepts P. 2001. Origins of plant agriculture and major crop plants. In *Our fragile world: challenges and opportunities for sustainable development*, Tolba MK, Ed., pp. 627–637. Oxford: EOLSS Publishers.
40. Gressel J. 1999. Tandem constructs: preventing the rise of superweeds. *Trends Biotechnol.* 17:361–366.
41. Gressel J. 2002. *Molecular biology of weed control.* London: Taylor and Francis. 520 pp.
42. Gu X-Y, Kianian SF, Foley ME. 2004. Multiple loci and epistases control genetic variation for seed dormancy in weedy rice (*Oryza sativa*). *Genetics* 166:1503–1516.
43. Halfhill MD, Millwood RJ, Weissinger AK, Warwick SI, Stewart CN Jr. 2003. Additive transgene expression and genetic introgression in multiple green-fluorescent protein transgenic crop x weed hybrid generations. *Theor. Appl. Genet.* 107:1533–1540.
44. Halfhill MD, Richards HA, Mabon SA, Stewart CN Jr. 2001. Expression of GFP and Bt transgenes in *Brassica napus* and hybridization with *Brassica rapa*. *Theor. Appl. Genet.* 103:659–667.
45. Halfhill M, Sutherland JP, Poppy GM, Warwick SI, Rufty TW, Weissinger AK, Raymer PL, Stewart CN Jr. 2004. Growth, fitness, and competitiveness of introgressed weedy *Brassica rapa* hybrids with a Bt transgene. ABIC Poster Abstracts, Cologne, Germany, p. 9.
46. Hansen LB, Siegismund HR, Jørgensen RB. 2001. Introgression between oilseed rape (*Brassica napus* L.) and its weedy relative *B. rapa* L. in a natural population. *Genetic Resources Crop Evol.* 48:621–627.
47. Harlan JR. 1965. The possible role of weedy races in the evolution of cultivated plants. *Euphytica* 14:173–176.
48. Harlan JR. 1992. *Crops and man.* 2nd ed. Madison, WI: American Society of Agronomy. 284 pp.
49. Harlan JR. 1992. Origin and processes of domestication. In *Grass evolution and domestication*, Chapman GP, Ed., pp. 159–175. Cambridge: Cambridge University Press.
50. Harlan JR. 1995. Barley. In *Evolution of crop plants,* Smartt J, Simmonds NW, Eds., 2nd ed., pp. 140–147. Harlow: Longman Group UK.
51. Harries HC. 1995. Coconut. In *Evolution of crop plants*, Smartt J, Simmonds NW, Eds., 2nd ed., pp. 389–398. Harlow: Longman Group UK.
52. Hayasaki M, Morikawa T, Tarumoto I. 2000. Intergenomic translocations of polyploid oats (genus *Avena*) revealed by genomic *in situ* hybridization. *Genes Genet. Syst.* 75:167–171.
53. Heap I. 2004. The International survey of herbicide resistant weeds. http://www.weedscience.com.
54. Hoang T, Liang GH. 1988. The genomic relationship between cultivated sorghum *Sorghum bicolor* (L.) Moench and johnsongrass *Sorghum halepense* (L.) Pers.: a re-evaluation. *Theor. Appl. Genet.* 76:277–284.
55. Holm LG, Plucknett DL, Pancho JV, Herberger JP. 1977. *The world's worst weeds: distribution and biology.* Honolulu: University Press of Hawaii. 609 pp.
56. Holm L, Pancho JV, Herberger JP, Plucknett DL. 1979. *A geographical atlas of world weeds.* New York: John Wiley & Sons. 391 pp.
57. Holm L, Doll J, Holm E, Pancho JV, Herberger JP. 1997. *World weeds: natural histories and distribution.* New York: John Wiley & Sons. 1152 pp.
58. James C. 2003. Global status of commercialized transgenic crops: 2003. ISAAA Briefs No. 30. Ithaca, NY: ISAAA (International Service for the Acquisition of Agri-Biotech Applications). http://www.isaaa.org/.
59. Jannoo N, Grivet L, Seguin M, Paulet F, Domaingue R, Rao PS, Dookun A, D'Hont A, Glaszmann JC. 1999. Molecular investigation of the genetic base of sugarcane cultivars. *Theor. Appl. Genet.* 99:171–184.

60. Jarvis DI, Hodgins T. 1999. Wild relatives and crop cultivars: detecting natural introgression and farmer selection of new genetic combinations in agroecosystems. *Mol. Ecol.* 8:S159–S173.
61. Jennings DL. 1995. Casava. In *Evolution of crop plants*, Smartt J, Simmonds NW, Eds., 2nd ed., pp. 128–132. Harlow: Longman Group UK.
62. Jørgensen RB, Andersen B. 1994. Spontaneous hybridization between oilseed rape (*Brassica napus*) and weedy *B. campestris* (Brassicaceae): a risk of growing genetically modified oilseed rape. *Amer. J. Bot.* 81:1620–1626.
63. Kanazawa A, Akimoto M, Morishima H, Shimamoto Y. 2000. Inter-and intra-specific distribution of *Stowaway* transposable elements in AA-genome species of wild rice. *Theor. Appl. Genet.* 101:327–335.
64. Keeler KH. 1989. Can genetically engineered crops become weeds? *Biotechnology* 7:1134–1139.
65. Keighery GJ. 1995. The ecological consequences of genetic engineering. *Search* 26:274–276.
66. Kermicle J. 1997. Cross compatibility within the genus *Zea*. In *Gene flow among maize landraces, improved maize varieties, and teosinte: implications for transgenic maize,* Serratos JA, Willcox MC, Castillo-González F, Eds., pp. 40–43. Mexico. DF: CIMMYT.
67. Khush GS. 1997. Origin, dispersal, cultivation and variation of rice. *Plant Mol. Biol.* 35:25–34.
68. Koinange EMK, Singh SP, Gepts P. 1996. Genetic control of the domestication syndrome in common bean. *Crop Sci.* 36:1037–1045.
69. Ladizinsky G. 1985. Founder effect in crop plant evolution. *Econ. Bot.* 39:191–199.
70. Ladizinsky G. 1998. *Plant evolution under domestication.* New York: Kluwer Academic Publishers. 254 pp.
71. Ladizinsky G, Zohary D. 1971. Notes on species delimitation, species relationships and polyploidy in *Avena. Euphytica* 20:380–395.
72. Langevin S, Clay K, Grace JB. 1990. The incidence and effects of hybridization between cultivated rice and its related weed rice rice (*Oryza sativa* L.) *Evolution* 44:1000–1008.
73. Lefol E, Danielou V, Darmency H. 1996. Predicting hybridization between transgenic oilseed rape and mustard. *Field Crops Res.* 45:153–161.
74. Lefol E, Fleury A, Darmency H. 1996. Gene dispersal from transgenic crops. II. Hybridization between oilseed rape and the wild hoary mustard. *Sexual Plant Reprod.* 9:189–196.
75. Li C-D, Rossnagel BG, Scoles GJ. 2000. Tracing the phylogeny of the hexaploid oat *Avena sativa* with satellite DNAs. *Crop Sci.* 40:1755–1763.
76. Linder CR, Taha I, Seiler GJ, Snow AA, Rieseberg LH. 1998. Long-term introgression of crop genes into wild sunflower populations. *Theor. Appl. Genet.* 96:339–347.
77. Lonsdale WM. 1994. Inviting trouble: introduced pasture species in northern Australia. *Aust. J. Ecol.* 19:345–354.
78. Marchais L. 1994. Wild pearl millet population (*Pennisetum glaucum*, Poaceae) integrity in agricultural Sahelian areas. An example from Keita (Niger). *Plant Syst. Evol.* 189:233–245.
79. Matsuoka Y, Vigouroux Y, Goodman MM, Sanchez JG, Buckler E, Doebley J. 2002. A single domestication for maize shown by multilocus microsatellite genotyping. *Proc. Natl. Acad. Sci. USA* 99:6080–6084.
80. McMullen MS, Phillips RL, Stuthman DD. 1982. Meiotic irregularities in *Avena sativa* x *Avena sterilis* hybrids and breeding implications. *Crop Sci.* 22:890–897.
81. McPartlan HC, Dale PJ. 1994. An assessment of gene transfer by pollen from field-grown transgenic potatoes to non-transgenic potatoes and related species. *Transgenic Res.* 3:216–225.
82. Moyes CL, Lilley JM, Casais CA, Cole SG, Haeger PD, Dale PJ. 2002. Barriers to gene flow from oilseed rape (*Brassica napus*) into populations of *Sinapis arvensis*. *Mol. Ecol.* 11:103–112.
83. Murphy JP, Frey KJ. 1984. Comparisons of oat populations developed by intraspecific and interspecific hybridization. *Crop Sci.* 24:531–536.
84. Nassar NMA. 1980. Attempts to hybridize wild *Manihot* species with cassava. *Econ. Bot.* 34:13–15.
85. Nassar NMA. 1984. Natural hybridization between *Manihot reptans* Pax and *M. alutacea* Rogers & Appan. *Canad. J. Plant Sci.* 64:423–425.
86. Olmo HP. 1995. Grapes. In *Evolution of crop plants*, Smartt J, Simmonds NW, Eds., 2nd ed., pp. 485–490. Harlow: Longman Group UK.
87. Panetsos CA, Baker HG. 1967. The origin of variation in "wild" *Raphanus sativus* (Cruciferae) in California. *Genetica* 38:243–274.

88. Paterson AH. 2002. What has QTL mapping taught us about plant domestication? *New Phytol.* 154:591–608.
89. Paterson AH, Schertz KF, Lin Y-R, Liu S-C, Chang Y-L. 1995. The weediness of wild plants: molecular analysis of genes influencing dispersal and persistence of johnsongrass, *Sorghum halepense* (L.) Pers. *Proc. Natl. Acad. Sci. USA* 92:6127–6131.
90. Pessel FD, Lecomte J, Emeriau V, Krouti M, Messean A, Gouyon PH. 2001. Persistence of oilseed rape (*Brassica napus* L.) outside of cultivated fields. *Theor. Appl. Genet.* 102:841–846.
91. Pollan M. 2001. The botany of desire: a plant's eye view of the world. New York: Random House. 271 pp.
92. Rawal KM. 1975. Natural hybridization among wild, weedy and cultivated *Vigna unguiculata* (L.) Walp. *Euphytica* 24:699–707.
93. Remington DL, Thornsberry JM, Matsuoka Y, Wilson LM, Whitt SR, Doebley J, Kresovich S, Goodman MM, Buckler IV ES. 2001 Structure of linkage disequilibrium and phenotypic associations in the maize genome. *Proc. Natl. Acad. Sci. USA* 98:11479–11484.
94. Renganayaki K, Amirthadevaranthinam A, Sadasivam S. 2000. Species relationship and hybrid identification in sorghum using RAPD, protein and isozyme techniques. *J. Genet. Breed.* 54:117–124.
95. Rieger MA, Potter TD, Preston C, Powles SB. 2001. Hybridization between *Brassica napus* L. and *Raphanus raphanistrum* L. under agronomic field conditions. *Theor. Appl. Genet.* 103:555–560.
96. Rieseberg LH, Kim MJ, Seiler GJ. 1999. Introgression between the cultivated sunflower and a sympatric wild relative, *Helianthus petiolaris* (Asteraceae). *Int. J. Plant Sci.* 160:102–108.
97. Rieseberg LH, Whitton J, Gardner K. 1999. Hybrid zones and the genetic architecture of a barrier to gene flow between two wild sunflower species. *Genetics* 152:713–727.
98. Roach BT. 1995. Sugar canes. In *Evolution of crop plants*, Smartt J, Simmonds NW, Eds., 2nd ed., pp. 160–166. Harlow: Longman Group UK.
99. Sakai AK, Allendorf FW, Holt JS, Lodge DM, Molofsky J, With KA, Baughman S, Cabin RJ, Cohen JE, Ellstrand NC, McCauley DE, O'Neil P, Parker IM, Thompson JN, Weller SG. 2001. The population biology of invasive species. *Ann. Rev. Ecol. Syst.* 32:305–332.
100. Seefeldt SS, Zemetra R, Young FL, Jones SS. 1998. Production of herbicide-resistant jointed goatgrass (*Aegilops cylindrica*) x wheat (*Triticum aestivum*) hybrids in the field by natural hybridization. *Weed Sci.* 46:632–634.
101. Simard M-J, Légère A, Séguin-Swartz G, Nair H, Warwick S. 2003. Fitness of double (imidazolinone+glufosinate) vs. single herbicide resistant canola (*Brassica napus*). *Weed Sci. Soc. Am. Abstract* 43:75–76.
102. Simard M-J, Légère A, Séguin-Swartz G, Nair H, Warwick S. 2004. Fitness of triple (imidazolinone+glufosinate+glyphosate) vs. single herbicide resistant canola (*Brassica napus*). *Weed Sci. Soc. Am. Abstract* 44:67.
103. Singh RJ, Hymowitz T. 1989. The genomic relationships among *Glycine soja* Sieb. and Zucc., *G. max* (L.) Merr. and '*G. gracilis*' Skvortz. *Plant Breed.* 103:171–173.
104. Small E. 1984. Hybridization in the domesticated-weed-wild complex. In *Plant biosystematics*, Grant WF, Ed., pp. 195–210. Toronto: Academic Press.
105. Song Z, Lu B, Zhu Y, Chen J. 2002. Pollen competition between cultivated and wild rice species (*Oryza sativa* and *O. rufipogon*). *New Phytol.* 153:289–296.
106. Song ZP, Lu B-R, Zhu YG, Chen JK. 2003. Gene flow from cultivated rice to the wild species *Oryza rufipogon* under experimental field conditions. *New Phytol.* 157:657–665.
107. Stewart CN Jr. 2004. Genetically modified planet: environmental impacts of genetically engineered plants. New York: Oxford University Press. 256 pp.
108. Stewart CN Jr, All JN, Raymer PL, Ramachandran S. 1997. Increased fitness of transgenic insecticidal rapeseed under insect selection pressure. *Mol. Ecol.* 6:773–779.
109. Stewart CN Jr., Excoffier L. 1996. Assessing population genetic structure and variability with RAPD data: application to *Vaccinium macrocarpon* (American cranberry). *J. Evol. Biol.* 9:153–171.
110. Stewart CN Jr, Halfhill MD, Warwick SI. 2003. Transgene introgression from genetically modified crops to their wild relatives. *Nature Rev. Genet.* 4:806–817.
111. Suh HS, Sato YI, Morishima H. 1997. Genetic characterization of weedy rice (*Oryza sativa* L.) based on morpho-physiology, isozymes and RAPD markers. *Theor. Appl. Genet.* 94:316–321.

112. Sun M, Corke H. 1992. Population genetics of colonizing success of weedy rye in northern California. *Theor. Appl. Genet.* 83:321–329.
113. Tal B, Rokem JS, Gressel J, Goldberg I. 1984. The effect of chlorophyll-bleaching herbicides on growth, carotenoid and diosgenin levels in cell suspension cultures of *Dioscorea deltoidea*. *Phytochemistry* 23:1333–1335.
114. Thomas H. 1995. Oats. In *Evolution of crop plants*, Smartt J, Simmonds NW, Eds., 2nd ed., pp. 132–137. Harlow: Longman Group UK.
115. U N. Genomic-analysis in *Brassica* with special reference to the experimental formation of *B. napus* and peculiar mode of fertilization. *Jpn. J. Bot.* 7:389–452.
116. Van Raamsdonk LWD, Van der Maesen LJG. 1996. Crop-weed complexes: the complex relationship between crop plants and their wild relatives. *Acta Bot. Neerl.* 45:135–155.
117. Van Slageren MW. 1994. Wild wheats: A monograph of *Aegilops* L. and *Amblyopyrum* (Jaub. & Spach) Eig (Poaceae). Wageningen, Netherlands: Wageningen Agricultural University. 513 pp.
118. Vaughan DA, Morishima H, Kadowaki K. 2003. Diversity in the *Oryza genus*. *Curr. Opin. Plant Biol.* 6:139–146.
119. Wang R-L, Stec A, Hey J, Lukens L, Doebley J. 1999. The limits of selection during maize domestication. *Nature* 398:236–239.
120. Wang T, Li Y, Shi Y, Reboud X, Darmency H, Gressel J. 2004. Low frequency transmission of a plastid-encoded trait in *Setaria italica*. *Theor. Appl. Genet.* 108:515–520.
121. Warwick SI, Beckie HJ, Small E. 1999. Transgenic crops: new weed problems for Canada? *Phytoprotection* 80:71–84.
122. Warwick SI, Beckie HJ, Simard MJ, Légère A, Nair H, Séguin-Swartz G. 2004. Chapter 24. Environmental and agronomic consequences of herbicide resistance (HR) canola in Canada. In *Introgression from genetically modified plants L(GMP) into wild relatives*, den Nijs HCM, Bartsch D, Sweet J, Eds., Wallingford UK: CABI Publishing, pp. 323–337.
123. Warwick SI, Simard MJ, Légère A, Beckie HJ, Braun L, Zhu B, Mason P, Séguin-Swartz G, Stewart CN. 2003. Hybridization between transgenic *Brassica napus* L. and its wild relatives: *B. rapa* L., *Raphanus raphanistrum* L., *Sinapis arvensis* L., and *Erucastrum gallicum* (Willd.) O.E. Schulz. *Theor. Appl. Genet.* 107:528–539.
124. Yang Q, Hanson L, Bennett MD, Leitch IJ. 1999. Genome structure and evolution in the allohexaploid weed *Avena fatua* L. (Poaceae). *Genome* 42:512–518.
125. Zemetra RS, Hansen J, Mallory-Smith CA. 1999. Potential for gene transfer between wheat (*Triticum aestivum*) and jointed goatgrass (*Aegilops cylindrica*). *Weed Sci* 46:313–317.
126. Zhou X, Jellen EN, Murphy JP. 1999. Progenitor germplasm of domesticated hexaploid oat. *Crop Sci.* 39:1208–1214.
127. Zohary D. 2000. *Domestication of plants in the old world: the origin and spread of cultivated plants in West Asia, Europe, and the Nile Valley.* 3rd ed. Oxford: Oxford University Press. 316 pp.

3 The Ecology and Detection of Plant Ferality in the Historic Records

Klaus Ammann, Yolande Jacot, and Pia Rufener Al Mazyad

3.1 GENERAL INTRODUCTION

The definitions of the terms *weed, weediness, feral crops,* and *domestication of crops* are discussed and defined in Chapter 2. From the ecologists' perspective, invasions of weeds into seminatural and natural plant communities are potentially risky (3,4). These problems are not focused on transgenic plants alone. It is just the bad luck of this new and elegant technique that it offers more clarity to the agroecologists, because transgenes can be precisely followed on a long-term basis. It is now emerging through new field research that, according to Sears et al. (45), the present day transgenes (e.g., Bt proteins to create insect resistance) do not have a dramatic impact on ecosystems in the field, while earlier lab studies of Losey et al. (30) and also Hilbeck et al. (21) showing toxic effect are questioned in the light of new studies (38). It will be a future chance to follow transgenes and their fate in agricultural systems. Thus, we will learn about crops and their wild relatives and feral populations.

But even crops themselves may cause a weed problem according to Schlink (43) and others (44), specifically in the last 2 decades, as studies on oilseed rape have shown. Volunteer oilseed rape (*Brassica napus*) has to be controlled by means of an adapted crop rotation system and tillage techniques, sometimes combined with herbicide applications. Furthermore, feral populations can often be observed in disturbed habitats outside agrosystems. How long such a population can survive needs to be checked. In addition, its potential for invading natural plant communities has to be analyzed. Appropriate long-term monitoring systems following potentially problematic weeds and their potential (natural) habitats and changes in the pattern of wild species have to be detected at an early stage. Following Hartmann et al. (20), it is an illusion that problematic weed types can easily be eradicated. Therefore, early detection of such weed types is essential.

Overall, the discussion shows that aggressive weeds are not directly related to the new technologies and not related to most feral populations, but, in worst-case scenarios, it is imaginable that the transgenes may contribute. It will be of great importance to follow up pertinent cases and make good use of the new precision the technology offers. The historic record provides considerable information on these long-term processes and much can be learned from it.

3.2 REVERSION OF CROPS TO WILD TYPES

According to Sukopp and Sukopp (46) there is no historic reference to a case where crops have totally been reverted to their wild type or where they would have lost all domestication characters. Centuries or even millennia of domestication obviously cannot be taken back easily. However, the example below shows that a single gene change may be sufficient to revert a crop toward a wild type. This has been shown by a photograph of Schwanitz given by Rauber (37) with the example of maize, which turned into "corn-grass" having a much smaller size in stem and leaf. Still, this is

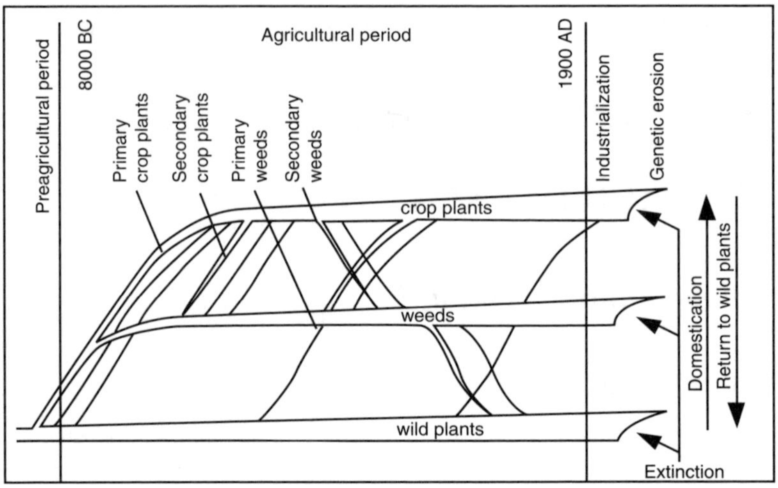

FIGURE 3.1 Complex, multiple origin of feral crops: feral crops are (or can be) of multiple origin. (Modified from Gladis and Hammer et al. (13,17). Used with permission.)

again not strictly a reversion into a wild or weedy grass. In cultivars that have a low degree of domestication, one mutation can cause the weedy form, which then successfully spreads (6,46). The loss of spikelet spindle toughness of cereals, for example, is sufficient for regaining the ability to spread seeds (e.g., *Avena*). Another example with *Avena sativa* with its fatuoid mutant: loss of a combination of genes which suppress awn, pubescence, and easy dehiscence of caryopses transform *Avena sativa* back into nearly wild plants (39). Gladis (13) and Hammer (17) have shown how complex the situation is Figure 3.1: whenever you find a feral population of any crop, one should be aware, according to Hammer and others, of the multiple possible origins of the taxon studied (17).

3.3 HISTORICAL ACCOUNTS OF FERAL CROPS

First, it has to be stated that the data for feral crops are still scanty. A thorough debate about morphological characters of domestication is given with many references (in German) (26). In a lucid account, the authors also refer to the confusing situation in nomenclature: modern names as defined on grounds of a genetic analysis are coined by MacKey and others (31,47). Nomenclature of hulled wheats, for instance, has bifurcated in modern names based on genetics as well as the traditional names in use in archaeobotany. Consequently, even on the simple basis of names, we always have to consider the history of how (and through whom) the name was assigned. Hanelt (18) gives a list of lesser known or forgotten cruciferous crops, most of them presently also surviving as feral populations: *Amaranthus* div. spec., *Mercurialis annua,* surprisingly *Bellis perennis, Cochlearia officinalis, Coronopus squamatus, Coronopus hortensis, Isatis tinctoria, Nasturtium officinale.*

3.3.1 Methods of Detection

3.3.1.1 Archaeobotanical Methods

Archaeobotanical methods are plentiful; they are extensively described with many examples in (26). One of the most important methods is washing (*elutriating*) the samples to purify and concentrate all plant remains. It is worthwhile to critically examine the details of such purifying methods, because details in the procedure can heavily influence the results and resulting statistics. After proper documentation, analysis can reveal many details, shown here with the example of hulled wheat.

Identification of archaeological wheat remains was discussed at a meeting of 25 archaeobotanists in London in 1992 and the published account is a useful source of more details (22,23). In summary, the reader of archaeobotanical reports should bear in mind three points:

1. All identifications should be critically examined. Identification criteria should be presented in detail and backed up by illustrations. Poorly documented identifications should be treated with even greater caution, and the older literature must always be used with care. Identifications of charred material are not absolute and even desiccated material can be problematic.
2. Glume wheat chaff can, if abundant and well preserved, be identified with greater certainty than grain.
3. Hulled wheat remains can be identified with some certainty to ploidy level, but the use of terms such as einkorn, emmer, or spelt does not imply full equivalency with current-day taxa.

The hulled wheats will be discussed as two groups. Einkorn and emmer were domesticated from wild ancestors growing in the Near East. These wild ancestors have been identified and much studied, and the area and time of domestication have been established with certainty. Spelt wheat, on the other hand, results from a hybridization that appears to have taken place after the origins of agriculture, under cultivation. It has no single wild ancestor, and the area and date of its domestication are still unclear.

There is some agreement among archaeobotanists that well-preserved assemblages of spikelets or spikelet forks and glume bases can be reliably separated into einkorn, emmer, and spelt (Figure 3.2 and Figure 3.3). This does involve the assumption that the morphological groups we identify in ancient material match modern taxa. This is undoubtedly true in general terms: the same character combinations that work on ancient material can distinguish current-day einkorn, emmer, and spelt from each other. However, archaeobotanists would not argue for complete similarity between modern einkorns and ancient einkorn. The ancient einkorn remains do belong to a hulled diploid wheat, but that does not necessarily imply that they all share ecological characteristics with their modern counterparts. In addition, forms of wheat may have existed in the past that are extinct today (33).

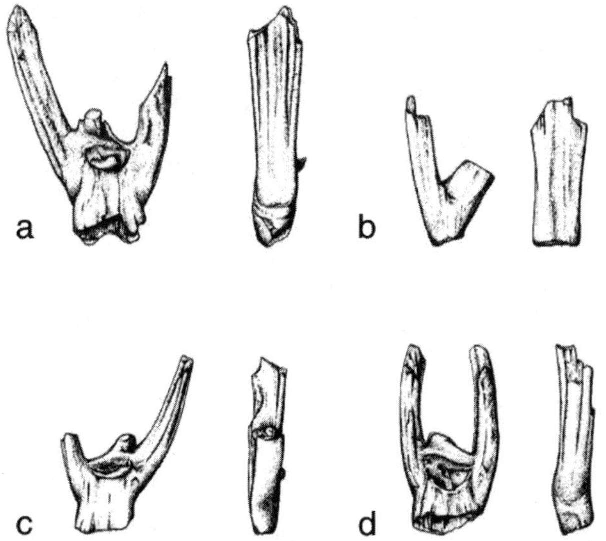

FIGURE 3.2 Charred hulled wheat chaff from Çayböyü, Turkey, dating to the Late Chalcolithic period (4th millennium BC). (A) Normal emmer spikelet fork, which originally came from the middle of the ear. (B) Terminal emmer spikelet fork — originally from the top of the ear. (C) Normal einkorn spikelet fork, with parallel glume bases. (D) Normal einkorn spikelet fork, with parallel glume bases. (From Padulosi et al. (34). Used with permission.)

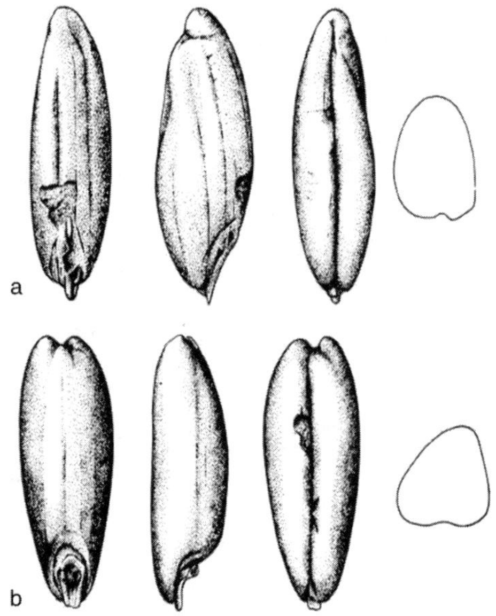

FIGURE 3.3 Charred hulled wheat grains from Çayböyü, Turkey dating to the Late Chalcolithic period (4th millennium BCE). The grains show the longitudinal grooves typical of hulled wheats, caused by the tightly investing chaff. (A) Einkorn grain, with the typical spindle shape, pointed apex, and pronounced ventral convex curve visible in the middle view. (B) Emmer grain, with a blunter apex and straighter sides in all views. Note the typical asymmetric triangular cross-section (34). (From Padulosi et al. (34). Used with permission.)

3.3.1.2 Pollen Analysis

Analysis of pollen grains and other microscopic remains is a well-known and still widespread method used in many laboratories (10). There is a plethora of literature and no need to go into details. For our specific question about feral crops, pollen morphology cannot offer many specific data, because it is not possible to discern single crop traits; it is even sometimes difficult to come down to species level. For specific archaeobotanical questions, see http://www.geo.arizona.edu/palynology/arch_pal.html. Maize pollen provides, just as other recognizable crop palynomorphs, reliable indicators of early settlement in soil profiles. Maize pollen, as soon as it is produced in adjacent fields, can be detected at all altitudes in the Alps (2).

3.3.1.3 Phytolith Analysis

Phytoliths are minute parts of silica in the cells of plants that are specific to certain parts of the plants (Figure 3.4). Phytoliths survive even after the plant decomposes or burns, which allows them

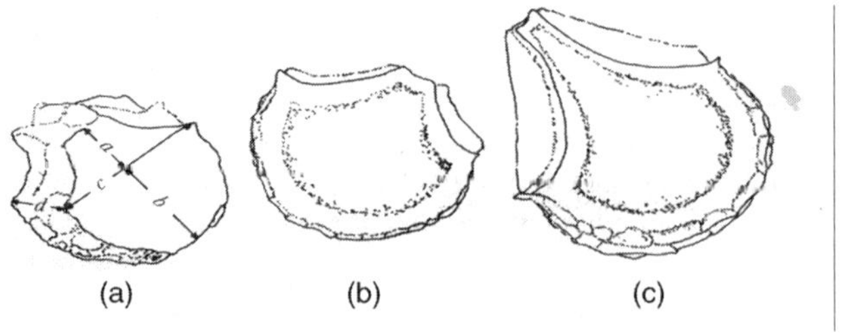

FIGURE 3.4 Measurements and types of fan-shaped phytoliths from rice: (A) parameters, (B) α-type, (C) β-type. (From Chen et al. (8). Used with permission.)

to serve as efficient clues to archaeologists on harvesting periods and techniques, as well as different characteristics of food production. Phytoliths can also help differentiate between wild and domestic plant species. Fujiwara discovered phytoliths of rice in walls of Jomon pottery of Japan (circa 500 BC), which proved that rice cultivation existed at that time. He then analyzed the phytoliths in soil samples and determined an estimated depth and areal extent of the fields and the total yield of rice (40,50–53).

Much needed data is often lacking in archeobotanical sites because little has been preserved and analysis is difficult. Many farm tools were excavated in many Chinese Neolithic sites without crops, a regrettable obstacle to the study of early cultivation. Phytoliths of crops and other plants were more easily preserved in sediments and other sites than sporopollens and more easily identified. For this reason, phytolith analysis has become an important technique in agricultural and environmental archaeology and will play an important role in future studies of the origin and propagation of cultivated rice (8).

3.3.1.4 Analysis of Herbarium Specimens

Methods have been developed to determine gene flow with morphometrics of hybrid specimens found in herbaria (5,12). These results, although derived from herbarium specimens, give a reliable picture on potential gene flow from crops toward their wild relatives, as long as hybrids can be analyzed successfully with morphometrical methods. The same methods can be used for the study of hybrid and feral crop populations. This remains to be done in a comprehensive style and considerable amounts of data could be derived from this method (27). Molecular analyses have confirmed those results (7,14–16).

In closely studying the hybridization records in herbaria and their morphometrics, the dynamics of gene flow can be estimated. The data set produced is valid for the decades the herbarium samples have been made; they can reliably give an impression of long-term hybridization dynamics, as long as one remains critical about the sampling. Those results can also show long-term dynamics in a more reliable way than short-term measurements of the marking transgenes in the field, because those measurements are limited by cost and time-consuming analysis. However, the herbarium results cannot reveal anything about the specific dynamics of the transgenes contained in the transgenic crops.

Some examples in the (non-transgenic) *Triticum aestivum* complex are given, all from (27) (Figure 3.5 through Figure 3.8). Wild *Triticum* species (or feral ones) are generally not weedy in the agronomic sense; they grow profusely in ruderal habitats adjacent to wheat fields (for the exception in Tibet, see Chapter 11). Hybrids between wheat and its wild relatives are generally rare, unlikely, and often sterile. There is little knowledge about the biology, the flowering behavior, and the reproductive systems of the wild relatives of wheat. The artificially obtained hybrids show intermediate characters and usually remain sterile, except in cases of chromosome duplication within the *Aegilops ovata–Triticum durum*-complex. Experiments also demonstrate low hybridization success and considerable differences between years and different conditions. In the same study, Jacot et al. (27) established herbarium-based morphometrical studies. Those morphometric studies take into account a high number of morphological characters; in the case of the study cited, we used some 50 to 75 characters (Table 3.1). Studies on herbarium samples collected in the field reveal long-term hybridization dynamics, but they cannot tell anything about the specific dynamics of transgenic crops and their wild relatives.

Two examples are presented from those morphometrical analyses, here:

1. The complex of hybrids between *Triticum aestivum* and *Aegilops ovata* (Figure 3.5 and Figure 3.6).
2. The complex of hybrids between *Triticum aestivum* and *Aegilops squarrosa* (Figure 3.7 and Figure 3.8).

TABLE 3.1
Morphological Characters Used to Distinguish Different Species and Hybrids of Wheat in Herbarium Collections

	Distinguishing Character	
Stem	1 — stem height *[cm]*	2* — stem form
Leaf	3 — leaf length *[0.5 cm]*	7 — leaf surface ventral
	4 — leaf width *[0.1 mm]*	8* — sheath form
	5* — leaf form	9* — auricle form
	6 — leaf surface dorsal	
Spike	10 — spike length *[mm]*	15 — spike: *nr* of fertile spikelets
	11 — spike width (at middle) *[mm]*	16 — *nr* of rudimentary spikelets at bottom
	12 — spike width (at top) *[mm]*	17 — spike: *nr* of sterile spikelets at top
	13* — spike form	18* — spike/plant color
	14 — spike density	
Internode	19 — internode length	20 — internode length at bottom
Rachis	21 — rhachis surface	23* — rhachis/spike fragility
	22* — rhachis edges	
Spikelet	24 — spikelet length *[mm]*	29 — spikelet: *nr* of fertile florets
	25* — spikelet width *[mm]*	30 — spikelet's highest flower
	26* — spikelet form	31 — spikelets: total *nr* of awns
	27 — spikelet surface	32 — spikelets (upper): total *nr* awns
	28 — spikelet: *nr* of florets	
Glume	33 — glume body length *[0.1 mm]*	44* — glume's 1. side t/a form
	34 — glume body width *[0.1 mm]*	45 — glume's 2. side t/a length
	35* — glume body form	46* — glume's 2. side t/a form
	36 — glume keeling	47 — glume's 4. t/a length
	37* — glume form at apex	48* — glume's 4. t/a form
	38 — glumes concealing lemma?	49 — glume-awns length low
	39* — glumes equilateral?	50 — glume-awns *nr* upper
	40 — glume: *nr* of awns	51 — glume-awns length upper
	41 — glume's middle t/a length	52* — glume awn position
	42* — glume's middle t/a form	53 — glume awn surface
	43 — glume's 1. side t/a length	
Lemma	54 — lemma body length *[0.1 mm]*	60 — lemma's middle t/a length
	55 — lemma body width *[0.1 mm]*	61* — lemma's middle t/a form
	56* — lemma body form	62 — lemma's 1. side t/a length
	57* — lemma nerves	63* — lemma's 1. side t/a form
	58* — lemma nerves convergency	64 — lemma's 2. side t/a length
	59 — lemma: *nr* of awns	65* — lemma's 2. side t/a form
	67 — lemma-awns length upp	66 — lemma-awns *nr* upper
Palea	68 — palea length *[0.1 mm]*	70* — palea form
	69 — palea width *[0.1 mm]*	
Fruit	71* — fruit length *[0.1 mm]*	73* — fruit form
	72* — fruit width *[0.1 mm]*	74* — fruit unit
Fertility	75* — fertility	

t/a = tooth or awn; *nr* = number; *excluded from statistics.

Source: From Jacot et al. (27). Used with permission.

FIGURE 3.5 Exoferal hybrids in herbaria: *Triticum aestivum*   *Aegilops ovata* (27). The hybrids show intermediate characters. (From Jacot et al. (27), with permission of the publisher.)

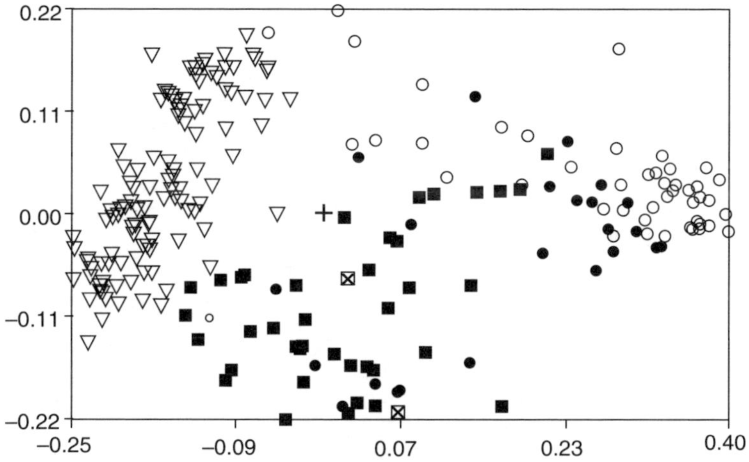

FIGURE 3.6 Scattergram of morphometrics in the *Aegilops ovata*   *Triticum aestivum complex* (27). The scattergrams of this figure and Figure 3.8 show a principal component analysis of some 60 characters (Table 3.1) involved in the measurements. Therefore, the dynamics of the hybridization are based on a full set of visible and measured (counted) characters of the herbarium specimens chosen over many decades and localities. It involves several hybridogene taxa as shown in the figure. *Aegilops ovata:* ▽, *Triticum aestivum:* ○, *Aegilops triticoides (F1):* ■, *Aegilops speltaeforme (BC)* ▲, *Aegilops triticoides (F2):* ⊠ (From Jacot et al. (27), with permission of the publisher.)

FIGURE 3.7 Exoferal hybrids between *Triticum aestivum* and *Aegilops squarrosa*, specimens from herbaria. (From Jacot et al. (27), with permission of the publisher.)

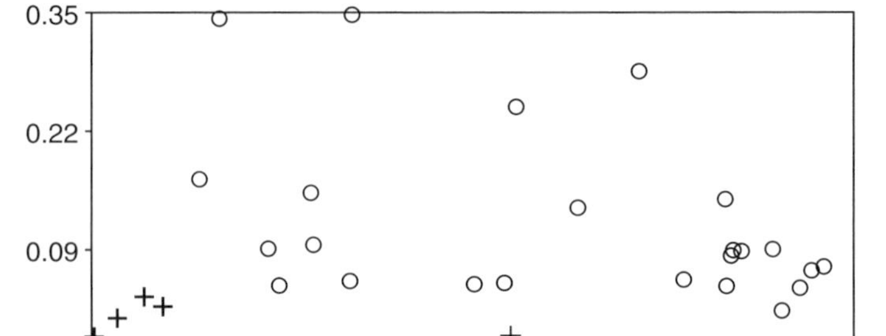

FIGURE 3.8 Scattergram of morphometrics of hybrid specimens *Triticum aestivum* *Aegilops ovata*. (From Jacot et al. (27), with permission of the publisher.)

3.3.2 Archaeobotanical Studies — The Example of Hulled Wheat

Once the wild ancestors of einkorn and emmer had been identified, it became clear that domestication most likely would take place in the same area where the wild ancestors grew. There are two main sources of information for the distribution of wild cereals during the period of domestication around 10,000 years ago (Figure 3.9): the current distribution of wild and feral cereals and archaeobotanical finds of wild (feral?) cereals from pre-agrarian sites (see Figure 3.2 and Figure 3.3).

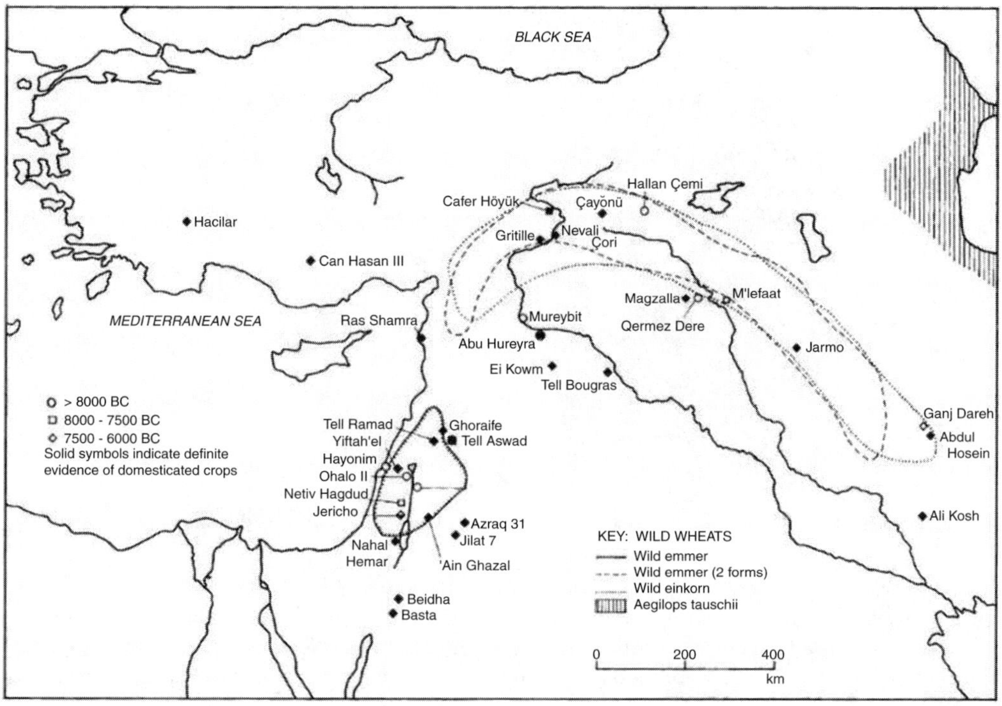

FIGURE 3.9 Distribution of archaeological sites in relation to the distribution of wild wheats. Only those Epipalaeolithic and Neolithic sites dating to before 6000 BCE with available archaeobotanical reports are shown. Solid symbols indicate definite evidence of domestication; empty symbols indicate sites that are nonagrarian or of uncertain status. Distribution of wild wheats in primary, truly wild, habitats is shown, but small populations of wild einkorn on Mount Hermon in southern Lebanon and of wild einkorn and emmer in Transcaucasia are not shown. Note that wild emmer in the Levant consists of pure *T. dicoccoides*; in the northern Fertile Crescent both *T. dicoccoides* and *T. araraticum* are present (indicated by the dashed line). The western extension of primary habitats of *Aegilops tauschii* is shown, around the Caspian Sea (34). (From Jacot et al. (27), with permission of the publisher.)

3.3.2.1 Einkorn

A relatively extensive discussion about one specific complex of ancient, modern, and feral taxa is given here to demonstrate that we are far from a final analysis of the situation. It is not enough to arrange a few scattered findings into a seemingly logical system of evolution; we need the population genetics approach and a multivariate data analysis in the whole range of those fascinating crop taxa. This situation is summarized in Padulosi (34): Today wild einkorn and wild emmer seem obvious candidates as wild ancestors of, respectively, einkorn and emmer wheat, because of their morphological similarity and ability to intercross. However, this has only been apparent for 100 years or so, after a series of botanical discoveries whose history is widely discussed (1,11,41,42) (Figure 3.10). Wild einkorn (*Triticum boeoticum*) was discovered in Greece and Turkey in the mid-19th century and by 1900, was widely accepted as the ancestor of domesticated einkorn wheat. *Triticum urartu*, the second diploid wild wheat, was named in 1938 by the Armenian botanist, Tumanian. It grows throughout the Fertile Crescent as a minor admixture of *T. boeoticum* on outcrops of basaltic soil (49). Unlike *T. boeoticum*, it has not spread outside the Fertile Crescent as a weed of disturbed ground. It is morphologically similar to *T. boeoticum*, but *T. urartu* can be consistently distinguished on the basis of anther length, the presence of a third lemma awn, and caryopsis color (29,32,48). Crosses between the two taxa result in sterile hybrids. Overall, the evidence points to *T. urartu* as a separate species. *T. urartu* is not a candidate species as a wild

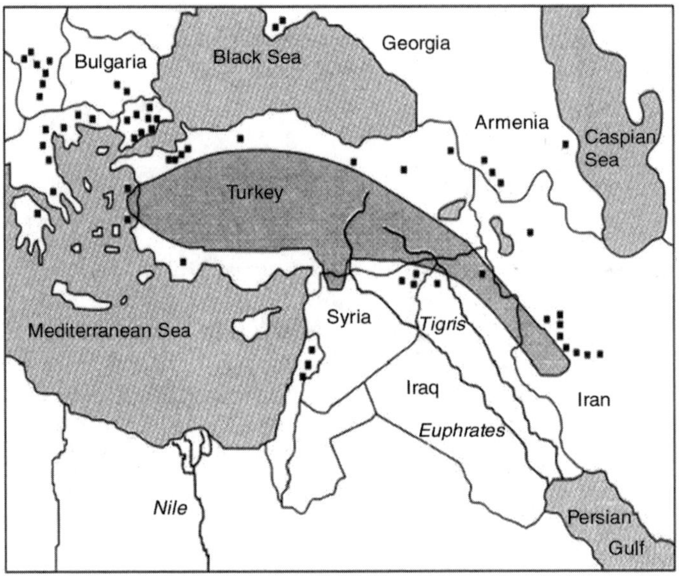

FIGURE 3.10 Wild einkorn, *Triticum boeoticum,* distribution map showing also isolated, maybe today also feral occurrences: shaded area represents the main center of origin of *Triticum boeoticum*, wild progenitor of *T. monococcum*; dots indicate areas of secondary importance where *T. boeoticum* also has been found Harlan et al. (19). (From Hunshal et al. (24), with permission of the publisher.)

ancestor for domesticated einkorn, according to Jaaska (25) and Waines (48), but it may be a parent of *T. dicoccoides* according to Dvorak (9).

3.3.2.2 Emmer

The wild ancestor of emmer was not identified until 1873, when Körnicke found part of a spike of *T. dicoccoides* in a collection of wild barley, *Hordeum spontaneum,* from Mount Hermon in southern Syria. However, it was Aaronson's discovery from 1906 (1) onwards of abundant wild emmer in Israel that led to general acceptance of its role as the wild ancestor of emmer. Two morphologically distinct forms of *T. dicoccoides* have been recognized by Poyarkova et al. (35,36), a narrow-eared, slender form native to the whole range of wild emmer, and a wide-eared, robust form of more restricted distribution (Figure 3.11). Although both forms are found in weedy habitats such as roadsides, both mostly grow in primary, undisturbed habitats. Unlike wild einkorn and barley, wild emmer has conspicuously failed to spread outside the Fertile Crescent. Its current distribution is therefore believed to be more representative of its early Holocene distribution than that of the other wild cereals. As with wild einkorn, wild emmer consists of two morphologically similar but reproductively isolated tetraploid species, *Triticum dicoccoides* and *T. araraticum*, the latter first recognized in the 1930s and named by Jakubziner (28). *T. araraticum* has been identified as the wild ancestor of *T. timopheevi*, a rare domesticated glume wheat found in the Republic of Georgia.

Whether the widely accepted theory that salinization of soils in Mesopotamia can be linked to the decline of emmer is put into question in the light of new scrutiny of the original data. Hunshal et al. (24) actually compared yields and found that one variety of Indian emmer wheat yielded much more than barley under a number of salinity levels. In view of modern-day adaptability of emmer to poor soil conditions, it is possible that some forms of ancient emmer were also resistant to saline conditions.

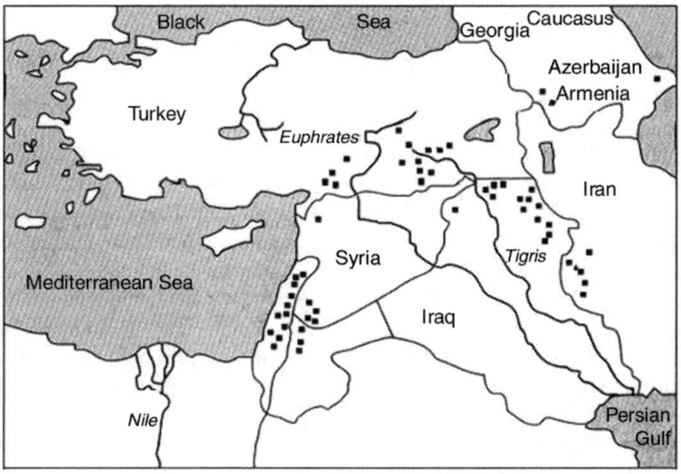

FIGURE 3.11 Distribution of *Triticum dicoccoides.* Dots indicate the presence of *Triticum dicoccoides*, progenitor of *T. dicoccon*. (From Hunshal et al. (24), with permission of the publisher.)

This combination of ambiguous texts, lack of archaeobotanical data, and lack of agronomic characterization of hulled wheats equally affects virtually all studies of ancient historical agriculture (34). From this short review on the ecogeographical distribution of hulled wheats, some considerations arise:

- Spelt, emmer, and einkorn are spread, in decreasing order of importance, in several countries, mainly of Europe, Near East, central Africa, and North America.
- Good information on the cultivation of these old crops is available for only a few countries, whereas for most of the others it is scarce or almost completely lacking.
- Only a few and rarely specific expeditions have been carried out to collect these species in their centers of diversity.
- In many cultivation areas, native populations are often extinct or mixed with other germplasm or replaced by modern cultivars, showing that the risk of genetic erosion is high.
- Characterization and evaluation data are available only for a part of the world collection.

This means in our case of feral crops that there is still a lot of work to be done to trace and understand feral wheat populations all over the world. Because each crop and its wild and feral relatives have multiple origins (compare with Figure 3.1), it will be necessary to conduct many regional and local studies. After having done this, it might be possible to find common ground for the definition of feral populations and individuals and proceed to some generalizations, keeping in mind that the complexity of history and migration, partly due to human activity, will only allow for a highly selective and thus insufficient data set.

LITERATURE CITED

1. Aaronson A. posthumous. 1994. Bibliography and diary. In *New Encyclopedia of Zionism and Israel*, Wigoder G, Ed., London: Associated University Press.
2. Ammann K. 1979. *Gletschernahe Vegetation der Oberaar, einst und jetzt*. Presented at Werden und Vergehen von Pflanzengesellschaften, Rinteln, 20–23. March 1978, Germany.
3. Ammann K. 1997. Botanists to blame? *Plant Talk* 8:4.
4. Ammann K, Jacot Y, Al Mazyad PR. 2000. Weediness in the light of new transgenic crops and their potential hybrids. *Z. Pflanzenkr. Pflanzenschutz* 17 (special issue):19–29.

5. Ammann K, Jacot Y, Rufener Al Mazyad P. 1996. Field release of transgenic crops in Switzerland: an ecological assessment of vertical gene flow. In *Gentechnisch veränderte krankheits- und schädlingsresistente Nutzpflanzen. Eine option für die Landwirtschaft?*, Schulte E, Käppeli O, Eds., pp. 101–157. Basel: Schwerpunktprogramm Biotechnologie, Schweiz. Nationalfonds zur Förderung der Wissenschaftlichen Forschung, BATS.
6. Bartsch D, Sukopp H, Sukopp U. 1993. Introduction of plants with special regard to cultigens running wild. In *Transgenic organisms*, Wöhrmann K, Tomiuk J, Eds., pp. 135–151. Basel: Birkhäuser.
7. Bianchi DS, Senften JK, Felber F. 2002. Isozyme variation of *Hordeum murinum* in Switzerland and test of hybridization with cultivated barley. *Weed Res.* 42:325–333.
8. Chen B, Zhang J, Lu H. 1995. Discovery of rice phytoliths in the neolithic site at Jiahu of Henan province and its significance. *Chin. Sci. Bull.* 40: issue 14, English scanning of translation by Leir G, Ed. by Leir and Gordon B on Web page. http://www.carleton.ca/~bgordon/Rice/papers/chen95.htm
9. Dvorak J, McGuire PE, Cassidy B. 1988. Apparent sources of the a genomes of wheats inferred from polymorphism in abundance and restriction fragment length of repeated nucleotide-sequences. *Genome* 30:680–689.
10. Faegri K, Iversen J, Kaland P, Krzywinski K. 1989. *Textbook of palynology, IV. Edition.* New York: John Wiley and Sons, Blackburn Press.
11. Feldman M, Millet E. 2001. The contribution of the discovery of wild emmer to an understanding of wheat evolution and domestication and to wheat improvement. *Israel J. Plant Sci.* 49:S25–S35.
12. Frietema DVF. 1996. Cultivated plants and the wild flora. Effect analysis by dispersal codes. Leiden: University of Leiden, 100 pp.
13. Gladis T. 1966. Unkräuter als genressourcen. *Z. Pflanzenkr. Pflanzenschutz* 15:39–43.
14. Guadagnuolo R, Bianchi D, Felber F. 2001. Specific genetic markers for wheat, spelt, and four wild relatives: comparison of isozymes, RAPDs, and wheat microsatellites. *Genome* 44:610–621.
15. Guadagnuolo R, Savova-Bianchi D, Felber F. 2001. Gene flow from wheat (*Triticum aestivum* L.) to jointed goatgrass (*Aegilops cylindrica* Host.), as revealed by RAPD and microsatellite markers. *Theor. Appl. Genet.*103:1–8.
16. Guadagnuolo R, Savova-Bianchi D, Keller-Senften J, Felber F. 2001. Search for evidence of introgression of wheat (*Triticum aestivum* L.) traits into sea barley (*Hordeum marinum* s.str. Huds.) and bearded wheatgrass (*Elymus caninus* L.) in central and northern Europe, using isozymes, RAPD and microsatellite markers. *Theor. Appl. Genet.* 103:191–196.
17. Hammer K, Arrowsmith N, Gladis T. 2003. Agrobiodiversity with emphasis on plant genetic resources. *Naturwissenschaften* 90:241–250.
18. Hanelt P. 1998. Lesser known or forgotten cruciferous vegetables and their history. *Acta Hort. Symposium on Brassicae* 549, ISHS:39–45.
19. Harlan J, Zohary D. 1966. Distribution of wild wheats and barley. *Science* 153:1074–1080.
20. Hartmann E, Schuldes H, Kübler R, Konold W. 1994. *Neophyten — Biologie, Verbreitung, Kontrolle ausgewählter Arten.* Landsberg, Germany: Ecomed. 302 pp.
21. Hilbeck A, Baumgartner M, Fried PM, Bigler F. 1998. Effects of transgenic *Bacillus thuringiensis* corn-fed prey on mortality and development time of immature *Chrysoperla carnea* (Neuroptera: Chrysopidae). *Environ. Entomol.* 27:480–487.
22. Hillman G, Hedges R, Moore A, Colledge S, Pettitt P. 2001. New evidence of late glacial cereal cultivation at Abu Hureyra on the Euphrates. *Holocene* 11:383–393.
23. Hillman G, Wales S, McLaren F, Evans J, Butler A. 1993. Identifying problematic remains of ancient plant foods — a comparison of the role of chemical, histological and morphological criteria. *World Archaeol.* 25:94–121.
24. Hunshal C, Balikai R, Viswanath D. 1990. *Triticum dicoccum* — its performance in comparison with barley under salinity. *J. Maharashtra Agric. Univ.* 15:376–377.
25. Jaaska V. 1993. Isoenzymes in the evaluation of germplasm diversity in wild diploid relatives of cultivated wheat. In *Biodiversity and wheat improvement*, Damania AB, Ed., pp. 247–257. Chichester: John Wiley.
26. Jacomet S, Kreuz A. 1999. *Archäobotanik, Aufgaben, Methoden und Ergebnisse vegetations — und agrargeschichtlicher Forschung.* Stuttgart, Germany: Eugen Ulmer. 368 pp.
27. Jacot Y, Ammann K, Rufener Al Mazyad P, Chueca C, David J, et al. 2004. Hybridization between wheat and wild relatives, a European Union research programme. In *Introgression from genetically modified plants into wild relatives*, den Nijs DBH, Sweet J, Ed., Oxford: Oxford University Press, pp. 63–74.

28. Jakubziner M. 1958. *New wheat species*. Presented at First International Wheat Genetics Symposium, Winnipeg, Canada: Wheat Information Service, pp. 207–220.
29. Johnson BL. 1975. Identification of apparent B-genome donor of wheat. *Can. J. Genet. Cytology* 17:21–39.
30. Losey JE, Raynor LS, and Carter ME. 1999. Transgenic pollen harms Monarch larvae. *Nature* 399:214.
31. MacKey J. 1966. Species relationships in *Triticum*. *Hereditas,* Suppl. 2:237–276.
32. Morrison L. 1993. *Triticum-Aegilops* systematics: taking an integrative approach. In *Biodiversity and wheat improvement*, Damania A, Ed., pp. 59–66. Chichester: John Wiley.
33. Nesbitt M, Samuel D. 1995. *From staple crop to extinction? The archeology and history of the hulled wheats*. Presented at 4. Proceedings of the First International Workshop on Hulled Wheats, Castelvecchio Pascoli, Tuscany, Italy.
34. Padulosi S, Hammer K, Heller Je. 1995. *4. Proceedings of the First International Workshop on Hulled Wheats*. Castelvecchio Pascoli, Tuscany, Italy. 262 pp.
35. Poyarkova H. 1988. Morphology, geography and infraspecific taxonomics of *Triticum dicoccoides* Korn — a retrospective of 80 years of research. *Euphytica* 38:11–23.
36. Poyarkova H, Gerechteramitai ZK, Genizi A. 1991. Two variants of wild emmer (*Triticum dicoccoides*) native to Israel — morphology and distribution. *Can. J. Bot.* 69:2772–2789.
37. Rauber R. 1977. Evolution von Unkräutern. Ergebnisse der 1. Deutschen Arbeitsbesprechung über Fragen der Unkrautbiologie und -bekämpfung. *Z. Pflanzenkran. Pflanzenschutz* 8:37–55.
38. Romeis J, Battini M, Bigler F. 2003. Transgenic wheat with enhanced fungal resistance causes no effects on *Folsomia candida* (Collembola: Isotomidae). *Pedobiologia* 47:141–147.
39. Rösler L. 1969. Zur Bestimmung und Beurteilung fon Fatuoiden und Bastardformen zwischen Saat- und Flughafer. *Saat-und Pflanzgut* 10:731–735.
40. Sato YI, Fujiwara H, Udatsu T. 1990. Morphological differences in silica body derived from motor cell of indica and japonica in rice. *Jpn. J. Breed.*40:495–504.
41. Schiemann E. 1951. New results on the history of cultivated cereals. *Heredity* 5:305–320.
42. Schiemann E. 1956. New dates for recent cultivation of *Triticum monococcum* and *Triticum dicoccum* in Jugoslavia1. In *Wheat information services* No. 3, Yokohama, Japan.
43. Schlink S. 1994. Ökologie der Keimung und Dormanz von Körnerraps (*Brassica napus* L.) und ihre Bedeutung für eine Überdauerung der Samen im Boden. Berlin, Stuttgart: Cramer. 194 pp.
44. Schönberger H, de Vries G. 1991. So bauen Sie den Rapsbestand gezielt auf. In Rapsanbau für Könner — Bestände gezielt aufbauen und schützen. *Top Agrar, Das Magazin für moderne Landwirtschaft. Landwirtschaftsverlag Münster-Hiltrup,* (special volume): 20–26.
45. Sears M, Hellmich R, Stanley-Horn D, Oberhauser K, Pleasants J, et al. 2001. Impact of Bt corn pollen on monarch butterfly populations: a risk assessment. *Proc. Natl. Acad. Sci. USA* 98:11937–11942.
46. Sukopp H, Sukopp U. 1993. Ecological long-term effects of cultigens becoming feral and of naturalization of nonnative species. *Experientia,* 49:210–218.
47. Szabó A, Hammer K. 1996. *Notes on the taxonomy of farro: Triticum monococcum, T. dicoccon and T. spelta*. Presented at Hulled wheats. Promoting the conservation and use of underutilized and neglected crops. 4. Proceedings of the First International Workshop on Hulled Wheats, Castelvecchio Pascoli, Tuscany, Italy.
48. Waines J, Barnhart D. 1990. Constraints to germplasm evaluation. In *Wheat genetic resources: meeting diverse needs,* Srivastava JP, Damania AB, Ed., Chichester: John Wiley. pp. 103–111.
49. Waines J, Rafi M, Ehdaie B. 1993. Yield components and transpiration efficiency in wild wheats. In *Biodiversity and wheat improvement*, Damania A, Ed., pp. 173–186. Chichester: John Wiley.
50. Wang C, Udatsu T, Fujiwara H. 1999. Effects of nitrogen levels on the morphology of silica bodies from motor cells in rice (*Oryza sativa* L.). *Jpn. J. Crop Sci.* 68:58–62.
51. Wang CL, Udatsu T, Fujiwara H. 1998. Genetic analysis of the morphology of silica bodies from motor cells in rice (*Oryza sativa* L.). *Breed. Sci.* 48:163–168.
52. Wang CL, Udatsu T, Tang LH, Zhou JS, Zheng YF, et al. 1998. Cultivar group of rice cultivated in Caoxieshan site (B.P.6000 similar to present) determined by the morphology of plant opals and its historical change. *Breed. Sci.* 48:387–394.
53. Zheng YF, Dong YJ, Matsui A, Udatsu T, Fujiwara H. 2003. Molecular genetic basis of determining subspecies of ancient rice using the shape of phytoliths. *J. Archaeol. Sci.* 30:1215–1221.

4 Feral Beets — With Help from the Maritime Wild?

Ulrich Sukopp, Matthias Pohl, Sarah Driessen, and Detlef Bartsch

4.1 HISTORY OF BEET DOMESTICATION

Beets have been cultivated for more than 2000 years in the eastern Mediterranean region, which is their center of origin. *Beta vulgaris* L. is an extraordinarily variable taxon, in which cultivated and wild forms are often difficult to distinguish (5,38). This is mainly due to the extensive use of sea beet (*B. vulgaris* ssp. *maritima* Arcang.) gene resources in conventional breeding programs (36). Sea beet is primarily a coastal plant growing in drift line vegetation with a wide distribution from Cape Verde and the Canary Islands in the west, northward along the Atlantic coast to the North and Baltic Seas. It also extends eastward from the Mediterranean region into Asia with occurrences in Asia Minor, the central and outer Asiatic steppes, and desert areas as far as western India (31,51). Sea beet varies from self-compatible annuals to self-incompatible, *polycarpic* (repeatedly fruiting) perennials with a life span of 1 to 11 years (41,73). Cultivated *B. vulgaris* forms, including Swiss chard, red garden beet, and sugar beet, are biennial. The latter is partially self-incompatible due to the extensive use of male sterility genes in sugar beet breeding (49). All cultivated and wild subspecies of *B. vulgaris* are mostly wind-pollinated, although some insect pollination has been noted (8).

Cultivated sugar beet varieties are either diploid or triploid. They originate from crossing diploid cytoplasmic male sterile (CMS) mother plants with diploid or tetraploid breeding lines. Although triploid varieties have certain yield advantages, they show a disturbed reproductive ability. Consequently, their offspring are limited and thus will be counter-selected naturally toward diploidy.

The historical development of beet cultivars has been intensively documented (38,54). Swiss chard and red beet are the oldest cultivated forms. Asian cultivars were introduced to western and central Europe as early as the Roman Age. Sugar beet is the most recent domesticated form, originating from wild beet ancestors of the north European channel coast at the end of the 18th century (35). In the course of domestication, the genetic diversity of certain cultivated forms was considerably constricted through breeding and is currently smaller than in wild forms (12). However, beets are allogamous so that they can easily recombine genetic information, and the overall variability among sugar beet cultivars is large compared to other crops. Domesticated traits of table (red) and sugar beets include leaf area, reduced *bolting* (premature flowering) behavior, small root in leaf beet, or large root in table and sugar beet. Weedy beets are bolting and flowering individuals within sugar beet fields. These bolters compete with the crop and decrease crop yield (as they themselves do not have the storage roots) and hamper harvest.

We discuss three biological processes that potentially contribute to volunteerism in beet — hybridization, ferality, and transgene introgression. Sukopp and Sukopp previously used the term *feral* in the context of dedomestication of cultigens and applied it to sugar beets and other crops (69). We present some of our unpublished data and discuss the results in the light of published evidence to illustrate these processes.

4.2 HYBRIDIZATION AND GENE FLOW IN BEET

Gene escape via hybridization is most likely to occur due to premature flowering (bolting) of plants during the first years of crop cultivation or at the commercial seed production stage. Unlike the wild sea beet progenitor of domesticated beets, cultivated forms of beet are mostly biennials. They are sown after spring frost and their roots and leaves are harvested at the end of the first cropping season. Vernalization of young beet plants at low temperatures (0°C to +10°C) for several days can result in bolting of biennial cultivars in the first year (29). This physiological response is widely used for commercial seed production in temperate climate regions of southwestern France and northeastern Italy, where beets for seed production are sown at the end of the summer, germinate in autumn, become vernalized during winter, and set seeds at the end of spring (50).

Four main factors can enhance gene flow between wild and cultivated beet:

1. Hybridization of sugar beet with flowering plants of sea beet in seed production areas (28)
2. Vernalization of volunteer vegetative beet parts remaining on the field after harvest (59)
3. Vernalization of young seedlings due to unexpected low temperatures below 10°C during the early spring of the growing season (16,74)
4. Cultivation of older varieties that have certain a percentage of bolting individuals, despite breeders' efforts to enhance biennial behavior of beets

Any factor that increases the incidence of premature bolting in cultivation areas leads to weediness. This includes beets that have regained feral characteristics from hybridization with wild sea beet. With regard to potential backcross hybridization of weedy beet in sugar beet plantation areas in Europe, two opposing parental sources hypothetically contribute to their genome:

1. Wild species as parents — Sea beets of France or Italy can act as the pollen parents of premature bolting progeny in seed production areas (hybridization between sugar beet and sea beet — Factor 1 described above). Naturalized weed beets from crosses between wild and domesticated beets are an additional source of the annual weed habit in seed production areas in Italy and France as well as in sugar beet fields in Germany (72).
2. Cultivars as parents — Although sugar beet is primarily the maternal source for weed beet, vernalization in spring (Factor 3) or hibernating beets (Factor 2) may increase the number of secondary maternal and paternal hybridization partners in sugar beet fields. This will add cultivar alleles to the gene pool of weed beet populations and is important in view of an additional source of gene flow. The backcross offspring will genetically be more related to the cultivar type. Even triploid sugar beet varieties occasionally hybridize with diploid weed partners.

We investigated which scenario is most likely to occur on a local scale by determining the genetic relatedness of weed beet offspring with their potential sea beet and sugar beet parents. RAPD-PCR (random amplified polymorphic DNA — polymerase chain reaction) analyses and neighbor dendrograms were used to determine whether a weed beet accession is more related to sea beet (Hypothesis 1) or a sugar beet ancestor (Hypothesis 2). We examined 13 sea beet, weed beet, and cultivated beet accessions of *Beta vulgaris* from Germany, the Netherlands, Ireland, France, Italy, Greece, and Portugal (Table 4.1). We also collected weed beet seeds from three sugar beet fields within a 20 km 20 km sugar beet cultivation area near Aachen, Germany. These seeds, along with cultivar seeds, were sown in a greenhouse and leaf samples of 10 germinated seedlings per accession of sea beet, sugar beet, and weed beet samples were used for RAPD-PCR analyses, as previously described (32,47), and as detailed in Table 4.1. We detected 7 to 13 fragments per primer resulting in an absolute number of 74 possible bands. Most (55%) of the weed beet fragments

TABLE 4.1
Genetic Similarity of Weed Beet to Sea and Sugar Beets. Diagnostic RAPD-PCR fragments according to primer and fragment combination number found in weed beets that are characteristic for sea beet and sugar beet (total number of weed beet individuals = 30; 10 per accession). A total of 13 accessions of sea beet, weed beet, and sugar beet from Germany, the Netherlands, Ireland, France, Italy, Greece, and Portugal were examined.[a]

	RAPD-PCR Fragments																		
	Sea Beet Specific														Sugar Beet Specific				
Primer no.	1		2		3	7		21		23		28			2	3	23	28	29
Fragment no.	2	6	16	21	29	35	42	45	48	54	60	62	64	67	18	28	55	66	72
Length (10 bp)	130	80	95	65	90	105	160	95	65	90	180	130	98	59	85	98	75	62	80
	Number of individuals (n = 30) where specific fragments were found:																		
Weed beet	6	11	18	0	14	3	6	13	30	0	3	0	14	8	25	27	8	10	13
Overall statistics	126 counts out of 420 possible = **30.0%**														83 counts out of 150 possible = **55.3%**				

[a] The PCR-amplified fragments were between 400 and 2200 bp long. Cultivar seeds were obtained from Kleinwanzlebener Saatgut AG (KWS) in Einbeck, Germany. Eight decamer primers with a GC-content of about 60% were used for DNA amplification. The amplification patterns were visually examined for the presence of fragments and the results were transformed into a binary matrix (0 for absence, 1 for presence of a specific DNA fragment). Genetic diversity was estimated both by Shannon's index for fragment frequency (34) as well as by the total number of RAPD-PCR fragments in a single population. The utility of a fragment for use in a differential diagnosis was scored by comparing the fragment pattern of sugar beet, sea beet, and weed beet. Genetic diversity data of the fragment frequency per accession were analyzed by Kruskal-Wallis-one-way analysis of variance on ranks with the SIGMASTAT software program (Jandel Scientific). Statistically significant differences were set for P-values < 0.05.

were similar to sugar beet specific fragments and 30% to sea beet specific fragments (Table 4.1). The quantitative distribution differed between the beet groups: 14 fragments were specific for individual sea beets and 5 fragments were specific for individual sugar beets. Sea beet and sugar beet could be distinctly separated from each other in the dendrogram (Figure 4.1) in which bootstrap values support most of the branch separation. Sea beets seemed to weakly subcluster according to their geographic origin into a Mediterranean group (Italy, Greece) and an Atlantic sea group (Portugal, Ireland, Netherlands, France). The frequency pattern of the PCR-amplified fragments demonstrated no clear picture of the coastal region where weed beets possibly originated from sea beet, as there is no evidence that weed beet is genetically more related to Mediterranean than to Atlantic sea beet. This suggests that weed beet has repeatedly evolved in different places. The genetic diversity based on Shannon's index was higher in weed beet (median = 0.211, 25% = 0.178, 75% = 0.218) than in sea beet (median = 0.151, 25% = 0.134, 75% = 0.158) and sugar beet (median = 0.071, 25% = 0.060, 75% = 0.081). This may be explained by the hybrid character of weed beet evolving from crosses between sugar beet and sea beet. Thus, weed beets have a higher genetic diversity than wild and cultivated forms. Sugar beet breeders increase cultivar diversity first by using many different beet genetic resources and second by producing seed material in several regions, with more distinct chances of unintentional hybridization with sea beet. Based on the total number of RAPD-PCR fragments, no difference was observed between sea beet (median = 53.5, 25% = 50.5, 75% = 54.5) and sugar beet (median = 53.5, 25% = 53, 75% = 54), but weed beet had a significantly greater diversity (median = 60, 25% = 59.25, 75% = 62.25). Other species may be involved in gene flow with sugar beet, both intentionally by breeders using interspecific crosses

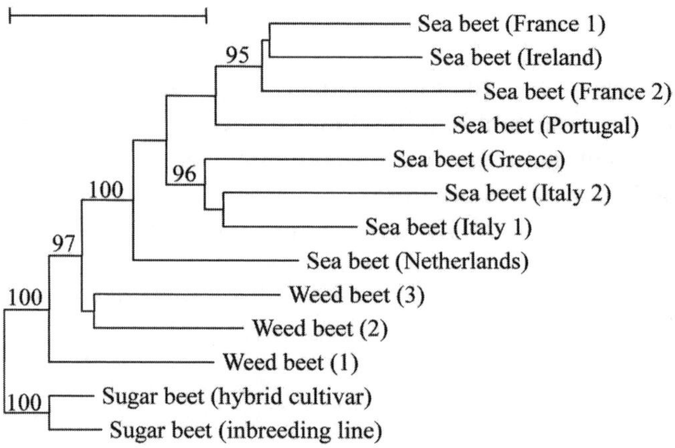

FIGURE 4.1 Weed beet is more closely related to sugar beet than to sea beet. Neighbor-joining-dendrogram of systematic relationships among 13 accessions of beet based on a genetic distance method (58) derived from allele frequency at 74 RAPD-PCR fragments. Bootstrap values >95% are given for branches (71). The scale bar represents a genetic distance of 0.1. The neighbor-joining method compares mean distance values of branch groups. Accessions are described in Bartsch et al. (12).

for crop improvement (e.g., *Beta patellaris*) (23) or unintentionally (e.g., *B. macrocarpa)* (see Section 4.3 below).

The sugar beet parents play a more important role for the population genetics of weed beet than the wild sea beet parents, as weed beet is significantly more related to sugar beet (overall identity = 0.86) than to sea beet (overall identity = 0.79). A similar result is presented in Desplanque et al. (28). An escape of transgenes in Germany far away from sugar beet breeding areas will be facilitated by frequent secondary local hybridizations in common sugar beet fields, such as between overwintering volunteer sugar beets and bolting weed beets. Another pathway will be the recurrent import of contaminated sugar beet seeds from the seed production areas. Hybridization and gene flow will add cultivar alleles to the gene pool of local weed beet populations. The fate of these alleles has to be closely observed.

4.3 FERALITY IN BEET CONNECTED TO THE BOLTING GENE "B"

Ferality of cultivated beet can have three origins (48):

1. Back mutations of the cultivar to the wild habit
2. Further evolution of the domesticated beets
3. Introgression from sea beet as the wild ancestor of sugar beet

Because there are no genetic barriers to crosses between sea beet and cultivars of *Beta vulgaris* (2), spontaneous hybridization occurs in the two main sugar beet seed production areas in France and Italy where both taxa grow in close proximity (14,20,57). Hybrids between sugar beet and sea beet bolt and flower in the first year, as the annual beet behavior is partly dominant over biennial behavior. These weedy hybrids flower within sugar beet crops and may cause tremendous yield losses in Europe and the U.S. (24,37,46,66,68).

Another interesting example for gene flow and ferality is found in California, where local weedy beets belong to two different taxa (*Beta vulgaris* and *B. macrocarpa*) and have at least three different origins (10,55). Weedy beet in California evolved either directly from escaped *B. vulgaris* cultigens (Swiss chard, red beet), or from wild *B. macrocarpa*, a beet species presumably introduced from Spain, or by hybridization of *B. vulgaris* cultigens with introduced *B. macrocarpa*. Although wild sea beet probably played some role in the origin of Californian weedy beets, genetic information is insufficient to determine the extent that hybridization with sea beets contributed to the contemporary *B. vulgaris*-type weedy beets in California. It is unknown whether the European sea beet has ever been introduced into California, but European hybrids between sea beet and sugar beet could have been transported directly to the U.S. To answer this question, it is suggested to check the CMS that belongs only to the crop and is found in hybrids and subsequent weedy populations in Europe but seldomly in sea beets.

The precise vernalization requirements for flowering in wild and cultivated beets widely differ among populations. Generally, cultigens require much more cold than wild beets, with evidence for adaptive significance of this trait according to regional climatic conditions (19). In contrast to cultivated forms, the trait of first year flowering is not evenly distributed among sea beet populations. In addition, the life span of wild *Beta vulgaris* is genetically fixed according to its distribution range, with up to 11 years maximum in north Brittany where individuals begin flowering in their second year (73). The shortest period observed is two growing seasons (germinating and flowering in the first year, overwintering, and second flowering period in the second year), as found in southwestern France. In southern Europe, wild *B. vulgaris* ssp. *maritima* beets behave as annuals: the plants start flowering in the first year and die after seed set due to unfavorable conditions during hot and dry summers. Flowering in the first season without vernalization (bolting) is mediated by genetic adaptation to environmental factors. Bolting is assigned to a single locus called B (B = bolting), but the offspring only partly follow Mendelian rules (1,21). Sugar beet has been strongly selected against cold sensitivity so that the lines hardly ever bolt (62). This behavior is based mainly, but not exclusively, on homozygosity of the recessive allele, that is, "bb" in diploid or "bbb" in triploid sugar beet varieties.

We studied the flowering phenology of different sea beet and weed beet accessions in relation to their geographic origin (the accessions referred to in Table 4.1). Seeds were germinated in a greenhouse. Twenty young seedlings from each accession were planted late spring and grown through the summer in an open field in Germany. Sugar beet is usually sown in early spring in this region. The delayed planting time was intended to prevent beets from unwanted bolting due to low temperature vernalization. The temperature was not lower than 3°C during the experiment, so there was still some risk of vernalization. Individuals were scored weekly for bolting and flowering. Weed and sea beet began flowering 50 days after planting in the open field. The geographic origin and genetic background clearly determined the bolting and flowering behavior of the beets. The flowering pattern of individual plants did not correlate with specific RAPD fragments. Some plants did not reach full anthesis, although they started bolting. This phenomenon was especially apparent in the Irish and French populations. The major differences between sea beet, weed beet, and sugar beet were the times of flowering. All individuals of the Portuguese and Greek sea beet populations flowered in the first year. Some vegetative individuals were detected in the Italian populations. A higher ratio of vegetative growth was observed among the Irish and Brittany populations. The Dutch sea beet population was strictly vegetative, similar to the cultivars. Approximately two-thirds of the individuals in the weed beet accessions flowered in the first year. Weed beets have indeed a high frequency of first year flowering habit indicating either the evolutionary importance of the potential sea beet ancestors or the possibility of mutation or back mutation toward bolting behavior. This phenomenon may be explained by the genetic background of the parental cultivar, which may have no genes favoring bolting other than the bolting allele "B". A general review of the genetic basis for flowering is given by Simpson and Dean (65). The geographic distribution pattern of the flowering phenology among the beet accessions examined supports the hypothesis that additional suppressing genes are present only in northern Europe and not in the Mediterranean region. It will be interesting to see how genetic and physiological mechanisms reported for *Arabidopsis* correspond with those in beet (e.g., whether the "FRI" vernalization locus is identical with the bolting locus). Mediterranean sea beets have an intermediate "B" frequency in France (73). Also, other Mediterranean accessions were not always fixed for "B" (L. Frese, 2001, personal communication). The occurrence of a few non-bolting individuals in Italian sea beet populations could either result from an intermediate frequency of the "B" allele or indicate gene flow from sugar beet seed production fields.

Vernalization occurs best when there are a few days at temperatures between 5°C and 10°C. The temperatures present for a few days during our field test could have stimulated vernalization. A test of the bolting behavior in the field provides better evidence of the natural variation among populations compared with the results of a greenhouse experiment. Although there is a strong selection pressure against vegetative "bb" weed plants in a conventional sugar beet field, various plants of weed beet offspring did not flower and may be assigned to the "bb" genotype. The obvious difference can be explained by the hybridization and introgression of genes from vernalized and flowering sugar beet cultivars with "bb" alleles, indicating the significant role of Hypothesis 2 in Section 4.2. Alternatively, if some other modifier genes interfere with "B" as suggested above, for instance a gene for sensitivity to vernalization, this could result in some "bb" plants flowering the first year, thereafter allowing selection for the secondary genes, and "bb" plants would be maintained within the population. Another explanation could be that "BB" plants may have some fitness cost compared to "Bb". Thus, there will be always enough "Bb" plants to produce "bb" seeds in weed populations. Also triploid beets are found as a result of crosses between tetraploid sugar beet and diploid sea beet, but the "BBb" and "Bbb" cases are likely to show intermediate phenotypes. In summary, the feral "B" allele enables the plants to flower without previous vernalization, but is not completely dominant. The molecular identification of "B" has attracted much attention, and gene mapping has made much progress (33).

Sugar beet, weedy beet, and sea beet form a highly dynamic crop-weed-wild complex, as gene flow also from the crop toward sea beets was detected in Italy (50) and in France (76). This complex has evolved at least over the past 2000 years under human impact, particularly under the changing influence of breeding techniques and agricultural practices. All three components may benefit from gene flow, the underlying processes of which are yet not fully understood within beets. In this respect, genetically engineered sugar beets can be an interesting study object. In quantitative terms, sugar beet genes have more in common with the weed beet genome than previously expected (75). We have probably underestimated the back mutation rate of sugar beet toward ferality in the past. This assumption is supported by our RAPD data analyses, which show that the weed beet cluster is genetically closer to the cultivated beet group than to the sea beet group.

Estimates of genetic variability among the 13 beet accessions from our study suggest that weedy *Beta vulgaris* is genetically more diverse than sugar beet or sea beet, although the local diversity of sea beet is high, supporting similar findings by Viard et al. (77). A reason for that can be the hybrid origin of weed beet from sea beet and sugar beet (as stated above). Reduced selection pressure or higher gene flow in cultivated fields may also be involved. Our results are the first to show that weed beet populations can be genetically more diverse than sea beet populations. As a result, weed beets would have a better potential for adaptation than sea beets.

Certainly, there is a greater likelihood to have weedy beets invade agricultural land and even semi-natural habitats compared to sea beets. Given the low fitness of beets outside strongly disturbed areas, conventional weed beets may only become invasive in managed land, in ruderal habitats, and in coastal sites where disturbances regularly occur. It is doubtful whether the beet gene pool presently harbors some hidden fitness genes that can increase the invasiveness of weed beet beyond the ecosystems mentioned above.

4.4 POTENTIAL IMPACT OF TRANSGENES ON FERALITY

In Europe, the cultivars of *Beta vulgaris* have so far not become invasive and have not displaced native plant species or native vegetation. However, based on the experience with unwanted effects of weed beet evolution, the introduction of transgenes into beet cultivars causes concern on the basis of a potential increase of ferality, for example, by genes that favor the spread and establishment

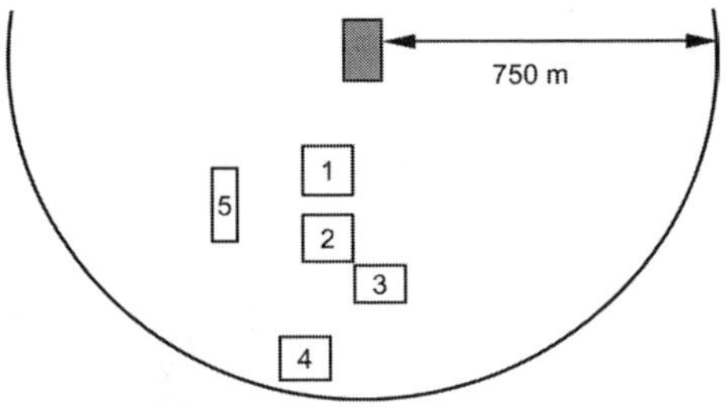

FIGURE 4.2 Weed-to-weed gene flow in *Beta vulgaris*. The rate of outcrossing was measured from a central transgene field consisting of F_1 (sea beet sugar beet) hybrids to all bolting weed beets in surrounding sugar beet fields. The F_2 offspring consisted of 1068 seedlings, of which 7 (= 0.7%) tested positively for transgene sequences (the number of transgenic offspring in detail: Field 1 = 1 of 30; Field 2 = 4 of 402, Field 3 = 2 of 43; none in Field 4 and Field 5).

in both managed and unmanaged habitats. Our results provide strong evidence that weed beet can act as an "avenue of escape" for transgenes from vernalized sugar beet bolters. In Italy, the geographic distance between sea beet populations and most sugar beet cultivation areas is too far, and sea beet pollen cannot have a considerable outcrossing effect (56). Still, flowering weed beets can link sugar beet breeding areas and sea beet habitats because of their distribution close to both habitats, such as in France (26). If transgenic sugar beet forms are placed on the market, outcrossing of the recombinant genes will occur in the near future (13,61,63). From a biosafety point of view, weed beet is considered both as a sink and a source for the flow of transgenes from sugar beet cultivars to sea beet.

We measured gene flow from 656 hemizygous transgenic virus-resistant weed beet individuals planted in a field experiment on a source plot of 800 m^2 area near non-transgenic beet bolters that spontaneously occurred in sugar beet fields within a 750 m radius of the source field (Figure 4.2). After pollination, all bolting recipient plants were harvested from an area of 3.6 ha. Seeds were transferred into the greenhouse for transgenic offspring testing. A total of 31 recipient plants yielded 1068 seedlings, of which 7 tested positive for transgenes (= 0.7% hybridization rate). The testing was laborious because no herbicide resistance marker was present in the transgenic plants. These data can be used for calculating the potential increment of increase in the number of transgenic beet plants within the following years. Starting with 7 transgenic weed beets (hemizygous for the transgene) and assuming a 3 year beet crop rotation, in the worst case, a finite population increase rate of 10 can lead to a population of 70,000 transgenic weed beet plants (allele frequency of 0.5, assuming neutral fitness effect) on 3.6 ha within 12 years (52,64). This figure ought to be compared to the number of 80,000 sugar beets usually planted per ha for cultivation. Therefore, bolter control is the key measure to minimize gene flow and adventitious transgenic weed beets in sugar beet production areas.

Considering the current progress in the field of developmental biology (18,70), specific flowering genes may be isolated in the near future and be used as transgenes to efficiently prevent bolting. Contrary to herbicide resistance genes, the direct ecological impact of genes providing weed beets with fitness enhancing traits has to be comprehensively addressed by future biosafety research (4,25). There are different opinions whether gene swamping can pose a threat to wild sea beet populations considering the future impact of various genetically engineered beet cultivars (39,43). Although non-bolting is potentially deleterious to sea beets, populations at the Adriatic coast in northeastern Italy seem not to be swamped and obliterated (12). Even with 10% outcrossing to sea beets — every generation — it will take 50 generations to have an effect (43). We examined the performance of rhizomania virus-resistant beet cultivars under various environmental conditions with regard to competitiveness, winter hardiness, and seed production. *Rhizomania* is a disease caused by the beet necrotic yellow vein virus (BNYVV). Greenhouse and field experiments were

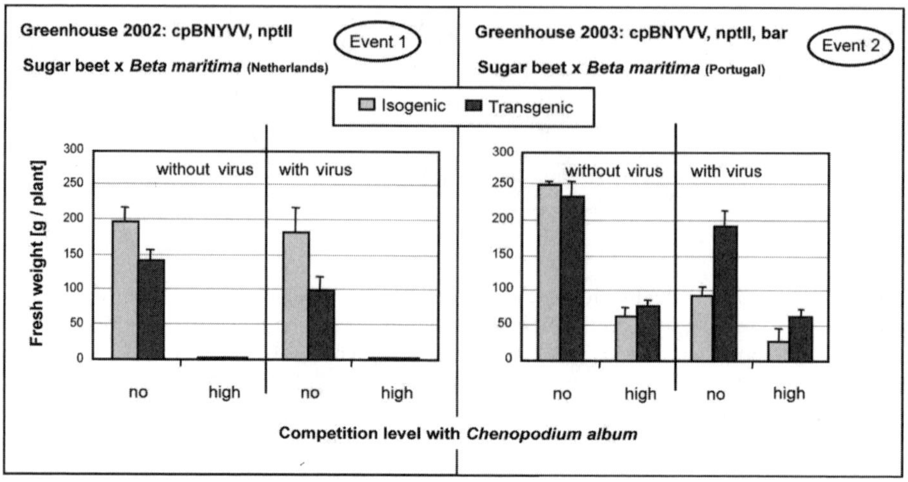

FIGURE 4.3 Competitive ability of transgenic virus-resistant sugar beet sea beet hybrids. The F_1 hybrids were grown in the greenhouse with or without virus infestation under two competition treatments with *Chenopodium album* as previously described (15). The performance clearly depends on the transformation event, which in case of Event 1 shows a "cost of resistance".

carried out between 1993 and 2001 and demonstrated that transgenic virus-resistant sugar beets performed better than virus susceptible beets, but only in soil infested by the virus (7,17,60). Some of our experiments focused on the overwintering of transgenic virus-resistant and non-transgenic sugar beet at different locations in Europe ranging from mild to cold winters of the years 1994 through 1999. We found no survival differences even under virus infestation conditions. Rhizomania is absent in saline wild beet habitats so that virus resistance is not an advantage there (6). Weed beets in infested sugar beet fields may benefit from virus resistance genes, whether they come from genetically engineered or conventionally bred cultivars (11).

In 2002 and 2003, greenhouse experiments were carried out with virus-resistant beet hybrids. Transgenic and near-isogenic hybrids of sugar beet sea beet were planted in two virus treatments (no or high pressure of infestation) at two competition levels (no or high density of *Chenopodium album*, a common weed in sugar beet crops). Biomass and seed production of beet genotypes strongly depend on virus and competition level. Transgenic virus resistance genes may offer a competitive advantage under certain field conditions, but the transgenic event and the conventional genetic background of the sea beet parent also determine this. In general, transgene mediated BNYVV resistance is unlikely to increase weediness of beet hybrids, *inter alia* due to the fact that many sea beet populations tend to be naturally virus tolerant. Thus, it can be hypothesized that wild beets have already passed tolerance genes via gene flow to weed beets. A typical example of transgene performance is shown in Figure 4.3. The transformation events — Event 1 cpBNYVV/nptII and Event 2 cpBNYVV/nptII/bar — have a distinct impact on plant performance. The difference between susceptible and resistant beets declined if more competing weeds were placed nearby. No differences were observed between transgenic and isogenic forms in case of Event 2 when the virus was absent. However, the case of Event 1 shows a "cost of resistance" effect under non infestation conditions.

4.5 CONCLUSIONS AND OUTLOOK

All forms of *Beta vulgaris* inhabit disturbed places, either agricultural land, field verges, ruderal sites, or coastal drift lines. Weed beets can invade fields with dicotyledonous crops, in particular with beet cultivars where they cannot be controlled by selective herbicides. Wild sea beet is a stress-tolerant colonizer of drift line vegetation on sea beaches, where competition is low but hard physical conditions prevail. In some Mediterranean regions, sea beets are also found inland growing in waste places and colonizing open soil along roadsides.

The most common hypothesis about the origin of European weed beets states that hybridization of wild sea beets and sugar beet cultivars is possible in the areas of seed production due to gene flow between flowering cultivars and the wild relatives. There are two hot spots of hybridization in southern Europe — the Italian seed production area in the northeastern Po valley and the seed production area around the town Agen in southwestern France (9). In both regions, inland weed beets occur close to sugar beet seed production fields whereas the non-weedy sea beet plants grow quite far away. This suggests that presently the local primary-type weed beet populations are the most likely source of the pollen contaminating sugar beet seed production, thus generating the seed of the secondary-type weed beet populations. In the latter process, pollen from wild sea beets might also rarely be involved. The weed beet populations in Germany are presumed to be imported with contaminated sugar beet seed from the seed production areas in southern Europe, as the large German sugar beet growing regions are far away from the coasts. There is still much discussion about the potential role of mutations or back mutations of sugar beet cultivars resulting in feral weed beets. It is most likely that such mutations would happen in the southern European seed production areas.

The weedy form does not require vernalization, hence it bolts, flowers, and produces seeds in the first growing season. It can quickly establish a seed bank, which remains viable in the soil for years. The infestation of sugar beet crops with weed beets is determined by two main factors:

1. Contamination of the sugar beet seeds
2. Cultivation practices that foster the growth and reproduction of annual weed beets

Considering that sugar beets are wind-pollinated obligatory outcrossers, and that wild sea beets or annual weedy beets are found close to seed production areas, continuous cross-fertilization probably occurs at low levels. Preferably, seed production should be transferred to places outside Europe, where sea beets and weed beets are absent. At present, farmers throughout Europe may be receiving contaminated sugar beet seeds.

Farmers must develop specific control strategies against weed beet infestation of sugar beet crops. These strategies may have a direct or indirect impact on farmland biodiversity (3,22,30,40,42,44,45). Comprehensive long-term environmental monitoring is needed that combines population genetic analysis with ecological field observations inside and outside cultivated land to keep such effects under surveillance. Because weed beets at present cannot be controlled by herbicides, mechanical removal of bolters is the only method available. Without control, the weed beet populations can quickly attain high densities rendering economically sound sugar beet cultivation impossible. Transgenic herbicide-resistant sugar beet cultivars were proposed to solve the weed problem (53), but herbicide-resistant weed beets would soon be generated through hybridization in seed production areas (27). Rotating glyphosate- and glufosinate-resistant cultivars would only open up the way to multitolerant weed beets, if gene flow in seed production areas is not inhibited. Even with herbicide-resistant cultivars, mechanical control of bolters will remain a critical part of agricultural practice in the future (67).

ACKNOWLEDGMENTS

Thanks are due to Jonathan Gressel and Herbert Sukopp for comments on this chapter.

LITERATURE CITED

1. Abe J, Guan G-P, Shimamoto Y. 1997. A gene complex for annual habit in sugar beet (*Beta vulgaris* L.). *Euphytica* 94:129–135.
2. Abe J, Yoshikawa H, Tsuda CH. 1986. Reproductive barriers in sugar beet and its wild relatives of the section *Vulgares* in the genus *Beta*. *J. Fac. Agric. Hokkaido Univ.* 63:40–48.
3. Andow DA. 2003. UK farm-scale evaluations of transgenic herbicide-tolerant crops. *Nature Biotech.* 12:1453–1454.
4. Arnaud JF, Viard F, Delescluse M, Cuguen J. 2003. Evidence for gene flow via seed dispersal from crop to wild relatives in *Beta vulgaris* (Chenopodiaceae): consequences for the release of genetically modified crop species with weedy lineages. *Proc. R. Soc. London B* 270:1565–1571.
5. Barocka KH. 1985. Zucker- und Futterrüben. In *Lehrbuch der Pflanzenzüchtung landwirtschaftlicher Kulturformen*, Fischbeck G, Plarre W, Schuster W, Eds., 2:245–287. Berlin/Hamburg: Paul Parey.
6. Bartsch D, Brand U. 1998. Saline soil condition decreases rhizomania infection of *Beta vulgaris*. *J. Plant Pathol.* 80:219–223.
7. Bartsch D, Brand U, Morak C, Pohl-Orf M, Schuphan I, Ellstrand, NC. 2001. Biosafety of hybrids between transgenic virus-resistant sugar beet and Swiss chard. *Ecol. Appl.* 11:142–147.
8. Bartsch D, Clegg J, Ellstrand NC. 1999. Origin of wild beet and gene flow between *Beta vulgaris* and *Beta macrocarpa* in California. *Gene flow and agriculture — relevance for transgenic crops, BCPC Symposium Proceedings 72*, pp. 269–274. Farnham, Surrey: British Crop Protection Council.
9. Bartsch D, Cuguen J, Biancardi E, Sweet J. 2003. Environmental implications of gene flow from sugar beet to wild beet — current status and future research needs. *Environ. Biosafety Res.* 2:105–115.
10. Bartsch D, Ellstrand NC. 1999. Genetic evidence for the origin of Californian wild beets (genus *Beta*). *Theor. Appl. Genet.* 99:1120–1130.
11. Bartsch D, Hoffmann A, Lehnen M, Wehres U. 2003. Ecological consequences of gene flow from cultivars to wild relatives, Rhizomania resistance genes in the genus *Beta*. In *Ecological impact of GMO dissemination in agro-ecosystems*, Lelley T, Balázs E, Tepfer M, Eds., pp. 115–130. Vienna: Facultas Verlags- und Buchhandelsgesellschaft AG.
12. Bartsch D, Lehnen M, Clegg J, Pohl-Orf M, Schuphan I, Ellstrand, NC. 1999. Impact of gene flow from cultivated beet on genetic diversity of wild sea beet populations. *Mol. Ecol.* 8:1733–1741.
13. Bartsch D, Pohl-Orf M. 1996. Ecological aspects of transgenic sugar beet: transfer and expression of herbicide resistance in hybrids with wild beets. *Euphytica* 91:55–58.
14. Bartsch D, Schmidt M. 1997. Influence of sugar beet breeding on populations of *Beta vulgaris* ssp. *maritima* in Italy. *J. Veg. Sci.* 8:81–84.
15. Bartsch D, Schmidt M, Pohl-Orf M, Haag C, Schuphan I. 1996. Competitiveness of transgenic sugar beet resistant to beet necrotic yellow vein virus and potential impact on wild beet populations. *Mol. Ecol.* 5:199–205.
16. Bartsch D, Schmitz G. 2002. Recent experience with biosafety research and role of post-market environmental monitoring in risk management of plant biotechnology derived crops. In *Biotechnology and safety assessment*, Thomas J, Fuchs R, Eds., 3rd ed. pp. 13–38. San Diego: Academic Press.
17. Bartsch D, Schuphan I. 2002. Lessons we can learn from ecological biosafety research. *J. Biotechnol.* 98:71–77.
18. Bastown R, Mylne JS, Lister C, Lippman Z, Martienssen RA, Dean C. 2004. Vernalization requires epigenetic silencing of FLC by histone methylation. *Nature* 427:165–167.
19. Boudry P, McCombie H, van Dijk H. 2002. Vernalization requirement of wild beet *Beta vulgaris* ssp. *maritima*: among population variation and its adaptive significance. *J. Ecol.* 90:693–703.
20. Boudry P, Mörchen M, Saumitou-Laprade P, Vernet P, van Dijk H. 1993. The origin and evolution of weed beets: consequences for the breeding and release of herbicide-resistant transgenic sugar-beets. *Theor. Appl. Genet.* 87:471–478.
21. Boudry P, Wieber R, Saumitou-Laprade P, Pillen K, van Dijk H, Jung C. 1994. Identification of RFLP markers closely linked to the bolting gene B and their significance for the study of the annual habit in beets (*Beta vulgaris* L.). *Theor. Appl. Genet.* 88:852–858.
22. Brooks DR, Bohan DA, Champion GT, Haughton AJ, Hawes C, et al. 2003. Invertebrate responses to the management of genetically modified herbicide-tolerant and conventional spring crops. I. Soil-surface-active invertebrates. *Phil. Trans. R. Soc. London B* 358:1847–1862.

23. Cai D, Kleine M, Kifle S, Harloff HJ, Sandal NN, et al. 1997. Positional cloning of a gene for nematode resistance in sugar beet. *Science* 275:832–834.
24. Champion GT. 2000. The biology of weed beet. *Br. Sugar Beet Rev.* 68:53–55.
25. Crawley MJ, Brown SL, Hails RS, Kohn D, Rees M. 2001. Transgenic crops in natural habitats. *Nature* 408:682–683.
26. Cuguen J, Arnaud JF, Delescluse M, Viard F. 2004. Crop/wild interaction within the *Beta vulgaris* complex: a comparative analysis of genetic diversity between sea beet and weed beet populations within the French sugar beet production area. In *Introgression from genetically modified plants into wild relatives and its consequences*, den Nijs H, Bartsch D, Sweet J, Eds., pp. 183–201. Wallingford: CABI.
27. Darmency H. 1997. Gene flow between crops and weeds: risk for new herbicide resistant weeds? In *Weed and crop resistance to herbicides*, De Prado R, Jorrin J, Garcia-Torres L, Eds., pp. 239–248. Dordrecht, Netherlands: Kluver Academic Publishers.
28. Desplanque B, Boudry P, Broomberg K, Saumitou-Laprade P, Cuguen J, van Dijk, H. 1999. Genetic diversity and gene flow between wild, cultivated and weedy forms of *Beta vulgaris* L. (Chenopodiaceae), assessed by RFLP and microsatellite markers. *Theor. Appl. Genet.* 98:1194–1201.
29. Desplanque B, Hautekeete N, van Dijk H. 2002. Transgenic weed beets: possible, probable, avoidable? *J. Appl. Ecol.* 39:561–571.
30. Dewar AM, May MHJ, Woiwood IP, Haylock LA, Champion GT, et al. 2003. A novel approach to the use of genetically modified herbicide tolerant crops for environmental benefit. *Proc. R. Soc. London B* 270:335–340.
31. Doney DL, Whitney ED, Terry J, Frese L, Fitzgerald P. 1990. The distribution and dispersal of *Beta vulgaris* L. ssp. *maritima* germplasm in England, Wales, and Ireland. *J. Sugar Beet Res.* 27:29–37.
32. Driessen S, Pohl M, Bartsch D. 2001. RAPD-PCR analysis of the genetic origin of sea beet (*Beta vulgaris* ssp. *maritima*) at Germany's Baltic Sea coast. *Basic Appl. Ecol.* 2:341–349.
33. El-Mezawy A, Dreyer F, Jacobs G, Jung C. 2002. High-resolution mapping of the bolting gene B of sugar beet. *Theor. Appl. Genet.* 105:100–105.
34. Elseth GD, Baumgardner KD. 1981. *Population biology.* New York: Van Nostrand.
35. Fischer HE. 1989. Origin of the 'Weisse Schlesische Rübe' (white Silesian beet) and resynthesis of sugar beet. *Euphytica* 41:75–80.
36. Ford-Lloyd BV. 1986. Infraspecific variation in wild and cultivated beets and its effect upon infraspecific classification. In *Infraspecific classification of wild and cultivated plants*, Styles BT, Ed., pp. 331–334. Oxford: Clarendon Press.
37. Ford-Lloyd BV, Hawkes JG. 1986. Weed beets, their origin and classification. *Acta Horticultura* 82:399–401.
38. Ford-Lloyd BV, Williams JT. 1975. A revision of *Beta* section *Vulgares* (Chenopodiaceae), with new light on the origin of cultivated beets. *Bot. J. Linn. Soc.* 71:89–102.
39. Gepts P, Papa R. 2003. Possible effects of (trans)gene flow from crops on the genetic diversity from landraces and wild relatives. *Environ. Biosafety Res.* 2:89–103.
40. Haughton AJ, Champion GT, Hawes C, Heard MS, Brooks DR, et al. 2003. Invertebrate responses to the management of genetically modified herbicide-tolerant and conventional spring crops. II. Within-field epigeal and aerial arthropods. *Phil. Trans. R. Soc. London B* 358:1863–1877.
41. Hautekeete N, Piquot Y, van Dijk H. 2002. Life span in *Beta vulgaris maritima*: the impacts of disturbance and of age at first reproduction. *J. Ecol.* 90:508–516.
42. Hawes C, Haughton AJ, Osborne JL, Roy DB, Clark SJ, et al. 2003. Responses of plants and invertebrate trophic groups to contrasting herbicide regime in the Farm Scale Evaluations of genetically modified herbicide-tolerant crops. *Phil. Trans. R. Soc. London B* 358:1899–1913.
43. Haygood R, Ives AR, Andow DA. 2003. Consequences of recurrent gene flow from crops to wild relatives. *Proc. R. Soc. London B* 270:1879–1886.
44. Heard MS, Hawes C, Champion GT, Clark SJ, Firbank LG, et al. 2003a. Weeds in fields with contrasting conventional and genetically modified herbicide-tolerant crops. I. Effects on abundance and diversity. *Phil. Trans. R. Soc. London B* 358:1819–1832.
45. Heard MS, Hawes C, Champion GT, Clark SJ, Firbank LG, et al. 2003b. Weeds in fields with contrasting conventional and genetically modified herbicide-tolerant crops. II. Effects on individual species. *Phil. Trans. R. Soc. London B* 358:1833–1846.

46. Johnson RT, Burtch LM. 1959. The problem of wild annual sugar beets in California. *J. Am. Soc. Sugar Beet Technol.* 10:311–317.
47. Jung C, Pillen K, Frese L, Fahr S, Melchinger AE. 1993. Phylogenetic relationships between cultivated and wild species in the genus *Beta* revealed by DNA fingerprinting. *Theor. Appl. Genet.* 86:449–457.
48. Lange W, Brandenburg WA, de Bock TSM. 1999. Taxonomy and cultonomy of beet (*Beta vulgaris* L.). *Bot. J. Linn. Soc.* 130:81–96.
49. Laporte V, Viard F, Béna G, Valero M, Cuguen J. 2001. The spatial structure of sexual and cytonuclear polymorphism in the gynodioecious *Beta vulgaris* ssp. *maritima* at a local scale. *Genetics* 157:1699–1710.
50. Lavigne C, Klein EK, Couvet D. 2002. Using seed purity data to estimate an average pollen mediated gene flow from crops to wild relatives. *Theor. Appl. Genet.* 104:139–145.
51. Letschert JPW. 1993. *Beta* section *Beta*: biogeographical patterns of variation and taxonomy. *Wageningen Agricultural University Papers 93.* 153 pp.
52. Longden P, May MJ, Fisher S. 2002. Weed beet control 2002. *Br. Sugar Beet Rev.* 70:38–41.
53. Madsen K, Jensen J. 1995. Weed control in glyphosate tolerant sugar beet (*Beta vulgaris* L.). *Weed Res.* 35:105–111.
54. Mansfeld R. 1986. *Verzeichnis landwirtschaftlicher und gärtnerischer Kulturpflanzen.* Hamburg: Springer.
55. McFarlane JS. 1975. Naturally occurring hybrids between sugarbeet and *Beta macrocarpa* in the Imperial Valley of California. *J. Am. Soc. Sugar Beet Tech.* 18:245–251.
56. Mücher T, Hesse P, Pohl-Orf M, Ellstrand NC, Bartsch D. 2000. Characterization of weed beets in Germany and Italy. *J. Sugar Beet Res.* 37(3):19–38.
57. Munerati O. 1932. Sull'incrocio della barbatiola coltivata con la *Beta* selvaggia della costa adriatica. *L'Indust. Saccarifera Ital.* 25:303–304.
58. Nei M. 1978. Estimation of average heterozygosity and genetic distance from a small number of individuals. *Genetics* 89:583–590.
59. Pohl-Orf M, Brand U, Driessen S, Hesse PR, Lehnen M, et al. 1999. Overwintering of genetically modified sugar beet, *Beta vulgaris* L. subsp. *vulgaris*, as a source for dispersal of transgenic pollen. *Euphytica* 108:181–186.
60. Pohl-Orf M, Morak C, Wehres U, Saeglitz C, Driessen S, et al. 2000. The environmental impact of gene flow from sugar beet to wild beet — an ecological comparison of transgenic and natural virus tolerance genes. *Proc. Int. Symp. on the Biosafety of Genetically Modified Organisms, 6th, Saskatoon,* pp. 51–55. Saskatoon, Canada: University of Saskatchewan.
61. Raybould AF, Gray AJ. 1993. Genetically modified crops and hybridization with wild relatives: a UK perspective. *J. Appl. Ecol.* 30:199–219.
62. Sadeghian SY, Becker HC, Johansson E. 1993. Inheritance of bolting in three sugar beet crosses after different periods of vernalization. *Plant Breed.* 104:328–333.
63. Saeglitz C, Bartsch D. 2002. *Gene flow from transgenic plants.* http://www.agbiotechnet.com.
64. Sester M, Delanoy M, Colbach N, Darmency H. 2004. Crop and density effects on weed beet growth and reproduction. *Weed Res.* 44:50–59.
65. Simpson GG, Dean C. 2002. *Arabidopsis*, the Rosetta Stone of flowering time? *Science* 296:285–289.
66. Slyvchenko O, Bartsch D. 2004. Introgression of cultivar beet genes to wild beet in the Ukraine. In *Introgression from genetically modified plants into wild relatives and its consequences*, den Nijs H, Bartsch D, Sweet J, Eds., pp. 173–182. Wallingford, UK: CABI.
67. Soukup J, Holec J. 2004. Crop/wild interaction within the *Beta vulgaris* complex: agronomic aspects of weed beet in the Czech Republic. In *Introgression from genetically modified plants into wild relatives and its consequences*, den Nijs H, Bartsch D, Sweet J, Eds., pp. 203–218. Wallingford, UK: CABI.
68. Soukup J, Holec J, Vejl P, Skupinova S, Sedlak P. 2002. Diversity and distribution of weed beet in the Czech Republic. *J. Plant Dis. Protect.* 18:67–74.
69. Sukopp H, Sukopp U. 1993. Ecological long-term effects of cultigens becoming feral and of naturalization of non-native species. *Experientia* 49:210–218.
70. Sung S, Amasino RM. 2004. Vernalization in *Arabidopsis thaliana* is mediated by the PHD finger protein VIN3. *Nature* 427:159–164.

71. van de Peer Y, de Wachter R. 1994. TREECON for Windows: a software package for the construction and drawing of evolutionary trees for the Microsoft Windows environment. *Comp. Appli. Biosci.* 10:569–570.
72. van Dijk H. 2004. Gene exchange between wild and crop in *Beta vulgaris*: How easy is hybridization and what will happen in later generations? In *Introgression from genetically modified plants into wild relatives and its consequences*, den Nijs H, Bartsch D, Sweet J, Eds., pp. 53–61. Wallingford, UK: CABI.
73. van Dijk H, Boudry P, McCombie H, Vernet P. 1997. Flowering time in wild beet (*Beta vulgaris* ssp. *maritima*) along a latitudinal cline. *Acta Oecol.* 18:47–60.
74. van Dijk H, Desplanque B. 1999. European *Beta*: crops and their wild and weedy relatives. *Plant evolution in man-made habitats, Proc. Int. Symp. IOPB, 7th, Amsterdam*, pp. 257–270. Amsterdam: Hugo de Vries Laboratory.
75. van Raamsdonk LWD, van der Maesen LJG. 1996. Crop-weed complexes: the complex relationship between crop plants and their wild relatives. *Acta Bot. Neerl.* 45:135–155.
76. Viard F, Arnaud JF, Delescluse M, Cuguen J. 2004. Tracing back seed and pollen flow within the crop-wild *Beta vulgaris* complex: genetic distinctiveness vs. hot spots of hybridization over a regional scale. *Mol. Ecol.* 13:1357–1364.
77. Viard F, Bernard J, Desplanque B. 2002. Crop-weed interactions in the *Beta vulgaris* complex at a local scale: allelic diversity and gene flow within sugar beet fields. *Theor. Appl. Genet.* 104:688–697.

5 Volunteer Oilseed Rape — Will Herbicide-Resistance Traits Assist Ferality?

Linda M. Hall, M. Habibur Rahman, Robert H. Gulden, and A. Gordon Thomas

5.1 INTRODUCTION

The production of *Brassica* crops grown for the production of oil and vegetables was recorded over 2000 years ago in China. Their culture has diversified widely, in both form and utilization. From their centers of origin in the Mediterranean, Europe, and Asia, they have been distributed around the world and are grown wherever temperate or Mediterranean climates occur. Their specific cropping systems have been selected for within the climatic conditions in each region. Along with these Old World crops, several wild relatives successfully colonized new continents and co-occur as weedy species within these agroecosystems. Near the center of origin, sympatric populations of several *Brassica* oilseed crops, crop volunteers, their wild progenitors, and weedy relatives co-occur (107). In other areas and climates, volunteers and feral populations (crop species that have established self-perpetuating populations) and weedy relatives have established in ruderal areas (108), while in other regions, volunteers occur, but form only short-lived populations (39,83) and weedy relatives may or may not co-occur. Therefore, considerable variation occurs between regions and continents in the presence and abundance of volunteer, feral, and wild and weedy relative populations in agroecosystems and ruderal or natural areas adjacent to agricultural ecosystems, which prohibits useful generalization. Case-by-case assessments of the impact of herbicide resistance and other traits on persistence of crop volunteers and feral populations must be made in each agroecosystem.

The introduction of herbicide-resistant *Brassica* crops provided the impetus, as well as convenient markers, to examine the demographics and ecological roles of volunteer crops and feral populations within and external to agroecosystems. The extent and significance of gene flow among domestic, feral, and wild and weedy relatives, and the impact of single and multiple herbicide-resistance genes on the biological attributes contributing to fitness can also be studied in this context.

Concern for the risk associated with herbicide-resistant crops predated their introduction (38,45,75,91,142,146). Would volunteer or feral transgenic crops be more fit and therefore increase within the agroecosystem? Would they become invasive of natural areas? Concerns were raised about the movement of transgenes to wild or weedy relatives. Would herbicide-resistance traits transfer to weeds or would the changes in cropping practices lead to the additional selection of resistant weeds? Resistant and multiple resistant weeds had been identified as a significant agricultural problem around the world (14,15,69). Therefore, considerable concern accrued to the first reports of incidences of multiple resistant volunteer oilseed rape *B. napus* (18,65).

In this review, we discuss the origins and biology of the common *Brassica* species grown for oil production. We identify the characteristics selected by crop breeding and speculate on their impact on volunteer crop demographics. We assess how cultivated *B. napus* and *B. rapa* (formerly *B. campestris*) conform to the characteristics associated with weeds found in agroecosystems.

Finally we review the evidence for an increase in persistence, abundance, or spread of volunteers or for dedomestication and evolution of feral populations since the introduction of herbicide-resistance traits. Because of the difference among regions and continents in the occurrence and ecological roles occupied by the *Brassica* species, we will focus the discussion on western Canada and identify divergence where reported in the literature. We do not address the potential impacts of transgene introgression into weedy relatives, but focus instead on the impacts of transgenes on volunteer and feral *Brassica* crop demographics within agroecosystems. We also consider evidence for the introgression of feral traits from wild or weedy species into volunteer crops and the effects this may have on ferality within the agroecosystem.

5.2 *BRASSICA RAPA* AND *B. NAPUS*, ORIGINS AND BIOLOGY

The genus *Brassica* contains a number of oilseed and vegetable crop species cultivated around the world. The three diploid species *B. rapa* (AA, 2n = 20), *B. nigra* (BB, 2n = 16), and *B. oleracea* (CC, 2n = 18) are elemental to the species *B. juncea* (AABB, 2n = 36), *B. carinata* (BBCC, 2n = 34); and *B. napus* (AACC, 2n = 38) are amphidiploids derived from them (10) (Figure 5.1). *Brassica* oilseeds, primarily *B. rapa*, *B. napus*, and *B. juncea*, are together currently the world's third most important source of vegetable oil, following soybean and palm oil.

The three *Brassica* genomes, A, B and C, share considerable homology. Cytogenetic studies support the hypothesis that the three diploid species are the secondary polyploid of a common prototype x = 6 (4,5,111,116). However, our knowledge on the origin of the three diploid species is limited. Molecular systematic studies by restriction fragment length polymorphism (RFLP) of nuclear DNA (127,128) and chloroplast and mitochondrial (mt) DNA (109,144) suggested two divergent pathways for origin of the three diploid species. One pathway gave rise to *B. nigra* (n = 8) and the other pathway gave rise to *B. oleracea* (n = 9) and *B. rapa* (n = 10). Cytological observations of a high frequency of homologous pairing between A- and C-genome chromosomes compared to pairing of the chromosomes of these two genomes with the chromosomes of the B-genome (6,7,8,35) support this hypothesis. Following emergence of the diploid species, natural interspecific crossing, followed by spontaneous chromosome doubling, led to the formation of the three amphidiploid species. This was verified by artificial synthesis of *B. napus* (2,31,50,100,126,139), *B. juncea* (49,126), and *B. carinata* (50,118,126). Cytoplasmic molecular analysis data of RuBISCO protein (140), cpDNA (102), and mtDNA (103) suggested *B. rapa* and *B. nigra*, respectively, functioned as female parent for *B. juncea* and *B. carinata*. However, no specific prototype has been proposed as the female parent of *B. napus*. Four different cytoplasms were found in diploid *Brassica* species and all types were also found in natural *B. napus*, suggesting multiple origin of this species. The cytoplasm of the majority of the cultivated *B. napus* differs from that of *B. rapa* and *B. oleracea*, but has the same chloroplast genome as that of *B. montana*

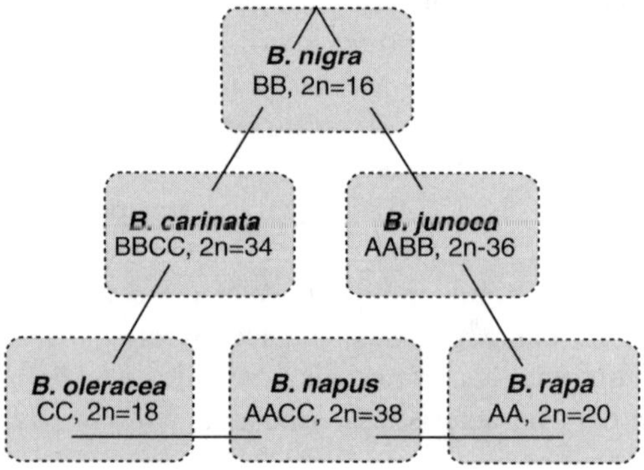

FIGURE 5.1 Phylogenic relationships between different *Brassica* species. (Adapted from N. U (139).)

(2n = 18) and a unique mitochondrial genome that is intermediate between *B. rapa* and *B. montana* suggesting *B. montana* or a close relative in some way has contributed cytoplasm in the origin of cultivated *B. napus* (125).

Brassica rapa is an annual, indigenous from western Europe to central Asia (76,122). *Brassica oleracea* is a low, shrubby perennial that grows along the coast of the Mediterranean from Greece through to the Atlantic coasts of Spain and France around the coast of England (123). It is found on limestone and chalk cliffs, below cliffs in scree among other shrubs, and some populations are found on steep grassy slopes or open rocky ground. It was speculated that *B. napus* originated in the Mediterranean, where these two species were sympatric (110).

Both *B. rapa* and *B. napus* have been domesticated and selected as oilseed crops. *B. rapa* was the predominant species grown in Canada in the early 1970s. Similar areas of *B. napus* and *B. rapa* were grown in the late 1980s but in the 1990s and 2000s the percentage of *B. rapa* has dropped to less than 10% (24). Development of *B. napus* cultivars with reduced days to maturity has permitted producers to utilize this more productive species.

Persistence of *B. napus* has been studied outside the agroecosystem in ruderal and natural areas. In Canada, anecdotal evidence suggests *B. napus* and *B. rapa* populations can occur where seed has been spilled, but that these populations are short lived and not invasive. In a study of patch dynamics of feral *B. napus* growing along a roadway in the U.K., Crawley and Brown (39) observed that patch extinction was common and rapid, except where patches may have had the seed bank replenished by spilled seed. In the absence of disturbance, *B. napus* was unable to compete with perennial grasses and patches became extinct in 3 years. However, Pessel et al. (108) reported that *B. napus* can persist for 8 to 9 years along a roadway in France, based on the presence of high erucic/high gluocosinilate used as a genetic marker. Several studies (38,39) support observations that *B. napus* is not invasive of natural areas.

Wild *B. rapa*, if it still exists (93), and weedy *B. rapa* occur throughout the center of origin and as an introduced weedy species in eastern Canada (51,145), and in the U.S. and Europe (130) coincident with cultivated *B. napus* and *B. rapa*. Weedy *B. rapa* is not commonly found in western Canada, the major *B. napus* growing region (97). Other weedy relatives, including *Raphanus raphanistrum, Erucastrum gallicum,* and *Sinapis arvensis*, are sympatric in some regions or continents.

5.2.1 Introgression between Crop and Wild Brassicaceae

The potential for introgression of *B. napus* with other species, including weedy and cultivated *B. rapa,* has been the subject of intense research scrutiny. As might be predicted from evolutionary and cytological similarities, hybridization is frequent between sympatric *B. napus* and *B. rapa*. Spontaneous hybrids have been located in field experiments and in wild populations in commercial fields, demonstrating a high potential for introgression where these species co-occur (74,145). In western Canada hybridization is unlikely because weedy *B. rapa* does not occur. Hybridization between *R. raphanistrum* and *B. napus* is rare (32,33; Powles, personal communication), and hybridization between *B. napus* and *E. gallicum* is even less likely. Hybridization with *S. arvensis* has not been demonstrated. The formation of hybrids suggests that introgression of crop genes, including herbicide resistance, may occur into the wild species genome. Gene flow from crops to wild species is beyond the scope of this review. Readers are directed to current research and review articles (33,59,66,82,98,114,115,145). Less is known about the introgression of genes from wild relatives including those that may increase ferality in weed/crop hybrids.

5.2.2 Crop Improvement Objectives, Domestication, and Ferality

Traits selected first in the process of domesticity have not been recorded, but similar to other crops (54,55), they probably include a loss of seed dormancy (Section 5.3), traits to increase seed yield, such as increased number and size of seeds, and changes in photoperiod sensitivity. Traditionally,

Brassica seed oil was used as edible oil in China and on the Indian subcontinent, but in the western world, it was used primarily as lamp oil and for lubricants. Rapeseed oil was also used as cooking oil in 16th century Europe, but only by the less affluent (70,76). Early breeding efforts in Europe focused on increased yield and on the development of both spring and winter seeded cultivars. Changes to the life cycle of volunteers, from spring annual to winter annual, impacted the crop volunteers' ability to compete and reproduce in winter and spring cropping systems.

After World War II, the oil was used for human consumption in the western world. However, nutritional quality was questioned in the early 1950s, because of high levels of erucic fatty acid. Laboratory studies with experimental animals concluded erucic acid might constitute a hazard to health and recommended removal in edible oil. Identification of zero eurcic acid *Brassica* sources (43,77,101,133) and incorporation into crop cultivars in the 1960s virtually eliminated erucic acid from edible *Brassica* seed oil. Conversion of Canadian fields to low or zero erucic acid type (0) for edible purpose occurred in early 1970s (132). Because winter seeded varieties require more time for breeding and seed increase, this was not accomplished in Europe until the late 1970s.

Although oil is the main product of *Brassica* crops, the seed meal remaining after extraction of the oil contains about 40% protein with an excellent amino acid composition (46). However, meal was seldom used as a protein source in animal feed due to the presence of glucosinolates. The low glucosinolate *B. napus* strain "Bronowski" was identified in the early 1970s (47) and was used extensively in breeding programs for the development of double low (00) (low glucosinolate and low erucic acid) cultivars. The first 00 *B. napus* cultivar Tower was released in Canada in 1974 (134). Conversion of Canadian crops to 00 type was accomplished in the late 1970s; however, another decade was required to convert European fall seeded areas to 00 type. To avoid any confusion between 0 and 00 types and to enhance market differentiation, Canadian rapeseed crushers adopted the name Canola (CANada Oil Low Acid) for the 00 rapeseed.

In wild species, glucosinolate in leaves has an impact on leaf feeders and hence on plant reproductive fitness. Giamoustaris and Mithen (57) reported that glucosinolate content in leaves decreased grazing by non-specialized arthropod herbivores, but increased the feeding and presence of adult flea beetles (*Psylliodes chrysocephala*) and larva of *Pieris rapae*. Hence, high glucosinolate levels in vegetative tissues are not always correlated with greater fitness potential (58,120,135). Seed glucosinolate level and composition in canola/00 cultivars are not directly related to the glucosinolates in leaves.

Recent crop improvement efforts in *B. napus* and *B. rapa* focused on specific objectives that could have direct effects on the fitness and persistence of volunteer crops (Table 5.1). We can speculate that increasing yield by increasing seed number and weight may confer a fitness advantage on volunteer crops as it might increase seedling numbers and vigor, respectively. Further crop improvement objectives, including winter hardiness and appropriate crop maturity for specific agroenvironments, may improve the fitness of the crop and similarly improve volunteer survival.

Breeding enhanced shattering resistance may facilitate domestication by reducing the volunteer seed bank (Section 5.4). However, with the exception of some synthetic lines (34), there is little variation for resistance to shattering within *B. napus*, and efforts to decrease shattering have not been successful to date (96). However, some variation in shattering does exists in *B. rapa* and *B. juncea* and introspecific introgression might play a role in future *B. napus* cultivar development. The identification of a shatterproof genes, *SHP1* and *SHP2,* that control dehiscence zone differentiation in *Arabidopsis thaliana* (84), may provide additional genetic resources for the reduction of shattering. Reducing shatter potential in *B. napus* is of greater priority in Europe, where the crops are generally direct harvested, than it is in western Canada, where they are usually windrowed prior to harvest and therefore seed losses are perceived to be lower.

Breeding for disease resistance has advanced rapidly. Cultivars have been conventionally bred and are now resistant (or less susceptible) to a range of fungal diseases, including blackleg (caused by *Leptosphaeria maculans*), white rust (caused by *Albugo candida*), etc. Disease resistance may provide a fitness advantage to crop cultivars and presumably to crop volunteers and feral populations.

TABLE 5.1
Some Recent Breeding Objectives and the Implication for Reproductive Fitness for Volunteer Crops

Breeding Objective	Fitness Implications for Volunteer Crops
Yield, increased seed number and seed weight	Potential increase in reproductive success
Oil content of seed	Potential increase in reproductive success
Protein content of seed	Unknown impact on reproductive success
Lodging resistance	Fitness neutral
Appropriate maturity	Enhanced fitness in environments where the cultivar is grown
Winter hardiness (winter crops)	Enhanced fitness in northern environments
Disease resistance	May increase fitness of volunteer crops
Shattering resistance	If successful, may decrease seed bank inputs and thus density of volunteers and feral crops
Oil and meal quality, may reduce glucosinolates in leaf tissue	May alter the feeding herbivores on volunteer crops, potentially changing fitness
Hybrid vigor/CMS breeding system	Enhanced fitness/decreased fitness
Herbicide resistance	Fitness benefit under specific agronomic conditions

To date, transgenic disease resistance has not been commercialized in *Brassica,* but transgenic *Cylindrosporium*/*Phoma*-resistant and *Sclerotinia*-resistant oilseed rape strains are currently being field tested in the U.S. (3). The contribution of disease resistance to volunteer fitness has not been established.

Hybrid *B. napus* cultivars have been recently introduced and adopted. Hybrids are notably more vigorous and produce more and sometimes larger seed, which usually results in increased productivity. However, compared to the F_1 hybrid parents, F_2 volunteers could experience a loss of heterosis and increased phenotypic diversity. Conversely, in the absence of linkage of favorable genes and the overdominance effect, part of the volunteer population might be as vigorous as the initial hybrid and will be self-selected. Hybrids are generated by several methods. In cultivars, where cytoplasmic male sterility (CMS) has been used to produce hybrids, the first generation of volunteers will be 75% fertile and 25% sterile plants. The proportion of fertile plants increases with subsequent generations. For other hybrid systems, termed genetic male sterility, 25% or less of the population are expected to be sterile. Overall, the effects on hybrid cultivars of *B. napus* on volunteer fitness is worthy of further study.

The primary crop improvement objectives (Table 5.1), including increased seed weight and seed oil content, agroenvironment specific maturity, cold tolerance, and disease resistance seem likely to improve fitness of volunteers and thereby increase potential ferality. Other changes, such as hybrid vigor and the reduction in glucosinolates in leaves may either increase or decrease fitness of volunteers, depending on the cultivar and environment.

Herbicide-resistant *B. napus* was commercially introduced in Canada in 1995 and now includes cultivars resistant to glyphosate, glufosinate, and imidazolinone herbicides. These varieties were rapidly adopted and currently occupy over 85% of the total annual oilseed rape growing area (29). Because of the spatial and temporal impact of herbicide-resistant *B. napus*, western Canada offers a unique opportunity to examine potential changes in volunteer canola demographics conferred by herbicide-resistance traits (17,65).

5.3 BIOLOGICAL CHARACTERISTICS INFLUENCING WEEDINESS

Weeds of agroecosystems (9,11,12) co-evolve with crops and many now have worldwide distributions within similar crop environments (71). For example, the weed *S. arvensis*, indigenous to

TABLE 5.2
Presence of Weed Characters in Oilseed Rape

Weedy Characteristic	Presence in *B. napus and B. rapa*
Extended seed viability and dormancy mechanism with germination requirements fulfilled in many environments	No primary dormancy, secondary dormancy a function of genotype and environment
Rapid growth development and reproductive maturation	Canola has rapid development and resource capture
High phenotypic plasticity often with the ability to reproduce as long as growing conditions permit	High individual plasticity
Unspecialized pollination mechanisms, including self-pollination, wind-pollination, or generalist insects	*B. napus* can self pollinate B. rapa is an obligate outcrossing species Both *B. napus* and *B. rapa* can also be pollinated by bees and other non-specialist pollinators
High reproductive capacity	Yes
Mechanisms for short- and long-distance dispersal of weeds	Requirements for dispersal mechanism negated by the human assisted dispersal. Seeds are dispersed over long distances either through transport or via planting
Genetic diversity to facilitate adaptation to a wide range of environmental conditions	Lack of genetic diversity, but cultivars are selected for specific local agroenvironments

Source: Adapted with permission from Armstrong and Keller (4), Barrett (13), and Holt (73).

temperate Europe, is now established in North America, South America, Australia, and New Zealand (143). This suggests that these weeds have common attributes that allow them to flourish within agroecosystems. Agroecosystems differ from other ecosystems because they are repeatedly and regularly physically and/or chemically disturbed. The type and periodicity of agronomic practices is predictable, favoring specific life cycles and biological adaptations. Plant nutrients, early in the growing season, are not usually limiting in this system because they are supplemented with chemical fertilizer. Competition is periodically depleted, either by application of herbicides or by tillage. During the early part of the cropping period there is an abundance of "safe sites" (67) for species assembled on the site (56). Several authors have attempted to delineate the underlying characteristics common to weeds, which facilitate success in agroecosystems (9,10,73,94). We will review these characteristics with reference to the demographics of volunteer *B. napus* and *B. rapa* (Table 5.2).

Extensive domestication of *B. napus* and *B. rapa* is relatively recent, compared to highly domesticated cereal crops. Hence, *B. napus* and *B. rapa* arguably may still retain more weedy characteristics than other crops. Seeds of many successful weeds of agroecosystems have prolonged viability and dormancy enabling them to survive periods unsuitable for plant growth or reproductive success (73). For example, seed longevity of up to 75 years has been reported for *S. arvensis* (143). Therefore, *seed banks* (the total viable seeds on the surface and buried in the soil at any depth) extend potential reproductive success temporally, just as seed dispersal (discussed below) extends it spatially.

Seed size and shape are important determinants of survival in the seed bank. In general, small seeds such as *B. napus* and *B. rapa* have the longest seed bank persistence (94). Oilseed rape seeds are small, relative to most other crops. In addition, the seeds are smooth and spherical, facilitating self-burial. This is also recognized as a weedy characteristic. Interactions between the soil micro-topography and germination among Brasicaceae with spherical seeds of different diameter have been reported (67).

Seed dormancy is regulated by both seed characteristics and environmental conditions. Crops, including *B. napus* and *B. rapa,* have virtually no primary seed dormancy (62,68) and when planted as crops, they have synchronous germination that facilitates rapid and predictable emergence.

However, *B. napus* seeds may enter secondary dormancy when exposed to conditions unfavorable for germination. Extended exposure (weeks) to cold, moisture-limiting conditions (90,95,104) or high temperatures (62) results in the development of secondary dormancy, particularly in the absence of light (e.g., seed burial). The development of secondary dormancy varies among cultivars (62,95,106) suggesting the trait is, at least in part, under genetic control. However, the genetics of seed dormancy in dicots are complex (48,53) and the genetic resources for low secondary dormancy *B. napus* appear to be limited in commercially available genotypes in western Canada (62). Hence, reducing secondary seed dormancy potential through conventional breeding seems unlikely to be rapid.

Volunteer crop seeds respond differently to light and nutrient stimuli than their wild and weedy counterparts. A comparison of feral *B. rapa* collected from Montana with domesticated *B. rapa* and hybrids showed higher germination in domestic seeds, irrespective of light quality and available nutrients (1). Feral seed collected from agricultural and non-agricultural areas showed differential response to light quality and available nutrients. Feral *B. rapa* found in crop agricultural areas was light quality and nutrient responsive, and seed from wild areas was responsive to light only when nutrients were limiting. Crop or wild hybrids responded similarly to the maternal parent. This suggests introgression of genes from feral *B. rapa* could increase seed bank persistence, seed germination responsiveness, and therefore ferality. A seasonal response of germination to light was also induced in buried European winter *B. napus* seeds where germination under light was lowest in seeds exhumed in the summer (119).

Successful weed species frequently have early and rapid establishment and resource capture. *Brassica napus* volunteers emerge more quickly from cold soils than other weedy species (23,41,141), but emerge even more rapidly when the soil warms (21,79,90,141) and are cold-tolerant when acclimated (131). In addition, *B. napus* has the ability to respond to nitrogen, thus capturing a disproportionate share of available resources, to the detriment of slower establishing neighboring species. In a comparison of nitrogen responsiveness of 23 weedy species commonly found with *B. napus*, only 10 exhibited a similar or greater response in shoot biomass to oilseed rape and only 5 had a similar or greater response in root biomass as nitrogen levels increased (20). Weedy *B. rapa* was not included in the study because it does not commonly occur in western Canada (97).

B. napus and *B. rapa*, like many successful weeds, have morphological and physiological flexibility in response to environmental variation. Individuals exhibit phenotypic plasticity, including indeterminate flowering, variability in branching, variable size, and seed production (80,92). Plasticity confers an ability to rapidly respond to changes and variability within the environment, thus broadening the environmental range favorable to reproductive success.

Many successful weeds also have unspecialized pollination mechanisms including self-compatibility, facilitating successful survival after long distance dispersal of an individual or a few plants. *Brassica napus* is self-compatible, but *B. rapa*, with the exception of the Indian yellow sarsan form, is an obligate outcrossing species. Successful fertilization in *B. rapa* depends on physical contact between plants and short distance pollen movement by wind or pollinators (72). Both species have yellow petals that attract pollinators. *B. napus* may be more capable than *B. rapa* of successful reproduction when plant densities are low.

Weed populations where high genetic diversity confers heterosis are generally considered to have an enhanced ability to respond to local selection pressures imposed by biotic and abiotic stress (9). Crop breeding has generally reduced population genetic diversity compared to the wild progenitors (13,54,55,127). One might assume crop species would be less able to respond to varying environments. However, in the case of *B. napus* and *B. rapa*, extensive selection within the crop species genotypes has occurred, including selection of genotypes with the ability to survive a predictable range of abiotic and biotic stress, such as short-term drought, temperature stress, and selection for insect and disease resistance. Increased resistance to soil pathogens greatly enhances invasiveness and abundance of a plant species (78). Cultivars have been bred for specific regions and therefore may have a high degree of fitness to the local environment.

Outcrossing can also influence genetic diversity of volunteer crops. *Brassica napus* is primarily self-compatible with up to 20% outcrossing between adjacent plants (40,112). The frequency of outcrossing diminishes with distance, but instances of long distance gene transfer have been demonstrated. Under field conditions, transfer from one field to another generally results in less than 1% outcrossing in the first 100 m (44). *Brassica rapa* is an obligate outcrossing species and would therefore have a higher level of gene flow via pollen and a higher level of genetic diversity than *B. napus* (24).

Weeds of agroecosystems are noted for their high reproductive capacity. For example, *S. arvensis* produces from 2000 to 35,000 seeds per plant (99) where as *B. napus* produces from 5000 to 25,000 seeds per plant, depending on environment and competition. Seed production in the crop *B. rapa* tends to be lower than *B. napus*. Cultivated *B. rapa*, can produce up to 5000 seeds (72). In one study, naturalized *B. rapa* produced between 593 to 1352 seeds per plant (124).

Weed distribution and inclusion in plant communities are determined, in part, by the ability of the weed species to disperse and the relative time of arrival at a site (22,56). Although weed dispersal mechanisms are numerous, for many weed species, dispersal is determined primarily by adaptability to movement with human activities, including transport on equipment and grain contamination through seed mimicry. In the case of crops, dispersal is widespread by intentional planting, seed spillage, and transport on equipment.

In agroenvironments, widespread dispersal facilitated by human intervention, high phenotypic plasticity, limited seed bank persistence, rapid resource capture, and abundant fecundity influence the persistence of volunteer *B. napus* and *B. rapa*.

5.4 PRESENCE AND PERSISTENCE OF VOLUNTEER *B. RAPA* AND *B. NAPUS*

Unlike many other crops, *B. napus* has significant *dehiscence* (shatter) prior to or during harvest (96). Additionally, this small seeded crop is dispersed by the harvest process due to incomplete removal of seeds from straw and chaff. Reported losses prior to harvest vary widely because total seed loss is influenced by many factors, including crop maturity, the use of cutting, or windrowing prior to harvest vs. direct combining, weather conditions, type of harvester, and the presence of fungal disease and insect damage. A recent study of 35 fields in western Canada showed an average seed loss of 107 kg ha^{-1} or 6% of the crop seed yield or 3000 viable seeds m^{-2}. However, seed losses ranged widely, from 1530 to 7130 seeds m^{-2} and can be as high as 14,000 seeds m^{-2} in isolated cases (60). Seed losses greatly exceed the recommended seeding rates of 70 to 180 seeds m^{-2}. In the U.K., seed loss is typically between 8 and 12%, but can increase to over 20% if harvesting is delayed, in which case loss can be greater than 10,000 seeds m^{-2} on the soil surface (90). Seed on the soil surface is vulnerable to predation by rodents, birds, and insects, but the importance of loss by predation has not been established. If even a portion of dehisced seed survives, inputs to the seed bank can still be significant.

Because of the lack of primary dormancy, spring seeded *B. napus* seed lost at harvest may germinate in the fall under favorable conditions and be killed by frost before seed set in western Canada. Winter kill and inappropriate germination decreases the available seed bank and subsequent volunteers. In western Canada, there are no reports of overwintering of oilseed rape seedlings (61). However, in eastern Canada, there have been reports of overwintering seedlings from spring seeded *B. napus* (121). In western Canada, subsequent volunteer oilseed rape seedling recruitment occurs primarily in spring prior to the typical time of preseeding or in-crop herbicide application (61).

Agronomic factors can influence induction of secondary dormancy and, consequently, seed bank persistence of *B. napus*. Seed burial immediately following harvest, either naturally or by cultivation, prolongs seed bank persistence in both *B. napus* and *B. rapa;* particularly in *B. napus* genotypes with a high secondary seed dormancy potential (63,86,88,104,106,117,129). Leaving seed on or near the soil surface reduces persistence of seed in the seed bank. Hence, the reduction of tillage in fall and spring can significantly reduce seed persistence.

The seed bank of volunteer *B. napus* and *B. rapa* tends to decline rapidly in agroecosystems. When Lutman et al. (87) compared the persistence of winter *B. napus* seed to 15 weedy species, the crop had the highest rate of decline in the seed bank. In general, seed bank decline of volunteer *B. napus* tends to be greater than 90% per annum under continuous cropping in European winter and Canadian spring genotypes in the first 2 to 3 years after seed bank establishment (61,88). Other short-term studies have given similar results, such as 1.5 and 2% seeds surviving after 1 and 2 years, respectively (64); 1% viability after only a few months; and 0% viability after 41 months in the seed bank (30). After approximately 3 years, however, rapid seed bank decline appears to cease with low, but stable levels of volunteer *B. napus* seeds remaining viable for up to 10 years (88). Much lower seed bank decline rates are observed when soil remains undisturbed after initial seed burial (62,88).

Weedy *B. rapa* may differ from domesticated *B. rapa* and *B. napus* in seed bank persistence. Linder and Schmitt (85), using spring seeded oilseed rape in California and Georgia, reported that the *B. rapa* cultivar Tobin has 7% viability after 1 year in the soil, and wild *B. rapa* exhibited 80% viability. In a second study on the persistence of high stearate *B. napus* compared to null segregants, Linder and Schmitt (85) also reported a low level of initial dormancy and a rapid decline in the seed bank (85). Gulden et al. (61) reported that in western Canada, rates of seed bank decline of volunteer *B. rapa* appeared to be similar or lower than those of *B. napus*, depending on *B. rapa* genotype. Taken together, evidence suggests that persistence in the seed bank maybe an important factor limiting self-perpetuating feral populations of both *B. rapa* and *B. napus*.

In addition to evidence from seed bank studies, surveys of volunteer *B. rapa/B. napus* (confounded) plants within typical western Canadian fields have been conducted. Even though surveys provide periodic assessments of occurrence and density of crop volunteers under normal management, the relative importance of seed bank persistence and volunteer seed bank replenishment can not be separated. Surveys conducted after in-crop herbicide applications in western Canada, show an average between 4.5 to 5.3 plants m^{-2} in the fields where volunteer *B. napus/B. rapa* occur when averaged over all crop types (Table 5.3) (81,136). The frequency of fields where volunteers occurred varied widely between crops and years. There is no apparent trend of increased frequency, density, or relative abundance in the 2000s following the introduction of herbicide-resistant varieties.

Several factors influence density and occurrence of volunteer *Brassica* crops in subsequent crops. Density and variability are highest in the year immediately following a *B. napus/B. rapa* crop (Figure 5.2) and decrease with time. Long-term persistence of volunteer *B. napus/B. rapa* has been reported, up to and including 4 years after the crop. However, after this time, densities were low (less than 1 plant m^{-2}) (138). Similar surveys in eastern Canada indicate that the first year following oilseed rape, there were 4.9 and 3.9 plants m^{-2} remaining after in-crop herbicide applications. Again, after 3 to 4 years, plant densities were reduced (0.2 plants m^{-2} in 1 of 3 fields sampled) (121). However, oilseed rape was not eradicated from fields within this 4 to 5 year time frame; instead, low densities of volunteers were recorded, presumably due to both long-term seed bank persistence and replenishment of the seed bank by surviving volunteers (83). Therefore, in both environments, volunteer oilseed rape does not maintain high populations and we must assume that volunteers do not produce sufficient seed to routinely replace themselves. Unless another oilseed rape crop is seeded, small populations appear to become locally extinct.

Other factors, including weather and agronomic controls also influence density and occurrence of volunteers in subsequent crops. In agroecosystems, survival is limited by agronomic practices of the grower. Control of volunteers includes tillage, herbicide application, and a host of agronomic practices designed to favor the seeded crop over weeds and volunteer crops (42). Because control by herbicides appears to be a significant factor in limiting volunteers, it is reasonable to question the implication of herbicide-resistance on volunteer populations.

TABLE 5.3
Frequency of Occurrence of Volunteer *B. napus/B. rapa* in Surveyed Fields, the Mean Density per Field and the Ranking of Volunteer Canola Based on a Relative Abundance Index from Provincial Weed Surveys of Common Cereal, Oilseed and Pulse Crops in Alberta (1987–1989, 1997, 2001), Saskatchewan (1986, 1995, and 2003), and Manitoba (1986, 1997, 2002)

Survey Period	Measurement	Wheat	Barley	Oats	Flax	Dry Peas	Lentils	Other Crops	All Non-Canola Crops
1980s	Frequency (%)	7.8	16.9	9.8	6.9	na	na	2.3	10.0
	Occurrence density (plants m^{-2})	5.4	3.6	4.5	7.3	na	na	0.3	4.6
	Relative abundance rank	19	17	30	18	na	na	37	19
	Number of surveyed fields[1]	1495	616	174	87	0	0	88	2460
1990s	Frequency (%)	13.3	12.5	11.7	19.8	40.0	14.7	3.6	13.9
	Occurrence density (plants m^{-2})	4.8	5.5	12.8	2.1	7.2	1.5	4.7	5.3
	Relative abundance rank	10	17	15	13	4	14	29	10
	Number of surveyed fields	1031	441	111	86	60	34	56	1819
2000s	Frequency (%)	11.0	12.6	13.2	6.9	24.3	5.5	9.5	11.9
	Occurrence density (plants m^{-2})	3.8	7.2	3.9	2.1	3.1	0.6	1.1	4.5
	Relative abundance rank	11	9	17	18	9	21	20	12
	Number of surveyed fields	1593	783	317	130	144	73	116	3158

[1] The relative abundance is a combination of the frequency, field uniformity (all fields), and mean field density (all fields).

Source: From Thomas and Leeson, unpublished data. Used with permission.

5.4.1 Influence of Herbicide-Resistance Traits on Persistence and Ferality

Triazine-resistant oilseed rape was introduced in Canada in the 1980s, but it was not commercially successful there, although it has been adopted and widely grown in Australia (19). Modern herbicide-resistant oilseed rape was first introduced in Canada in 1995 and herbicide-resistant cultivars have been widely and rapidly adopted. Given an average crop rotation of one *B. napus* crop in 3 years, most fields in Alberta have grown herbicide-resistant *B. napus* and harbor some herbicide-resistant volunteers. Cultivars with varying types of herbicide-resistance are frequently grown in adjacent fields. Because of low frequency outcrossing, gene flow via pollen generates resistant and multiple resistant volunteer oilseed rape (18,65,113). Gene flow also may be facilitated by contamination of certified seed (52) and presumably by admixtures of replanted farm-saved seed. Therefore, western Canadian producers should assume that volunteer *B. napus* may be resistant to one, two, or three types of herbicides. Most have adjusted management practices to reflect this assumption. Practices include the use of 2,4-D prior to seeding and mixing of herbicides with different modes of action for in-crop applications.

Additionally, although important in reducing weed abundance in *B. napus*, the use of non-selective herbicides, glyphosate, glufosinate, and imazethapyr/imazamox was not the only agronomic change to occur over the time period from 1995 to 2001. The amount of tillage and the tillage intensity was significantly reduced (Table 5.4) during this period, which may be directly related to the reduced application of soil incorporated herbicides previously used in conventional oilseed rape. Reduced tillage leads to a shorter seed bank half-life in volunteer oilseed rape, which contributes to reduced ferality (61).

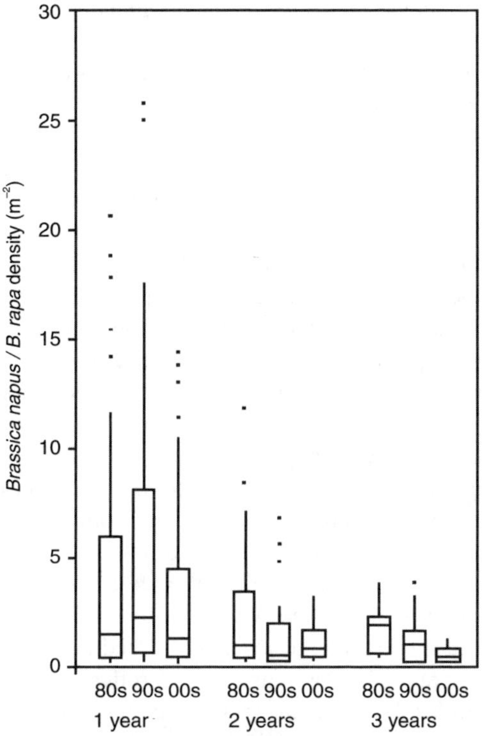

FIGURE 5.2 Box plot of density of volunteer *B. napus/B. rapa* 1, 2, and 3 years after a crop was seeded. Data are taken from provincial weed surveys of common cereal, oilseed, and pulse crops (excluding canola) in Alberta (1987–1989, 1997, 2001), Saskatchewan (1986, 1995, and 2003), and Manitoba (1997, 2002). Extreme outliers (more than three interquartile ranges) are not illustrated on the graph. (From Thomas and Leeson, unpublished data. Used with permission.)

TABLE 5.4
Time of Tillage Practices in Alberta in 1997, When Herbicide-Resistant *B. napus* Was Introduced and in 2001, after It Had Been Widely Adopted by Growers

	%	
Time of Tillage	**1997**	**2001**
None	6.7	26.0
Spring only	2.1	17.5
Fall only	41.6	21.1
Spring and Fall	49.5	29.6
Other	0.1	5.8

Source: From Thomas and Leeson, unpublished data. Used with permission.

Most weeds, including herbicide-resistant weeds, are managed using herbicides or mixtures of herbicides with multiple modes of action and by agronomic practices that reduce dispersal and reproductive potential (15). Many of these practices have been adopted to control herbicide-resistant and multiple resistant feral oilseed rape (17,42).

Weeds, including volunteer oilseed rape, are subject to repeated and redundant control throughout the growing season. Based on a grower's survey in Alberta, less than 3% of growers did not use herbicide application in 2001. Because volunteer oilseed rape is one of the earliest weeds to emerge, many of the seedlings can be controlled prior to crop seeding. Prior to seeding, the majority of fields are either tilled or a herbicide is applied. In Alberta in 2001, over 19% of fields received a preseeding herbicide application and 47% received a preseeding tillage treatment (Table 5.4). The preseeding herbicide of choice has traditionally been glyphosate, but changes have occurred to include the use of premixed products specifically for preseeding control of herbicide-resistant volunteer oilseed rape. Herbicide mixes have been widely adopted by producers (Hall and Thomas, unpublished data).

Following seeding, in-crop herbicide applications were applied to 85% of all Alberta fields in 2001 (Hall and Thomas, unpublished data). Although not all volunteer oilseed rape is killed by the herbicide application, most survivors are affected by the combination of crop competition and partial herbicide control that reduces seed set. In all crops except field peas, in-crop herbicides control herbicide-resistant oilseed rape because glyphosate and glufosinate are not used in crops other than oilseed rape at this time in western Canada. Herbicides normally used in field peas do not control single- or multiple-imidazolinone-resistant oilseed rape, and in this instance, volunteer *B. napus* populations may increase in the year of pea production. Imidazolinones are premixed with 2,4-D ester to control imidazolinone-resistant oilseed rape volunteers and acetolactate synthase (ALS)-resistant broadleaf weed species in imidazolinone-resistant wheat. In balance, redundant controls, coupled with the low seed bank persistence reduce volunteer oilseed rape to low levels after 2 years. Where *B. napus* and *B. rapa* are grown in western Canada, non-selective herbicides are rarely used along roadsides, as a perennial grass cover is preferred.

Elsewhere in the world, *B. napus* and *B. rapa* exist within ruderal or natural areas in stable or short-lived metapopulations (39,108) or there may be higher potential for gene flow from weedy relatives like *B. rapa* to cultivated *Brassica* oilseed crops. In these locations, there may be a higher potential for the formation of populations with feral traits. Should non-selective herbicides such as glyphosate, glufosinate, or ALS inhibitors be used for weed control where these populations exist, they would have a selective advantage and may increase. Fortunately, most roadside weed control is confined to mowing or using of selective herbicides to preserve grassy species.

Although it is interesting to speculate about the influence of insect- and disease-resistant oilseed rape on volunteer and feral reproductive fitness at this point, we have little information on the impact of disease and insect predation on volunteer fecundity. It appears volunteer populations are limited primarily by the lack of seed bank persistence and redundant agronomic controls with high efficacy that limit seed production. Clearly each trait, each crop, and each environment must be independently assessed.

5.4.2 Anticipated and Unintended Consequences

Prior to the release of herbicide-resistant *B. napus* varieties, crop competition due to weeds was a significant factor affecting crop yield. Selective herbicides for broadleaf weed control in *B. napus* were limited to ethametsulfuron-methyl and the dinitroanaline herbicides trifluralin and ethalfluralin. Dinitroanaline herbicides are soil applied and require incorporation by tillage. Field trials conducted prior to release had shown weed control would increase with the use of non-selective or broad-spectrum herbicides and that yield increases could be expected. Other consequences of the introduction of the technology were less well understood or communicated to growers.

Following the introduction of (glyphosate or glufosinate) herbicide-resistant *B. napus*, the Canola Council of Canada conducted a paired comparison in 2000 (29) and found that the transgenic oilseed rape varieties yielded over 10% more than the conventional varieties. Furthermore, contamination with weed seeds (dockage) was significantly lower in the transgenic system resulting in a higher value crop.

Growers of transgenic *B. napus* also reported fewer tillage operations than growers of conventional varieties with the majority indicating that they practiced minimum or zero till production. Transgenic growers report that they were able to seed earlier thus benefiting from soil moisture conservation. In 2001, over 1.05 million ha in oilseed rape rotations in western Canada had been positively impacted by increased conservation tillage practices since the introduction of the technology. These benefits include increased organic matter and water retention, reduced erosion by wind and water, and increased soil microorganism biodiversity. Energy consumption (fuel used) was lower for transgenic production systems due to fewer field operations. Minimum till and zero tillage are more available options with the herbicide regime used on transgenic varieties. As a result, fuel savings attributed to growing transgenic *B. napus* have grown from 9.5 million liters in 1997 to 31.2 million liters in 2000. Contrary to the popular belief that herbicide-resistant crops use more herbicides, the report estimated a reduction in herbicides of 1500 tons in 1997 and 6000 tons in 2000. They summarized that the transgenic *B. napus* systems had a positive economic and agronomic impact when compared to the conventional *B. napus* systems in western Canada for the 4-year period, 1997 to 2000.

Environmental risk assessments were conducted, including the potential for herbicide-resistant *B. napus* to become a weed of agricultural systems or to become invasive of natural habitats. The potential for gene flow to wild relatives, whose hybrid offspring may become weedier or more invasive, the potential impact on non-target organisms, and the potential impact on biodiversity were also assessed (25–28). Outcrossing and stacking of resistance traits to produce multiple herbicide-resistant volunteers were predicted (112). However, information on the disadvantages of herbicide-resistant oilseed rape did not receive as much dissemination as information on advantages. Therefore, some growers were unprepared for the lack of control of volunteer oilseed rape in the initial period following its introduction. Changes in agronomic practices, including the adoption of herbicide mixtures rapidly followed.

5.4.3 A Simple Scenario for Population Demographics in Western Canada

Currently, a herbicide-resistant *B. napus* cultivar is seeded in most (>85%) oilseed rape fields in western Canada. Seed is usually obtained from a certified seed grower, but may contain a low frequency of seeds with resistance to more than one herbicide. During flowering of these self-compatible plants, the primary source of pollen is the maternal plant, followed by close neighbors. Some flowers may be fertilized by pollen from adjacent fields, but at a lower frequency. Thus, both resistant and multiple-resistant traits are likely present in the seed produced. Pollen flow from weedy relatives is possible, but in western Canada weedy *B. rapa* is not present (97) and gene flow with other weedy relatives is highly constricted by the low natural occurrence of hybridization.

Seed loss at harvest initiates a high input to the seed bank. Through predation, fall germination, or loss of viability over winter, the seed bank is diminished, but large numbers of volunteers germinate in the first year following the oilseed rape crop (61). Seedling recruitment of volunteer oilseed rape is limited to the spring and most volunteers are controlled or are limited in seed production through agronomic practices including tillage or redundant herbicide use. Redundant herbicide use usually controls both resistant and multiple-resistant volunteers, but in some fields, selection for herbicide-resistant individuals occurs. Survey data suggests that in most fields, few seeds are returned to the seed bank in the years following the oilseed rape crop and the seed bank is diminished before the next oilseed rape crop is seeded.

TABLE 5.5
Percentage of Growers Reporting Volunteer *B. napus* or *B. rapa* as One of the Top Five Troublesome Weeds in Their Fields in Comparison to Selected Species

	% of Growers Who Responded		
Species	**1980s**	**1990s**	**2000s**
B. napus/B. rapa	0.5	3.3	8.0
Avena fatua L.	70.0	78.1	68.7
Polygonum convolvulus L.	42.2	41.1	44.6
Setaria viridis (L.) P. Beauv.	35.9	39.4	24.1
Cirsium arvense (L.) Scop.	45.5	45.4	33.7
Sinapis arvensis L.	11.7	13.6	12.2
Thlaspi arvense L.	41.6	27.2	26.4

Data are derived from management questionnaires returned in conjunction with provincial weed surveys of common cereal, oilseed, and pulse crops in Alberta (1987–1989, 1997, 2001), Saskatchewan (1986, 1995, and 2003), and Manitoba (1997, 2002).

Source: From Thomas and Leeson, unpublished data. Used with permission.

This simple scenario reflects the number of oilseed rape volunteers found in subsequent crops where the highest volunteer densities occur in the year following seeding. It is also confirmed by the level of concern expressed by growers. Most producers did not consider *B. napus/B. rapa* a significant weed problem. When asked to identify the most troublesome weeds in fields, few chose volunteer *B. napus/B. rapa* compared to other weeds such as wild oat (Table 5.5). The perception of concern increased slightly following the introduction of herbicide-resistant *B. napus,* presumably because some varieties were not controlled without modification of previous weed control practices or because of extension efforts to heighten awareness. Producer perception is critical to identification of post release problems. Around the world, producers have been responsible for identifying herbicide-resistant weeds. Should volunteers increase or feral populations establish, it will probably be producers who bring this information to the attention of researchers.

In western Canada, where herbicide-resistant oilseed rape has been grown extensively for 7 years, there is no evidence at present that volunteer *B. napus* has increased or is more prevalent because of herbicide-resistance traits (Table 5.3, Figure 5.2) (16). Extensive pre- and post release weed population monitoring has been conducted in thousands of fields and will continue to play an important role in assessing populations of herbicide-resistant volunteers, weed population shifts, and changes to weed biodiversity due to herbicide-tolerant crops.

Although crop-breeding efforts have generally increased reproduction and local adaptation of crop seed through increases in yield, disease resistance, cold tolerance, and hybrid vigor, volunteer *B. napus* is not a weed that can routinely maintain a population within the western Canada agroecosystem. Population dynamics merit further study and modeling, but an important aspect limiting ferality may be the lack of a persistent seed bank. Conventional management practices appear to be limiting seed set of volunteer *B. napus*, including herbicide-resistant types, by use of herbicide mixtures and redundant controls that strongly limit ferality.

In other locations, including Europe, where:

- Ruderal areas may harbor volunteer or feral oilseed rape.
- Set aside areas with little alternative herbicide use allow for the increase in seed banks in the years following *B. napus* crops.
- Both winter and spring seeded oilseed rape are grown.
- Weedy relatives may be more frequent and wild *B. rapa* is a source of secondary seed dormancy traits.
- Field size is smaller and pollen flow more likely.

the scenario delineated above will not be appropriate. Should secondary dormancy be obtained through introgression with weedy *B. rapa*, we anticipate populations with increased ferality. Modeling efforts are currently underway in several countries to address specific agroecological realities (36,37). Biological and agricultural differences between regions reinforce the concept that decisions to introduce transgenic crops must be made on a case-by-case basis.

5.5 CONCLUSIONS

Brassica crops have been domesticated since recorded history, circa 1500 BC (10). Domestication included a reduction in genetic diversity and a loss of secondary seed dormancy. Recent breeding objectives and methods do not appear to have enhanced the domestication of this crop. Increases in yield, disease resistance, hybrid vigor, and regionally adapted fitness may increase the abilities of volunteer crops. Unlike many crops, *B. napus* seed is not readily retained and planted in subsequent years. An increase in shattering resistance would reduce the volunteer crop concerns by decreasing the seed inputs to the weed seed bank and should be considered as a significant breeding objective to lessen the impacts of transgenic *B. napus*.

Brassica crops share many of the common characteristics of weedy species. There is abundant production of small round seeds. The crop is morphologically and physiologically plastic, with rapid site capture and luxury use of nutrients. The lack of an effective long-term seed bank persistence mechanism in *B. napus* and *B. rapa* and a susceptibility to broadleaf herbicides allow it to be effectively controlled within a crop rotation. The loss of these critical traits could increase persistence and weediness in *B. napus,* should introgression from weedy relatives occur.

In western Canada, herbicide-resistant volunteer *B. napus* is not currently at risk of becoming feral, because of the lack of persistence in the seed bank, redundant and repetitive control of volunteer weeds in subsequent crops, absence of persistent populations in ruderal areas, and limited weedy relatives with a significant potential for hybridization. However, post release monitoring is being carried out on a large scale. Caution must be exercised prior to the release of herbicide-resistant oilseed rape especially where feral populations and weedy *B. rapa* persist in the agroecosystem. In these cases, the enhanced ferality of herbicide-resistant oilseed rape is more probable.

LITERATURE CITED

1. Adler LS, Wikler K, Wyndham FS, Linder CR, Schmitt J. 1993. Potential for persistence of genes escaped from canola: germination cues in crop, wild, and crop-wild hybrid *Brassica rapa*. *Funct. Ecol.* 7:736–745.
2. Akbar MA. 1989. Resynthesis of *Brassica napus* aiming for improved earliness and carried out by different approaches. *Hereditas* 111:239–246.
3. Animal and Plant Health Inspection Service. http://www.aphis.usda.gov/. Accessed June 10.

4. Armstrong KC, Keller WA. 1981. Chromosome pairing in haploids of *Brassica campestris*. *Theor. Appl. Genet.* 59:49–52.
5. Armstrong KC, Keller WA. 1982. Chromosome pairing in haploids of *Brassica oleracea*. *Can. J. Genet. Cytol.* 24:735–739.
6. Attia T, Busso C, Röbbelen G. 1987. Digenomic triploids for an assessment of chromosome relationships in the cultivated diploid *Brassica* species. *Genome* 29:326–330.
7. Attia T, Röbbelen G. 1986. Cytogenetic relationship within cultivated *Brassica* analyzed in amphihaploids from the three diploid ancestors. *Can. J. Genet. Cytol.* 28:323–329.
8. Attia T, Röbbelen G. 1986. Meiotic pairing in haploids and amphidiploids of spontaneous versus synthetic origin in rape, *Brassica napus* L. *Can. J. Genet. Cytol.* 28:330–333.
9. Baker HG. 1965. Characteristics and modes of origin of weeds. In *The genetics of colonizing species*, Baker HG, Stebbins GL, Eds., pp. 147–168. New York: Academic Press. 588 pp.
10. Baker HG. 1974. The evolution of weeds. *Ann. Rev. Ecol. Syst.* 5:1–24.
11. Baker HG. 1991. The continuing evolution of weeds. *Econ. Bot.* 45:445–449.
12. Barrett SCH. 1988. Genetics and evolution of agricultural weeds. In *Weed management in agroecosystems: ecological approaches*, Altieri MA, Liebman M, Eds., pp. 57–75. Boca Raton, FL: CRC Press. 354 pp.
13. Barrett SCH. 1992. Genetics of weed invasions. In *Applied population biology*, Jain SK, Botsford LW, Eds., pp. 91–119. Dordrecht, Netherlands: Kluwer. 304 pp.
14. Beckie HJ, Hall LM, Tardif FJ. 2001. Impact and management of herbicide-resistant weeds in Canada. In *Proc. Brighton Crop Protection Conf.-Weeds 2001*, pp. 747–754. London: British Crop Protection Council.
15. Beckie HJ, Hall LM, Tardiff FJ. 2001. Herbicide resistance in Canada — where are we today? In *Integrated weed management: Explore the potential*, Blackshaw RE, Hall LM, Eds., pp. 1–36. Sainte-Anne-de-Bellevue, Quebec: Expert Committee on Weeds. 136 pp.
16. Beckie HJ, Hall LM, Warwick SI. 2001. Impact of herbicide-resistant crops as weeds in Canada. In *Proc. Brighton Crop Protection Conf.-Weeds 2001*, pp. 135–142. London: British Crop Protection Council.
17. Beckie HJ, Séguin-Swartz G, Nair H, Warwick SI, Johnson E. 2004. Multiple herbicide–resistant canola can be controlled by alternative herbicides. *Weed Sci.* 52:152–157.
18. Beckie HJ, Warwick SI, Nair H, Seguin-Swartz G. 2003 Gene flow in commercial fields of herbicide-resistant canola (*Brassica napus*). *Ecol. Appl.* 13:1276–1294.
19. Beversdorf WD, Kott LS. 1987. Development of triazine resistance in crops by classical plant breeding. *Weed Sci.* 35:9–11.
20. Blackshaw RE, Brandt RN, Janzen HH, Ent T, Grant CA, Derksen DA. 2003. Differential response of weed species to added nitrogen. *Weed Sci.* 51:532–539.
21. Blackshaw RE. 1991. Soil temperature and moisture effects on downy brome vs. winter canola, wheat, and rye emergence. *Crop Sci.* 31:1034–1040.
22. Booth BD, Swanton CJ. 2002. Assembly theory applied to weed communities. *Weed Sci.* 50:2–13.
23. Bullied WJ, Marginet AM, Van Acker RC. 2003. Conventional- and conservation-tillage systems influence emergence periodicity of annual weed species in canola. *Weed Sci.* 51:886–897.
24. Canadian Food Inspection Agency. 1999. The biology of *Brassica rapa* L. Regulatory Directive Dir 1999-02. http://www.inspection.gc.ca/english/plaveg/bio/dir/dir9902e.shtml. Accessed on April 14, 2004.
25. Canadian Food Inspection Agency. 2004. Decision Document DD95-01: Determination of environmental safety of Agrevo Canada Inc.'s glufosinate ammonium-tolerant canola. http://www.inspection.gc.ca/english/plaveg/bio/dd/dd9501e.shtml. Accessed June 1, 2004.
26. Canadian Food Inspection Agency. 2004. Decision Document DD95-02: Determination of environmental safety of Monsanto Canada Inc.'s Roundup® herbicide-tolerant *Brassica napus* canola line GT73. http://www.inspection.gc.ca/english/plaveg/bio/dd/dd9502e.shtml#A7. Accessed June 1, 2004.
27. Canadian Food Inspection Agency. 2004. Decision Document DD95-03: Determination of environmental safety of Pioneer Hi-Bred International Inc.'s imidazolinone-tolerant canola. http://www.inspection.gc.ca/english/plaveg/bio/dd/dd9503e.shtml. Accessed June 1, 2004.

garden plantings as propagule sources. The forest populations then start to set fruits, allowing dispersal to continue without further seed influx from gardens (157). Birds also disperse yew (*Taxus baccata*) from gardens into managed forests (139). Growth anomalies such as columnar forms of yew have been observed in Germany. They descend from domesticated forms of *Taxus*, which were widely distributed by nurseries (74,126,139). Dedomestication is also evident in poplars. Hybrid swarms are found in the German Ruhrgebiet, originating from *Populus nigra* cv. "Italica" and *Populus maximowiczii*.

In general, only a few ornamentals spread over large areas; these, however, often led to substantial ecological and economic impacts (see species listed in Table 7.4). A good example of wide-reaching dispersal in cultivated species is the European buckthorn *Rhamnus frangula* in southern Ontario. Wild populations of introduced *Rhamnus frangula* were confined, until the 1930s, to three urban centers. By 1950, they had spread up to 40 km and by 1970 up to 150 km, and they still appear to have a largely urban distribution in southern Ontario. Twenty years later, *Rhamnus frangula* is widespread and common in certain regions (27).

Flowing water may move the propagules of ornamentals from adjacent gardens for long distances in seminatural or natural ecosystems. Many populations of giant hogweed (*Heracleum mantegazzianum*), policeman's helmet (*Impatiens glandulifera*), and knotweed (*Fallopia*) species in central European riparian ecosystems result from long-distance dispersal from gardens (5,55,77,115,134). For example, all populations of giant hogweed in the catchment of the Auschnippe river (northern Germany) probably descended from a single plant that was cultivated in 1982 in a village garden next to the river (134).

7.4 UNDERLYING PROCESSES

Why do many ornamental plants only begin to disperse after a long period? There has been remarkable recent progress in understanding the mechanisms that underlie invasion processes and their subsequent impacts (83,88). Genetic changes in introduced species are believed to function as the driving forces in subsequent invasions (39). A well-documented case is the cordgrass *Spartina anglica,* which evolved in Europe as a hybrid between the American *S. alternifolia* and the European (or possibly west African) *S. maritima* as parental species (50).

The examples described here demonstrate that dedomestication processes clearly play a role in ornamental plants leaving intentional cultivation. However, the extent of this phenomenon, especially with marginally domesticated plants, is not exactly known. Causal processes are also little understood.

The *post hoc* nature of research on plant invasions generally remains one of the weak links in the field (118). Most studies focus on invasions that have already started and have proved worthy of attention. There is, however, little information on the role of dedomestication in ornamentals as a precursor of processes that facilitate spread. In many cases, escapes proceed through a combination of diverse mechanisms. The frequency or number of plantings may be decisive or species begin to spread as a response to environmental changes. A precise assessment of the role of genetic changes as drivers in invasions is not yet feasible.

7.4.1 Ecological Role of Cultivation

The success of ornamentals in plant invasions has been explained by the ecological role of cultivation in enhancing the naturalization and spread of species (78,86,89). Cultivation integrates different driving forces:

- Maintenance of the cultivated plants may help protect populations from detrimental environmental effects including those with stochastic expression. Extreme drought periods or competition by other species, for example, may prevent the establishment of new populations. Irrigation and weeding may counteract such negative impacts in cultivated populations. Cultivation thus emerges as a potential counterforce to environmental stochasticity and thus may well facilitate the establishment of ornamentals (89).

- Cultivating species in large numbers leads to a high propagule pressure, which has been recognized as a powerful driving force in invasions. Several studies have found a correlation between the frequency of cultivated individuals and the invasiveness of a species (123,166).
- Secondary releases by planting or sowing may facilitate the establishment of sustainable populations and facilitate their spread by overcoming spatial separation from adequate but inaccessible sites (78). A conspicuous example is the North American skunk cabbage, *Lysichiton americanus.* It was cultivated in Germany as an ornamental for a long time, but it was not seen outside of cultivation before the 1980s. Then, in the 1980s, it was planted by a single gardener on some natural sites in the Taunus region, which served as invasion foci for subsequent spread (6).

Decisions to plant ornamentals in greater or lesser quantities, however, are mostly based on human preferences and biases that normally change significantly over time. The North American black cherry (*Prunus serotina*) was introduced as an ornamental into Europe between 1623 and 1635. The spread, however, only began about 200 years later when the species had been planted abundantly as a forest crop and for functional landscaping (143). A considerable proportion of the problematic populations of this and other highly invasive non-native species can be traced directly back to plantings or sowings in the vicinity of the current sites in northern Germany. Invasions of wetlands and managed pine forests by hybrid swarms of cultivated blueberry can be traced back to commercial plantings less than 1000 m away. Deliberate sowings were shown to directly function as foci of a quarter of subsequent invasions by *Heracleum mantegazzianum* (134,136,178).

The probability of a species becoming naturalized may be a direct function of the number of plantings. In four Australian urban areas, a total of 289 naturalized woody ornamental species were identified. Of these, 106 taxa had spread to the extent that they were dominant in at least one city over an area of at least 400 m^2. There was a significant relationship between naturalized occurrence and the number of times a species had been planted (98).

Future escapes from cultivation may result from North American prairie species that were introduced early to Europe without spreading vigorously. Now, a greater potential for proliferating and spreading is expected, due to the new fashion of using prairie species in European gardening (58).

7.4.2 Spread as a Response to Environmental Changes

Changes in site conditions are broadly recognized as a driving force in plant invasions. The role of human-induced environmental changes is illustrated here through the example of the effects of climate change on the spread of ornamentals.

One of the major constraints to ornamentals introduced to central Europe is late flowering, associated with the inability to set seeds, as well as frost intolerance, caused by an imperfect climatic match between the source and target areas. This is usually reflected in the decisive influence of altitude on the performance of introduced species and the role of cities as "heat islands" (95,116,145). One example involves the frequently cultivated East Asian tree of heaven (*Ailanthus altissima*). Its spread across central Europe started in larger cities. The distribution pattern in Berlin corresponds to the extent of the urban heat island. Recent studies show that the growth of *Ailanthus* is supported by warmer temperatures (77,79,133). With global climate change, it is predicted that the edges of species ranges or the boundaries of biogeographical regions may adjust (e.g., review in Walther (158)). Because the formation of urban heat islands anticipates some effects of the expected global warming, changes should be expected in the ecological behavior of ornamentals that until now have not escaped cultivation or that only occur infrequently.

Palm trees in European forests are no longer a fantasy. Palms are already making headway, namely, in the lake region at the border of Italy and Switzerland. Here, a new type of warm-temperate forest is about to be established. A minor climatic shift has been sufficient to open new ecological niches for warm-temperate species that form evergreen broad-leafed vegetation instead of the previous deciduous broad-leafed forests (*laurophyllisation*) (70). The local flora is poor in evergreen species. Thus, the long history of cultivating evergreen ornamentals has established a rich source of warm-temperate species in the vicinity of gardens and parks. Changes in climatic conditions the last 3 decades have facilitated the spread and naturalization of the palm tree *Trachycarpus fortunei* and other evergreen ornamentals (*Cinnamomum glanduliferum, Laurus nobilis, Lonicera japonica, Prunus laurocerasus, Elaeagnus pungens, Mahonia aquifolium*) (48,156,159). Elevated CO_2 may also enhance the spread of cherry laurel (*Prunus laurocerasus*) in Swiss forests (56). The spread of other thermophilic ornamentals in the wild has recently been reported in Spain and southern France (102,140).

7.4.3 Dedomestication Processes

Ornamental plants may spread from cultivation with or without alterations in their genetic constitution. Studying the genetic basis of plant invasions is a relatively young field (39,63). As a consequence, the information on the genetics of ornamentals growing in the wild is scarce. There are two ways domesticated organisms can become feral (Chapter 1):

1. In endoferality, a species dedomesticates on its own, using mutations to achieve the feral forms.
2. In exoferality, relatives contribute ferality genes and hasten the evolutionary process.

The following examples illustrate that dedomestication also plays a role in ornamental plants. In contrast to agricultural crops, the significance of this process for successful spread of ornamentals is little understood.

7.4.3.1 Endoferality — Some Examples

The North American goldenrods *Solidago canadensis* and *S. gigantea* were introduced into Europe as ornamentals in 1645 and 1758, respectively. They are now widespread throughout Europe, but have not yet reached the limits of their potential ranges (161). Differences in morphological and life history characteristics were found across a gradient from northern to southern regions within wild European populations of both species. This variation reflects rapid adaptive population differentiation after introduction of both species to Europe (163). European populations of *S. canadensis* differ from populations native to North American in various traits, but have no increased competitive ability in response to simulated herbivory (150). Scholz (137) proposed distinguishing newly evolved European *Solidago* taxa from the North American parents as *S. anthropogena.*

Dedomestication processes are most evident in hybrids of North American blueberry species, which are cultivated in diverse forms on approximately 600 ha in northern Germany (84,101). Hybridization with native *Vaccinium* species has not been observed in the wild. Among the parental species, the North American *V. corymbosum* and *V. angustifolium* are most important (53). Both species were introduced into Europe as ornamentals in the late 18th century. The spread of non-native blueberries only started in the mid-20th century. It is unknown whether this lag time was due to insufficient propagule pressure by the rarely planted parental species or whether the recent commercially grown hybrids were better adapted than the parental species to their new range. The offspring of the cultivated forms vary in growth pattern, as well as in leaf and fruit characteristics (e.g., larger leaves and fruits than their wild ancestors). Currently, the spreading hybrids, combined as *Vaccinium corymbosum angustifolium,* have become established in pine plantations and also natural ecosystems in the vicinity of the commercial plantations (77,135,136).

Mahonia aquifolium, an evergreen shrub native to western North America, was introduced to Europe in 1822 for horticultural purposes. *Mahonia* has been one of the most popular urban ornamentals since the mid-20th century. It frequently escapes from cultivation and naturalizes in urban, silvicultural, and natural ecosystems (10,11,74,85). However, the cultivated and naturalized forms in Europe are mostly hybrids of a North American *Mahonia* species closely related to them (3). The invasive forms of *Mahonia* probably have a higher reproductive potential compared to wild North American relatives because of horticultural efforts to breed plants with more flowers and fruits (141). A higher seed production may generally enhance a species' invasion success (104,122).

The North American Jerusalem artichoke (*Helianthus tuberosus*) was first introduced into Europe in 1607 as a crop and as an ornamental (165). It is morphologically variable (146,155) and a broad variety of domesticated and possibly also wild forms have been used for the production of edible tubers or as ornamentals (late-flowering, tall-growing perennial). In central Europe, plants that have escaped from cultivation grow on urban sites as well as in seminatural vegetation along streams and rivers (77,85). Feral populations differ from cultivated taxa by their smaller, spindle-shaped tubers at the end of long rhizomes and their often strongly branched stems (72,85). *H. tuberosus* was thought to mainly reproduce vegetatively in central Europe. A recent study, however, confirmed the production of mature seeds in established central European populations (72).

7.4.3.2 Exoferality

Hybridization between an introduced and a native species or between two introduced species can lead to important evolutionary consequences. Hybridization and introgression may create varieties that adapt to novel habitats or improve the competitive ability of the recipient species (1,39,62).

Hybridization and introgression occurred in California, between the introduced South African Hottentot fig *Carpobrotus edulis* and the non-invasive putative native *Carpobrotus chilensis* (4,47,164). Both the introduced species and the hybrids are successful invaders in coastal ecosystems. Hybridization is thought to accelerate the spread of *Carpobrotus* due to the formation of additional genotypes that promote dispersal. The introduced species and hybrids produce more fruits per clone primarily due to larger clone sizes. The fruits are preferred by frugivores over those of the native *Carpobrotus* species. The hybrids also grow faster than both parents and have a higher survival rate than the native species (152,164). Hybridization between *Carpobrotus* species has also occurred in southern France (144).

Two introduced knotweed species hybridized in their new range. Both parental species, *Fallopia japonica* and *F. sachalinensis* are regarded as problematic invaders in several parts of Europe (28,77,162). A single clone of *Fallopia japonica*, introduced in 1848 from Japan to Britain is thought to be the origin of all populations in Britain (14,30). The introduced forms grow taller than many of the Japanese plants. This is understood to be an effect of selection for tall-growing clones for introduction to Europe. In 1878, *Fallopia japonica* was highly praised in English garden literature for its strong growth (132). The hybrid *F. bohemica* described in 1983, emerged in the new European range of both *Fallopia* species, and spread widely across central Europe and Britain (5,13). It grows taller and regenerates better from rhizome segments than both parents and shows stronger growth during mowing, herbivory, and competition experiments. The hybridization increased the competitiveness and, therefore, also the effort needed for control (5,22,23). A hybrid of *F. japonica* and *F. sachalinensis* was first described in Japan in 1997. Bailey (13) suggests that plantings of *F. japonica* within the distribution range of *F. sachalinensis* encouraged this hybridization.

F. bohemica occurs in Europe at three different ploidy levels, which vary in their degree of fertility. The tetraploid form is almost totally fertile and is capable of forming euploid progeny in the few localities where it occurs with other tetraploid taxa of *Fallopia*. Octoploid *F. bohemica* is known on one British site, but is more widely dispersed across central Europe (13). There are also other, less frequently found *Fallopia* taxa established in Europe (Table 7.6). Surprisingly,

TABLE 7.6
Spread of Japanese *Fallopia* in Europe and Hybridization with Introduced Congenors

Taxon	Category	Distribution	Cytology	Comments
F. japonica var. *japonica*	Variety	Northern Europe, Hungary, U.S., Canada, Australia, New Zealand	2n = 88	One male-sterile clone in U.K. — probably also in most of introduced range
F. japonica var. *compacta*	Variety	U.K., Czech Republic	2n = 44	Both sexes occur
F. sachalinensis	Species	N. Europe, U.S. elsewhere	2n = 44,66	Both sexes occur
F. *bohemica* 8x	Hybrid	U.K., Czech Republic, France, and Germany	2n = 88	Both sexes occur
F. *bohemica* 6x	Hybrid	N. Europe, U.S., Australia, and New Zealand	2n = 66	Both sexes occur
F. *bohemica* 4x	Hybrid	U.K. and Czech Republic	2n = 44	Both sexes occur, reciprocal crosses also found
F. *conollyana* *F. jap.* var. *jap.* *F. baldschuanica*	Hybrid	U.K., Germany, Hungary, Norway, and France	2n = 54	Frequently found as seed on *F. jap.* var. *jap.* throughout its range
F. japonica var. *japonica* *F. japonica* var. *compacta*	Intraspecific hybrid	U.K. and Germany	2n = 66	Morphologically similar to *F. japonica* var. *japonica*
F. japonica *F.* *bohemica* *(6x)*	Backcross	Wales	2n = 76 – 110	Two established seedlings
F. *bohemica (8x)* *F. sachalinensis*	Backcross	Wales	2n = 66	Several established putative plants

Source: Bailey (13), with permission.

F. japonica has also hybridized with its climbing woody relative *F. baldschuanica*, a species not native to Japan. The latter is frequently planted as an urban ornamental. This hybrid, *F.* *conollyana* was first discovered in the early 1980s (12).

Hybridizations between frequently cultivated introduced and native species occur in *Rosa, Crataegus* (38,41,57), and *Narcissus* (110), but little is known about evolutionary consequences. Examples of hybridizations between two introduced ornamental species include *Echinops* (62) and *Rhododendron* taxa. The exchange of genes among *Rhododendron* spp. may have led to an increase in the ecological amplitude of the recipient taxa.

Rhododendron ponticum, a native to the Caucasian region, Asia Minor, and the Iberian peninsula, was introduced into Britain as an ornamental about 1763. The first reports of self-sowing date back to 1849. *R. ponticum* is now widespread across the British Isles and is regarded as an environmental weed with great economic damage (29,33,37,110,168). The present invasion appears to be moving eastward with actual occurrences in the Netherlands, Belgium, and northern Germany (40,103).

A comparison of native and naturalized populations of *Rhododendron ponticum* revealed an increased growth rate and seedling recruitment for the Irish populations (40). Genetic analyses indicate an Iberian origin for the British material (96). However, evidence of genetic introgression from the North American *R. catawbiense* was found in 27 of 260 naturalized British accessions of *R. ponticum.* Such accessions were significantly more abundant in cold, eastern Scotland, indicating that introgression from the North American congener may have improved cold tolerance, which may be a selective advantage in other cold areas. Milne and Abbott (96) discuss three possible underlying mechanisms:

1. The observed increase in frequency of introgressed individuals may be due to horticultural selection, that is, the preference of growers in colder regions for "hardy hybrids" (cultivars of *R. ponticum* crossed with *R. catawbiense, R. maximum,* and sometimes *R. arboretum*) in their gardens.
2. Natural selection may have acted on the *R. ponticum* populations during and after naturalization, favoring individuals with a higher level of introgression from *R. catawbiense* in colder regions.
3. The observed pattern simply results from varying tastes or breeding programs in different parts of Britain.

Introgression associated with a broadened habitat tolerance has also been observed in other feral ornamentals such as purple loosestrife (*Lythrum salicaria*), which is widely established in North America, although in no case has a direct link been proven (96).

7.5 CAN WE PREDICT THE SPREAD OF INTRODUCED ORNAMENTALS?

Predicting which species will spread has been a long-standing goal of ecologists, at least since the late 1980s. Estimates on the progress in predicting invasions, however, vary from awareness of shortcomings (167,169) to promising attempts at identifying potential invaders (34,123). In a recent review, Rejmanek (123) sketched five groups of non-exclusive, sometimes complimentary approaches to predict invasions:

1. Stochastic approaches that allow probabilistic predictions about potential invaders based on initial population size, residence time, and number of introduction attempts.
2. Empirical taxon-specific approaches that are based on previously documented invasions of particular taxa.
3. Evaluations of the biological characters of non-invasive taxa and successful invaders that give rise either to general or to habitat-specific screening procedures.
4. Evaluation of environmental compatibility that helps to predict whether a particular plant taxon can invade specific habitats.
5. Experimental approaches to attempt to tease apart intrinsic and extrinsic factors underlying successful invasion.

Predictions are possible to a certain extent, by focusing on smaller subsets of potential invaders (e.g., taxonomical groups, life forms). In pine (*Pinus*) species, for example, distinctions between invasive and non-invasive species were achieved by analyzing species traits. Small seed masses, a high relative growth rate of seedlings, and short generation times, for example, characterize invasive pine species. However, such findings cannot be generalized over different taxonomical groups (51,124). There is obviously no consistent set of species traits shared by invasive species (31,169).

The progress in predicting biological invasions is obvious, but has to be contrasted with remaining uncertainties. After discussing 10 limiting reasons, Williamson (167) concludes, "prediction of the ecological behavior of a species in a new environment may be effectively impossible." Even if considerable progress in understanding biological mechanisms as drivers of invasions will be achieved, predictions will be constrained by one remaining obstacle. Human agency may mimic, as the decisive factor, the action of biological mechanisms in the growth, establishment, and expansion of population ranges of non-indigenous species. The underlying cultural processes may, however, not be explained or predicted by applying ecological rules (78). The remaining uncertainties are well illustrated by the considerable time lag between the introduction of a species and the break out of subsequent invasions (see Section 7.3.4). These are at least partly due to cultural processes that affect, for example, propagule pressure or creation or availability of adequate sites. This holds true especially for ornamentals because their use depends strongly on fickle human preferences.

7.6 CONCLUSIONS

The use of introduced ornamentals will continue to increase as global commerce grows. Only a small proportion of the total of cultivated species is likely to spread, but these species may cause far-reaching ecological and economic impact. Introduced ornamentals escaped from cultivation predominate the weeds of urbanized areas, managed forests, flood plains, some managed grasslands, and natural ecosystems. The spread of non-native species is mostly apparent within the urban ornamental species pool. More emphasis should be placed on a usually clandestine part of such invasion processes: the spreading of "cryptic non-native species," that is, hardly domesticated forms or introduced provenances of native species. This group is usually neglected, but may have considerable ecological and evolutionary consequences due to large plantations (e.g, missing adaptation to habitats, losses in the regionally evolved genetic diversity). Regional provenances of native species should be preferred for conservational purposes outside of settlements instead of (cryptic) non-native species. Such regional provenances are certified in Germany and produced by specialized nurseries (81). However, non-native species often have beneficial functions in urban environments. They should be only regulated if they spread vigorously to non-urban sites.

As the examples described here show, dedomestication processes can play a large role in the spread of cultivated ornamentals. Exoferal hybrids can enhance the ability to establish, as exemplified by *Rhododendron* or *Fallopia*. However, the start and the course of invasion processes can be directed by other than genetic mechanisms. In times of global change, habitat qualities will change faster than they did in the past. They are subject to increasing human impact in most parts of the world. This and an additional cultural factor complicates predictions of invasions: choosing a species as an ornamental, deciding where to plant it, and in what quantities depends on cultural values. Human agency cannot be predicted by applying ecological rules, but it has considerable ecological consequences (e.g., propagule pressure, accessibility to new habitats).

What should be done? Should the introduction and use of all non-native species be opposed to prevent negative impacts of introduced ornamentals? This strategy disregards the beneficial functions of introduced species and neglects the fact that most of the introduced species will not spread. The following complementary approaches may be promising to reduce remaining risks and take, at the same time, advantage of the aesthetic and functional benefits of introduced species:

- Risk assessments should be performed prior to introducing a species into a new region. The predictive power of the *invades-elsewhere* criterion should be used (167). This means that the introduction or cultivation of species should be omitted in a given area if these species are known to be invasive in other similar areas. This criterion can be used and combined with other biological criteria in preparing decision trees for admissions of introduced species (120).
- Risk assessments should be also performed for species that already have been introduced to a region. In Germany, for example, the planting or sowing of non-native species outside of settlements requires a license issued by the nature conservation authorities. A case-by-case procedure (80) allows individual decisions to be made based on the species, the ecosystem, and the specific area affected. This is reasonable because even those plants that can damage plants and animals in certain situations may be completely harmless in

others. This case-by-case approach allows for continued traditional uses of non-native species and reduces the danger of overreaction. An overall ban on the use of specific non-natives is justified only when the species exert large impacts and cannot be contained.
- Prevention and early detection are the most effective strategies that can be used against harmful introduced species. Legal regulations, voluntary codes of conduct, and best management practices in using native and non-native plants should complement one another. One example is the draft of a voluntary code of conduct for nursery professionals, which has been developed in St. Louis (20). It fosters:
  - Risk assessments prior to introducing and marketing, also by integrating regional knowledge on invasive species.
  - Identification of plants that could be suitable alternatives and to develop and promote alternative plant material by plant selection and breeding.
  - Phasing out existing stock of invasive species in regions where they are considered as a threat.
  - Encouraging customers to use, and garden writers to promote, non-invasive plants.

Significant research efforts are still necessary to better understand the causal structure of genetic, ecological, and cultural mechanisms that underlie invasion processes. Methodological tools in risk assessments should be further developed, and existing databases should be strengthened.

ACKNOWLEDGMENTS

Thanks are due to Jonathan Gressel, Ulrich Sukopp, and Uwe Starfinger for comments on the chapter, to Moritz von der Lippe for preparing Figure 7.2, to Kelaine Vargas for the translation, and to Ines Grabarse, Janneke Westermann, and Kerstin Matz for technical support.

LITERATURE CITED

1. Abbott RJ, James JK, Milne RI, Gillies ACM. 2003. Plant introductions, hybridization and gene flow. *Phil. Trans. R. Soc. London* B 358:1123–1132.
2. Adolphi K. 1995. Neophytische Kultur- und Anbaupflanzen als Kulturflüchtlinge des Rheinlandes. *Nardus* 2:1–272.
3. Ahrendt LWA. 1961. *Berberis* and *Mahonia*: a taxonomic revision. *J. Linn. Soc. Bot.* 57:1–410.
4. Albert ME, D'Antonio CM, Schierenbeck KA. 1997. Hybridization and introgresssion in *Carpobrotus* spp. (Aizoaceae) in California. 1. Morphological evidence. *Am. J. Bot.* 84:896–904.
5. Alberternst B. 1998. Biologie, Ökologie, Verbreitung und Kontrolle von *Reynoutria*-Sippen in Baden-Württemberg. *Culterra* 23:1–198.
6. Alberternst B, Nawrath S. 2002. *Lysichiton americanus* in Kontinental-Europa. Bestehen Chancen für die Bekämpfung in der Frühphase ihrer Einbürgerung? In: Kowarik I, Starfinger U. (Eds.): Biologische Invasionen: Herausforderung zum Handeln? *Neobiota* 1:91–99.
7. Altan T, Fahrenhorst B. 1988. Der Export von Geophyten aus der Türkei und seine Folgen. *Natur Land.* 63:21–25.
8. Aragon R, Morales JM. 2003. Species composition and invasion in NW Argentinian secondary forests: effects of land use history, environment and landscape. *J. Veg. Sci.* 14:195–204.
9. Asmus U. 1981. Der Einfluß von Nutzungsänderung und Ziergärten auf die Florenzusammensetzung stadtnaher Forste in Erlangen. *Ber. Bayer. Bot. Ges.* 52:117–121.
10. Auge H. 1997. Biologische Invasionen: Das Beispiel *Mahonia aquifolium*. In: Feldmann R, Henle K, Auge H, Flachowsky J, Klotz S, Krönert R. (Eds.): *Regeneration und nachhaltige Landnutzung: Konzepte für belastete Regionen*. Berlin: Springer Verlag, pp. 124–129.
11. Auge H, Brandl R. 1997. Seedling recruitment in the invasive clonal shrub, *Mahonia aquifolium* Pursh (Nutt.). *Oecologia* 110:205–211.

12. Bailey JP. 1988. Putative *Reynoutria japonica* Houtt. x *Fallopia baldschuanica* (Regel) Holub hybrids discovered in Britain. *Watsonia* 17:163–164.
13. Bailey JP. 2003. Japanese knotweed *s.l.* at home and abroad. In: Child L, Brock JH, Brundu G, Prach K, Pysek P, Wade PM, Williamson M. (Eds.): *Plant invasions: ecological threats and management solutions.* Leiden: Backhuys Publishers, pp. 183–196.
14. Bailey JP, Conolly AP. 2000. Prize winners to pariahs — a history of Japanese knotweed *s.l.* (Polygonaceae) in the British Isles. *Watsonia* 23:93–110.
15. Bakker PA. 1986. Erhaltung von Stinzenplanten (Zwiebel- und Knollengewächse an alten Burgen). *Schrift. Stift. zum Schutz gef. Pflanz.* 4:105–116.
16. Bakker PA, Boeve E. 1985. *Stinzenplanten.* Uitgeverij Terra, Zutphen, 168 pp.
17. Balder H, Ehlebracht K, Mahler E. 1997. *Straßenbäume. Planen, Pflanzen, Pflegen am Beispiel Berlin.* Berlin: Patzer Verlag, 240 pp.
18. Bartels H, Bärtels A, Schroeder FG, Seehann GR. 1991. Erhebung über das Vorkommen winterharter Freilandgehölze. Die Gehölze und ihre Verbreitung in den Gärten und Parks. *Mitt. Dt. Dendrol. Ges.* 74:1–377.
19. Barton AM, Brewster LB, Cox AN, Prentiss NK. 2004. Non-indigenous woody invasive plants in rural New England town. *Biol. Invasions* 6:205–211.
20. Baskin Y. 2002. The greening of horticulture: new codes of conduct aim to curb plant invasions. *Bioscience* 52:464–471.
21. Bell CE, Wilen CA, Stanton AE. 2003. Invasive plants of horticultural origin. *HortScience* 38:14–16.
22. Bimova K, Mandák B, Pysek P. 2001. Experimental control of *Reynoutria* congeners: a comparative study of a hybrid and its parents. In: Brundu G, Brock J, Camarda I, Child L, Wade M. (Eds.): *Plant invasions. Species ecology and ecosystem management.* Leiden: Backhuys Publishers, pp. 283–290.
23. Bimova K, Mandak B, Pysek P. 2003. Experimental study of vegetative regeneration in four invasive *Reynoutria* taxa (Polygonaceae). *Plant Ecol.* 166:1–11.
24. Binggeli P. 1996. A taxonomic, biogeographical and ecological overview of invasive woody plants. *J. Veg.* Sci. 7:121–124.
25. Boag B, Yeates GW. 2001. The potential impact of the New Zealand flatworm, a predator of earthworms, in western Europe. *Ecol. Appl.* 11:1276–1286.
26. Carroll-Spillecke M (Ed.). 1995. *Der Garten von der Antike bis zum Mittelalter.* 2nd ed. Mainz: von Zabern, 293 pp.
27. Catling PM, Porebski ZS. 1994. The history of invasion and current status of glossy buckthorn, *Rhamnus frangula*, in southern Ontario. *Can. Field-Nat.* 108:305–310.
28. Child LE, Wade PM. 2000. *The Japanese Knotweed manual.* Chichester, U.K.: Packard Publishing Limited, 123 pp.
29. Chocholousková Z, Pysek P. 2003. Changes in composition and structure of urban flora over 120 years: a case study of the city of Plzen. *Flora* 198:366–376.
30. Conolly AP. 1977. The distribution and history in the British Isles of some alien species of *Polygonum* and *Reynoutria. Watsonia* 11:291–311.
31. Crawley MJ. 1987. What makes a community invasible? In: Gray AJ, Crawley MJ, Edwards PJ (Eds.) *Colonization, succession and stability.* Oxford: Blackwell, pp. 429–453.
32. Crosby AW. 1986. Ecological imperialism. The biological expansion of Europe, 900–1900. Cambridge, U.K.: Cambridge University Press.
33. Cross JR. 2002. The invasion and control of *Rhododendron ponticum* L. in native Irish vegetation. In: Kowarik I, Starfinger U. (Eds.): Biologische Invasionen. Herausforderungen zum Handeln? *Neobiota* 1:329–337.
34. Daehler CC, Carino DA. 2000. Predicting invasive plants: prospects for a general screening system based on current regional models. *Biol. Invasions* 2:93–102.
35. Dehnen-Schmutz K. 2000. Nichteinheimische Pflanzen in der Flora mittelalterlicher Burgen. *Diss. Bot.* 334, 119 pp.
36. Dehnen-Schmutz K. 2004. Alien species reflecting history: medieval castles in Germany. *Diversity Distrib.* 10:147–151.
37. Dehnen-Schmutz K, Perrings C, Williamson M. 2004. Controlling *Rhododendron ponticum* in the British Isles: An economic analysis. *J. Environ. Man.* 70:323–332.

38. Eigner A, Wissemann V. 1999. *Rosa* x *mangii*, eine neue intersektionelle Hybride charakterisiert durch morphologische und genetische Untersuchungen. *Haussknechtia* 7:35–40.
39. Ellstrand NC, Schierenbeck KA. 2000. Hybridisation as a stimulus for the evolution of invasiveness in plants? *Proc. Natl. Acad. Sci. USA* 97:7043–7050.
40. Erfmeier A, Bruelheide H. 2004. Comparison of native and invasive *Rhododendron ponticum* populations: growth, reproduction and morphology under field conditions. *Flora* 199:120–133.
41. Feuerhahn B, Spethmann W. 1995. Kreuzungen bei Wildrosenarten. *Gehölzforschung* 3:1–123.
42. Fischer W. 1975. Vegetationskundliche Aspekte der Ruderalisation von Waldstandorten im Berliner Gebiet. *Arch. Natursch.. Landschaftsforsch.* 15:21–32.
43. Fischer W. 1993. Zur Einbürgerung von Parkpflanzen in Brandenburg (Teil 1). Ein Beitrag zur Neophytenflora. *Verh. Bot. Ver. Berlin Brandenb.* 126:190–200.
44. Fischer W. 1997. Zur Einbürgerung von Parkpflanzen in Brandenburg (Teil 2). *Verh. Bot. Ver. Berlin Brandenb.* 130:159–184.
45. Forman J. 2003. The introduction of American plant species into Europe: issues and consequences. In: Child L, Brock J, Brundu G, Prach K, Pysek P, Wade PM, Williamson M (Eds.): *Plant invasions: ecological threats and management solutions,* Leiden, Netherlands: Backhuys Publishers, pp. 17–39.
46. Gaddum M. 1999. *Gaddum's plant finder 2000.* Gisborne, New Zealand, 290 pp.
47. Gallagher KG, Schierenbeck KA, D'Antonio CM. 1997. Hybridization and introgression in *Carpobrotus* spp. (Aizoaceae) in California. 2. Allozyme evidence. *Am. J. Bot.* 84:905–911.
48. Gianoni G, Carraro G, Klötzli F. 1988. Thermophile, an laurophyllen Pflanzenarten reiche Waldgesellschaften im hyperinsubrischen Seenbereich des Tessins. *Ber. Geobot. Inst. ETH, Stiftung Rübel* 54:164–180.
49. Goeze E. 1916. Liste der seit dem 16. Jahrhundert bis auf die Gegenwart in die Gärten und Parks Europas eingeführten Bäume und Sträucher. *Mitt. Dt. Dendrol. Ges.* 26:160–188.
50. Gray AJ, Marshall DF, Raybould AF. 1991. A century of evolution in *Spartina anglica. Adv. Ecol. Res.* 21:1–62.
51. Grotkopp E, Rejmanek M, Rost TL. 2002. Toward a causal explanation of plant invasiveness: seedling growth and life-history strategies of 29 pine (*Pinus*) species. *Am. Nat.* 159:396–419.
52. Groves RH. 1998. *Recent incursions of weeds to Australia 1971–1995.* Technical Series No. 3, Canberra: Cooperative Research Centre for Weed Management Systems.
53. Hancock JF, Siefker JH. 1982. Levels of inbreeding in highbush blueberry cultivars. *HortScience* 17:363–366.
54. Hanske C. 1991. Verfügbarkeit, Herkünfte und Qualität von heimischem Gehölzsaatgut. Unpublished diploma thesis, University of Hannover.
55. Hartmann E, Schuldes H, Kübler R, Konold W. 1995. *Neophyten. Biologie, Verbreitung und Kontrolle ausgewählter Arten.* Landsberg, Germany: Ecomed, 301 pp.
56. Hattenschwiler S, Körner C. 2003. Does elevated CO_2 facilitate naturalization of the non-indigenous *Prunus laurocerasus* in Swiss temperate forests? *Funct. Ecol.* 17:778–785.
57. Hellwig FH. 2000. Ist die Flaumeiche in Thüringen noch zu retten? Genetische Diversität und introgressive Hybridisierung. *Schrift. Vegetat.* 32:157–165.
58. Hitchmough J, de la Fleur M, Findlay C. 2004. Establishing North American prairie vegetation in urban parks in northern England. *Lands. Urban Plan.* 66:75–90.
59. Hodkinson DJ, Thompson K. 1997. Plant dispersal: the role of man. *J. Appl. Ecol.* 34:1484–1496.
60. Holm LG, Doll J, Holm E, Pancho JV, Herberger J. 1997. *World weeds: natural histories and distribution.* New York: J. Wiley & Sons.
61. Hubbes T. 1999. The American elm and Dutch elm disease. *For. Chron.* 75:265–273.
62. Hügin G, Lohmeyer W. 1993. Bastardbildung und intraspezifische Sippengliederung bei *Echinops sphaerocephalus* (Asteraceae, Cardueae) in Mitteleuropa. *Willdenowia* 23:83–89.
63. Hurka H, Bleeker W, Neuffer B. 2003. Evolutionary processes associated with biological invasions in the Brassicaceae. *Biol. Invas.* 5:281–292.
64. Jäger EJ. 1973. Zur Verbreitung und Lebensgeschichte der Wildtulpe (*Tulipa sylvestris* L.) und Bemerkungen zur Chorologie der Gattung *Tulipa* L. *Hercynia N.F.* 10:429–448.
65. Jäger EJ. 1989. *Ornithogalum nutans* L. und *O. boucheanum* (Kunth) Aschers. Heimatareal, synanthrope Ausbreitung und Lebensgeschichte. *Bot. Rundbr. Bez. Neubrand.* 21:13–18.

66. Jones AT, Evans PR. 1994. A comparison of the growth and morphology of native and commercially obtained European *Crataegus monogyna* Jacq. (Hawthorn) at an upland site. *Watsonia* 20:97–103.
67. Jones AT, Hayes MJ, Sackville Hamilton NR. 2001. The effect of provenance on the performance of *Crataegus monogyna* in hedges. *J. App. Ecol.* 38:952–962.
68. Kloot PM. 1987. The naturalised flora of South Australia. 3. Its origin, introduction, distribution, growth forms and significance. *J. Adelaide Bot. Gard.* 10:99–111.
69. van Kleunen M, Schmid B. 2003. No evidence for an evolutionary increased competitive ability in an invasive plant. *Ecology* 84:2816–2823.
70. Klötzli F, Walther GR. 1999. Recent vegetation shifts in Switzerland. In: Klötzli F, Walther GR (Eds.): *Recent shifts in vegetation boundaries of deciduous forests, especially due to general global warming*. Basel: Birkhäuser, pp. 19–25.
71. Kolar CS, Lodge DM. 2001. Progress in invasion biology: predicting invaders. *Trends Ecol. Evol.* 19:199–204.
72. Konvalinková P. 2003. Generative and vegetative reproduction of *Helianthus tuberosus*, an invasive plant in central Europe. In: Child L, Brock J, Brundu G, Prach K, Pysek P, Wade PM, Williamson M (Eds.): *Plant invasions: ecological threats and management solutions*. Leiden: Backhuys Publishers, pp. 289–299.
73. Kowarik I. 1990. Some responses of flora and vegetation to urbanization in Central Europe. In: Sukopp H, Hejni S, Kowarik I. (Eds.): *Urban ecology*. The Hague: SPB Academic Publ. pp. 45–74.
74. Kowarik I. 1992. Einführung und Ausbreitung nichteinheimischer Gehölzarten in Berlin und Brandenburg. *Verh. Bot. Ver. Berlin Brandenb., Beiheft* 3, 188 pp.
75. Kowarik I. 1995. Time lags in biological invasions with regard to the success and failure of alien species. In: Pysek P, Prach K, Rejmànek M, Wade PM. (Eds.): *Plant invasions. General aspects and special problems*. Amsterdam: SPB Academic Publ., pp. 15–38.
76. Kowarik I. 1995. On the role of alien species in urban flora and vegetation. In: Pysek P, Prach K, Rejmanek M, Wade M. (Eds.): *Plant invasions. General aspects and special problems*. Amsterdam: SPB Academic Publ., pp. 85–103.
77. Kowarik I. 2003. Biologische Invasionen. Neophyten und Neozoen in Mitteleuropa. Stuttgart: Ulmer, 380 pp.
78. Kowarik I. 2003. Human agency in biological invasions: secondary releases foster naturalisation and population expansion of alien plant species. *Biol. Invas.* 5:293–312.
79. Kowarik I, Böcker R. 1984. Zur Verbreitung, Vergesellschaftung und Einbürgerung des Götterbaumes (*Ailanthus altissima* (Mill.) Swingle) in Mitteleuropa. *Tuexenia* 4:9–29.
80. Kowarik I, Heink U, Schmitz G, Starfinger U. 2003. Development of assessment criteria for the release of alien plants. Effects on plant communities and selected animal species. Unpubl. research report, UFOPLAN 299 812 02 of the Federal Ministry for Environment by order of the Federal Environmental Agency, Berlin, 171 pp.
81. Kowarik I, Seitz B. 2003. Perspektiven für die Verwendung gebietseigener ("autochthoner") Gehölze. In: Seitz B, Kowarik I. (Eds.): Perspektiven für die Verwendung gebietseigener Gehölze in Brandenburg. *Neobiota* 2:3–25.
82. Kunick W. 1982. Zonierung des Stadtgebietes von Berlin (West). Ergebnisse floristischer Untersuchungen. *Landsch. Umweltforsch.* 14:1–164.
83. Levine JM, Vilá M, D'Antonio CM, Dukes JS, Grigulis K, Lavorel S. 2003. Mechanisms underlying the impacts of exotic plant invasions. *Proc. R. Soc. London B* 270:775–781.
84. Liebster G. 1961. *Die Kulturheidelbeere*. Berlin, Hamburg: Paul Parey, 229 pp.
85. Lohmeyer W, Sukopp H. 1992. Agriophyten in der Vegetation Mitteleuropas. *Schr. Vegetationskde.* 25:1–185.
86. Mack RN, Erneberg M. 2002. The United States naturalized flora: largely the product of deliberate introductions. *Ann. Missouri Bot. Gard.* 89:176–189.
87. Mack RN, Lonsdale WM. 2001. Humans as global plant dispersers: getting more than we bargained for. *Bioscience* 51:95–102.
88. Mack RN, Simberloff D, Lonsdale M, Evans H, Clout M, Bazzaz F. 2000. Biotic invasions: causes, epidemiology, global consequences, and control. *Ecol. Appl.* 10:689–710.
89. Mack RN. 2000. Cultivation fosters plant naturalization by reducing environmental stochasticity. *Biol. Invas.* 2:111–122.

90. Mack RN. 2001. Motivations and consequences of the human dispersal of plants. In: McNeely JA (Ed.): *The great reshuffling: human dimensions in invasive alien species.* Gland: IUCN, pp. 23–34.
91. Mack RN. 2003. Plant naturalizations and invasions in the Eastern United States: 1634–1860. *Ann. Missouri Bot. Gard.* 90:77–90.
92. Maurer U. 2002. Pflanzenverwendung und Pflanzenbestand in den Wohnsiedlungen der 1920er und 1930er Jahre in Berlin. Ein Beitrag zur historischen Pflanzenverwendung. *Diss. Bot.* 353:1–221.
93. Merkel J, Walter E, Rebhahn H. 1991. Naturschutz in Oberfranken. Zur Problematik der Einbürgerung von Pflanzen und Tieren. *Heimatbeil. Amtl. Schulanz. RegBez. Oberfranken* 178:1–64.
94. Meyer JY, Florence J, Tchung V. 2003. The endemic *Psychotria* (Ruiaceae) of Tahiti (French Polynesia) threatened by the invasive *Miconia calvescens* (Melastomataceae): status, distribution, ecology, phenology and conservation. *Rev. d'Ecol. — Terre Vie* 58:161–185.
95. Mihulka S. 1998. The effect of altitude on the pattern of plant invasions: a field test. In: Starfinger U, Edwards K, Kowarik I, Williamson M. (Eds.): *Plant invasions: ecological mechanisms and human response.* Leiden, Netherlands: Backhuys Publishers, pp. 313–320.
96. Milne RIM, Abbott RJ. 2000. Origin and evolution of invasive naturalized material of *Rhododendron ponticum* L. in the British Isles. *Mol. Ecol.* 9:541–556.
97. Moran MA. 1984. Influence of adjacent land use on understory vegetation of New York forests. *Urban Ecol.* 8:329–340.
98. Mulvaney M. 2001. The effect of introduction pressure on the naturalization of ornamental woody plants in south-eastern Australia. In: Groves RH, Panetta FD, Virtue JG, (Eds.): *Weed Risk Assessment.* Melbourne: CSIRO, pp. 186–193.
99. Murchie AK, Moore JP, Walters KFA, Blackshaw RP. 2003. Invasion of agricultural land by the earthworm predator, *Arthurdendyus triangulatus* (Dendy). *Pedobiologia* 47:920–923.
100. Nath M. 1990. Historische Pflanzenverwendung in Landschaftsgärten. Auswertung für den Artenschutz., Worms, Germany: Wernersche Verlagsgesellschaft, 236 pp.
101. Naumann WD. 1993. Overview of the *Vaccinium* industry in western Europe. *Acta Horticult.* 346:53–55.
102. Neff C. 2003: Les Corbières maritimes — forment-elles un étage de végétation méditerranéenne thermophile masqué par la pression humaine? In: Fouache, E. (Ed.): *The Mediterranean world environment and history.* Paris: Elsevier, pp. 191–202.
103. Niinemets U, Valladares F, Ceulemans R. 2003. Leaf-level phenotypic variability and plasticity of invasive *Rhododendron ponticum* and non-invasive *Ilex aquifolium* co-occurring at two contrasting European sites. *Plant Cell Environ.* 26:941–956.
104. Noble IR. 1989. Attributes of invaders and the invading process: terrestrial and vascular plants. In: Drake JA, Mooney HA, di Castri F, Groves RH, Kruger F, Rejmánek M, Williamson M. (Eds.): *Biological invasions. A global perspective.* New York: John Wiley & Sons, pp. 301–313.
105. Perrings C, Williamson M, Dalmazzone S. (Eds.) 2000. *The economics of biological invasions.* Cheltenham, U.K.: Edward Elgar, 249 pp.
106. Pimental D, Lach L, Zuniga R, Morrison D. 2000. Environmental and economic costs of nonindigenous species in the United States. *BioScience* 50:53–65.
107. Poppendieck HH. 1996. Stinzenpflanzen in Schleswig-Holstein und Hamburg. In: Buttlar A, Meyer MM. (Eds.): *Historische Gärten in Schleswig-Holstein.* Heide, Germany: Boyens & Co., pp. 612–613.
108. Prach K, Hadinec J, Michalek J, Pysek P. 1995. Forest planting as a way of species dispersal. *For. Ecol. Manag.* 76:191–195.
109. Prange W. 1996. Das Kleine Immergrün (*Vinca minor* L.) in Westdeutschland: eine Kulturreliktpflanze aus römischer Zeit. *Schrift. Naturwiss.Vereins Schleswig-Holstein* 66:71–96.
110. Preston CD, Pearman DA, Dines TD. (Eds.) 2002. *New atlas of the British and Irish flora.* Oxford: Oxford University Press.
111. Prinzing A, Durka W, Klotz S, Brandl R. 2002. Which species become aliens? *Evol. Res.* 4:385–405.
112. Pysek P. 1993. Factors affecting the diversity of flora and vegetation in central European settlements. *Vegetatio* 106:89–100.
113. Pysek P. 1997. Clonality and plant invasions: can a trait make a difference? In: de Kroon H, van Groenendal J. (Eds.): *The ecology and evolution of clonal plants.* Leiden, Netherlands: Backhuys Publishers, pp. 405–427.

114. Pysek P. 1998. Alien and native species in Central European urban floras: a quantitive comparison. *J. Biogeogr.* 25:155–163.
115. Pysek P, Prach K. 1993. Plant invasions and the role of riparian habitats: a comparison of four species alien to Central Europe. *J. Biogeogr.* 20:413–420.
116. Pysek P, Sádlo J, Mandák B. 2002. Catalogue of alien plants of the Czech Republic. *Preslia* 74:97–186.
117. Pysek P, Sádlo J, Mandák B, Jarosik V. 2003. Czech alien flora and the historical pattern of its formation: what came first to Central Europe? *Oecologia* 135:122–130.
118. Pysek P, Sádlo J, Mandák B. 2003: Alien flora of the Czech Republic, its composition, structure and history. In: Child L, Brock J, Brundu G, Prach K, Pysek P, Wade PM, Williamson M. (Eds.): *Plant invasions: ecological threats and management solutions.* Leiden, Netherlands: Backhuys Publishers, pp. 113–130.
119. Rauer G, von den Driesch M, Lobin W, Ibisch PL, Barthlott W. 2000. *Beitrag der deutschen Botanischen Gärten zur Erhaltung der Biologischen Vielfalt und Genetischer Ressourcen. Bestandsaufnahme und Entwicklungskonzept.* Bonn, Germany: Bundesamt für Naturschutz, 246 pp.
120. Reichard SH, Hamilton CW. 1997. Predicting invasions of woody plants introduced into North America. *Conserv. Biol.* 11:193–203.
121. Reichard SH, White P. 2001. Horticulture as a pathway of invasive plant introductions in the United States. *BioScience* 51:103–113.
122. Rejmanek M. 1996. A theory of seed plant invasiveness: The first sketch. *Biol. Conserv.* 78:171–181.
123. Rejmanek M. 2000. Invasive plants: approaches and predictions. *Aust. Ecol.* 25:497–506.
124. Rejmanek M, Richardson DM. 1996. What attributes make some plant species more invasive? *Ecology* 77:1655–1661.
125. Richardson DM, Pysek P, Rejmánek M, Barbour MG, Panetta FD, West CJ. 2000. Naturalization and invasion of alien plants: concepts and definitions. *Diversity Distrib.* 6:93–107.
126. Ringenberg J. 1994. *Analyse urbaner Gehölzbestände am Beispiel der Hamburger Wohnbebauung.* Hamburg: Kovac, 220 pp.
127. Rouget M, Richardson DM. 2003. Understanding patterns of plant invasion at different spatial scales: quantifying the roles of environment and propagule pressure. In: Child L, Brock J, Brundu G, Prach K, Pysek P, Wade PM, Williamson M. (Eds.): *Plant invasions: ecological threats and management solutions.* Leiden, Netherlands: Backhuys Publishers, pp. 3–15.
128. Rozefelds ACF, Cave L, Morris DI, Buchanan AM. 1999. The weed invasion in Tasmania since 1970. *Aust. J. Bot.* 47:23–48.
129. Rumpf H. 2002. Phänotypische, physiologische und genetische Variabilität bei verschiedenen Herkünften von *Viburnum opulus* L. und *Corylus avellana* L. PhD thesis, University of Hannover, 176 pp.
130. Sachse U, Starfinger U, Kowarik I. 1990. Synanthropic woody species in the urban area of Berlin. In: Sukopp H, Hejny S, Kowarik I. (Eds.): *Plants and plant communities in the urban environment.* The Hague: SPB Academic Publishing, pp. 233–243.
131. Sachse U. 1989. Die anthropogene Ausbreitung von Berg- und Spitzahorn. Ökologische Voraussetzungen am Beispiel Berlins. *Landsch. Umweltforsch.* 63, 132 pp.
132. Salisbury E. 1964. *Weeds and aliens.* 2nd ed., London: Collins, 384 pp.
133. Säumel I, Kowarik I. 2003. Non-native tree species in Berlin. Does urban climate influence on the spatial patterns of phytodiversity? *Bol. Soc. Argentina Bot.* 38 (Suppl.):198.
134. Schepker H. 1998. *Wahrnehmung, Ausbreitung und Bewertung von Neophyten. Eine Analyse der problematischen nichteinheimischen Pflanzenarten in Niedersachsen.* Stuttgart: Ibidem Verlag, 246 pp.
135. Schepker H, Kowarik I, Garve E. 1997. Verwilderungen nordamerikanischer Kultur-Heidelbeeren (*Vaccinium* subg. *Cyanococcus*) in Niedersachsen und deren Einschätzung aus Naturschutzsicht. *Nat. Landschaft* 72:346–351.
136. Schepker H, Kowarik I. 1998. Invasive North American blueberry hybrids (*Vaccinium corymbosum x angustifolium*) in northern Germany. In: Starfinger U, Edwards K, Kowarik I, Williamson M. (Eds.): *Plant invasions. Ecology and human response.* Leiden, Netherlands: Backhuys Publishers, pp. 253–260.
137. Scholz H. 1993. Eine unbeschriebene anthropogene Goldrute (*Solidago*) aus Mitteleuropa. *Florist Rundbr.* 27:7–12.
138. Schroeder FG. 1969. Zur Klassifizierung der Anthropochoren. *Vegetatio* 16:225–238.
139. Seidling W. 1999. Spatial structures of a subspontaneous population of *Taxus baccata* saplings. *Flora* 194:439–451.

140. Sobrino Vesperinas E, Gonzales Moreno A, Sanz Elorza M, Dana Sanchez E, Sanchez Mata D, Gavilan R. 2001. The expansion of thermophilic plants in the Iberian Peninsula as a sign of climatic change. In: Walther GR, Burga CA, Edwards PJ. (Eds.): *"Fingerprints" of climate change — adapted behaviour and shifting species ranges.* New York: Kluwer Academic/Plenum Publishers, pp. 163–184.
141. Soldaat LL, Auge H. 1998. Interactions between an invasive plant, *Mahonia aquifolium*, and a native phytophagous insect, *Rhagoletis meigenii*. In: Starfinger U, Edwards K, Kowarik I, Williamson M. (Eds.): *Plant invasions: ecological mechanisms and human response*. Leiden, Netherlands: Backhuys Publishers, pp. 347–360.
142. Spethmann W. 1995. In-situ/ex-situ-Erhaltung von heimischen Straucharten. In: Kleinschmit J, Begemann F, Hammer K. (Eds.): *Erhaltung pflanzengenetischer Ressourcen in der Land- und Forstwirtschaft. Schriften zu Genetischen Ressourcen* (Bonn, ZADI,) 1:68–87.
143. Starfinger U, Kowarik I, Rode M, Schepker H. 2003. From desirable ornamental plant to pest to accepted addition to the flora? The perception of an alien plant species, *Prunus serotina,* through the centuries. *Biol. Invas.* 5:323–335.
144. Suehs CM, Affre L, Medail F. 2004. Invasion dynamics of two alien *Carpobrotus* (Aizoaceae) taxa on a Mediterranean island: I. Genetic diversity and introgression. *Heredity* 92:31–40.
145. Sukopp H, Wurzel A. 2000. Changing climate and the effects on vegetation in central European cities. *Arboricult. J.* 24:1–24.
146. Swanton CJ, Cavers PB, Clements DR, Moore, MJ. 1992. The biology of Canadian weeds. 101. *Helianthus tuberosus* L. *Can. J. Plant Sci.* 72:1367–1382.
147. Thellung A. 1912. La flore adventice de Montpellier. *Mém. Soc. Sci. Nat. Cherbourg* 38:622–647.
148. Trepl L. 1990. Research on the anthropogenic migration of plants and naturalization. Its history and current state of development. In: Sukopp H, Hejny S, Kowarik I. (Eds.): *Urban ecology. Plants and plant communities in urban environments.* The Hague: SPB Academic Publishing, pp. 75–97.
149. Tutin TG, Heywood VH, Burgess NA, Moore DM, Valentine DH, Walters SM, Webb DA. 1980. *Flora Europaea.* Cambridge, U.K.: Cambridge University Press.
150. van Kleunen M, Schmid B. 2003. No evidence for an evolutionary increased competitive ability in an invasive plant. *Ecology* 84:2816–2823.
151. van Wilgen BW, Richardson DM, le Maitre DC, Marais C, Magadlela D. 2001. The economic consequences of alien plant invasions: examples of impacts and approaches to sustainable management in South Africa. *Environ. Develop. Sustain.* 3:145–168.
152. Vila M, D'Antonio CM. 1998. Fruit choice and seed dispersal of invasive vs. noninvasive *Carpobrotus* (Aizoaceae) in coastal California. *Ecology* 79:1053–1060.
153. Viola H, Margolis C. 1991. *Seeds of change: a quincentennial commemoration.* Washington, DC: Smithsonian Institution Press.
154. Vitousek PM, D'Antonio CM, Loope LL, Rejmanek M, Westbrooks R. 1997. Introduced species: a significant component of human-caused global change. *N.Z. J. Ecol.* 21:1–16.
155. Wagenitz G. 1979. Compositae [*Helianthus*]. In: Hegi G. (Ed.): *Illustrierte Flora von Mitteleuropa.* Hamburg: Paul Parey, 2nd ed., Vol 6/3, pp. 244–257.
156. Walther GR. 2001. Laurophyllisation — a sign for a changing climate? In: Burga CA, Kratochwil A. (Eds.): *Biomonitoring: general and applied aspects on regional and global scales.* Dordrecht, Netherlands: Kluwer, pp. 207–223.
157. Walther GR. 2003. Wird die Palme in der Schweiz heimisch? *Bot. Helv.* 113:159–180.
158. Walther GR. 2003. Plants in a warmer world. Plant Ecol., *Evol. Syst.* 6:169–185.
159. Walther GR, Carraro G, Klötzli F. 2001. Evergreen broad-leaved species as indicators for climate change. In: Walther GR, Burga CA, Edwards PJ. (Eds.): *"Fingerprints" of climate change — adapted behaviour and shifting species ranges.* New York: Kluwer Academic/Plenum Publishers, pp. 151–162.
160. Webb SL, Kaunzinger CK. 1993. Biological invasion of the Drew-University (New Jersey) forest preserve by Norway maple (*Acer platanoides* L.). *Bull. Torrey Bot. Club* 120:343–349.
161. Weber E. 2001. Current and potential ranges of three exotic goldenrods (*Solidago*) in Europe. *Conserv. Biol.* 15:122–128.
162. Weber E. 2003. *Invasive plant species of the world: a reference guide to environmental weeds.* Wallingford, UK: CABI Intl., 548 pp.
163. Weber E, Schmid B. 1998. Latitudinal population differentiation in two species of *Solidago* (Asteraceae) introduced into Europe. *Am. J. Bot.* 85:1110–1121.

164. Weber E, Vila M, Albert M, D'Antonio CM. 1998. Invasion by hybridization: *Carpobrotus* in coastal California. In: Starfinger U, Edwards K, Kowarik I, Williamson M. (Eds.): *Plant invasions: ecological mechanisms and human response.* Leiden, Netherlands: Backhuys Publishers, pp. 275–281.
165. Wein K. 1963. Die Einführungsgeschichte von *Helianthus tuberosus* L. *Die Kulturpflanze* 11:43–91.
166. Williamson M. 1996. *Biological invasions.* London: Chapman and Hall, 244 pp.
167. Williamson M. 1999. *Invasions. Ecography* 22:5–12.
168. Williamson M. 2002. Alien plants in the British Isles. In: Pimentel D. (Ed.): *Biological invasions. Economic and environmental costs of alien plant, animal, and microbe species.* Boca Raton, FL: CRC Press, pp. 91–112.
169. Williamson M, Brown KC. 1986. The analysis and modelling of British invasions. *Philos. Trans. R. Soc. London* B 314:505–522.
170. Williamson M, Fitter A. 1996. The characteristics of successful invaders. *Biol. Conserv.* 78:163–170.
171. Wu SH, Chaw SM, Rejmanek M. 2003. Naturalized Fabaceae (Leguminosae) species in Taiwan: the first approximation. *Bot. Bull. Acad. Sin.* 44:59–66.
172. Zerbe S, Maurer U, Schmitz S, Sukopp H. 2003. Biodiversity in Berlin and its potential for nature conservation. *Landscape Urban Plan.* 62:139–148.

8 Sorghum and Its Weedy Hybrids

Gebisa Ejeta and Cécile Grenier

8.1 INTRODUCTION

Genetic exchange between wild and cultivated sorghums occurs in nature. The appearance of weedy traits such as shattering and *rhizomatousness* (development of plants that propagate from rhizomes) in crop-wild hybrids of sorghum has often been used as evidence for such exchanges. The impact of unusual traits resulting from these introgressions in the nature of the resultant hybrids including the appearance of feral forms has not been widely studied and documented. Sorghum is one of the five most important grain crops in the world and is mainly cultivated in the developing world including Africa, the ancestral home of the crop. In many of these areas, volunteer, wild, and weedy forms exist both together with the crop and in nearby ruderal areas. Intermediary forms such as shattercanes are ubiquitous and exist in a continuum of forms ranging from those that closely resemble the wild to those nearly indistinguishable from the cultivated members of the same genus. Although there is no clear evidence for the existence of feral types emerging as a result of mutational dedomestication (*endoferality*), the existence of intermediary forms in most sorghum growing areas offers an empirical evidence for noxious weedy forms arising from continued introgression (*exoferality*) among different sorghum types. For instance, the most widely recognized noxious weed in the sorghum taxa, *Sorghum halepense* (johnsongrass), is a natural hybrid between the cultivated *Sorghum bicolor* and wild rhizomatous species *S. propinquum*.

With the advent of the transgenic era, genetic transformation of sorghums has been approached with some apprehension among sorghum research scientists. If developed, the deployment of transgenic sorghums is likely to be met with some trepidation primarily due to the concern that feral transgenic hybrid forms that are more fit and difficult to eradicate may arise as a result of natural and accidental cross-pollination. Research is needed to assess the extent of spontaneous gene flow among the cultivated-wild-weed complex of the sorghum taxa and, should there appear some danger, to devise means for its mitigation. In this chapter, we describe the sorghum taxa, the nature of the profile of the crop-wild-weed complex in sorghum, and the existence of intermediary forms of varying ferality, relative to the ease of genetic exchange among the species in the taxa. The influence of the agroecosystem in modifying the natural habitat, in catalyzing effective genetic exchange, and in disseminating the resultant hybrid forms is also discussed using anecdotal evidence from three major sorghum growing environments around the world.

8.2 THE SORGHUM TAXA

The genus sorghum belongs to the Poaceae (Gramineae) family. Along with maize (*Zea mays*), sugarcane (*Saccharum* spp.), and all the millets (*Pennisetum*, *Eleusine*, *Eragrostis*, *Setaria,* etc.), it falls into the tribe Andropogoneae. The genus sorghum is divided into five subgenera — *Eu-Sorghum*, *Parasorghum*, *Heterosorghum*, *Chaetosorghum,* and *Stiposorghum* (Figure 8.1). Three species are recognized in the subgenera *Sorghum* (*Eu-Sorghum*) including the two rhizomatous weedy taxa, *Sorghum halepense* (L.) Pers. and *S. propinquum* (Kunth.) Hitchc., as well as cultivated sorghum

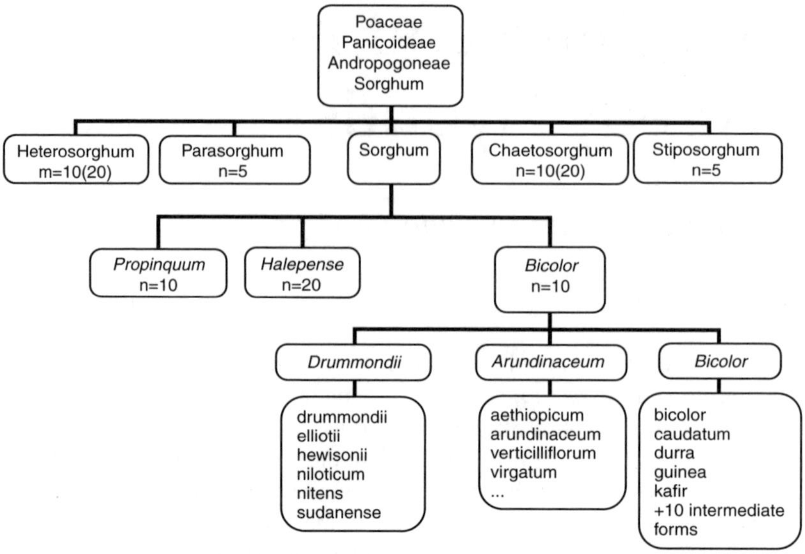

FIGURE 8.1 The sorghum taxa as per de Wet (8): the sorghum genus is a member of five major subgenera in the Poaceae family. Species bicolor is the cogenitor of the crop-wild-weed complex in the sorghum genus.

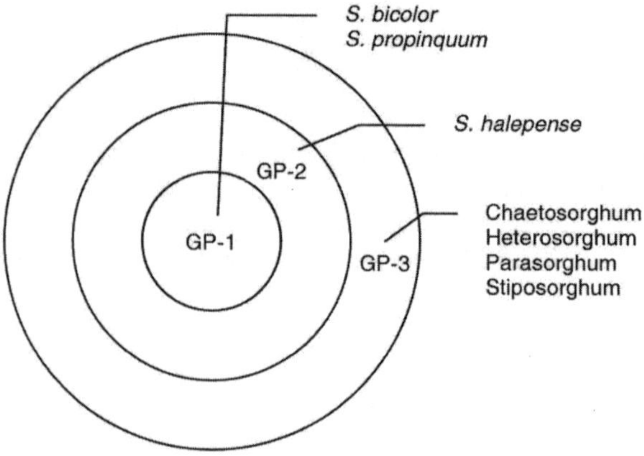

FIGURE 8.2 The sorghum gene pool based on classification of a "biological species" as per Harlan and de Wet (21).

S. bicolor (L.) Moench, which also has weedy forms. According to Harlan and de Wet (21), *S. bicolor* and *S. propinquum* are found in the primary gene pool (GP-1) of sorghum, which contains biological species of both the cultivated and spontaneous (wild or weedy) races. The species in this primary gene pool intercross readily and produce fertile hybrids. *Sorghum halepense* is found in the secondary gene pool (GP-2), which includes all species that can be crossed with GP-1 with at least some fertility in the hybrids. This suggests that gene transfer between these two gene pools is possible, though it may be difficult in some situations. The other sections of the sorghum taxa are encompassed in the tertiary gene pool (GP-3), where hybrids with GP-1 are anomalous, lethal, or nearly completely sterile, and gene transfer is not readily possible or requires radical techniques (Figure 8.2).

Sorghum bicolor is indigenous to Africa. It includes the weedy type sorghums classified under *S. bicolor drummondii*, the wild progenitor *S. b. arundinaceum,* and the cultivated crop, *S. b. bicolor* (Figure 8.1). Members of the *S. bicolor* subspecies are *diploid* (2n = 20). Cultivated sorghum in much of the world is sympatric with other sexually compatible, cultivated, and feral relatives (2). *Sorghum propinquum* is a strongly rhizomatous perennial indigenous to Asia, which freely introgresses with its congener, *S. bicolor. Sorghum halepense,* also a perennial, is a *tetraploid* (2n = 40) species. Both its rhizomatous nature and its geographical distribution suggest that *S. halepense* is an interspecific hybrid derived as a descendant of *S. bicolor* and *S. propinquum* (34,35).

8.2.1 Cultivated Sorghums

The pattern of distribution and history of domestication of sorghum has been carefully delineated in a series of insightful observations by early botanists (22,31,40,42). Cultivated sorghum arose from the wild *Sorghum bicolor* ssp. *arundianaceum* (28). There is no evidence suggesting that presently cultivated sorghums have evolved from the rhizomatous diploid or tetraploid wild species (10). Domestication of the sorghum crop is presumed to have started in the northeast quadrant of Africa (22) some 5000 years ago, perhaps in the expanse of arable land area recognized in contemporary geography as Ethiopia and Sudan. The great genetic variability of the crop in this geographical area, the wide range of ecological habitats there, and the long history of human selection efforts in the region have given sufficient credence to the theory of the origin and early domestication of sorghum in eastern Africa. Several routes have been recognized for the later movement of sorghum into other parts of Africa and beyond (Figure 8.3). Early movement between 2000 and 4000 years ago has taken the crop to west, central, and southern Africa leading to further domestication and appearance of distinct forms (races) in each of these regions. Harlan and de Wet (22) recognize 5 major races and 10 intermediate hybrid forms having arisen from these early movements of *S. bicolor.* Of the sorghum races, bicolor was domesticated first (11). Bicolors are grown across the range of sorghum cultivation in Africa. Race caudatum is an old race of grain sorghum, still widely grown in present Chad, Sudan, northeastern Nigeria, and Uganda. The race guinea, although occasionally cultivated in several places around the world is uniquely predominant in the west African savannah. The durra race is preponderant in the Ethiopian and Yemen highlands. Durras are widely grown along the fringes of the southern Sahara, across arid west Africa, whereas race kafir is distinctly eastern and southern African from Tanzania southward. Beyond Africa, early domestication of sorghum took place also in the Indian subcontinent about 3000 years ago having moved there early both over the Indian Ocean as well as over the Arabian peninsula. The small-panicled, dryland durras commonly found in India today are distinctly different in form, adaptation, stress tolerance, as well as overall productivity from those found in the east African highlands. Distinctly later than in these early movements, sorghum was also moved from India to China over the Himalayas as evidenced by appearance of new races of sorghum. Chinese kaoliangs uniquely make up the only sorghums in the world that have evolved in the temperate parts of the world. As a result, they have become an invaluable source of early season cold tolerance in sorghum. Even much more recently, sorghum was moved from Africa in the last 3 centuries, via the slave trade, to the Americas. Evidence for distinct evolutionary changes is limited in the Americas, yet sorghum received its greatest transformation as a crop through modern selection efforts and mechanized cultivation. Today, cultivated sorghum is one of the most economically important crops of the world with an annual production of approximately 60 million metric tons (15). It is used in a wide array of forms and for a range of purposes, constituting an essential portion of the diet of people living in the semi-arid tropics. The U.S., India, and Nigeria together produce nearly half of the grain sorghum harvested annually around the world. Among African countries, Nigeria (8.1 million tons), Sudan (4.4 million tons), and Ethiopia (1.6 million tons) are the largest producers of grain sorghum. Sorghum is most widely grown in the semi-arid tropics. Of the total land area of 95.2 million ha annually committed to cereal production in Africa, over 24.3 million ha of the arable land is cultivated with sorghum (15). In some nations, such as the Sudan, sorghum is important and widely cultivated providing the nutritional backbone of the population with grain harvested on more than 65% of the total area allocated to cereals (18).

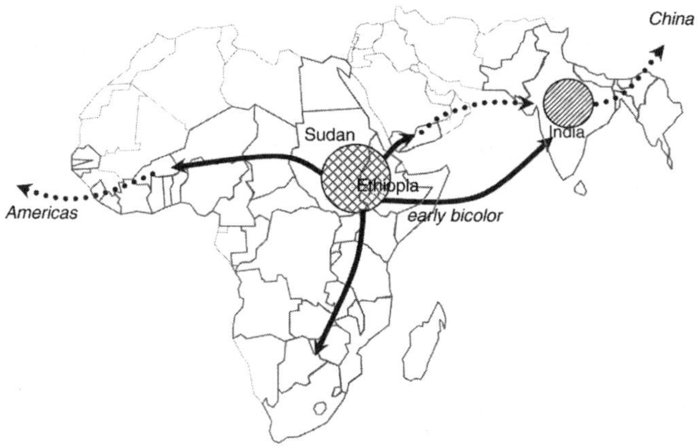

FIGURE 8.3 Pattern of domestication and spread of the genus *Sorghum*.

Sorghum is well adapted to a wide range of environmental conditions. It is generically drought and flood tolerant, and often does well in both heavy and light sandy soils over a pH range of 5.5 to 8.5, and can also withstand high salinity levels. Sorghum has a number of morphological and physiological characteristics that contribute to its adaptation to contrasting moisture, temperature, and as well as varying soil fertility and ecological conditions. In spite of its broad adaptation and being fit to a range of harsh environments, sorghum production can be constrained by biotic and abiotic stresses. Insects, diseases, parasitic weeds, as well as severe deficits and erratic distribution of rainfall in drier areas often devastate sorghum crops. However, the great wealth of genetic diversity in the species has allowed both natural and artificial selection efforts to result in continual progress in deriving hardier variants of the species.

Several plant breeding programs have successfully identified sources of agronomically valuable genes within the sorghum germplasm and effectively introduced them into improved varieties adapted to cultivation under various environmental conditions. The use of molecular marker assisted selection offers promise for faster and even more efficient crop breeding. Though no sorghum cultivar has yet been developed and released through marker assisted selection, techniques have been developed and molecular markers identified that promise a more deliberate and rapid incorporation of traits into preferred germplasm backgrounds. Furthermore, new genetic engineering tools promise an even faster rate of progress and a more precise mode of incorporating new genes in existing cultivars in a crop improvement program. Genetic transformation of sorghum has been pursued, in recent years, to genetically engineer special purpose sorghums. Although stable transformations were obtained through biolistic bombardment (5,24) and *Agrobacterium* mediated transformation (44), further improvements are needed to achieve more reliable transformation protocols before transgenic sorghums can be more readily exploited. With efficient transformation systems in sorghum and other cereal species, valuable genes selected from an organism in the same or different species, family, tribe, or even from an organism belonging to a different kingdom could more precisely and readily be transferred into a crop, a feat unimaginable in natural selection or traditional selection efforts of the past.

8.2.2 Wild and Weedy Sorghums

Domestication of sorghum began with wild members of *Sorghum bicolor* ssp. *arundinaceum*. Ecological and geographical isolation probably gave rise to the following four races of wild sorghums from *S. b. arundinaceum* (21):

1. Race aethiopicum formed part of the grass vegetation across the drier parts of the west African savannah, extending from western Ethiopia to Mauritania. This race can be locally abundant appearing in large tracts of land and occasionally invades cultivated fields.
2. Race arundinaceum is a forest grass widely distributed in moist tropical forests of the Guinea coast and the Congo. It appears as a common grass along stream banks and forest paths, and occasionally establishes in cultivated fields.
3. Race verticilliflorum is a common grass across most of the African savannah, between Sudan and Nigeria, found in weedy patches along roadsides, and often found established in cultivated fields. It is the most widely distributed wild sorghum, having been introduced to tropical Australia, parts of India, and the New World (8). Along with race aethiopicum, it is found most everywhere sorghum is grown in Sudan.
4. Race virgatum is a desert grass distributed along irrigation ditches and stream banks in central Sudan and extends along the length of the Nile northward to Cairo.

Weedy sorghums exist parallel to the wild sorghums, as either the perennial rhizomatous forms derived from *Sorghum propinquum* or the annual grassy weeds that resulted from hybridization between cultivated and wild sorghums within the *Sorghum bicolor* species. *Sorghum propinquum* is a strongly rhizomatous, perennial species indigenous to Asia. Ecologically, *S. propinquum* thrives as a forest grass well adapted to moist habitats such as riverbanks. The pattern of evolutionary changes for weedy sorghum types suggests an ancestral diploid species common to the wild grass *S. bicolor arundinaceum* and the diploid perennial *S. propinquum* (10). It is suspected that one prototype spread into southeast Asia where the humid climates and high rainfall favored the development of rhizomes, and a second form spread into northeast Africa in a habitat of open country with long dry seasons, where the annual growth habit of the plant was favored (Figure 8.3). The primary geographical area of occurrence for *S. propinquum* is the Indian subcontinent, and includes the area from Burma eastward to the islands of southeastern Asia. This species readily crosses with the introduced grain sorghums in the Philippines and derivatives of such crosses are recognized as noxious weeds.

S. halepense is a perennial, tetraploid sorghum native of southern Eurasia, east of India. It occupies a continuous area from southern and eastern India to the Mediterranean littoral, about halfway between the region of distribution of the *Sorghum bicolor* ssp. *verticilliflorum* of tropical Africa and the southeast Asian populations of *S. propinquum* (Figure 8.3). According to Doggett and Prasada Rao (11), *S. halepense* probably arose from chromosome doubling after a natural cross between these two species. *S. halepense* has also been introduced as a weed to all the warmer temperate regions of the world (8). Many ecotypes of *S. halepense* have been reported (29). Though not widely acknowledged, two morphologically distinct complexes were reported within *S. halepense*: a Mediterranean tetraploid ($2n = 40$ chromosomes) ecotype and a tropical diploid with $2n = 20$ chromosomes (9). To our knowledge, no further research has been carried out to confirm this. Within the *Sorghum bicolor* species, ssp. *drummondii* is the primary group of weedy sorghums that are strictly African. Sorghum shattercanes often appear in sympatry with the crop as hybrids resulting from crosses between cultivated and wild sorghums. Morphologically stabilized derivatives of shattercane often infest fields of grain sorghum and occur widely in farmers' fields in India and the highlands of Ethiopia (8).

8.3 WEEDY SORGHUMS IN AGROECOSYSTEMS

In general, weeds are harmful in most agroecosystems (43). *S. halepense* is reported as among the world's 10 worst weed pests (23). *S. halepense* is an aggressive perennial grass described as a serious weed from the Mediterranean through the Middle East to India, Australia and the nearby islands, central South America, and the Gulf coast of the U.S. It has been reported as a troublesome weed of numerous other important crops. *S. halepense* was introduced into the U.S. early in the 19th century as a fodder crop before eventually becoming a pest, known in the Americas as johnsongrass (30). *S. halepense* is now classified as a noxious weed in 22 states in the U.S. and 1 province in Canada (http://invader.dbs.umt.edu). It is wind-pollinated, often found sympatric with grain sorghum virtually everywhere the crop is cultivated, and its flowering time often overlaps that of grain sorghum making genetic exchange likely (23). *Sorghum almum* (Parodi), also called Columbus grass, does not present as much of a threat to agricultural systems as *S. halepense,* although it is widespread (10). This perennial weed is classified as noxious in 7 U.S. states. Weedy *S. bicolor* (shattercane) is an aggressive annual weed, which is well equipped with survival traits that allow it to readily out compete its cultivated progenitor. The shattering ability of the panicles of this weedy intermediate facilitates dispersal of seeds and increases the likelihood for its survival in crop fields. Unlike *S. halepense,* however, shattercanes lack rhizomes and are annual weeds. In highly mechanized agriculture, annual weeds are generally not hand pulled and thus their spread through seed dispersal is not controlled when they are able to overwinter.

Weedy sorghums are also detrimental to agriculture not only in terms of direct cost due to crop losses and chemical treatments, but also because they can serve as alternate hosts for pests and viruses that harm crops. For instance, sorghum ergot caused by *Claviceps africana* greatly reduces quality of grain in infected fields. Ergot can be a serious disease of seed parents in hybrid sorghum seed production fields, and the presence of ergot can, therefore, impact international seed markets. *C. africana*, a specialized fungus that parasitizes only the flowers of specific grasses, survives in the conidial stage on feral sorghums and alternate hosts such as *S. halepense* (1). Furthermore, once established, *C. africana* could become endemic on *S. halepense,* thus enhancing the risk. *S. halepense* is also a host for *Colletotrichum graminicola*, the anthracnose fungus. Isolates collected from *S. halepense* are highly pathogenic to sorghum (33).

8.4 GENE FLOW AMONG SORGHUMS

The natural outcrossing rate is variable among *Sorghum bicolor* strains and varieties. Although preferentially self-pollinated, outcrossing rates among sorghums can reach 26% for a grain-type sorghum with compact panicle typical of commercial hybrids and 61% for an open grasslike panicle such as sudangrass (*S. sudanensis*) (Piper) Stapf (36). Gene flow naturally exists between individuals that belong to different sorghum species and within or between gene pools. The most widely recognized interspecies sorghum weed, shattercane, arises from the hybridization of ssp. *bicolor* and all wild relatives (20). Though not widely adopted, some also describe shattercanes as hybrids between sorghum crop and *S. halepense* (32). Outcrossing can also involve individuals with different ploidy levels. *Sorghum almum* is a tetraploid grass that arose from a natural cross between the tetraploid *S. halepense* and the diploid *S. bicolor* (10). The same weed is also described as an allopolyploid perennial weed resulting from the cross between *S. propinquum* and *S. bicolor* (14). Hybrids between grain sorghum ($2n = 20$) and *S. halepense* ($2n = 40$) can include highly sterile 30-chromosome and relatively fertile 40-chromosome types (19).

8.4.1 WEED-TO-CROP GENE FLOW

The flow of genes from wild and weedy types to landraces and improved crop cultivars is hard to detect in agricultural systems. However, landrace sorghum varieties maintain superiority over most improved sorghum cultivars in their arsenal of survival traits that may have been introgressed in them through time from their wild and weedy kin. Adaptation to harsh environment and competitiveness in agroecosystems are often important features of weeds, and they appear to be true for weedy sorghums as well. The prevalence of such traits in wild and weedy types has drawn some attention in modern plant breeding to their genetic potential as sources of agronomically useful genes for the improvement of the sorghum crop. Several studies were undertaken in recent years to search for genes of interest among wild and weedy types of sorghum.

A high level of allelopathic activity was found in root exudates of *S. halepense* (7). Among traits of agronomic importance in sorghum, resistance to a highly virulent biotype E of the greenbug (*Schizaphis graminum*) was found in *S. halepense* (12). Resistance to ergot was found in the weedy type *S. b. drummondii* (37). The broad adaptation and perennial growth habit of *S. halepense* was reported as a source of valuable genes for improving sorghum (39). We have recently reported a unique source of resistance in *S. b. drummondii* to the parasitic weed *Striga,* which plagues African sorghum (38). Wild sorghums are also a significant source of genes for crop improvement providing resistance sources to various biotic and abiotic stresses in the cultivated sorghum germplasm. The collection of wild sorghum germplasm maintained at International Crops Research Institute for the Semi-Arid Tropics (ICRISAT) (461 accessions) has been evaluated for resistance to various pathogens and pests (25,26). Wild sorghums with resistance to downy mildew, stem borer, shoot fly, and midge were isolated. Furthermore, most of the greenbug (Biotype C)-resistant sorghum hybrids currently grown in the U.S. were originally derived from a wild sorghum source that belonged to race virgatum (12). Resistance to ergot was also found in several accessions of wild sorghum belonging to the *S. arundinaceum* ssp. (37).

Because few, if any, organized studies have been undertaken on assessing weed-to-crop gene flow in sorghum, there is insufficient information on the rate and the extent of genetic exchange *in situ.* It is probable, based on the ease of cross-pollination and the overlap in natural habitats, that there is a continual transfer of an array of fitness genes from weedy types to cultivated sorghum.

8.4.2 Crop-to-Weed Gene Flow

Natural gene flow between cultivated and wild and weedy sorghums in areas where they are sympatric has also led to gene exchange between the cultivated crop and wild relatives. Several studies (3,39) showed that under natural conditions, crop-to-weed gene exchange is likely in sorghum. Success in moving genes between crop and wild relatives depends on several factors including crossability, spontaneous hybridization, fertility, and fitness of the resultant hybrids. Potential hybridization of cultivated sorghum (*S. bicolor*) with sudangrass (*S. sudanense*) and its feral relatives (*S. almum* and *S. halepense*) was assessed for three congeners commonly growing in natural habitats near sorghum fields (2). However, although the potential for gene flow among this group of plants was recognized to be high, no deliberate study has been carried out, except for *S. halepense,* to characterize the extent of crossing and nature of hybrid progenies among these weedy species.

Sangduen and Hanna (39) conducted chromosome and fertility studies on reciprocal crosses between tetraploid *S. bicolor* (autotetraploid induced by chemical treatment) and *S. halepense.* A relatively higher (71 to 83%) outcrossing rate was reported when *S. halepense* was the female parent compared to only 0 to 33% hybrids produced when the tetraploid *S. bicolor* was used as the seed parent. The study also showed that when *S. halepense* was used as the female parent, the average seed set on both selfed and open pollinated panicles were similar and high (from 82 to 95%). In contrast, when *S. bicolor* was used as the female parent, the average seed set was only 18% on selfed panicles, and as high as 74% on open pollinated panicles. Arriola and Ellstrand (3) devised an experiment to measure the incidence and rate of hybrid formation in the field. Their study revealed that hybrid seeds could be detected on panicles of *S. halepense* plants at distances up to 100 m from the crop. Regardless of apparent variability among fields, years, and distance, hybrid formation occurred at a range of 0 to 12% overall. Arriola and Ellstrand (4) conducted a fitness study of crop-weed hybrids of sorghum grown under field conditions to estimate the likelihood of persistence of hybrids in the wild. From their experiments, crop-weed hybrids were determined to be as fit as either of the parent plants for sexual characters as well as vegetative traits. However, the authors tempered their results in suggesting an evaluation of hybrid fitness in wild conditions, without added irrigation. Fitness of hybrids obtained from crosses between grain sorghum (*S. bicolor*) and *S. halepense* was evaluated for potential propagation of the weed-gene in subsequent generations (19). The results showed that the highly sterile 30-chromosome hybrids have extremely vigorous rhizomes, and present a threat through vegetative reproduction. Conversely, the 40-chromosome hybrids, whether self-fertile or male-sterile, were found to be weakly rhizomatous and did not constitute a threat other than being a source of grassy weed seedlings.

8.4.3 Consequences of Recurrent Gene Flow

8.4.3.1 In Conventional Agriculture

As briefly described earlier, distant relatives of sorghum, both wild and weedy types, can be valuable sources of important genes for crop improvement. Wild and weedy species can also serve as a threat to modern agriculture. Agroecosystems can be severely affected by unmonitored gene flow in which genes from landraces and elite germplasm are moved into the wild relative of the crop. The prevalence of such a gene exchange in sorghum and its potential effect on increasing the evolution of feral forms has only been studied in association with johnsongrass. However, intermediate forms, such as the shattercanes, readily appear in the agroecosystems in much of Africa where sorghum cultivation is practiced without sufficient isolation from wild forms and in sympatry with weedy relatives. Progenies from an interspecific cross between sorghum and *S. propinquum* were genotyped with molecular markers and several quantitative trait loci associated to weediness were discovered (34). In the Americas, *S. halepense* has introgressed with grain sorghum to produce the widely distributed noxious weed. Derivatives of such introgression were described in Argentina as *S. almum*, a rhizomatous tetraploid grass that appears as a weed of sorghum. Enhanced weediness due to genetic exchange between wild and cultivated types is also illustrated by the ubiquitous and well-defined stable intermediates, the shattercanes, particularly in Africa (20).

Several cases, among which *S. almum* was also listed, have been reported where invasive taxa have evolved after intertaxon hybridization (14). In some cases, such as in *S. almum*, invasiveness could possibly be a result of the fitness benefits conferred by heterosis. In evaluating the potential risk of genetic introgression into weedy types, therefore, fitness data are critical and need to be determined.

The other consequence of recurrent gene flow from crops to wild relatives is demographic swamping, which can occur following two different paths. A case of so-called “outbreeding depression” from detrimental gene flow can lead to reduced fitness of a locally rare species and eventually to its extinction (13). It is expected that domestication genes transferred into a weed might cause reduction in fitness as they might decrease the potential for weediness and lead to maladaptation (41). The other cause leading to demographic swamping is when a locally rare species loses its genetic integrity and becomes assimilated into a locally common species as a result of repeated bouts of hybridization (13,27). Domesticated species have been implicated in the extinction or increased risk of extinction of wild species for 2 of the world’s 13 most important crops (13). The risk of extinction by hybridization depends on patterns of mating, and it is expected that the time to extinction by outbreeding depression or swamping will double if assortative mating is imposed more frequently over random mating. In sorghum, there may have been cases of genetic erosion, but only as a result of habitat change, despite cultivated sorghum growing adjacent to wild species, suggesting that genetic swamping is unlikely.

We believe that consequences of recurrent gene flow between cultivated sorghum and its wild relatives need not be generalized, but should be considered on a case-by-case basis and in individual countries where the crop is grown. Sorghum offers an excellent example of the sympatric association and interaction of a crop-wild-weed complex of a species in an agroecosystem. The nature of genetic interaction among forms of the taxa and the consequences of these exchanges depend not only on the power of the genes involved but also on several other associated factors. Prevalence of wild relatives naturally varies from region to region based on extent of inherent genetic diversity, existence of selective pressures, and farming systems in the region. Based primarily on knowledge of the agroecosystems and the history of genetic resource management and conservation, we have selected Ethiopia, Sudan, and the U.S. to demonstrate how different consequences may arise from gene flow between wild and cultivated sorghums. Although the empirical evidence we describe here is circumstantial and not experimental, it offers some basis for recognizing the differential effects of gene flow in contrasting ecological, demographic, and farming practices.

Sudan and Ethiopia are the birthplaces of the crop and have witnessed the evolution of wild and primitive forms of sorghum. The U.S. is the largest producer of sorghum and is likely to be one of the first places transgenic sorghums may be grown in commercial agriculture. Even though wild sorghums are not widely present in the Americas, the introduced weed *S. halepense* has been reported in more than half the states of the U.S. and has been described as a severe noxious weed in several states.

8.4.3.2 Introgression Disrupted by Genetic Erosion

Ethiopian sorghums are cultivated over a large range of environments from 400 to 2400 m above sea level. Landraces have been selected for their specific adaptation and use for each of 18 major agroecological zones that characterize the country. Obviously, genetic exchange and interactions between wild and cultivated sorghum have taken place in Ethiopia. The recurrent gene flow that has existed for thousands of years, most probably acted upon by natural and deliberate selection, accounts for the present makeup of highly adapted and diverse forms of sorghum landraces in Ethiopia. Growing population pressure in recent years, recurrent drought, political strife, and persistent famine have forced farmers to practice extensive agriculture and to cultivate even the most marginal of land areas that formerly served as the natural habitat for wild and weedy species. Consequently, wild plant populations of sorghum have become a rarity in some regions of the sorghum growing areas of Ethiopia. As a result, few populations and patches of wild sorghums can be found. The prevalence of wild sorghums has been so severely reduced that the apparent potential gene flow between cultivated and wild sorghum has been significantly diminished. Genetic erosion in diversity of Ethiopian sorghums is apparent in both the wild and cultivated forms, but is due to habitat loss and not to genetic swamping. In patches where wild and weedy types are seen, hybrid forms are found that represent intermediate types with favorable traits from repeated events of hybridization that have taken place since the beginning of domestication of the crop.

8.4.3.3 *In situ* Introgression in a Natural Habitat

The scenario in the Sudan is different. Sudan has the largest land area in Africa, yet the population base is low, only a third of the population in Ethiopia. Per capita holdings of arable land are higher in the Sudan than all other African countries. In northern Sudan, where human settlement has been historically light, the genetic identity of wild sorghums may have been further protected by their isolation from human disturbance. In the central clay plains of the country where sorghum farming is practiced under irrigation and in rotation with other crops, wild sorghums have also survived as weeds in cotton and wheat fields and along irrigation ditches (20). In both the rain fed and irrigated Sudanese agriculture, genetic exchange between sorghum and its wild relatives has resulted in formation of two widely recognized forms of crop-wild hybrids. Aggressive forms of weedy *S. bicolor* have evolved that are readily identified and recognized by most everyone as feral weeds and known under a local name, "adar." This form of shattercane is widely distributed and almost accepted as unavoidable. In spite of continual weeding and selective rouging, this weedy *S. bicolor* has not been easy to eradicate in Sudan. The second form of intermediate is equally feral, but appears more similar to cultivated sorghum and produces grains that only slowly shatter. Continued introgression of cultivated sorghum genes into wild forms has resulted in this hybrid form called "kerketita." Farmers selectively harvest these types and encourage their continued existence, for they rely on them as feed and food depending on the harvest prospect. In bad years, these fast growing intermediates provide the only harvest possible, particularly for fodder. We have recently started a study in Sudan to investigate the extent of gene introgression among these types and hope to assess the differential fitness of these introgressed intermediates in contrast to the cultivated and wild progenitors.

8.4.3.4 Introgression Enhanced by Modern Farming Practices

Sorghum and maize production in the U.S. has been severely affected by the widely recognized, highly competitive, and noxious *S. halepense*. Mechanization favors perennial weed propagation as it stimulates rhizome growth. Because mechanized farming is practiced, farm implements used in plowing and cultivation in the U.S. crop fields resulted in fortified rhizomes that spread readily and enhance the aggressiveness of the perennial weed. Aggressiveness of weeds can further be enhanced by the introgression of genes from highly improved cultivars as long as these genes confer an advantage. Continuous crossing, however low the frequency, also generates hybrid progeny that are ever more vigorous. These resultant hybrids possess additional fitness benefits conferred on to them via heterosis.

A successful transfer of fitness from cultivated sorghum into weedy types is best illustrated in the creation of a now famous forage crop, sudangrass (*Sorghum sudanense*). This forage grass resulted from the introgression of cultivated sorghum genes into the weedy germplasm of *Sorghum sudanense*. It probably evolved during the millennia of genetic exchange among sorghum species in the Sudan. It combines excellent attributes of cultivated and wild sorghums that allow effective regeneration of seeds as well as fast crop development and regrowth after repeated cuts as a fodder. Sorghum–sudan forages are the most widely commercialized forage crop in the world, perhaps next to alfalfa.

8.4.4 Gene Flow in the Transgenic Era

The expectation of reduced fitness brought about by domestication genes transferred to wild recipient genome may not hold for transgenic traits that confer herbicide resistance or other fitness enhancing traits. In most cases, transgenes added to highly domesticated crops are not anticipated to survive in nature without human intervention, as shown for potato, oilseed rape, maize, and sugar beet (6). However, partially domesticated crops may constitute an exception for the dependency on humans. Assessment of the survival of transgenes in nature requires considering the impact of the performance of transgenic plants under field conditions, that is, tolerance to abiotic stresses or pest resistance, as well as the feral tendency of crops. When transgenic traits are not expected to increase plant fitness in natural habitats, such as herbicide resistance, the risk of its persistence in the wild is lessened. However, when transgenes are transferred into forage crops to confer herbicide resistance, increased weediness may occur by reducing the opportunities to control feral populations (16).

Many domestication genes are presumed to represent a loss rather than a gain of function, as indicated by their recessiveness. Exception is found in sorghum where action of *S. propinquum* allele for reduced seed size were mainly (67% of the alleles) recessive to the corresponding *S. bicolor* alleles (35). For a species with a relatively high outcrossing rate such as sorghum, dominant mutations may have been easily selected during the domestication process, mostly for a trait such as seed size, which is under strong selective pressure. In contrast to domestication genes favored in agroecosystems, transgenes frequently represent gains of function that might release wild relatives from constraints that limit their fitness, but often make crops less adapted to natural ecosystems (17).

Whether gene flow from a transgenic crop to its wild relative can lead to transgene escape and long-term survival, implies asking whether the transgene will effectively introgress and add to fitness of the wild population. The question is broader than the elements we addressed in this review. Undoubtedly, gene flow exists between cropped sorghum and its wild relatives, and transfer of transgenes from a genetically engineered sorghum is likely. The prevalence of wild relatives near the transgenic crop, as would be true in centers of crop diversity, significantly increases this likelihood. For a gene to introgress into a wild relative, it is necessary that repeated backcrosses and stabilization of the transgene occur in the new host. The likelihood of transgene introgression into a wild relative will depend on its dominance, absence of association with deleterious crop alleles or traits, and location in the genome. The speed with which the transgene introgresses and stabilizes in the population will depend on selective pressure and population size. Furthermore, the nature of the allele that is introgressed and whether or not the function gained because of the introgession contributes to fitness and ferality are key considerations.

Important questions remain to be considered to assess the risk of transgene introgression into the wild gene pool. What is the actual magnitude of gene flow (both through pollen and seed) in different agroecosystems? Will the interactive effects of the introduced alleles and the management practices in the agroecosystems force selection of aggressive hybrids? How do transgenes behave (expression and stability) in wild genetic backgrounds? Will transgenes be preferentially selected in the wild populations and will they persist? These and perhaps many other similar questions are currently under investigation in several laboratories, including our own. The lessons we learn should offer empirical evidence on the degree of gene flow within and among related taxa as well as what can be devised to mitigate serious consequences that may arise from the evolution of feral forms.

LITERATURE CITED

1. Alderman S, Frederickson D, Milbrath G, Montes N, Narro-Sanchez J, Obdvody G. 1999. A laboratory guide to the identification of *Claviceps purpurea* and *Claviceps africana* in grass and sorghum seed samples. www.oregon.gov./ODA/PLANT/docs/pdf/Ergotmanual2.pdf.
2. Arriola PE. 2002. *Gene flow and hybrid fitness in the Sorghum bicolor — Sorghum halepense complex.* Presented at Gene Flow Workshop, Ohio State University, March 5 and 6, 2002. http://www.bio-sci.ohio-state.edu/~asnowlab/Proceedings.pdf.
3. Arriola PE, Ellstrand NC. 1996. Crop-to-weed gene flow in the genus Sorghum (Poaceae): spontaneous interspecific hybridization between johnsongrass, *Sorghum halepense*, and crop sorghum, *S. bicolor. Am. J. Bot.* 83:1153–1160.
4. Arriola PE, Ellstrand NC. 1997. Fitness of interspecific hybrids in the genus *Sorghum*: persistence of crop genes in wild population. *Ecol. Appl.* 7:512–518.
5. Casas A, Kononowicz A, Zehr U, Tomes D, Axtell J, et al. 1993. Transgenic sorghum plants via microprojectile bombardment. *Proc. Natl. Acad. Sci. USA* 90:11212–11216.
6. Crawley MJ, Brown SL, Hails RS, Kohn DD, Ress M. 2001. Transgenic crops in natural habitats. *Nature* 409:682–683.
7. Czarnota MA, Rimando AM, Weston LA. 2003. Evaluation of root exudates of seven sorghum accessions. *J. Chem. Ecol.* 29:2073–2083.
8. de Wet JMJ. 1978. Systematics and evolution of sorghum sect. Sorghum (Gramineae). *Am. J. Bot.* 65:477–484.
9. de Wet JMJ, Huckabay JP. 1967. The origin of *Sorghum bicolor*. II. Distribution and domestication. *Evolution* 21:787–802.
10. Doggett H. 1988. *Sorghum*. New York:Wiley & Sons. 403 pp.
11. Doggett H, Prasada Rao KE. 1995. Sorghum. In *Evolution of crop plants*, Smartt J, Simmonds NW, Eds., pp. 173–180. New York: Longman.

12. Duncan RR, Bramel-Cox PJ, Miller FR. 1991. Contributions of introduced germplasm to hybrid development in the USA. In *Use of plant introductions in cultivar development*, Shands HL, Weisner L, Eds., pp. 69–102. CSSA Special Publication no. 17.
13. Ellstrand NC, Prentice HC, Hancock JF. 1999. Gene flow and introgression from domesticated plants into their wild relatives. *Ann. Rev. Ecol. Syst.* 30:539–563.
14. Ellstrand NC, Schierenbeck KA. 2000. Hybridization as a stimulus for the evolution of invasiveness in plants? *Proc. Natl. Acad. Sci. USA* 97:7043–7050.
15. FAOSTAT. 2003. http://faostat.fao.org.
16. Gepts P. 2002. A comparison between crop domestication, classical plant breeding, and genetic engineering. *Crop Sci.* 42:1780–1790.
17. Gepts P, Papa R. 2003. Possible effects of (trans)gene flow from crops on the genetic diversity from landraces and wild relatives. *Environ. Biosafety Res.* 2:89–103.
18. Grenier C, Bramel PJ, Dahlberg JA, El-Ahmadi A, Mahmoud M, et al. 2004. Sorghums of the Sudan: analysis of regional diversity and distribution. *Genet. Resour. Crop Evol.* 51:489–500.
19. Hadley HH. 1958. Chromosome numbers, fertility and rhizome expression of hybrids between grain sorghum and johnsongrass. *J. Agron.* 50:278–282.
20. Harlan JR. 1992. *Crops and man*. Madison, WI: American Society of Agronomy. 284 pp.
21. Harlan JR, de Wet JMJ. 1971. Toward a relational classification of cultivated plants. *Taxon* 20:509–517.
22. Harlan JR, de Wet JMJ. 1972. A simplified classification of cultivated sorghum. *Crop Sci.* 12:172–177.
23. Holm LG, Plucknett DL, Pancho JV, Herberger JP. 1977. *The world's worst weeds, distribution and biology*. Honolulu: University Press of Hawaii. pp. 54–61.
24. Jeoung JM, Krishnaveni S, Muthukrishnan S, Trich HN, Liang GH. 2002. Optimization of sorghum transformation parameters using genes for green fluorescent protein and β-glucuronidase as visual markers. *Hereditas* 137:20–28.
25. Kamala V, Singh SD, Bramel PJ, Rao DM. 2002. Sources of resistance to downy mildew in wild and weedy sorghums. *Crop Sci.* 42:1357–1360.
26. Kameswara Rao N, Reddy LJ, Bramel PJ. 2003. Potential of wild species for genetic enhancement of some semi-arid food crops. *Genet. Resour. Crop Evol.* 50:707–721.
27. Levin DA, Francisco-Ortega J, Jansen RK. 1996. Hybridization and the extinction of rare plant species. *Conserv. Biol.* 10:10–16.
28. Mann JA, Kimber CT, Miller FR. 1983. The origin and early cultivation of sorghums in Africa. *Tex. Agric. Exp. Stn. Bull.* 1454. 21 pp.
29. McWhorter CG. 1971. Growth and development of johnsongrass ecotypes. *Weed Sci.* 19:141–146.
30. McWhorter CG. 1971. Introduction and spread of johnsongrass in the United States. *Weed Sci.* 19:496–500.
31. Murty BR, Arunachalam V, Saxena MBL. 1967. Classification and catalogue of a world collection of sorghum. *Indian J. Genet. Plant Breed.* 27:1–394.
32. OTA. 1993. *Harmful non-indigenous species in the United States*. Washington, D.C.: Office of Technology Assessment, U.S. Government Printing Office.
33. Pastor-Corrales MA, Frederiksen RA. 1978. *Sorghum anthracnose*. Presented at International Workshop on Sorghum diseases — A world review. Hyderabad, India: ICRISAT.
34. Paterson A, Schertz K, Lin Y, Liu S, Chang Y. 1995. The weediness of wild plants: molecular analysis of genes influencing dispersal and persistence of johnsongrass, *Sorghum halepense* (L.) Pers. *Proc. Natl. Acad. Sci. USA* 92:6127–6131.
35. Paterson A, Schertz K, Lin Y, Li Z. 1998. Case history in plant domestication: sorghum, an example of cereal evolution. In *Molecular dissection of complex traits*, Paterson A, Ed., pp. 187–195. Boca Raton, FL: CRC Press.
36. Pedersen J, Toy J, Johnson B. 1998. Natural outcrossing of sorghum and sudangrass in the central great plains. *Crop Sci.* 38:937–939.
37. Reed JD, Ramundo BA, Claflin LE, Tuinstra MR. 2002. Analysis of resistance to ergot in sorghum and potential alternate hosts. *Crop Sci.* 42:1135–1138.
38. Rich PJ, Grenier C, Ejeta G. 2004. Striga resistance in wild relative of sorghum. *Crop Sci.* 44: 2221–2229.
39. Sangduen N, Hanna WW. 1984. Chromosome and fertility studies on reciprocal crosses between two species of autotetraploid sorghum. *J. Hered.* 75:293–296.

40. Snowden JD. 1936. *The cultivated races of sorghum*. London: Adlard & Son, Ltd.
41. Stewart CN, Halfhill MD, Warwick SI. 2003. Transgene introgression from genetically modified crops to their wild relatives. *Nature Rev. Genet.* 4:806–817.
42. Vavilov NI. 1951. The origin, variation, immunity and breeding of cultivated plants. *Chron. Bot.* 13:1–366.
43. Westbrooks RG. 1998. *Invasive plants: changing the landscape of America: fact book.* Federal Interagency Committee for the Management of Noxious and Exotic Weeds, Washington, D.C. 109 pp.
44. Zhao Z-Y, Cai T, Tagliani L, Miller M, Wang N, et al. 2000. *Agrobacterium*-mediated sorghum transformation. *Plant Mol. Biol.* 44:789–798.

9 Multidirectional Gene Flow among Wild, Weedy, and Cultivated Soybeans

Bao-Rong Lu

9.1 INTRODUCTION

The cultivated soybean (also referred to as soy or soyabean), *Glycine max* (L.) Merrill, is an important food crop grown widely in the world on about 76 million ha (22). The major producers of soybeans are the U.S. (annual production of more than 74 million tons), Brazil (51 million tons), Argentina (35 million tons), and China (16 million tons), accounting for more than 90% of the total world soybean production. Soybean has a multitude of uses, but mainly edible oil and high-quality proteins for human consumption and animal feed. Fractions and derivatives of soybean seeds have substantial economic significance in a wide range of industrial (e.g., yeast substrates, varnish, and soap), food, pharmaceutical, and agricultural products (35,40). Soybean food products are popular and are consumed in daily life in Asian countries, not only for its food use, but also for its medicinal values. There is an increasing knowledge that soybean-rich diets provide significant health benefits. Population studies have shown that people living in Asian countries have a lower incidence of menopausal symptoms, osteoporosis, heart disease, and certain cancers (breast and prostate). In the other parts of the world, besides main food use of soybean as oil and hydrogenated fat, it is used in various products, including artificial milk and meat products.

With the rapid development of transgenic biotechnology and increasing demands of soybean production, a large number of transgenic soybean varieties carrying different herbicide-resistant genes (e.g., resistant to glyphosate and glufosinate), insect-resistant genes (e.g., *Bt* gene), quality traits (e.g., high oleic acid soybeans), and genes for suppressing allergens for infant formulae have been produced. Some of these varieties have already entered commercial markets or are released to the environment for field testing. The International Service for the Acquisition of Agribiotech Applications estimates the global transgenic soybean cultivation area has reached 42 million ha since 2003, which accounts for more than 55% of the global area of soybean and 61% of the global area of total transgenic crops (22). The massive cultivation of transgenic crops, including transgenic soybeans, has aroused concern of environmental biosafety worldwide. Among these perceived concerns, transgene escape and its potential ecological consequences have attracted the attention of the international community, particularly the issue of possible spread of transgenic crops to different agricultural and natural habitats becoming weeds through volunteerism and ferality, and potential consequences of unintended transgene escape in the centers of crop origin and species diversification.

The concerns and consequences of transgene movement to wild relatives mainly include the potential change in fitness of weedy and wild relatives and the concomitant risk of increased weediness, the loss of herbicide resistance as a tool to protect the crop from closely related weeds, and the effects on the genetic identity and diversity of sexually compatible relatives (12). The controversial debate on the massive import of transgenic herbicide-resistant soybean (with an average of 15 million tons per annum) and the possible release of transgenic soybean in China in the future is a good example of such concerns. Will the release of transgenic soybean cause any ecological disturbance in the environment? Will transgenic soybean get away from fields into the wild habitats and become agricultural weeds through volunteerism or ferality? Will transgene spread to weedy or wild soybeans and cause severe weed problems from the wild species or hybrids with them? Will the spread of transgenes to soybean and its weedy types increase the ferality of this crop? Will the release of transgenic soybean into its center of origin have an impact on the diversity of wild relatives of soybean?

9.2 SOYBEAN AND ITS WEEDY AND WILD RELATIVES

Transgene movement is essentially associated with gene flow. Therefore, the first step in the assessment of transgene escape and its potential consequences is to determine which weedy and wild species exist near the cultivated species and can hybridize with the crop through gene flow to produce fertile offspring. If there is no wild relative that can sexually cross with the crop and produce viable offspring, there will be no scientific concern of transgene escape through gene flow. This is particularly true for a highly self-pollinated crop such as soybean, because the opportunity of horizontal gene transfer between soybean and unrelated species is extremely low, if any. Knowledge of the general biology of soybean and its weedy or wild relatives will help to understand the possibility of transgene flow and its potential ecological consequences and to address the above questions.

The cultivated soybean, *G. max*, is a diploidized tetraploid (2n = 40) species. This crop is an erect and bushy herbaceous annual about 1 m above ground level (up to 2 to 3 m) and 2 m below ground level. The nodulated root system is intermediate between a taproot type and a diffuse type. The foliage leaves are alternate, pinnately trifoliolate, with pulvini, stipels, and stipules. The soybean flower is a standard papilionaceous flower with calyx of 5 united sepals; zygomorphic corolla of carina, alae, and vexillum; androecium of 10 diadelphous (9 + 1) stamens; and gynoecium of a single carpel. Two to four seeds develop in the pods. The seeds have 2 large cotyledons and scant endosperm. The anthers mature in the bud and shed their pollen directly onto the stigma of the same flower, thus ensuring a high degree of self-pollination. Still, cross-pollination is usually less than 1%. Soybean plants are thus virtually pure breeding homozygous lines, although manual cross-pollination is practiced routinely in breeding programs.

Cultivated soybean does not usually form volunteers in agricultural habitats in China because of the intensive agricultural practices and prevailing manual planting, harvesting, and weed control systems. Occasionally, there are a few escaped soybean plants outside of farmers' fields, but farmers or children collecting grasses for feed eventually remove them.

Taxonomically, *G. max* is included in the legume family (*Fabaceae*), subfamily *Faboideae*, tribe *Phaseoleae*, subtribe *Glycininae*, and the genus *Glycine* Willd. The subtribe to which soybean belongs contains about 16 genera, but none of these, except for soybean (*Glycine*) and kudzu (*Pueraria*), are commonly known outside of botanical science. The genus *Glycine* is unique within the subtribe for several morphological and chromosomal characters, and does not seem to bear an intimate relationship with any of the other genus in the subtribe (26). No fertile hybridization events were achieved between soybean and species in related genera of the same tribe (including kudzu), using all the genetic methods of the breeders to facilitate such crosses (such as embryo rescue), so that useful traits could be moved into soybean.

Based on the taxonomic treatment by Hymowitz et al. (21) and Shimamoto (32), the genus *Glycine* is divided into two distinct subgenera (Table 9.1). These are subgenus *Glycine* Willd. and subgenus *Soja* (Moench) F. J. Herm. Confusingly, soybean (*Glycine max*) is in subgenus *Soja*. The subgenus *Glycine* currently contains 16 perennial species, primarily indigenous to Australia, although some species are also found on the South Pacific Islands. The subgenus *Soja* contains three annual species, *G. max*, *G. soja*, and *G. gracilis*, originally from eastern Asia. The wild *G. soja* is an herbaceous annual that spreads vinelike with many branches densely covered with hairs,

TABLE 9.1
Taxonomy, Cytogenetics, and Geographic Distribution of the Genus *Glycine*

Subgenus Species	Habit	2n =	Genome	Geographic Distribution
			***Glycine* Willd.**	
G. albicans Trin. *et* Craven	Perennial	40	II	Australia
G. arenaria Tind.	Perennial	40	HH	Australia
G. argyrea Tind.	Perennial	40	A_2A_2	Australia
G. canescens F. J. Herm.	Perennial	40	AA	Australia
G. clandestina Wendl.	Perennial	40	A_1A_1	Australia, South Pacific Islands
G. curvata Tind.	Perennial	40	C_1C_1	Australia
G. cyrtoloba Tind.	Perennial	40	CC	Australia
G. falcata Benth.	Perennial	40	FF	Australia
G. hirticaulis Tind. *et* Craven	Perennial	40, 80	H_1H_1	Australia
G. lactovirens Tind. *et* Craven	Perennial	40	I_1I_1	Australia
G. latifolia (Benth.) Newell *et* Hymowitz	Perennial	40	B_1B_1	Australia
G. latrobeana (Meissn.) Benth.	Perennial	40	A_3A_3	Australia
G. microphylla (Benth.) Tind.	Perennial	40	BB	Australia
G. pindanica Tind. *et* Craven	Perennial	40	H_2H_2	Australia
G. tabacina (Labill.) Benth.	Perennial	40, 80	B_2B_2 $AAB_2B_2BBB_2B_2$	Australia; West, Central, and South Pacific Islands; Japan; China
G. tomentella Hayata	Perennial	40, 80	DD, EE DDEE, AADD	Australia, Papua New Guinea, Indonesia, Philippines, China
			***Soja* (Moench) F.J.Herm.**	
G. soja Sieb. *et* Zucc.	Annual	40	GG	China, Far East of Russia, Korea, Japan
G. gracilis Skvortzow	Annual	40	GG	China
G. max (L.) Merrill	Annual	40	GG	Cultigen (Worldwide)

Source: Modified based on Hymowitz and Singh (20) and Shimamoto (32).

climbing and entwining on itself. There are no predominant main stems and the vinelike stems usually climb over other objects or creep on the ground. The leaves are alternate, pinnately trifoliolate, ovate, oblong lanceolate, or narrow-lanceolate shaped, with pulvini, stipels, and stipules. The flower of *G. soja* is also a standard papilionaceous flower like the cultivated soybean, but with a great variation in size (usually smaller than that of cultivated soybean) and color (from dark purple to white). The anthers are believed to mature in the bud and shed their pollen directly onto the stigma of the same flower. One to four seeds (rarely five) develop in the pods. The pods explode as they mature, shattering the seeds. Usually, the seeds are black, or sometimes brown or brownish-yellow, flat-ellipse or long-ellipse shaped, with a 100-grain weight of 1 to 2.5 g, having yellow cotyledons.

Glycine soja is recognized as the immediate ancestor of the cultivated soybean, and *G. gracilis* (sometimes referred to as semi-wild soybean) is the weedy form of the soybean (26). *Glycine gracilis* probably originated exoferally by repeated spontaneous hybridization events between cultivated soybean and its wild ancestor where the two species have sympatric distribution. The hypothetical domestication procedure of the cultivated soybean from its wild ancestral *G. soja* is illustrated in Figure 9.1, along with the formation of the weedy *G. gracilis* through hybridization and introgression between *G. soja* and *G. max*, after the domestication of the cultivated soybean. These events reflect the especially close genetic relationships among these three soybean species.

The weedy form is usually found in soybean fields and disturbed semi-wild habitats, mostly in the northeastern parts and Yellow River region of China. Its morphology varies in a continuum

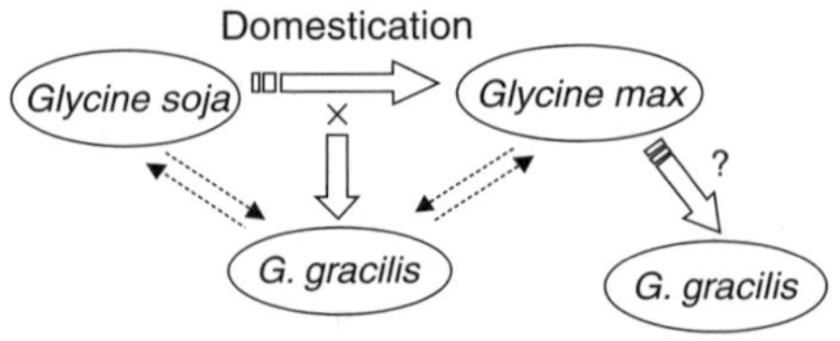

FIGURE 9.1 Hypothetical domestication procedure of cultivated soybean (*Glycine max*) from its wild ancestor *G. soja* and the possible pathways by which the weedy *G. gracilis* may have been formed.

between weedy *G. gracilis* and *G. soja*. Therefore, there are no good morphological characteristics that can be used to accurately distinguish between *G. gracilis* and *G. soja*, indicating significant introgression between the weedy and wild soybeans. However, 100-grain weight is artificially used by breeders and geneticists to differentiate between the wild and the weedy types. The grain weight of weedy soybean is considered to be heavier than the wild, although with considerable variation. In breeding and germplasm management practices, the 100-grain weight of *G. soja* is artificially determined as less than 2.5 g and that of weedy *G. gracilis* is between 2.6 to 5 g or even heavier. The morphological variation of weedy soybean found in different habitats is significant, with some phenotypic characteristics intermediate between *G. max* and *G. soja*, such as the size and shape of leaves, size of pods, and grain weight. The weedy types that occur in China in soybean fields are generally termed as "semi-cultivated soybean" and the weedy types that occur in the wild habitats are called "semi-wild soybean." This reflects the difficulty of accurate classification of the weedy types. The occurrence of weedy types of soybean seems to be associated with the occurrence of wild soybean (*G. soja*), from which the weedy forms originated. However, little is understood about the weedy *G. gracilis* in terms of its distribution patterns, seed dormancy, flowering habit, breeding system, and dispersion. Therefore, some breeders and taxonomists only recognize *G. soja* in a broad sense that includes the weedy types (40). More studies should be conducted regarding this aspect.

9.2.1 Distribution and Relationships of Soybean and Its Wild Relatives

Information on geographical distribution of soybean and its wild relatives is important for the estimation of transgene flow from soybean to its wild relatives. This is because spatial contact (sympatric distribution) is the prerequisite for regular gene exchange between soybean and its wild relatives, provided that a certain frequency of cross-pollination occurs.

Most species in the subgenus *Glycine* are distributed in Australia and South Pacific Islands (10,20,32) and only two species, that is, *G. tabacina* (A, B genomes) and *G. tomentella* (A, D, E genomes), are found in the Philippines, Miyako Island of Japan, and southern China (20,40).

Glycine soja is distributed mainly in China and its adjacent areas of the Far East region of Russia, Korea, and Japan (20,32,40) at the latitudes between approximately 24ϒ and 53ϒ N and longitudes between approximately 97ϒ and 143ϒ E (Figure 9.2). *Glycine gracilis* is usually found in areas where the cultivated soybean and its wild ancestor have a sympatric distribution. It was mostly reported from the northeast part of China (18). However, herbarium specimen collections indicate a wide range of geographical distribution. Nevertheless, weedy soybean has not been reported growing naturally outside its center of origin in other parts of the world such as the Americas and Europe where only the cultivated soybean is grown. This suggests an exoferal nature of weedy soybean.

The annual (subgenus *Soja*) and perennial (subgenus *Glycine*) species are rather distantly related (10). Attempts to hybridize between species of the subgenera *Soja* and *Glycine* were unsuccessful. In these studies, the pods resulting from interspecific hybridization eventually aborted although pod development could be initiated (2,16,27). Intersubgeneric hybrids of *G. max* *G. clandestina*, *G. max* *G. tomentella*, and *G. max* *G. canescens* were obtained *in vitro* either through embryo

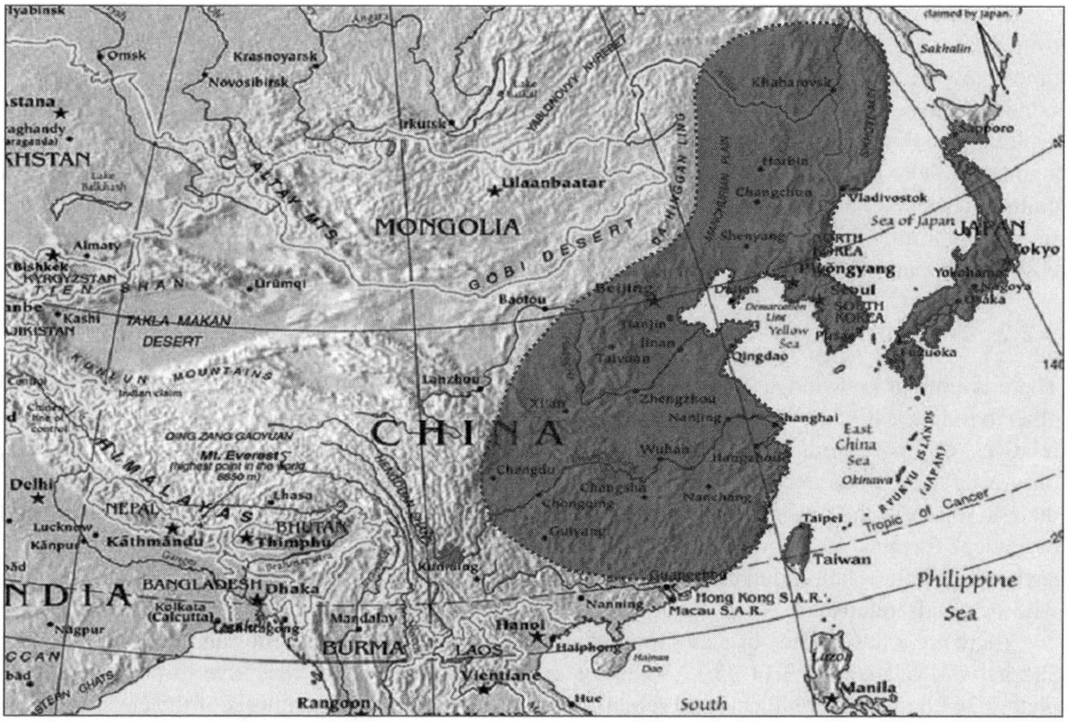

FIGURE 9.2 Geographic distribution of wild soybean (*Glycine soja*). Scattered populations are found throughout the area shaded on the map.

rescue (33,34) or use of transplanted endosperm as a nurse layer (6). Thus, soybean can only hybridize with members of the subgenus *Glycine* with extreme technical assistance. All the progeny of such inter-subgeneric hybrids were completely male and female sterile, and no backcrosses with genetic recombination were achieved. Transgene flow from soybean to wild species of the subgenus *Glycine* thus seems completely improbable under natural conditions.

In contrast, cultivated soybean can easily hybridize with *G. soja* and *G. gracilis*. Interspecific hybridization between *G. max* and *G. soja* (2,15), and between *G. max* and *G. gracilis* (24) produced fertile hybrids without much difficulty. Hybrids between *Glycine* species usually have normal meiosis (40). In fact, wild and weedy (semi-wild) soybeans are important genetic resources in the primary gene pool and have been extensively explored and used in soybean breeding in China through sexual interspecific hybridization. This indicates a significantly close genetic relationship among the three species. The utilization of wild and weedy soybeans in breeding became widespread since the beginning of the 1980s when more wild soybean germplasm was collected, characterized, and studied (40). To date, there are more than 6200 accessions of wild soybean germplasm being stored in the Chinese National Genebank. This genebank serves as a valuable genetic resource for soybean improvement. Wild and weedy soybeans were applied in breeding programs in China mainly:

- To improve quality of the cultivated soybean by offering germplasm with high protein content (>55%), high oil content (>9%), and high amino acid content
- To develop resistance to biotic stresses such as soybean aphid, soybean cyst nematode, soybean leaf mosaic virus, and other fungal diseases
- To increase tolerance to abiotic stresses such as drought and salt

In addition, utilization of soybean heterosis was actively attempted by searching for male sterility genes in wild soybean genetic resources (40).

The cultivated *G. max* was domesticated by farmers in China about 3000 to 4000 years ago from *G. soja*, and brought to the New World in the 1700s through Europe (19). It has had more time to evolve in China than in other parts of the world, to hybridize with its progenitor and its weedy relative, and to mutate. Such a limited period of diversification is insufficient, in terms of evolution, to develop a perfect reproductive isolation among the three taxa. In addition, gene flow among the cultivated and wild soybeans since the domestication of soybean has kept the three taxa united. Therefore, the possibility of transgene escape from transgenic soybean to *G. soja* and *G. gracilis* through gene flow could be comparatively high, with ecological consequences, if there is sufficient outcrossing between the cultivated and wild or weedy soybeans.

9.2.2 The Potential of Transgene Flow in Soybean Species

There is limited knowledge about the flowering habit, mating systems, population structures, and other baseline data that will allow prediction of gene flow from soybean to its weedy and wild relatives. *Glycine soja* is believed to be inbreeding with almost completely cleistogamous flowers, like its cultivated counterpart. Such information about weedy *G. gracilis* is not available. Nevertheless, to assess the probability of gene flow, the most important key factors are the fundamental biological features of the wild soybeans, understanding gene flow frequency between different soybean varieties, and understanding gene flow frequency between soybean and its interbreeding wild or weedy relatives.

There are a few studies of gene flow between different soybean varieties and between soybean and its wild relatives (1,3,11,25,29,30,39). The stigma of soybean is receptive to pollen approximately 24 hours before anthesis and remains receptive 48 hours after anthesis of the cleistogamous flowers. Many consider soybean to be an almost completely self-pollinated species in the field (7,28). Weber and Hanson (39) found the natural cross-pollination rate in adjacent rows of different soybean varieties to be between 0.5 and 1%, in a field survey. Caviness (8) reported that the outcrossing rate of soybean in field conditions is usually less than 1%. Ahrent and Caviness (3) as well as Gumisiriza and Rubaihayo (14) observed natural outcrossing rates as high as 2.5 and 4.5%, respectively. These data demonstrate that natural outcrossing rates fluctuate significantly among different soybean varieties and can be a function of the presence of pollinating insects. The natural cross-pollination rates in conventionally sown soybean were followed in the Mississippi Delta by Ray et al. (30) using different soybean varieties with dominant purple or recessive white flowers as genetic markers. An average of 1.8% natural cross-pollination rates ranging from 0.7 to 6.3% was recorded, suggesting the potential of significant within-crop gene flow in soybean. The results indicate that considerably higher outcrossing rates can be found in soybean varieties under particular environmental conditions such as the favorable climate for pollination and abundance of pollinators.

There are fewer studies relating to gene flow to wild and weedy soybeans. The natural outcrossing rates of the annual wild or weedy soybeans are generally expected to be low, but some genetic diversity studies of *G. soja* showed surprisingly high cross-pollination rates of this species. This is a determining factor for predicting transgene flow from cultivated soybean to the wild relative. Fujita et al. (11) analyzed the genetic structure of four natural wild soybean populations along the Omono River in Akita Prefecture in Japan by examining allozyme variation to evaluate the extent of natural cross-pollination rate in *G. soja*. They found that at some collection sites there was higher within-population genetic variation and low genetic divergence among populations than would be expected for an inbreeding species. Furthermore, they estimated the mean natural outcrossing rate by multilocus analysis to be 13%, ranging from 9.3 to 19% among the four populations. These values are much higher than outcrossing rates previously reported for both *G. soja* and *G. max*. Frequent visits by honeybees and carpenter bees to the wild soybean flowers were observed in this study, which may account for the high outcrossing rate.

From an earlier study of genetic structure of wild soybean (*G. soja*) populations in Iwate Prefecture of Japan, Kiang et al. (25) analyzed seeds systematically sampled from 4 natural populations growing along banks of the Kitakami River. Based on the scores of 15 enzymes and 1 non-enzyme protein representing 42 loci, they observed an average outcrossing rate of 2.3% among the wild soybean populations. In a fine scale genetic structure study of a wild *G. soja* population near Shanghai, we found a relatively high heterozygosity of a wild soybean population and occasional long-distance (60 m) pollen travel, as estimated by the intersimple sequence repeat analysis (23). Our unpublished data based on heterozygosity estimation by microsatellite SSR (simple sequence repeats) indicated about 4% outcrossing for a few natural *G. soja* populations in China.

There were also two reports on natural hybridization or introgression between wild (*G. soja*) and cultivated soybeans by actually measuring pollen-mediated gene flow to wild soybean in a designed population. Natural hybridization using molecular tools was examined (1,29). A maximum of 5.8% natural hybridization between cultivated and wild soybeans was reported (1). More studies on actual gene flow from soybean to its weedy and wild populations need to be conducted to more appropriately understand the possibilities of transgene escape to the weedy and wild soybean populations. The present data all demonstrate a high possibility of gene movement from soybeans to the wild relatives, which would be equally possible for transgene flow, if the transgenic soybean is grown in close contact with the wild relatives.

9.3 POSSIBLE CONSEQUENCES OF GENE FLOW FROM TRANSGENIC SOYBEAN

Gene flow between soybean and its wild or weedy relatives must have occurred since the commencement of farmers' domestication of the cultivated soybean. This gene flow led to the formation of weedy *G. gracilis*, in addition to the generation of rich varietal diversity of the cultivated soybean in and near its center of origin. The introduction of transgenes into cultivated soybean will lead to the movement of these genes among soybean and its wild to weedy relatives through gene flow and introgression. Some transgenes, for example, those with a natural selective advantage that can change the fitness of weedy or wild soybean individuals, may promote or reduce the wild or weedy populations, with possible ecological consequences.

The major consequences and concerns of transgene escape include the potential change in genetic diversity and ecological fitness of the weedy and wild relatives of soybean, in addition to other unexpected ecological impacts. The massive introgression of transgenes into wild soybeans could potentially alter genetic integrity or biodiversity of the wild soybean gene pool at the population level, particularly when these transgenes are subject to natural (e.g., insect) and human (e.g., herbicide) selection. There is a concern that the addition of transgenes could lead to a change in allelic frequencies through a selective advantage conferred by transgenes in wild populations (12). This would be particularly so if soybeans with multiple transgenes are extensively grown in the center of soybean origin. The successive and extensive outflow of transgenes from soybean to its wild relatives could lead to changes in biodiversity patterns (evenness and richness) of the wild soybeans. In fact, studies have demonstrated that the favored alleles could spread and fix faster than neutral alleles in a natural population (30). This phenomenon can also be evidenced in plant breeding procedures in which an agronomically advantageous gene (allele) or trait can rapidly increase its frequencies in a breeding population by human and natural selection. This also would have occurred if native genes of the Australian soybeans would have produced fertile hybrids with cultivated soybeans.

Some transgenes (with better fitness) may quickly disseminate when they spread to natural populations. If an allelic frequency change accumulates in wild soybean populations to a significant degree, or if common alleles in wild soybean populations are significantly replaced by transgenes through selection over time, the genetic integrity of wild soybean in the origin center might be changed gradually by transgenic wild soybean. The currently available transgenes, such as insect-resistant genes and herbicide-resistant genes, certainly have selective advantages in different habitats.

If these transgenes introgress to and persist in wild and weedy soybean populations, the wild and weedy individuals carrying the transgenes may become more fit (insect- and herbicide-resistant) and more invasive to out compete their wild and cultivated counterparts, leading to more weed problems in agricultural ecosystems, and change of diversity in wild soybean populations in natural habitats. Conversely, such a fit transgene may spread throughout the diverse genotypes in the population. The future anticipated transgenes, such as drought- and salt-resistant genes and other abiotic stress tolerant genes, might even have a greater impact on this aspect.

The effect of some transgenes on the ecological fitness of wild soybean would significantly influence the dynamics of the wild soybean population, if these transgenes, such as drought resistance and pest resistance could significantly increase the frequency of wild soybean individuals carrying these types of transgenes. Hybrids or their progeny with transgenes were found to have higher vegetative or reproductive productivity in sugar beet (5), *Brassica napus* (38), and sunflower (36). If weedy and wild soybean populations contain transgenes that could largely increase their fitness, the wild populations may become more productive or aggressive under natural conditions, and ultimately become difficult weeds in agroecosystems.

Weedy soybeans are already a problem in some soybean fields, and the situation will be even more complicated with some transgenes such as herbicide resistance. If the ecological fitness of weedy soybean can sometimes be enhanced by picking up particular transgenes through outcrossing with transgenic soybean, the weedy soybean may become weedier, out competing its cultivated counterpart. Although there has not been report of such an event yet in soybean species, the lessons from the Canadian experience of herbicide-resistant gene-stacked volunteers of oilseed rape having three different herbicide resistances demonstrates the potential risk of transgene escape, although no problem is perceived by the Canadians (Chapter 5).

With the extensive use of herbicide-resistant transgenic soybean varieties, the concerns of transgene escape extend to the loss of herbicide resistance as a tool to protect the crop from closely related weeds in the center of origin. When the traits of herbicide resistance introgress from transgenic soybeans and accumulate in wild and weedy soybean populations through gene flow, it might transform them into herbicide-resistant wild or weedy soybeans (with multiple herbicide-resistant traits by gene stacking) that make the weed management by herbicides more problematic, causing further unwanted agroecological consequences.

Consequently, proper biosafety measures with adequate spatial isolation distances in combination with different transgene containment methods (9,13,37) should be taken into consideration to avoid or minimize the significant amount of gene flow between transgenic soybean varieties and the weedy and wild relatives where they exist sympatrically. Altering gene flow by interfering with pollination to confine transgenes of a transgenic soybean from its wild relatives can be done by genetic engineering of chloroplasts (or other cytoplasmic organelles) to promote maternal inheritance of transgenes, because soybean is a diploid annual species with only one genome. This might significantly reduce outflow of pollen-mediated transgenes, because pollen usually does not carry (or only carries an extremely low number of) transgenes, but will not prevent the wild or weedy species from being the pollen parent. It is not clear whether this would be effective with soybeans, as traits encoded on chloroplasts are carried by pollen in many other legume species.

If the transgene escape to its weedy and wild populations is inevitable and the escaped transgenes may possibly provide a fitness advantage to the wild populations causing ecological consequences, transgenic mitigation (TM) technology proposed by Gressel (13) and Al-Ahmad et al. (4) could be considered to restrict the spread of the transgenes and to minimize its potential ecological consequences. An example is if a gene such as "non-exploding pod" (dominant) were available and could be constructed in tight linkage with a transgene and then transferred to cultivated soybean. Thus, even if the transgene can move to wild or weedy soybeans through outcrossing, it will not be able to effectively spread out in these populations. This is because the wild or weedy individuals picking up the transgene cannot self-reproduce and disseminate normally due to the non-exploding pod gene that constrains the free seed shattering.

ACKNOWLEDGMENTS

The support of Shanghai Commission of Science and Technology (Grant Nos. 02JC14022 and 03DZ19309) is kindly acknowledged.

LITERATURE CITED

1. Abe J, Hasegawa A, Fukushi H, Mikami T, Ohara M, Shimamoto Y. 1999. Introgression between wild and cultivated soybeans of Japan revealed by RFLP analysis from chloroplast DNAs. *Econ. Bot.* 53:285–291.
2. Ahmad QN, Britten EJ, Byth DE. 1977. Inversion bridges and meiotic behaviour in species hybrids of soybeans. *J. Hered.* 68:360–364.
3. Ahrent DK, Caviness CE. 1994. Natural cross-pollination of twelve soybean cultivars in Arkansas. *Crop Sci.* 34:376–378.
4. Al-Ahmad H, Galili S, Gressel J. 2004. Tandem constructs to mitigate transgene persistence: tobacco as a model. *Mol. Ecol.* 13:697–710.
5. Bartsch D, Schmidt M, Pohl-Orf M, Haag C, Schuphan I. 1996. Competitiveness of transgenic sugar beet resistant to beet necrotic yellow vein virus and potential impact on wild beet populations. *Mol. Ecol.* 5:199–205.
6. Brouè P, Gouglass J, Grace JP, Marshall DR. 1982. Interspecific hybridization of soybeans and perennial *Glycine* species indigenous to Australia via embryo culture. *Euphytica* 31:715–724.
7. Carlson JB, Lersten NR. 1987. Reproductive morphology. In: *Soybeans: improvement, production, and uses.* Wilcox JR, Ed., 2nd ed., pp. 95–134, Madison, WI: American Society of Agronomy.
8. Caviness CE. 1966. Estimates of natural cross pollination in Jackson soybeans in Arkansas. *Crop Sci.* 6:211–212.
9. Daniell H. (2002) Molecular strategies for gene containment in transgenic crops. *Nature Biotechnol.* 20:581–586.
10. Doyle JJ, Doyle JL, Rauscher JT, Brown AH. 2003. Diploid and polyploid reticulate evolution throughout the history of the perennial soybeans (*Glycine* subgenus *Glycine*). *New Phytol.* 161:121–132.
11. Fujita R, Ohara M, Okazaki K, Shimamoto Y. 1997. The extent of natural cross-pollination in wild soybean (*Glycine soja*). *J. Hered.* 88:124–128.
12. Gepts P, Papa R. 2003. Possible effects of (trans)gene flow from crops on the genetic diversity from landraces and wild relatives. *Environ. Biosafety Res.* 2:89–103.
13. Gressel J. 1999. Tandem constructs: preventing the rise of superweeds. *Trends Biotechnol.*17:361–366.
14. Gumisiriza G, Rubaihayo PR. 1978. Factors that influence outcrossing in soybean. *Acker Pflanzenbau* 147:129–133.
15. Hadley HH, Hymowitz T. 1973. Speciation and cytogenetics. In: *Soybeans: production and uses.* Caldwell BE, Ed., pp. 97–116. Madison, WI: American Society of Agronomy.
16. Hood MJ, Allen FL. 1980. Interspecific hybridization studies between cultivated soybean, *Glycine max* and a perennial wild relative, *G. falcate*. Agronomic Abstract, pp. 58. Madison, WI: American Society of Agronomy.
17. Huang J, Rozelle S, Pray C, Wang Q. 2002. Plant biotechnology in China. *Science* 295:671–676.
18. Hymowitz T. 1970. On the domestication of the soybean. *Econ. Bot.* 24:408–412.
19. Hymowitz T, Harlan JR.1983. Introduction of soybean to North America by Samuel Bowen in 1765. *Econ. Bot.* 39:371–379.
20. Hymowitz T, Singh RJ. 1987. Taxonomy and speciation. In: *Soybeans: improvement, production, and uses.* Wilcox JR, Ed., pp. 23–48. Madison, WI: American Society of Agronomy.
21. Hymowitz T, Singh RJ, Kollipara KP. 1998. The genomes of the *Glycine*. *Plant Breed. Rev.* 16:289–317.
22. James C. 2003. Global status of commercialized transgenic crops: 2003. ISAAA Briefs No. 30. Ithaca, NY: ISAAA. www.isaaa.org.
23. Jin Y, He TH, Lu BR. 2003. Fine scale genetic structure in wild soybean population (*Glycine soja* Sieb. et Zucc.) and the implication for conservation. *New Phytol.* 159:513–519.

24. Karasawa K. 1952. Crossing experiments with *Glycine soja* and *G. gracilis*. *Genetica* 26:357–358.
25. Kiang YT, Chiang YC, Kaizuma N. 1992. Genetic diversity in natural populations of wild soybean in Iwate Prefecture. *Jpn. J. Hered.* 83:325–329.
26. Lackey JA. 1981. Systematic significance of epihilum in *Phaseoleae* (Fabaceae, Faboideae). *Bot. Gaz.* 142:160–164.
27. Ladizinsky G, Newell CA, Hymowitz T. 1979. Wild crosses in soybean: prospects and limitations. *Euphytica* 28:421–423.
28. McGregor SE. 1976. *Insect pollination of cultivated crop plants*. Washington, DC: U.S. Department of Agriculture, Agriculture Handbook No. 496, pp. 411.
29. Nakayama Y, Yamaguchi H. 2002. Natural hybridization in wild soybean (*Glycine max* ssp. *soja*) by pollen flow from cultivated soybean (*Glycine max* ssp. *max*) in a designed population. *Weed Biol. Manag.* 2:25–30.
30. Ray JD, Kilen TC, Abel CA, Paris RL. 2003. Soybean natural cross-pollination rates under field conditions. *Environ. Biosafety Res.* 2:133–138.
31. Rieseberg LH, Church SA, Morjan CL. 2003. Integration of populations and differentiation of species. *New Phytol.* 161:59–69.
32. Shimamoto Y. 1999. Research on wild legume with an emphasis on soybean germplasm. In: Proc. 7th International Workshop on Genetic Resources (MAFF ed.), pp. 5–17. Tsukuba, Japan.
33. Singh RJ, Hymowitz T. 1985. An intersubgereric hybrid between *Glycine tomentella* Hayata and the soybean, *G. max* (L.) Merr. *Euphytica* 34:187–192.
34. Singh RJ, Kollipara KP, Hymowitz T. 1987, Intersubgeneric hybridization of soybeans with a wild perennial species, *Glycine clandestina* Wendl. *Theor. Appl. Genet.* 74:391–396.
35. Smith KJ, Huyser W. 1987. World distribution and significance of soybean. In *Soybeans: improvement, production, and uses*. Wilcox JR, Ed., pp. 1–22. Madison, WI: American Society of Agronomy.
36. Snow A, Pilson D, Rieseberg L, Paulsen M, Pleskac N, Reagon M, Wolf D, Selboa S. 2003. A *Bt* transgene reduces herbivory and enhances fecundity in wild sunflowers. *Ecol. Applic.* 13:279–286.
37. Stewart CN, Halfhill MD, Warwick SI. 2003. Transgenic introgression from genetically modified crops to their wild relatives. *Nature Rev. Genet.* 4:806–817.
38. Stewart AN, All JN, Raymer PL, Ramachandran S. 1997. Increased fitness of transgenic insecticidal rapeseed under insect selection pressure. *Mol. Ecol.* 6:773–779.
39. Weber CR, Hanson WD. 1961. Natural hybridization with and without ionizing radiation in soybeans. *Crop Sci.* 1:389–392.
40. Zhuang BC. Ed. 1999. *Biological studies of wild soybeans in China.* Beijing: Science Press (in Chinese).

QUESTIONS AND ANSWERS

S. Powles: What is the potential for U.S. soybeans imported to China to be "illegally" grown in China, leading to gene flow of resistance gene to weedy soybean?

Answer: There is a possibility that the transgene from imported U.S. glyphosate-resistant soybeans can leak to the environment in China and grow as volunteers in agricultural ecosystems if no weeding measures are taken. There is also a possibility that these soybeans would be grown by farmers in China without knowledge of their transgenic nature, leading to flow of the resistance gene to wild or weedy soybeans. Therefore, strong control measures to keep imported soybean transgenes from escaping is absolutely necessary.

S. Morris: Is there any evidence of herbicide resistance occurring in any non-herbicide controlled soybean in Japan or elsewhere? If so, can that trait be established in weedy population?

D. Vaughan responds: I am not aware of herbicide tolerance present in soybeans in Japan. Currently, Japan does not grow any transgenic soybeans. In Japan, natural weeds are controlled mainly by cutting rather than herbicide; hence even if herbicide resistance is present, it would not provide a selective advantage.

I do not know the situation in China where a large proportion of the imported soybean crop is glyphosate resistant. Although wild soybeans are sympatric with soybeans over large areas of China, constant weeding and intensive land use probably would mitigate any effect of gene flow from the crop to wild populations.

Bao-Rong Lu responds: As far as I know, no transgenic soybean has been approved for commercial production in China. The imported transgenic soybeans from the U.S. are only used for oil pressing. So far, I have not heard of the control of natural weeds by using herbicides, and the weeding is essentially done manually in China, so there is no selective advantage for this trait.

J. Gressel: Is an alternative hypothesis possible that *G. soja* was first domesticated to the intermediate weedy *G. gracilis*, which was then further domesticated to *G. max*?

Answer: It might possibly be, but so far there is no proof of such hypothesis, because the weedy soybean was mainly found in the northeastern part and Yellow River region of China, and cultivated and wild soybeans are found nearly all over China.

Zaida Lentini: Have molecular markers been used to elucidate whether *G. gracilis* is a hybrid between *G. soja* and *G. max*? Or if it is a continuum?

Answer: No molecular markers have been used so far to study whether *G. gracilis* is a hybrid between *G. soja* and *G. max.* It is a good suggestion to try such markers to study the origin of *G. gracilis*.

D. Gealy: We should keep in mind that the maximum outcrossing distances from crops to weeds or crop to crop are probably not an absolute (e.g., 60 m), but rather a detection limitation phenomenon. If testing 10^6, 10^9, 10^{12} seeds, etc., we probably would find a few hybrids at much greater distances. But, the presence of $1/10^9$ or $1/10^{12}$ seeds tested is likely "insignificant" compared to $1/10^3$ or $1/10^6$.

Answer: Yes, there is no absolute determination for outcrossing. That is why we usually use an outcrossing rate to measure the pollen-mediated gene flow. A larger sample size would provide you a more accurate frequency than the situation when you include a small number of samples. However, I think that any gene flow frequency less than $1/10^3$ would be considered as insignificant in terms of ecological consequences.

D. Vaughan: The soybean crop complex appears to be different from other legume crop complexes, such as the azuki bean (*Vigna angularis*) complex and common bean (*Phaseolus vulgaris*) crop complexes. In Japan, morphologically distinguishable weedy or intermediate types of soybean are uncommon even though wild and cultivated soybeans are sympatric across much of the country. It is not clear if gene flow from cultivated to wild soybean results in hybrids that closely resemble wild soybean and therefore outcrossing occurs more often than can readily be visually detected or if outcrossing between wild and cultivated soybean is rare.

Answer: It is indeed a good comment that our understanding of the domestication process in soybean is extremely limited. Knowledge on weedy or intermediate types of soybean is sparse. We do not even know the current distribution of weedy soybeans. Therefore, it is important to conduct fundamental biological studies of the annual soybeans including their origin, evolutionary process, and diversification patterns to address these questions. It is the same situation for the actual gene flow from cultivated to wild soybean through hybridization.

10 Maize and Soybeans — Controllable Volunteerism without Ferality?

Micheal D.K. Owen

10.1 INTRODUCTION

The issues surrounding recent developments in crops, specifically the introduction of traits by biotechnology that are intended to improve pest management, have been and are likely to continue to be a source of contentiousness in the world community. Early in the debate, the scientific community expressed a number of concerns. The message was that the benefits of the *transgenic* (often colloquially referred to as genetically modified or GM) crop, presumably better management of pernicious pests, should be carefully evaluated against the risks (e.g., increased cost of production or the escape of the trait into the wild plant community) (51). Unfortunately, an effective assessment of benefits and risks has not been possible in most agronomic situations, in part due to the inability to assess economic gains attributable to the transgenic trait, a lack of definitive knowledge about the potential environmental and ecological impacts, and the global political climate (47). Significant differences in how these traits are regulated by governments (49), the concerns about food safety issues (44), and the perceptions on how transgenic crops will impact global food supply (21) have led to much confusion within urban communities about the utility of biotechnology. It has polarized communities, agriculture, companies, and environmentalists to the point that an objective scientific debate, with the potential to effect change, does not appear likely in the near future (12,13).

The concern surrounding the use of biotechnology in plant breeding is an interesting conundrum. Traditional plant breeding tactics have been effective and have improved crop productivity and food quality impressively during post-World War II agriculture (23). The change from open-pollinated cultivars to inbred lines for hybrid maize seed production has contributed to increased maize yields from approximately 1.5 metric tons ha^{-1} to almost 8.0 metric tons ha^{-1} (23). Recent developments in soybean breeding reflect opportunities to improve productivity (31). However, none of the changes attributable to traditional plant breeding have made the impact, nor been adopted as quickly as transgenic traits (e.g., glyphosate resistance) introduced via biotechnology. Plant breeders may not see relevant differences between resistance to herbicides or insecticides engineered into plants compared to those transferred by traditional breeding (14). In either case, hybridization between crops and compatible plants within the weed community could occur and the resultant hybrid could gain an ecological opportunity (improved fitness) compared to the native population (32,57).The major difference is that with transgenic crops, the traits may come from unrelated species, genera, families, or kingdoms and possibly not from traditional breeding. Thus, these transgenic traits may cause a greater impact on the plant community associated with the agroecosystem if introgressed from transgenic crops than traits from traditionally bred crops might.

10.1.1 Maize and Soybean

Maize and soybean are different from other crops insofar as genetically compatible wild or weedy species are not a major factor in the major production areas for these important crops (1). Thus, the major consideration with regard to the potential for maize and soybean ferality and volunteerism is not the movement of traits to the wild plant community, but rather the impact of transgenic traits on the ability of the producer to manage the resultant volunteers from the previous crop. However, on the surface, this also seems to be a non-issue. Reference searches for pertinent information about the management of volunteer maize and soybean found only 19 and 1 citations, respectively, and the research was typically conducted prior to the adoption of transgenic crops.

However, with the introduction of herbicide resistance, volunteer maize and soybean represent a different problem than the management of non-transgenic cultivars. Furthermore, maize is open pollinated and thus the movement of the transgenic trait is not easily controlled (59). Consider that gene flow can occur via pollen and seeds (56). Given the inability of the current grain handling system to segregate grain (47) and the potential for movement of traits over great distances (1), crop-to-crop transgene flow is likely to be an issue of greater importance and urgency in the weed community than the movement of traits from crops to compatible plants (15). The transgenic maize variety StarLink® continues to be a worldwide issue (26) and anecdotal reports from farmers of glyphosate-resistant volunteer maize in glyphosate-resistant soybean illustrate the potential for this type of volunteerism and gene flow to become a greater economic and environmental issue than the movement of traits to wild species. The increasing occurrence of these transgenic volunteers also illustrates the problems in regulating transgenic plants and the inability of the regulators to foresee these issues (49,64).

An assessment of benefits attributed to transgenic crops includes improving the environment (19), increasing populations of organisms that break down crop residue (27), improving and making safer crop production (62), and improving yield (5). It can be extrapolated that the lack of diversity found in typical maize and soybean production systems may also lessen the risk of problems with invasive species in those agroecosystems (37). The objective of this chapter will be to review maize and soybean production, assess the extent of volunteerism for these crops, and describe the implications of transgene movement in these agroecosystems.

10.2 CURRENT MAIZE AND SOYBEAN PRODUCTION

Maize and soybean are major crops in the world and rank first and fourth, respectively, with wheat and rice ranking second and third, respectively, in million metric tons (MMT) produced in 2003 (Table 10.1). World maize production represented an estimated 137 million ha in 2003 and had an average yield of 4.4 metric tons (MT) ha^{-1} (Table 10.2). World soybean production involved an estimated 81 million ha with an average yield of 2.4 MT ha^{-1} (Table 10.3). The U.S., China, Mexico, and Europe planted most of the maize and rank first through fourth in the world for area planted, respectively. However, the European Union (E.U.) had a greater average yield than the U.S. (Table 10.2). Maize yield in China was close to the world average, but Africa and Mexico were considerably lower than the world average.

The U.S., Brazil, Argentina, and China are the primary soybean production areas of the world (Table 10.3). The U.S. plants more soybeans than Brazil, Argentina, or Paraguay, but the average yield of the latter three countries was 10% greater than the soybean yield produced in the U.S. India has a considerable area planted to soybeans, but average yields were about 30% of the world average.

TABLE 10.1
World Grain Production in 2003 and Projected Production in 2004 (MMT)

Crop		2003	2004
Maize	World	602	611
	U.S.	228	257
Wheat	World	566	548
	U.S.	44	64
Rice	World	378	391
	U.S.	7	6
Soybeans	World	197	199
	U.S.	75	65

Source: Adapted from Anonymous (6).

TABLE 10.2
World Maize Production in 2003

Country	Area Planted (million ha)	Average Yield (MT per ha)
World	137.5	4.4
U.S.	28.1	8.1
Argentina	2.45	6.3
South Africa	3.65	2.6
China	24.6	4.9
European Union	4.3	9.2
Mexico	7.0	2.7

Source: Adapted from Anonymous (6).

TABLE 10.3
World Soybean Production in 2003

Country	Area Planted (million ha)	Average Yield (MT per ha)
World	81.4	2.4
U.S.	29.3	2.6
Brazil	18.4	2.8
Argentina	12.6	2.8
Paraguay	1.6	2.9
China	8.7	1.9
India	5.7	0.7

Source: Adapted from Anonymous (6).

10.2.1 North and South America

North America and South America account for approximately 66% of the world maize production and 85% of the soybean production (6). Brazil and Argentina are the primary producers of maize and soybeans in South America. Generally, production is based on high inputs and weed control is with herbicides. In the midwestern U.S., maize is often grown continuously or in rotation only with soybeans. In South America, soybeans are often grown continuously. In 2003, approximately 28 million ha of maize were grown in the U.S. Mexico grew 7 million ha and Canada grew an estimated 1.3 million ha of maize (6). The U.S. planted almost 30 million ha of soybeans in 2003 and the area planted has not changed markedly for several years (6). Canada produced 1 million ha and Mexico grew less than 0.1 million ha of soybeans in 2003. Most of the maize and soybeans are grown in the midwestern U.S., particularly in Iowa and Illinois, which account for approximately one-third of maize and soybean production in the U.S.

10.2.2 China

China is the only country other than the U.S. that planted significant areas of both maize and soybean (6). Maize was produced on almost 25 million ha in 2003, but the yield was approximately 60% of that reported for the U.S. Soybeans were produced on approximately 9 million ha. Soybean yields were approximately 74% of the U.S.

10.2.3 Status of Biotechnologically Derived Traits in Maize and Soybean

The worldwide rate of adoption of biotechnology by agriculture, specifically transgenic crops, is unprecedented. Furthermore, the impact of transgenic crops on production agriculture is probably greater than any other advancement experienced thus far. The breeding of hybrid maize significantly improved yield; however, the adoption of this technology occurred over 30 years (23). Adoption of transgenic crops by agriculture increased 40-fold in 7 years and was estimated to represent 68 million ha of transgenic crops planted worldwide in 2003 (21). In the U.S., commercial transgenic crops typically have 1 of 6 traits introduced into the genome via biotechnology; resistance to insects, resistance to herbicides, resistance to viruses, or genes to delay ripening, alter oil content, or control pollen. Crops that are primarily included in the transgenic technology shift are maize, soybeans, cotton, and canola and more than 99% of the transgenic traits in these crops are for herbicide- or insect-resistance (21). Resistance to insects resulting from the incorporation of optimized genes from *Bacillus thuringiensis* (*Bt*) genes into the genomes of crops could result in a savings of $2.6 billion expended annually on insecticides. Transgenic crops may have a greater yield potential than non-transgenic crops, particularly in developing nations where the use of pesticides is less frequent. For example, the use of transgenic cotton increased production by 5 to 80% when compared to non-transgenic cotton (21). Transgenic cultivars of soybeans and maize accounted for an estimated 61 and 23%, respectively, of the area planted to those crops in 2003 (21). The U.S. had approximately 63% of the world area planted, followed by Argentina with 21%, Canada with 6%, China with 4%, and South Africa with 1%. Brazil was estimated to have 4% of the transgenic hectares; however, the author suggests that this figure is not representative of reality and the value indicated was politically expedient (21). Interestingly, an estimated 5.5 million farmers planted transgenic cultivars in 2002 and 90% of the hectares planted to transgenic cultivars was in the U.S., Argentina, and Canada. Approximately 133 million ha of crop land was planted in the U.S. in 2003 (8). Transgenic herbicide-resistant soybeans, oilseed rape, cotton, and maize represented 81, >58, 32, and 11% of the area planted, respectively. Most of these transgenic herbicide-resistant crops were resistant to glyphosate, although resistance to glufosinate also is registered and commercially available in oilseed rape and maize cultivars. Glyphosate-resistant cultivars were grown on approximately 25% of the total crop area of the principal crops grown in 2003 (8).

Differences in yield potential for the transgenic herbicide-resistant cultivars compared to the non-transgenic cultivars were observed early in the adoption of the technology. The differences were attributed to yield lag (suggesting the herbicide-resistant trait was not yet incorporated into the highest yielding cultivars) and yield drag (implying a pleiotropic effect of the herbicide resistance trait that negatively impacted cultivar yield potential). When sister lines of glyphosate-resistant and susceptible soybeans were compared with the high yielding cultivars, the high yielding non-herbicide-resistant cultivars yielded 5% more than the susceptible sister lines and 10% more than the glyphosate-resistant sister lines (17). The application of glyphosate to the glyphosate-resistant sister lines did not affect yield (18). Earlier introduced cultivars demonstrated a greater affect on yield attributable to glyphosate resistance than later cultivars (54). Current transgenic glyphosate-resistant soybean cultivars are reported to have the same yield potential and do not differ in protein or oil content (58). Early glyphosate-resistant maize hybrids were observed to be somewhat sensitive to glyphosate depending on the application timing, rate, and subsequent environmental conditions, and did not have the same yield potential as the best commercial hybrids available (Owen, personal observation). However, *Bt* transgenic maize hybrids are suggested to represent a significant advantage when compared to non-transgenic maize, particularly in developing countries. Advantages are attributable to protection the *Bt* affords insect feeding, but also improved food safety due to lower levels of the mycotoxin fumonisin, but not aflatoxin (5,55).Transgenic *Bt* maize cultivars are projected to be planted on 40 to 45 million ha, or about 30% of the global maize. It is not anticipated that transgenic herbicide-resistant maize will achieve that high level of adoption and acceptance in the U.S. has been slow.

10.3 ORIGIN OF MAIZE AND SOYBEANS

It is well documented that maize (*Zea mays* L.) is derived from the annual grass *Zea mays* ssp. *May* L. and is a highly important crop originating in Mexico (16). The closest wild relatives of maize are the teosintes (*Zea* spp.) and include annual and perennial species (53). The distribution of teosinte includes the tropical and subtropical areas of Guatemala, Honduras, Mexico, and Nicaragua. Although closely related to maize, there is considerable debate about the extent that teosinte participated in the eventual development of maize races in Mexico and thus contributed to prominence of maize as a leading world grain. The origins of cultivated soybeans and the relationship cultivars have with weedy and wild relatives, solely in the center of origin, are discussed in Chapter 9.

10.3.1 POTENTIAL FOR INTROGRESSION WITH COMPATIBLE PLANT SPECIES IN THE AGROECOSYSTEM

The movement of genes from domesticated plant populations to compatible plant populations is well documented, but not for all crops in all locations (16). It has been suggested that the introgression of traits is an inevitable consequence of crop husbandry (32) and may actually be increasing in frequency due to current production practices that have resulted in significant landscape modification and habitat disturbances, which may lessen barriers to introgression (1). The literature is replete with numerous examples of introgression from domesticated plants and related plants that occur in the agroecosystem (15). Furthermore, there is nothing to suggest that domesticated plant populations that contain transgenes should behave any differently than other plant populations, and many researchers suggest that it has become more important to monitor the introgression of crop genes into genetically compatible plant populations (28).

10.3.1.1 Maize

Maize is reported to cross with relatives of maize (*Zea* spp.) (teosinte) (16,53). Movement of traits from diploid teosinte to maize is somewhat successful although the reciprocal crosses are often unsuccessful (36). However, the relationship between teosinte and maize is not well understood and thus the implications of introducing transgenic maize cultivars into regions where teosinte populations exist cannot be defined (34). Research suggests that transgene movement is possible, although this work has been refuted as flawed (10,12,50). Other scientists suggest the introgression of the transgenic *Bt* trait into land races would be a good thing and is of no major consequence (39,40). Regardless, the introgression of traits from maize to teosinte should be considered a possibility and thus the attempts to preserve teosinte populations are justified to maintain the genetic opportunity for future maize improvement (34,53).

Research suggests that there are isolation distances that will effectively curtail the movement of traits via pollen and thus serve to protect the genetic integrity of maize and teosinte populations (38). However, even though the probability of gene flow may approach 0, this goal is not likely achievable given the current economic and sociological constraints (30).

Although introgression of traits between teosinte and maize appears to be possible, there is no compelling evidence that this means anything with regard to increased weediness for the resulting putative hybrids. Importantly, in the major maize production areas, wild *Zea* spp. are not found and thus there is little likelihood of teosinte becoming more pernicious and difficult to manage in maize production fields (20).

10.3.1.2 Soybean

The potential movement of transgenes between soybeans and its wild and weedy relatives is addressed in Chapter 9. Soybeans are described as an autogamous species, and although commercial hybrid cultivars have been available, hybridization requires manual cross-pollination and is difficult and time-consuming (48). The potential for natural outcrossing in cultivated soybeans is considerably lower than in wild species. Estimates of spontaneous outcrossings in cultivated soybeans range from 0.02 to 5% and were accommodated by small insects like *Thrips tabaci* (thrips) and *Apis mellifera* L. (honeybees) (48). The occurrence of spontaneous outcrossing in cultivated soybean raises the question as to the potential for transgene movement between transgenic and non-transgenic soybean cultivars. Movement of glyphosate resistance was observed between two soybean cultivars, but the frequency of natural cross-pollination was quite low and dependent on the distance that the receptor flowers were from the transgenic donor flowers (2). The frequency of cross-pollination was 0.5% when plants were 1 m apart and the frequency dropped dramatically from 0.01% with a 2 m separation to 0.005% with a 10 m separation. The implications of the limited opportunity for the introgression of genetic traits between cultivated and wild soybean species are that it is unlikely that the wild soybeans will achieve an ecological advantage and become more problematic weeds.

10.4 EXTENT OF VOLUNTEERISM IN MAIZE AND SOYBEANS

Volunteer maize and soybeans as weeds are not a new concern and are a relatively consistent problem in production systems where maize and soybeans are grown. Interestingly, there is little information in the scientific literature describing the problem of maize or soybean volunteerism, be it the potential management of the problem or the economic impact. Farmers typically have used various herbicides and mechanical tactics to control volunteer maize in soybeans and volunteer soybeans in maize. Although the potential for volunteerism of the crops increases as tillage decreases, typically growers do not see volunteer maize or soybean as a serious problem. However, with transgenic traits for herbicide resistance, management could become more difficult and costly.

Given the extent that maize and soybeans are grown worldwide, the importance that these crops have with regard to global food production and the fact that most of the transgenic crops grown are either maize or soybean, the lack of scientific information is surprising. Based on the lack of information, one might assume then that volunteerism for maize and soybeans, whether transgenic or conventional cultivars, either is not a significant problem or is an issue that has been resolved. The author suggests that both may indeed be the case in most agroecosystems in which maize or soybeans are grown. However, some prudence must be recommended; failure to predict a problem has greater negative implications than predicting a problem that does not occur (35). For example, the occurrence of volunteer maize in soybeans grown in rotation could potentially increase maize insect or plant pathogen problems by allowing populations to increase during the year soybeans were grown. The other important consideration is whether the inclusion of transgenes increases the weediness of the volunteer crop or weed. Hypothetically, several characteristics would need to be acquired for maize and soybeans to become weeds that are more virulent. These include the acquisition of dormancy and possibly shattering of the grain. It is unlikely that the cultivar genome could change that dramatically in the short period that volunteer crops exist within an agroecosystem.

Generally, if the phenotype of the rotational crop is similar to the volunteer crop weed (e.g., monocot crop grown in rotation with maize), management may be more difficult. This problem is exacerbated if the rotational crops have similar transgenic traits. The control of volunteer glyphosate-resistant wheat in glyphosate-resistant maize required that herbicides other than glyphosate be used (46). Similarly, additional herbicide applications were necessary to control volunteer herbicide-resistant crops in the crop rotations of the Pacific Northwest U.S. (52). However, if the transgenic herbicide-resistant maize volunteers into a conventional cultivar, or if the transgene is for resistance to insects or plant pathogens, the volunteer crop weed will have minimal impact on the management tactics.

Another consideration is the planting configuration; crops grown in relatively narrow row spacing will not have mechanical cultivation as an option to manage the volunteer crop weed. This then places all management opportunities on the availability of an herbicide that has selectivity for the crop and efficacy on the volunteer crop weed. The requirement for an herbicide specifically to control the volunteer crop weed may represent a significant additional expense to the grower. Furthermore, the use of an additional herbicide may also increase the risk of crop injury, thus further negatively impacting profitability.

10.4.1 Weedy Characteristics of Volunteer Maize

Volunteer maize has been considered to be an economic problem as a volunteer weed in soybeans. Thus, problems were greatest for the Midwest maize belt in the U.S. Recent increases in maize and soybean production in China and Argentina suggest that volunteer maize could also be a weed problem in those countries. Some maize is grown in rotation with small grains and forages, but there is no information to suggest that volunteer maize is an important volunteer crop or weed in these production systems. Maize is a competitive crop and thus exhibits the potential to become a serious economic problem if the amount of the volunteerism is high. Maize grown for direct human consumption (sweet corn) is more subject to plant pathogens and insect damage, increasing harvest losses and potentially representing a greater volunteer problem in regions where these cultivars are grown. However, these maize cultivars are less robust and not considered a serious volunteer issue.

10.4.1.1 Factors Involved with Volunteerism in Maize

Factors that influence volunteer maize populations include pest infestations, harvesting, and tillage. Preharvest losses reflect dropped maize ears and are attributable to insect or disease infestations. Harvesting impacts volunteer maize populations directly by gather losses, where ears are missed or dropped by the harvest machinery, and cylinder and separator losses, which reflect the efficiency of grain shelling by the harvest machinery (24). Harvest losses can be as great as 206 kg ha^{-1} of maize grain and represent individual grain and whole ears that are not collected in the grain tank. Consider that in the midwestern U.S., maize is seeded at approximately 74,100 seeds ha^{-1}, which corresponds to approximately 92 kg ha^{-1}. Thus, harvest losses can be as great as 224% of the seeding rate. These losses occur when the harvest machinery is adjusted incorrectly to gather the ears into the machinery or efficiently shell the grain from the cob, or if weeds are a significant problem at harvest and the grain sieves become plugged. Maize ears break from the stalk or seeds pass through the equipment and are deposited on the ground, thus becoming a volunteer weed problem. Tillage affects volunteer maize populations directly by placing the seeds in an unfavorable position for germination. Volunteer maize problems are a greater concern in conservation tillage systems where high residue and volunteer maize seeds remain on or near the soil surface. Generally, volunteer maize infestations are characterized by clumps of maize rather than individual plants. Soybean yield losses attributable to volunteer maize were as great as 25% with 5380 maize clumps ha^{-1} (10 plants clump^{-1}) (9). When volunteer maize plants in a clump ranged from 2 to 7 plants, soybean losses varied from 6 to 21%. Timing of control of volunteer maize populations also impacted soybean losses; and if the volunteer maize population was allowed to coexist with the soybean population for 10 weeks after soybean emergence, 27% soybean yield loss was recorded (9).

Management of volunteer maize has not been a major concern in soybeans. Traditionally, graminicides (e.g., diclofop) or glyphosate, applied selectively, has been an effective management tactic (4). The current predominant tactic for weed control in soybean is the topical application of glyphosate. This tactic has been relatively successful in controlling a broad spectrum of weeds including volunteer maize. However, the evolution of weed species resistant to glyphosate, as well the problems controlling other common weeds, has reinforced the assumption that appropriate selection pressure on agroecosystems will result in weed population shifts that do not respond to previously effective management tactics (61,66). Regardless, control of volunteer maize has continued to be effective with glyphosate. However, with the increased adoption of glyphosate-resistant maize hybrids, growers in the midwestern U.S. are experiencing problems managing volunteer maize with glyphosate even if the transgenic maize was not planted the year before the transgenic glyphosate-resistant soybean. The glyphosate resistance transgene has been distributed widely in pollen and has resulted in a considerable problem controlling volunteer maize. Concerns about parentage of volunteer maize were prophetically expressed 2 decades prior to the adoption of transgenic maize (3). Growers now often include graminicides to control volunteer maize based on the assumption that the glyphosate resistance transgene was transferred via pollen to a previously glyphosate-sensitive maize crop.

10.4.2 Weedy Characteristics of Soybeans

Soybeans are not generally considered a serious volunteer weed problem as exemplified by the lack of published research. Soybean seeds that are lost during harvest do not overwinter particularly well in the midwestern U.S.; and if volunteer plants develop in the rotational crop, losses attributable to interference are minimal. In warmer climates, it is likely that seeds would imbibe and rot quickly, thus precluding the chances for volunteerism in the following growing season. In soybean/maize rotations, a number of herbicides typically used for weed control in maize (e.g., atrazine) effectively control volunteer soybeans. Other crop rotations where volunteer soybeans might be considered a problem include rice, cotton, and continuous soybeans. In the rotations that alternate crops, management practices preclude the likelihood that volunteer soybeans would become a significant issue, whether or not the volunteer population included a transgene for herbicide resistance. However, there is a question as to possible management tactics to effectively control volunteer soybeans in a soybean monoculture. It is suggested that soybean seeds do not survive in the agroecosystem and are lost due to predation, rotting, germination resulting in death during the winter, or due to management practices prior to planting the current soybean crop. However, there is no information in the literature that describes any issue with volunteerism in soybeans. Thus, regardless of the reason, volunteerism of soybeans is not an economic problem.

10.4.3 Implications for Ferality and Volunteerism

To assess the ecological risks of transgenic crops, the occurrence of spontaneous crosses with compatible weedy relatives must be evaluated, the fitness of the transgenic crops and transgene hybrids determined, and unforeseen interactions of the transgene hybrid in the environment considered (33). Based on these assessment criteria, ferality in maize and soybeans is not a serious agricultural issue. Although there are genetically compatible species for the potential introgression of transgenic traits from maize and soybeans, the species are not known to be weedy (i.e., teosinte and maize), do not occur in important production areas (i.e., wild soybean and soybean production in the Western Hemisphere), or do not have a reported competitive ability to represent an agronomic threat (i.e., wild soybean in China). There are no indications in the literature that either maize or soybean will revert to aggressive weedy types like other crops (i.e., sunflowers). Thus, the only issues to consider are the implications of volunteer transgenic maize and transgenic soybeans in the agroecosystem.

It is important to note that transgenic traits for pest resistance do not contribute to the difficulties in managing the volunteer crops. Furthermore, herbicide resistance only contributes to weediness if the herbicide used for weed control in the rotational crop is the one for which the volunteer crop has the resistance transgene. Even in this scenario, the volunteer transgenic crop does not generally represent a management problem, as there are effective management tactics available for most rotational crops, if the presence of the transgene is anticipated. In the case of volunteerism in soybean, given that self-pollination predominates the crop, and the fact that volunteer soybeans are not particularly competitive, volunteerism is not an economic issue.

The volunteerism of maize represents a different problem. Non-transgenic maize volunteers are easily controlled with glyphosate, glufosinate, and a number of graminicides. The economic issue reflects the inability for the producer to determine whether the transgenic herbicide-resistant trait for glyphosate has introgressed into a non-transgenic crop, thus making the planned glyphosate applications to the rotational transgenic soybean crop ineffective for the control of volunteer transgenic maize. Thus, growers can either include a graminicide in anticipation that the volunteer maize has the transgene conferring resistance to glyphosate, or are faced with a second herbicide application if they guess wrong. Both decisions represent an added cost of production, and possibly an unneeded herbicide application, if the transgene did not move into the previous maize crop. Furthermore, if growers decide not to include a graminicide with the glyphosate application and the transgene is there, there is an increased potential for reduced soybean yield due to volunteer maize interference prior to the second remedial herbicide application.

10.5 VOLUNTEERS NEED NOT BE IN FIELDS — IDENTITY PRESERVATION AND DERIVED TRAITS

It is interesting to note that the concept of genetic identity preservation (IP) in maize has not historically been a major agronomic or economic issue due to guidelines established to isolate the seed fields, thus minimizing the introgression of traits by controlling, to a great degree, pollen from other genetic sources (26). It is suggested, however, that while the percentage of successful pollination by outside sources was quite low, the actual introgression of the traits occurred at a greater frequency than desired, and thus the preservation of genetic identity was poor. The reason that this was not considered to be a serious problem was the difficulty in actually identifying the trait that was introduced. This problem, identifying the introgressed trait, has been greatly simplified and the accuracy dramatically improved with the use of transgenic traits and biotechnological methods. Furthermore, IP has received considerable notoriety given the polarization of societal acceptance of transgenic traits (e.g., StarLink contamination of maize intended for human consumption) (47). Recently, the inability of the industry to isolate experimental pharmacrops (transgenic maize grown for use in the manufacture of pharmaceuticals) has not strengthened the position for the acceptance of transgenic crops and the presumption that genetic identity can be preserved (42). It is clear that the industry must address three issues to resolve concerns expressed by the public about the development and production of transgenic maize:

1. Pollen drift
2. Containment of plant products during the production year
3. Volunteer plants in following years

The introgression of transgenic traits in soybean is considerably less of a problem when compared to open pollinated maize. However, the movement of transgenes is still an issue, as gene flow also can occur via seeds (56). Currently, many growers receive a premium of $0.015 to $0.022 kg^{-1} for non-transgenic soybeans. However, the grain handling industry does not have the ability to segregate grain effectively, thus placing the growers at risk of forfeiting any premium based on the genetic identity of the soybean (47). As the requirement for IP becomes more stringent, and the relative tolerance of contamination approaches 0%, the cost of maintaining IP potentially becomes prohibitive (30). Thus, a non-transgenic soybean can be inadvertently contaminated with transgenic soybean during harvest, transport, and storage. This problem can escalate quickly and the resultant costs become extremely high. For example, in 2002, soybeans grown in rotation with nonregistered transgenic maize were infested with volunteer maize, presumably containing a pharmaceutical transgene. Furthermore, maize residues from the previous year were also in evidence. The 13,608 kg of soybeans harvested from the area that potentially included the nonregistered transgenic maize were comingled with 13,608,000 kg of soybeans that were impounded by the United States Department of Agriculture (USDA) Animal and Plant Health Inspection Service (APHIS) and ultimately destroyed at considerable cost to the grain handler and grower (42). It is important to recognize that movement of transgenes via grain can occur over greater distances and perhaps more consistently than that via pollen movement.

10.5.1 Implications of Volunteerism on Labeling Requirements and Regulations

The implication of regulations is, in theory, to improve the consumers' trust of the quality of the food and to provide an understanding of the vigorous nature of the testing, handling, and manufacturing requirements for foods. However, in the current global environment, the implications of regulating food, as it relates to transgenic crops, are presumed by producer organizations to limit fair trade opportunities in key markets such as the E.U. and Japan. Again, this issue relates, in part, to presumed benefits and risks of the transgene and specifically focuses on food safety issues and the ability to regulate the process. A majority of people (57%) from the U.S. indicated that they are not concerned about health problems attributable to farming practices in the U.S., although 80% of Americans expressed concern about the practices in other countries (65). With regard to food safety and foods with transgenic traits, 28% of the sample population from the U.S. indicated that these foods were unsafe. Given the strict regulation requirements and the governmental agencies that control the development of transgenic crops, this concern is surprising. It is presumed that the percentage would be considerably higher if people from the E.U. were surveyed. For those who are frightened by transgenic maize and soybeans, volunteer transgenic material, whether in the field or in grain processed for food, is an issue.

One of the key issues that has been a significant factor of the testing of transgenic maize and soybean is field trial stewardship. Researchers are required to make sure that no transgenic residues remain in the environment upon completion of a field test. Although a number of tactics are used to achieve confinement of the transgene, including using isolation distances as proposed by the Association of Official Seed Certifying Agencies (AOSCA), it is impossible to achieve complete containment. Maize represents the most common transgenic crop tested in the U.S. and transgenic traits that are most frequently tested are for herbicide resistance (28%), quality characteristics (25%), and insect resistance (23%) (49).

It seems clear that the strict regulations that are in effect for the development of transgenic crops in the U.S. are intended to ensure the safety of the environment. However, they are still a function of various government agencies and thus decisions may be affected, not so much by the science, but by the political will. The incident with the transgenic StarLink maize provides sufficient evidence to indicate the fallibility of the system (47). Even though the criticisms of the science that developed the transgenic crops may not be accurate, the concerns of the public must still be alleviated, and even the stringent requirements implemented in the U.S. have not been able to overcome global opinions based on public fears.

10.5.1.1 Pollen Drift

The concern about IP and transgenic trait containment in soybean does not necessarily preclude a discussion about pollen drift. However, from a practical perspective, it has already been established that movement of transgenic traits via pollen is not a significant issue with soybeans (2,48). However, for maize, this represents a significant economic issue and can arguably be considered a major economic problem. Although research on the effect of distance from the pollen source suggests that the greater the distance, the lower the probability of successful fertilization, containment of pollen that includes transgenes is still considered a serious threat. The European Union Scientific Committee on Plants holds the position that transgene contamination is an inevitable consequence of transgenic maize production (26). In Denmark, pollen drift is the legal responsibility of the farmer who causes the pollen trespass, and consequently the potential for transgene introgression (59). Another consideration is that even if pollen drift could be managed, other factors would still create a great potential for transgene contamination (42). The question is whether the management of transgenic maize and soybeans will result in movement of the transgene at levels currently allowable by the E.U.

Models that predict the potential for pollen trespass have been developed, but have limitations on the accuracy of the prediction, the simplicity of the model that hampers predictions of the small amounts of pollen that move to distances greater than 250 m, and impact of variable climatic conditions (7,60). The objective of the models was to provide agriculture with a tool that rationally and economically allows for the isolation of fields and thus the containment of pollen drift. Importantly, the development of more robust models might improve the management of transgenic pollen trespass into non-transgenic crops.

A number of factors influence pollen-mediated gene flow. Distance from the pollen source is a critical factor (38). Pollen shed density is an important consideration, more so for maize yield, but with significant considerations for transgene movement (63). Reports of the quantitative relationship between pollen density and maize grain yield suggested that a minimum pollen density of 3000 pollen grains silk^{-1} were required for maximum yield potential (63). Time also influences the potential for pollen movement. Both of these factors influence the probability of whether pollen will move and successfully transfer traits. Also, influencing the distance pollen will move and the time pollen will remain viable are the environmental conditions, notably air temperature and relative humidity (38). Pollen viability declines quickly with desiccation. The maximum distance for cross-pollination was measured to be 200 m from the pollen source (38). In other studies, only 0.75 to 0.5% of viable maize pollen was estimated to move 500 m from the source plant (20). Researchers suggested that although pollen clouds are found to occur at great distances from the pollen source, this does not mean that transgenic trait movement occurs over that distance. Similarly, it can be suggested that the occurrence of transgenic pollen in a non-transgenic maize field does not mean the non-transgenic maize is fertilized by the transgenic pollen (20). However, the implications of this must reflect the tolerance of transgene allowed. Research suggested that increasing isolation

FIGURE 10.1 Movement of traits in maize as a result of pollen. Picture demonstrates the movement of pollen from purple popcorn to yellow maize at a distance of 500 m. Dark kernels indicate the successful pollination with the purple popcorn pollen. (Photo from Jean McGuire, Iowa State University, 2003. Used with permission.)

distance between transgenic maize and non-transgenic maize is a useful tool in containing gene flow (38). Other research suggests that isolation distances required to maintain a near 0% transgenic contamination are too large to be realistic (41). Research recently conducted at Iowa State University demonstrated clearly that distances approaching 500 m were not sufficient to keep transgene movement in maize contained (Figure 10.1) (45). Given the systems currently used for maize production, it is not possible to effectively argue that field isolation will serve to control pollen trespass. Fields occur side by side or at best, are only somewhat remote from each other. However, given the potential for transgene introgression as a result of pollen movement, and the desire from major markets that require a low level of transgene contamination, movement of transgenic traits via maize pollen will continue to be an issue of contention (22,30).

10.5.1.2 Grain Marketing Issues

The implications of transgenic maize and soybeans on grain marketing issues do not reflect how the movement of the transgene occurred and may not be a factor of the relative amount of transgene found. Global markets for non-transgenic maize and soybeans are strong and growers are paid a premium for the crop, particularly in the U.S. (58). However, the market for non-transgenic crops in the E.U., particularly transgenic soybeans, may be softening. The market demands, in many instances, that there is a zero or near-zero tolerance of transgenic crop contamination, whether the occurrence of the transgene is attributable to pollen movement or the inability to segregate grain resulting in the comingling of transgenic and non-transgenic grain. The problem of transgenic maize and soybeans in the marketplace reflects consumer preferences and often does not give credence to the science (29). Thus, countries entertain circular debates about the political and social agenda of food supply and quality. The scientific community may not be able to bring useful discussion to the debate, and governmental agencies have not clearly demonstrated that they have the ability to effectively regulate transgene development and utilization. Recent decisions of considerable economic consequence made by the private industry are a testament to vigorous and often contentious debate that surrounds transgenic crops. Monsanto has transgenic glyphosate-resistant sugar beets registered, but has thus far refused to commercially introduce the cultivars; and recently, they removed the prospect of introducing transgenic wheat due to concerns expressed by the general public, primarily in the E.U. (11).

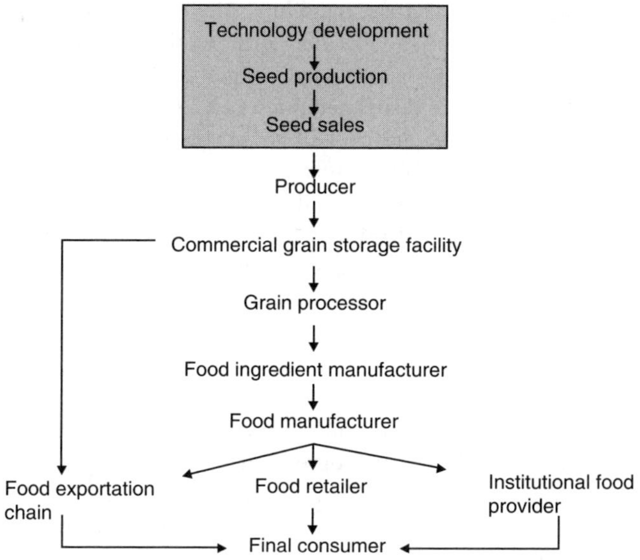

FIGURE 10.2 Development and movement of transgenic grain through the marketplace. The current business model reflects the trend that the same company that develops the transgenic technology, markets the technology. (Adapted from Ginder (22).)

A general understanding of the global grain marketing system is useful in appreciating this issue (Figure 10.2) (22). Using the problems surrounding StarLink as an example, the current issues encompassing the acceptance of transgenic maize and soybeans in the market can be better understood. The basis for the StarLink problems reflected the capital investment made in the development of the trait (a *Bt* ssp. *tolworthi* insecticidal protein Cry9C) developed by AgrEvo and the rush to get the trait registered and in the commercial trade (25). The Environmental Protection Agency (EPA) granted a limited and conditional registration of the transgenic trait that allowed the grain to be used only as animal feed and not in the human food channels. The agreement also required that growers adopt IP practices such as field isolation to prevent introgression of the trait into other fields. Apparently, no consideration was given to the inability of the marketplace to segregate grain (47). Although there was considerable discussion about potential problems of the EPA-restricted registration during 1998 and 1999, the first complaint occurred in September 2000 when taco shells manufactured by Kraft Foods and marketed by Taco Bell were found to contain the *Bt* protein Cry9C (25). A cascade of events quickly followed that resulted in Aventis CropScience purchasing all StarLink maize including maize grown within 200 m of the transgenic fields, a number of lawsuits and legal arrangements with individual states, and eventually the withdrawal of the registration by the EPA.

To put the issues of transgenic crops and grain marketing in perspective, StarLink accounted for only 0.5% of the corn production in the U.S. The protein is detected by an immunoassay that requires about 10 minutes and is accurate at 0.25% (1 kernel in 400). The North American Millers Association (NAMA) suggested guidelines for the detection of the StarLink protein in grain intended to be used in processing, but the guidelines gave only a 95% confidence to detect the protein if present at 0.38% or less. This level of detection was not acceptable to most foreign markets. Even though StarLink hybrids have not been planted for several years, concerns continue. For example, a shipment of 1200 MT of maize was received in Japan in December 2002 and was found to contain the StarLink transgene. The question arises as to whether the recent detections were attributable to transgene movement via pollen into other hybrids presumed not to contain the StarLink protein or via contamination from stored maize.

Given the apparent inability of the grain handling system to segregate grain, the comingling of transgenic and non-transgenic grain is likely. Furthermore, the movement of transgene traits via pollen or other vectors increases the probabilities of contamination even when efforts are made to preserve the genetic identity of maize and soybean grain (22,29). Markets that require low tolerance levels of contamination by transgenic grains will make future grain marketing difficult if not impossible, and importantly, as the transgene tolerance requirements approach 0, the expense to ensure that tolerances are met increase dramatically (30). Interestingly, it has been proposed that seed for non-transgenic cultivars be produced in areas free of any transgenic crop production (26). Whether this is a viable concept remains to be seen. However, there can be no question that the identity of food (e.g., IP) is a critically important global issue.

10.6 CONCLUSIONS

Maize and soybeans uniquely illustrate the complexity of imparting transgenic traits into major crops. Even though the concerns for the introgression of the transgene into the wild relatives resulting in a more difficult to manage hybrid is a serious consideration for many crops (e.g., rice), it is suggested that this is not a significant issue for maize or soybean. Although these crops have wild relatives that present some potential for introgression, the wild relatives either do not function as effective weeds or do not exist in major maize and soybean production areas (exceptions would be Mexico for maize and China for soybeans). Furthermore, there is no indication that either maize or soybeans have a great potential for ferality. However, the perspective that maize and soybeans have the potential for "controllable volunteerism" is misleading. With regard to soybeans, the movement of transgenic traits via pollen is minimal (31,43). However, the issue of comingling grain from transgenic and non-transgenic cultivars is real and economically important (29,30). Furthermore, although transgenic soybeans may occur as volunteer weeds in rotational crops, management tactics do not need modification from those used to manage most indigenous weed communities.

The concern about volunteer transgenic maize cultivars is a much different issue. Transgene movement via pollen is not controllable in current agroecosystems, nor does there appear to be near-future biotechnological fixes for this problem. Teosinte does exist in limited areas where maize production occurs as an important component of the local economy, and teosinte maize hybrids are reported to exist (34). However, teosinte is not indigenous to most of the maize production areas and there is no indication that hybridization with transgenic maize cultivars increases the weediness of teosinte. Thus, the introgression of transgenes from transgenic maize cultivars to compatible wild plants is presumed not to be an important economic issue. However, the movement of the transgene from transgenic cultivar to non-transgenic cultivar is a serious issue, whether the movement is from pollen drift or from comingling of grain. Pollen drift of a transgene for herbicide resistance (e.g., glyphosate resistance) results in added difficulty and expense in controlling volunteer maize in rotational crops. The potential for the pollen drift to occur and the success of the introgression of the transgenic trait are not predictable. Thus, many growers decide to include a graminicide that will control the herbicide-resistant volunteer maize, whether the introgressed transgenic trait is known to be present or not. This decision potentially represents a needless expense, but growers rationalize that it is cheaper than making a second herbicide application and may also prevent loss of crop yield attributable to maize interference (9). A more important potential issue of pollen drift is the loss of genetic identity of the maize (22,30). Given the premiums paid for non-transgenic maize, maize with specialty traits, and organic maize, contamination by transgenic maize pollen or by comingling with transgenic grain could result in significant economic losses for the grower, depending on the allowable percentage of the transgene grain (25). This is a current issue of considerable importance in global maize production and marketing.

Unfortunately, there does not appear to be a strategy to manage the movement of transgenic traits into non-transgenic grain in the near term or the longer term. Unless consumer fears are alleviated, particularly in the European Union, the author suggests that the problem of transgene contamination of food grains will escalate.

LITERATURE CITED

1. Abbott RJ, James JK, Milne RI, Gillies ACM. 2003. Plant introductions, hybridization and gene flow. *Phil. Tran. R. Soc. London* 358:1123–1132.
2. Abud SPIMS, Moreira CT, Farias Neto AL, Vianna GR, Andrade SRM, Nunes J, Guerzoni RA, Monteiro PMFO, Assuncao MS, Rech EL Aragao FJL, Eds. 2004. *Distance of flow in transgenic BROO-69515 and the nontransgenic soybean in the Cerrado Region, Brazil.* Presented at VII World Soybean Research Conference, Foz do Iguassu, Brazil.
3. Andersen RN, Geadelmann JL. 1982. The effects of parentage on the control of volunteer corn (*Zea mays*) in soybeans (*Glycine max*). *Weed Sci.* 30:127–131.
4. Andersen RN, Ford JH, Lueschen WE. 1982. Controlling volunteer corn (*Zea mays*) in soybeans (*Glycine max*) with diclofop and glyphosate. *Weed Sci.* 30:132–136.
5. Anonymous. 2003. Biotech corn can boost yields. In *Dealer and applicator*, Vol. 12, pp. 30–31.
6. Anonymous. 2004. World agricultural production. FAS USDA. http://www.fas.usda.gov/commodity.html.
7. Ireland DS. 2004. The influence of cultural practice, biology, and environment on maize pollen travel. PhD thesis. Ames, IA: Iowa State Univesity, 101 pages.
8. Banks PA, Branham B, Harrison K, Whitson T, Heap I. 2004. Determination of the potential impact from the release of glyphosate- and glufosinate-resistant *Agrostis stolonifera* L. in various crop and non-crop ecosystems, *Weed Science Society of America Abstracts.* p. 65.
9. Beckett TH, Stoller EW. 1988. Volunteer corn (*Zea mays*) interference in soybeans (*Glycine max*). *Weed Sci.* 36:159–166.
10. Blancas LD, Arias M, Ellstrand NC. 2002. Patterns of genetic diversity in sympatric and allopatric population of maize and its wild relative teosinte in Mexico: evidence for hybridization. Presented at Ecological and agronomic consequences of gene flow from transgenic crops to wild relatives, Columbus, OH. pp. 31–38. http://www.biosci.ohio-state.edu/~asnowlab/Proceedings.pdf.
11. Brasher P. 2004. Wheat decision hardly a surprise. In *Des Moines Sunday Register*, May 16, 2004, pp. 1M–2M. Des Moines, IA.
12. Butler D. 2002. Alleged flaws in gene-transfer paper spark row over genetically modified maize. *Nature* 415:948–949.
13. Dalton R, Dalton S. 2002. Superweed study falters as seed firm deny access to transgene. *Nature* 419:655.
14. Duvick DN. 1999. *Consequences of classical plant breeding for pest resistance.* Presented at Ecological effects of pest resistance genes in managed ecosystems, Bethesda, MD.
15. Ellstrand NC. 2002. *Gene flow from transgenic crops to wild relatives: what have we learned, what do we know, what do we need to know?* Presented at Ecological and agronomic consequences of gene flow from transgenic crops to wild relatives, Columbus, OH. pp. 39–46. http://www.biosci.ohio-state.edu/~asnowlab/Proceedings.pdf.
16. Ellstrand NC. 2003. Dangerous liaisons? When cultivated plants mate with their wild relatives. Baltimore, MD: Johns Hopkins University Press. 244 pp.
17. Elmore RW, Roeth, FW Nelson LA, Shapiro CA, Klein RN, Knezevic SZ, Martin A. 2001. Glyphosate-resistant soybean cultivar yields compared with sister lines. *Agron. J.* 93:408–412.
18. Elmore RW, Roeth FW, Klein RN, Knezevic SZ, Martin A, Nelson LA, Shapiro CA. 2001. Glyphosate-resistant soybean cultivar response to glyphosate. *Agron. J.* 93:404–407.
19. Fawcett RS. 2001. *Environmental impacts of transgenic crops.* Presented at North Central Weed Science Conference, Milwaukee, WI.

20. Feil B, Schmid JE. 2002. Dispersal of maize, wheat and rye pollen. A contribution to determining the necessary isolation distances for the cultivation of transgenic crops, Zurich: Institute of Plant Sciences, Swiss Federal Institute of Technology.
21. Fish AC, Rudenko L. 2004. A look at biotechnology and world hunger. Pew Initiative on food and biotechnology. www.pewagbiotech.org.
22. Ginder RG. 2001. Channeling, identity preservation and the value chain: lessons from the recent problems with StarLink corn. Iowa State University. http://www.exnet.iastate.edu/Pages/grain/publications/buspub/0103channel.pdf.
23. Hallauer AR, Russell WA, Lamkey KR. 1988. Corn breeding. In *Corn and corn improvement*, pp. 473–564. Madison, WI: ASA-CSSA-SSSA.
24. Hanna M, Van Fossen L. 1995. *Profitable corn harvesting. Rep. ISU ExtensionPM-574*, Iowa State University, Ames, IA. http://www.abe.iastate.edu/machinery/pm574.asp.
25. Harl NE, Ginder RG, Hurburgh CR, Moline S. 2003. The StarLink™ situation. Iowa State University. http://www.exnet.iastate.edu/Pages/grain/publications/buspub/0010star.pdf.
26. Haslberger A. 2001. GMO contamination of seeds. *Nature Biotechnol.* 19:613.
27. Haughton AJ, Champion GT, Hawes C, Heard MS, Brooks DR, et al. 2003. Invertebrate responses to the management of genetically modified herbicide-tolerant and conventional spring crops. II within-field epigeal and aerial arthropods. *Phil. Trans. R. Soc. London* 358:1863–1877.
28. Haygood R, Ives AR, Andow DA. 2003. Consequences of recurrent gene flow from crops to wild relatives. *Proc. R. Soc. London* 270:1879–1886.
29. Hurburgh CR Jr. 2000. The GMO controversy and grain handling for 2000. Iowa State University. http://www.exnet.iastate.edu/Pages/grain/gmo/99gmoy2k.pdf.
30. Hurburgh CR Jr. 2003. Constraints for isolation and traceability of grains. Personal communication. Iowa State University.
31. Hymowitz T. 2004. *Diversity within the perennial Glycine spp.* Presented at VII World Soybean Research Conference, Foz do Iguassu, Brazil.
32. Jordan N. 1999. *Escape of pest resistance transgenes to agricultural weeds: relevant facets of weed ecology.* Presented at Ecological effects of pest resistance genes in managed ecosystems, Bethesda, MD,
33. Jorgensen RB, Andersen B, Snow A, Hauser TP. 1999. Ecological risks of growing genetically modified crops. *Plant Biotechnol.*16:69–71.
34. Kato TA. 1997. *Review of introgression between maize and teosinte.* Presented at Gene flow among maize landraces, improved maize varieties, and teosinte: implications for transgenic maize, Mexico City.
35. Keeler KH, Turner CE, Bolick MR. 1996. Movement of crop transgenes into wild plants. In *Herbicide-resistant crops — agricultural, environmental, economic, regulatory, and technical Aspects*, Duke SO, Ed., pp. 303–330. Boca Raton, FL: CRC Press.
36. Kermicle J. 1997. *Cross compatibility within the genus Zea.* Presented at Gene flow among maize landraces, improved maize varieties, and teosinte: implications for transgenic maize, Mexico City.
37. Levine JM. 2000. Species diversity and biological invasions: relating local process to community pattern. *Science* 288:852–854.
38. Luna S, Figueroa J, Baltazar B, Gomez R, Townsend R, Schoper JB. 2001. Maize pollen longevity and distance isolation requirements for effective pollen control. *Crop Sci.* 41:1551–1557.
39. Martinez-Soriano JPR, Bailey AM, Lara-Reyna J. 2002. Transgenics in Mexican maize. In *Nature Biotechnol.*, p. 19.
40. Martinez-Soriano JPR, Leal-Klevezas DS. 2000. Transgenic maize in Mexico: no need for concern. *Science* 287:1399.
41. Matus-Cadiz PH, Horak MJ, Blomquist LK. 2004. Gene flow in wheat at the field scale. *Crop Sci.* 44:718–727.
42. McCabe D, Swoboda R. 2003. Where will we farm pharmacrops? In *Wallaces Farmer*, Vol. 1, pp. 12–13.
43. Muniz FRS, Di Mauro AO, Oliveira JA, Barbaro IM, Uneda-Trevisoli SH, Silveira GD, Goncalves ECP, Costa MM. 2004. *Heritability and genetic gain in soybean segregant populations.* Presented at VII World Soybean Research Conference, Foz do Iguassu, Brazil.

44. Noteborn HPJM, Peijnenburg AACM, Zeleny R. 2002. Strategies for analyzing unintended effects in transgenic food crops. In *Genetically modified crops: assessing safety*, Atherton KT, Ed., pp. 74–93. London: Taylor and Francis.
45. Olsen T, Rossiter L, McGuire J. 2003. Research project examines corn pollen drift. Iowa State University Extension. http://www.extension.iastate.edu/newsrel/2003/sep03/sep0317.html.
46. Oltmans SM, Zollinger RK, Henson RA. 2001. *Management of volunteer glyphosate-resistant wheat in glyphosate-resistant corn*. Presented at North Central Weed Science Society, Milwaukee, WI.
47. Owen MDK. 2000. Current use of transgenic herbicide-resistant soybean and corn in the USA. *Crop Protect.* 19:765–771.
48. Palmer RG, Gai J, Sun H, Burton JW. 2001. Production and evaluation of hybrid soybean. In *Plant Breeding Reviews*, Janick J, Ed., pp. 263–308. New York: John Wiley and Sons.
49. Payne J. 1997. Regulating transgenic plants: the experience of USDA in field testing, wide-scale production, and assessment for release in centers of origin. Presented at Gene flow among maize landraces, improved maize varieties, and teosinte: implications for transgenic maize, Mexico City.
50. Quist D, Chapela IH. 2001. Transgenic DNA introgressed into traditional maize landraces in Oaxaca, Mexico. *Nature* 414:541–543.
51. Radosevich SR, Ghersa CM, Comstock G. 1991. Concerns a weed scientist might have about herbicide-tolerant crops. *Weed Technol.* 6:635–639.
52. Rainbolt CR, Thill DC, Young FL. 2002. *Managing volunteer crops following herbicide-resistant crops*. Presented at Weed Science Society of America, Reno, NV.
53. Sanchez Gonzalez JJ, Ruiz Corral JA. 1997. *Teosinte distribution in Mexico*. Presented at Gene flow among maize landraces, improved maize varieties, and teosinte: implications for transgenic maize, Mexico City.
54. Santos DJ, Sandras VO Andrade FH. 2004. *Genetically modified soybean in Argentina: a yield–glyphosate tolerance trade off*. Presented at VII World Soybean Research Conference, Foz do Iguassu, Brazil.
55. Shelton AM, Zhao JZ, Roush RT. 2002. Economic, ecological, food safety, and social consequences of the deployment of Bt transgenic plants. *Annu. Rev. Entomol.* 47:845–881.
56. Snow AA. 2002. *Possible phenotypic effects of genetically modified pathways on gene flow from field tests*. Presented at Criteria for field testing of plants with engineered regulatory, metabolic, and signaling pathways, Washington, D.C.
57. Snow AA. 2002. Transgenic crops — why gene flow matters. *Nature Biotechnol.* 20:20.
58. Swoboda R. 2002. Bean quality bonus. In *Wallaces Farmer*, Vol. 2, pp. 16–17.
59. Swoboda R. 2003. Corn pollen trespass. In *Wallaces Farmer*, Vol. 7, p. 12.
60. Thompson S. 2002. A cloud of pollen. In *Wallaces Farmer*, Vol. 2, pp. 40–41.
61. VanGessel MJ. 2001. Glyphosate-resistant horseweed from Delaware. *Weed Sci.* 49:703–705.
62. Virgin I, Frederick R. 1997. *Searching for a balance: environmental concerns and potential benefits of transgenic crops in centers of origin and diversity*. Presented at Gene flow among maize landraces, improved maize varieties, and teosinte: implications for transgenic maize, Mexico City.
63. Westgate ME, Lizaso J, Batchelor W. 2003. Quantitative relationship between pollen shed density and grain yield in maize. *Crop Sci.* 43:934–942.
64. White JL. 2002. *U.S. regulatory oversight for the safe development and commercialization of plant biotechnology*. Presented at Ecological and agronomic consequences of gene flow from transgenic crops to wild relatives, Columbus, OH. http://www.biosci.ohio-state.edu/~asnowlab/Proceedings.pdf.
65. Wimberley RC, Reynolds WN. 2003. Food from our changing world: the globalization of food and how Americans feel about it. Southern Rural Development Center. http://srdc.msstate.edu/cred/.
66. Zelaya I, Owen MDK. 2002. *Amaranthus tuberculatus* (Mq. ex DC) J. D. Sauer: potential for selection of glyphosate resistance. Presented at 13th Australian Weeds Conference, Perth, Australia.

11 Wheat Domestication and Dedomestication — What Are the Odds?

Sharon Ayal and Avraham A. Levy

11.1 INTRODUCTION

How different is a wild progenitor from its domestic progeny? This kind of question has been at the heart of heated debates among geneticists, such as the Beadle–Mangelsdorf controversy on the origin of corn. A famous experiment by Beadle showed that five major genes distinguish the wild corn progenitor, teosinte, from its domestic derivative (5). It is also necessary to understand how many mutations are needed to reverse the process and turn a domestic crop into a feral, partially wild, self-propagating species. Here, we have addressed this question for wheat, using the new tools of genomics.

11.2 WHEAT DOMESTICATION

Most scholars agree that the domestication of wheat, as well as that of barley, several legumes, flax, and others occurred approximately 11,000 years ago in the western arch of the Fertile Crescent (8,11) (now Israel, Jordan, Syria, Lebanon, and southern Turkey). Wild emmer wheat, *Triticum turgidum* ssp. *dicoccoides,* is a tetraploid species ($2n = 4x = 28$; genome BBAA). This species, still living in the wild, was discovered by Aaronsohn at the beginning of this century in the vicinity of Mt. Hermon (1). This wild taxon is the progenitor of most domesticated wheats (Figure 11.1). It is the direct progenitor of domestic tetraploid hulled wheat, *Triticum turgidum* ssp. *dicoccum* (genome BBAA), and of domestic tetraploid free-threshing wheat, *Triticum turgidum* ssp. *durum* (genome BBAA). It is also the progenitor of the A and B genomes of hexaploid bread wheat (genome BBAADD). Interestingly, there are no known cases of ferality for tetraploid or hexaploid wheat in the regions where it is extensively grown, such as Europe or the American continent, and where there is no wild progenitor. The only reported case of ferality in wheat is in Tibet where feral hexaploid wheat was found (14).

Back to the question, how different are wild progenitors of wheat from domestic wheat? The obvious differences that can be noticed by analysis of visible phenotypes are as follows:

- Spike fragility (shattering), an essential trait in the wild, imposed too much effort for the harvest by early farmers. Thus, selection for variants with non-fragile spikes was one of the first consequences and may be one of the first priorities in the domestication process.
- The domestic has free-threshing grains (naked grains) vs. hulled grains in wild wheat.
- Seeds are non-dormant with uniform and rapid germination.
- Plants are erect with increased number of grains per spikelet.

All these traits, which are harmful in the wild, present obvious advantages to the early farmer, such as ease of harvest, higher yield, and high percentage of germination. There might be many additional

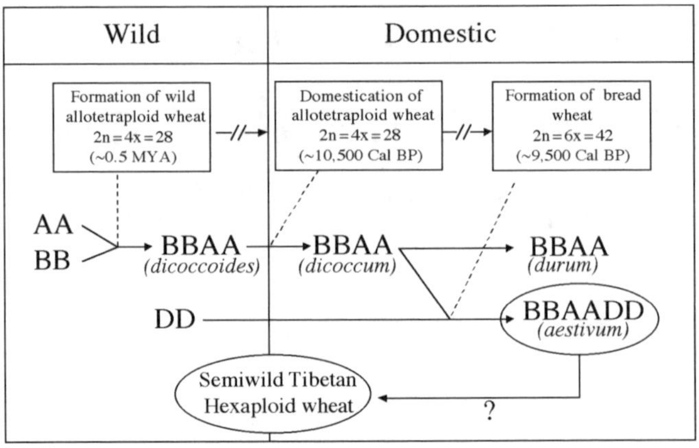

FIGURE 11.1 Evolutionary history of allotetraploid and allohexaploid wheat. Interspecific hybridization between the diploid *Triticum urartu* (genome AA, 2n = 2x = 14) as male and the donor of the B genome (an unknown species similar to *Aegilops speltoides*) as female, followed by chromosome doubling, gave rise, approximately 0.5 million years ago (MYA) (7,9), to wild allotetraploid wheat, *Triticum turgidum,* ssp. *dicoccoides* (2n = 4x = 28, genome BBAA). This allotetraploid is considered to be the direct progenitor of durum and bread wheat (6). Domestication of allotetraploid wheat took place approximately 10,500 calibrated years before present (Cal. BP) giving rise to a hulled domestic wheat *Triticum turgidum,* ssp. *dicoccum* (2n = 4x = 28, genome BBAA), and later on to free-threshing types including 2000 to 3000 Cal. BP to pasta wheat *Triticum turgidum* ssp. *durum* (2n = 4x = 28, genome BBAA) (6). Domestication of tetraploid wheat was rapidly followed (approximately 9500 Cal. BP) by a second round of intergeneric hybridization and chromosome doubling between domesticated allotetraploid wheat and the donor of the D genome, *Ae. tauschii* (2n = 2x = 14, genome DD), giving rise to bread wheat, an allohexaploid with 2n = 6x = 42 chromosomes (genome BBAADD). There is no known hexaploid wheat wild progenitor; however, a semiwild wheat discovered in Tibet has been proposed as a case of feral, semi-wild wheat. (Based on information in Feldman (6), Huang et al. (7), and Mori et al. (9).)

traits, not visible to the eye, that are essential for survival in the wild, such as resistance to a broad range of abiotic and biotic stresses, loss of secondary metabolites that might have been used by the wild progenitor for preserving the grain in the soil or for symbiosis with microbes or efficiency in the uptake of nutrients that are less abundant in nature than in the farmer's field, etc. Some of these genes might become dispensable in the domestic types and thus might become mutated or silenced. The new tools of genomics offer a new opportunity to estimate the extent of the genetic distance between the wild progenitors and their domestic progenies.

11.3 GENOMICS OF WHEAT DOMESTICATION

11.3.1 The Approach and the Materials

The purpose of genomics is to understand how all the genetic components work together in cells and organisms. Traditionally, knowledge on how genes are regulated was derived from several techniques that allowed the analysis of a limited number of genes simultaneously. During the last few years, complementary DNA (cDNA) array technologies have been developed (12) that allow the parallel and comparative analysis of the expression of many genes. Expression profiling using cDNA microarrays is emerging as a powerful method to identify genes of interest in plants like the SAAT gene involved in strawberry flavor (2), or genes that are involved in response to different kinds of stresses (10,13).

We have used cDNA microarrays to compare the genetic makeup of wild and domesticated tetraploid wheat. For this purpose, we chose 23 different lines of tetraploid (*turgidum*) wheat,

11 wild ecotypes (*Triticum turgidum* ssp. *dicoccoides* (TTD)) from different parts of the Fertile Crescent, 12 domesticated cultivars, 8 free-threshing (*Triticum turgidum* ssp. *durum* (TTR)), and 4 non-free-threshing (*Triticum turgidum* ssp. *diccocum* (TTC)). Seeds from these lines were kindly provided by Professor M. Feldman. The lines were analyzed first for DNA polymorphism, using amplified fragment length polymorphism (AFLP). Based on this analysis, 10 lines representing different phylogenetic taxa were chosen for gene expression analysis: 5 wild ssp. *dicoccoides* ecotypes and 5 domestic lines — 3 of ssp. *durum* and 2 of ssp. *dicoccum*.

11.3.2 Design and Analysis of cDNA Microarrays

We have used suppression subtraction hybridization (SSH) to enrich for cDNA clones that are differentially expressed between wild and cultivated tetraploid wheat. Two subtractive cDNA libraries were obtained: one library was enriched with the genes from domesticated wheat (SSH1) and the second with genes from wild wheat (SSH2). Sequencing of approximately 400 cDNA clones showed that the libraries were of high quality and had little redundancy. A total of 2492 cDNA clones were spotted in two replicas on superamine slides: 192 expressed sequence tags (ESTs) were from the SSH library that was enriched for domesticated genes (SSH1), 2108 ESTs were from the SSH library enriched for wild genes (SSH2), and 192 ESTs were derived from bread wheat seedlings that underwent a cold treatment. The bread wheat clones were kindly provided by the lab of Dr. Eliot Herman (USDA — Danford Lab., St. Louis). Total RNA (ribonucleic acid) was extracted, subjected to reverse transcription, and amplified by *in vitro* transcription with T7 polymerase (according to Ambion's protocol with some adjustment). Amplified RNA (aRNA) products were subjected to reverse transcription and then labeled with Cy3 and Cy5 by the indirect aminoallyl method. For each biological repetition, at least two hybridization reactions with swapped dye labeling were performed. Separate images for each fluorescence type (Cy3 or Cy5) were acquired using ScanArray® 4000 software (Packard BioScience, Meridan, CT) at a resolution of 10 μm per pixel, adjusting the photomultiplier and laser power to achieve an optimal distribution of signals without minimal saturation. Initial image analysis was performed using the QuantArray™ version 3 software using the histogram method. Data analysis was performed applying per-tip, per-spot, and per-chip normalization using GeneSpring® 6 software from Silicon Genetics.

11.3.3 Differential Expression of Genes in Young Spikes of Wild vs. Domesticated Wheat

RNA was extracted from a mix of different lines of either wild or domestic young spikes at the stage of 1 week after anthesis. The rationale for choosing this stage is that we expect some of the genes that were involved in domestication might already be active at this stage (e.g., fragility, free-threshing). Other tissues might also have been subjected to unconscious domestication-driven selection (e.g., roots or leaves) and would be interesting to study, but this is beyond the scope of this work. A first pool was made out of 5 wild wheat accessions and a second pool was prepared from 5 domesticated lines. The pools of wild vs. domesticated aRNA were labeled against each other in 6 replicas with dye swap. Genes that were up or down regulated by a factor of at least 1.8, in at least 4 of 6 replicas were identified and sent for sequencing and annotation. The genes for which sequences are available are listed in Table 11.1 and Table 11.2. The expression levels of 63 genes, which represent 2.5% of all genes tested, were up or down regulated: 38 genes (1.5%) were up regulated and 24 genes (1%) were down regulated in the wild compared to the domesticated lines (Figure 11.2). Most of the genes that were up regulated in the domestic varieties are correlated with sugar metabolism, such as the Rubisco large and small subunit and the *sucrose synthase* gene. Genes up regulated in the wild corresponded to storage proteins and to ribosomal RNA genes and to stress-related genes.

TABLE 11.1
Expression Levels of Selected Transcripts[a] that Were Up Regulated in Young Wild Spikes Compared to Domestic Spikes

Clone	D/W[b]	Putative Annotation[c]
WCC-2-23g	0.617	3-isopropylmalate dehydrataselike protein
SSH2-u3b	0.628	ribosomal protein
WCC-2-23d	0.602	30S ribosomal protein S3
SSH2-l5a	0.621	40S ribosomal protein S24
SSH2-u5g	0.605	60S ribosomal protein L34
SSH1-23e	0.54	alpha/beta-gliadin precursor
SSH1-211h	0.441	alpha/beta gliadin precursor
SSH1-23g	0.505	alpha-amylase inhibitor
SSH1-21d	0.517	ESTs AU032852
SSH1-211a	0.592	gamma-gliadin
SSH1-22h	0.669	gamma-gliadin precursor
SSH1-212a	0.577	glutenin low molecular weight subunit
SSH2-a9h	0.608	histone H2A
SSH1-24e	0.612	low-molecular-weight glutenin subunit group 6 type IV
WCC-2-14d	0.595	permease 1
WCC-2-18g	0.603	glyceraldehydes 3-phosphate dehydrogenase
SSH2-t2e	0.647	unspecific monooxygenase

[a] A more complete list and description is in preparation.
[b] The average induction level in domesticated spikes is compared to wild spikes.
[c] Annotations are based on the TIGR (The Institute for Genomic Research) database.

TABLE 11.2
Expression Levels of Selected Transcripts[a] that Were Down Regulated in Young Wild Spikes Compared to Domestic Spikes

Clone	D/W[b]	Putative Annotation[c]
SSH2-q3a	1.872	caffeic acid *O*-methyltransferase
SSH2-s10g	2.286	Rubisco small subunit
SSH2-s12g	3.314	phenylalanine ammonia lyase
SSH2-d11b	2.525	peroxidase
SSH2-j9g	1.801	bifunctional nuclease
SSH2-i6g	1.992	ribulose 1 5-bisphosphate carboxylase
SSH2-i8h	2.171	Rubisco chain precursor
SSH2-K10d	1.91	Rubisco activase A
WCC-2-18e	1.786	Rubisco large subunit
SSH2-d2d	2.708	Rubisco small subunit
SSH2-u3g	1.799	sucrose synthase

[a] A more complete list and description is in preparation.
[b] The average induction level in domesticated spikes is compared to wild spikes.
[c] Annotations are based on the TIGR database.

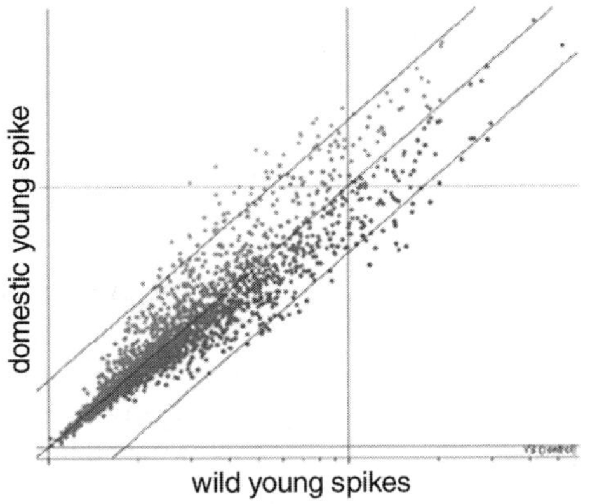

FIGURE 11.2 Microarray analysis of differential gene expression in wild vs. domestic wheat in young spikes. A scatterplot diagram is shown for the expression of 2493 genes in young spikes of domesticated vs. wild durum wheat. Expression levels are in the same normalized scale for the X- and Y-axes. The points beyond the diagonal lines represent genes that are up or down regulated genes by a significant factor greater than 1.8 in wild or domestic wheat.

11.4 ASSESSING THE ODDS OF DEDOMESTICATION

We have studied the alterations in gene expression, in the early stages of spike development that occurred during wheat domestication using novel tools of expression profiling via cDNA microarrays. One of our major findings is that approximately 2.5% of the genes are differentially expressed between wild vs. domestic wheat. Of the genes, 1.5% were up regulated in the wild; that is, they were lost, mutated, or silenced or their expression was significantly reduced during the 11,000 years of wheat domestication. Assuming that the tetraploid wheat genome contains approximately 40,000 genes, this means that approximately 1000 genes are differentially expressed between wild and domestic wheat and out of these, the expression of approximately 600 genes was reduced (i.e., lost, mutated, or silenced) by the domestication process. This figure is much higher than Beadle's 5 gene difference between wild and domesticated corn. Obviously, it is probable that not all 600 less-expressed genes are essential for survival in the wild. Conversely, it is also likely that it takes more than the few major genes that control visual phenotypic features (e.g., spike fragility) to make a species that can thrive in the wild. Many other genes (e.g., dormancy, nutrient uptake, hardiness) are probably involved. Moreover, another factor not taken into account in the analysis presented here is the variance in gene expression (the analysis shown here relates to mean effects from pooling of several genotypes). We have calculated the variation in expression between ecotypes for each of the 2492 genes studied. On the average, the variation is approximately 4-fold higher for a locus from the wild compared to the same locus in domestic wheat (data not shown). This enhanced biodiversity is probably the "insurance policy" of a wild species. It is the raw material for facing future natural selection pressures. We assume that dedomestication involves a genetic bottleneck as strong, if not stronger than domestication. Regaining genetic variation might be essential for the long-term survival of the feral plants and for it turning into a wild species.

11.5 SEMIWILD WHEAT FROM TIBET — A CASE OF DEDOMESTICATION?

Despite our assessment that the odds of dedomestication for wheat must be extremely low, becoming feral (e.g., becoming a weed in the crop of origin) might be easier than becoming a true wild plant. It might be achieved by only a few dominant mutations of reversion to the wild type because the habitat of the feral plant is similar to that of its crop of origin. We review below the possibility that wheat became feral in Tibet. Semi-wild hexaploid (2n = 6x = 42) wheat has been found in Tibet where it grows as a weed in barley and wheat (14). It is tall, weedy wheat with a broken rachis that facilitates shattering. Seven of eight accessions studied had red grain color, three had curved awns, two had short awns, and three were awnless. In crosses with wheat, the feral traits of broken rachis and red grain were dominant, as is the non-threshing hulled character (3,4). The Tibetan wheat crosses freely with bread wheat (*Triticum aestivum* ssp. *aestivum*) and the more primitive, hulled, spelt wheat (*T. aestivum* ssp. *spelta*). The greater shattering and the hulled traits led researchers to believe that it came from spelt wheat. A more careful analysis showed that the shattering phenotype was morphologically different from spelt wheat, and a random amplified polymorphic DNA (RAPD) analysis clinched that the Tibetan weedy wheat is much more closely related to bread wheat than spelt wheat (15). In support of the dedomestication origin of the Tibetan semi-wild wheat are the following facts: the lack of wild wheat relatives in the region of Tibet, and the lack of a wild hexaploid progenitor in the region of origin of wheat (the Fertile Crescent). One could therefore envision a reversion of the mutation controlling spike fragility to the dominant wild type. However, if the Tibetan wheat is indeed phylogenetically close to bread wheat and is different both morphologically and molecularly from spelt wheat, then we have to assume a reversion of yet another set of genes controlling the free-threshing trait that is present in ssp. *aestivum*. This is possible, given sufficient time (number of generations), large populations, and increase in fitness of the revertant. Nevertheless, the odds of double reversion are quite low and therefore one should keep in mind the possibility of an alternative theory to the dedomestication origin of the Tibetan semi-wild wheat. One may consider that when bread wheat was introduced to Tibet, some of the seeds might have been brought as hybrids with a wild relative. The progeny of such hybrids might have been backcrossed to local bread wheat, thus the phylogenetic and cytogenetic similarity to ssp. *aestivum*, and might have found favorable grounds in Tibet and succeeded to turn into semi-wild wheat and currently as a weed. This is a formal possibility; however, it has no support "on the ground": a wild relative growing in China (but absent in Tibet) that could have hybridized with wheat is *Aegilops tauschii*. The type of shattering of *Ae. tauschii* (barrel type), however, differs from that of the Tibetan feral wheat (wedge type).

It is thought that wheat was introduced to Tibet 3000 to 4000 years ago (6). Unfortunately, the details are not known. The reports on the appearance of the semi-wild Tibetan wheat do not go beyond the technical differences and do not attempt to provide agrohistorical data on wheat cultivation in Tibet — if known. The feral origin theory for Tibetan wheat should be further examined. In particular, an essential point is to establish a robust phylogeny of the semi-wild Tibetan wheat, using markers more reliable that RAPD.

11.6 CONCLUSIONS

In summary, the odds of returning to the wild for a species that has been seriously domesticated (and not mildly tamed), such as bread or durum wheat, and to thrive in the wild, seem extremely low because domestication has favored so many traits that are deleterious in the wild. However, becoming a weed might be much easier through a few mutations of reversion to the dominant wild type. In the case of wheat, isolating the genes that gave rise to domestic wheat, as we have initiated in this work, remains a major goal. It should shed light not only on the domestication process but also on the analysis of dedomestication events.

ACKNOWLEDGMENTS

We are thankful to Professor Moshe Feldman for providing seeds, support, and inspiration to this work and to Professor Jonathan Gressel for fruitful discussions, editing of the manuscript, and contribution to the paragraph related to feral wheat. This work was supported by an ISF Bikura grant to Abraham A. Levy.

LITERATURE CITED

1. Aaronsohn A. 1910. Agricultural and botanical explorations in Palestine. *Bull. Plant Industry* 180:1–63.
2. Aharoni A, Keizer LC, Bouwmeester HJ, Sun Z, Alvarez-Huerta M, et al. 2000. Identification of the SAAT gene involved in strawberry flavor biogenesis by use of DNA microarrays. *Plant Cell* 12:647–662.
3. Chen QF, Yen C, Yang JL. 1998. Chromosome location of the gene for brittle rachis in the Tibetan weed race of common wheat. *Genet. Resour. Crop Evol.* 45:407–410.
4. Chen QF, Yen C, Yang JL. 1999. Chromosome location of the gene for the hulled character in the Tibetan weedrace of common wheat. *Genet. Resour. Crop Evol.* 46:543–546.
5. Doebley J. 1992. Mapping the genes that made maize. *Trends Genet.* 8:302–307.
6. Feldman M. 2001. The origin of cultivated wheat. In *The world wheat book*, Bonjean A, Angus W, Eds., Paris: Lavoisier Tech. & Doc.
7. Huang S, Sirikhachornkit A, Faris JD, Su X, Gill BS, et al. 2002. Phylogenetic analysis of the acetyl-CoA carboxylase and 3-phosphoglycerate kinase loci in wheat and other grasses. *Plant Mol. Biol.* 48:805–820.
8. Lev-Yadun S, Gopher A, Abbo S. 2000. Archaeology. The cradle of agriculture. *Science* 288:1602–1603.
9. Mori N, Liu YG, Tsunewaki K. 1995. Wheat phylogeny determined by RFLP analysis of nuclear DNA. 2. Wild tetraploid wheats. *Theor. Appl. Genet.* 90:129–134.
10. Reymond P, Weber H, Damond M, Farmer EE. 2000. Differential gene expression in response to mechanical wounding and insect feeding in *Arabidopsis*. *Plant Cell* 12:707–720.
11. Salamini F, Ozkan H, Brandolini A, Schafer-Pregl R, Martin W. 2002. Genetics and geography of wild cereal domestication in the near east. *Nat. Rev. Genet.* 3:429–441.
12. Schena M, Heller RA, Theriault TP, Konrad K, Lachenmeier E, Davis RW. 1998. Microarrays: biotechnology's discovery platform for functional genomics. *Trends Biotechnol.* 16:301–306.
13. Schenk PM, Kazan K, Wilson I, Anderson JP, Richmond T, et al. 2000. Coordinated plant defense responses in *Arabidopsis* revealed by microarray analysis. *Proc. Natl. Acad. Sci. USA* 97:11655–11660.
14. Shao Q, Li C, Basang C. 1983. *Semi-wild wheat from Xizang (Tibet).* Presented at 6th International Wheat Genetics Symposium, Kyoto, Japan.
15. Sun QX, Ni ZF, Liu ZY, Gao JW, Huang TC. 1998. Genetic relationships and diversity among Tibetan wheat, common wheat and European spelt wheat revealed by RAPD markers. *Euphytica* 99:205–211.

12 Feral Rye — Evolutionary Origins of a Weed

Jutta C. Burger and Norman C. Ellstrand

12.1 INTRODUCTION

Feral, or weedy, cereal rye (Poaceae: Triticeae, *Secale cereale* L.) is a widely distributed weed of agricultural and ruderal habitats. It persists in agricultural systems under strong selection pressure for crop mimicry as well as in non-agricultural systems under strong selection pressure for several non-cultivated traits. The diversity of weedy rye has led to much confusion about its origins as well as their potential impact. In the case of cereal rye, we cannot even be sure that all weedy rye is indeed feral (i.e., derived from a crop either directly or via hybridization) and not merely an ancestral form of cereal rye. Cereal rye is itself believed to have originated as a secondary crop from an ancestral weedy form of *S. cereale* in wheat and barley fields (26,47). Clearly, determining the origin of weedy rye and its relationship to cultivated rye is critical to understanding the evolutionary origins and adaptations of weedy rye populations. We will therefore first review taxonomic relationships within *Secale*. We will then provide an overview of domestication, followed by an introduction to dedomesticated and predomesticated forms of *Secale cereale*. Finally, we will present preliminary data from an investigation of the origin and divergence of feral rye in the western U.S. In this chapter, the term *weedy rye* will be used to refer to persistent annual cereal rye that could be ancestral or only distantly related to cultivated cereal rye, whereas *feral rye* will be used when we know that weedy plants are at least in part derived from cultivated cereal rye.

Cultivated cereal rye (*Secale cereale cereale*), weedy rye, and related species in the genus *Secale* have 7 pairs of chromosomes like all other diploid Triticeae (24). With the exception of *S. strictum africanum* and *S. sylvestre*, *Secale* species are highly outcrossing and have a complex gametophytic incompatibility system (28). Cultivars as well as weedy forms of rye are highly polymorphic and are morphologically poorly distinguishable from one another. Rye ranks economically among the 20 most important crops worldwide. It has, however, been steadily decreasing in status since the advent of modern agricultural technology, the development of cold-tolerant wheat varieties, and the decline in demand for cereal rye (28). According to FAO (Food and Agriculture Organization of the United Nations) statistics, rye production over the last decade has decreased in 49 countries, whereas it has increased only modestly in 11 (18a). Despite its diminishing popularity, rye has valuable properties as a crop. It can be grown across a wider range of climates than any other cereal grain. Winter rye can survive temperatures of –25 to –35YC even in the absence of a protective snow cover (35). Rye can also germinate and grow to maturity with as little as 130 mm of rain per year. Cereal rye, its weedy relatives, and its perennial sister species, mountain rye (11), all have extensive and deep root systems that help to maximize water uptake in dry climates. Cereal rye, for instance, can have roots that extend downward to 1.5 m and laterally to 0.9 m (36). This species is a strong competitor for water and purportedly also has allelopathic properties that make it useful as a cover crop and a weed-suppressing rangeland grass.

Despite having highly beneficial characteristics such as drought and cold tolerance, rye has long held a dubious reputation as a cereal crop because of its tendencies to volunteer in years following harvest, as was already documented by Pliny the Elder nearly 2000 years ago (36). This

TABLE 12.1
Traits Likely to Be Associated with Weedy Rye in Different Environments

	Environment	
Trait	**Cultivated**	**Uncultivated**
Flowering/ripening	Uniform	Opportunistic
Phenology	Convergent with crop	Independent of crop and potentially associated with environment (winter in north, spring in south)
Seed dormancy	No	Yes
Seed enclosure	Free	Enclosed in floret
Seed size	Large	Small
Seed spike	Stiff	Freely shattering (brittle)
Vegetative growth	Limited	Extensive

phenomenon is of special concern when wheat is planted after rye. Because rye grains cannot be sorted from wheat after harvest, they remain either as a seed contaminant that is resown or as an impurity that reduces leavening properties of wheat flour. Rye is grown in every continent where agriculture is possible, but there is little published about the distribution of its cultivated or weedy forms outside of Eurasia and North America. Therefore, we will restrict our analysis of the origins of weedy rye to its native range in Europe and Asia and its introduced range in the U.S.

12.1.1 The Many Faces of Weedy Rye

Weedy rye is a weed of cultivated fields, roadsides, ruderal habitats, and, occasionally, undisturbed wildlands. In contrast to rye persisting outside of cultivation, weedy forms not exposed to annual cultivation, harvesting, and re-seeding will experience selection for dispersing or shattering seed heads (Table 12.1, Figure 12.1). Strong selection for traits beneficial to either cultivated or uncultivated environments has probably at least in part maintained high genetic diversity in feral rye and produced ecological divergence in cross-compatible forms of rye. Spike fragility and perenniality are both considered to be useful taxonomic characters for identifying species, subspecies, and races of rye in Eurasia, whereas they have been generally disregarded as being evolutionarily or even ecologically important to weedy rye in North America.

Based on available data, it appears that the origins of weedy rye are unknown and may vary across continents. We generalize that weedy rye has one or more of three origins.

1. Wild and not derived from the crop cereal rye (*non-feral*)
2. Derived directly from one or more cereal rye cultivars (*endoferal*)
3. Derived from hybridization between cereal rye and a related wild species (*exoferal*)

In short, weedy traits can originate either from rare alleles in cultivars, from back mutations of cultivated traits to feral traits, or from hybridization with a species or race that has the weedy traits.

12.1.2 Taxonomy of Rye and Its Relatives

Divergent opinions exist about the exact taxonomy and phylogeny of *Secale*. A recent taxonomic revision of *Secale* recognizes only three species encompassing a total of five subspecies: *Secale cereale* (*S. c. ancestrale* the progenitor *and S. c. cereale* the crop), *S. strictum (S. s. africanum,* and *S. s. strictum*)*,* and *S. sylvestre* (21). For consistency, we will use the species definitions of this revision. Historically, and even still presently, however, many more species and subspecies have been recognized (for a partial list, see Table 12.2). Most notably, the self-compatible "vavilovii"

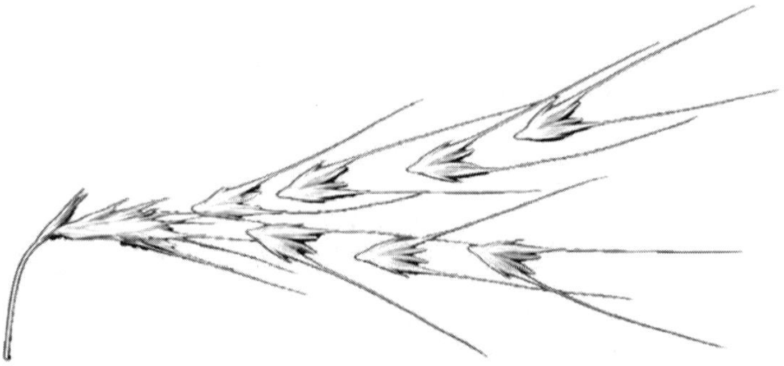

FIGURE 12.1 Brittle seed head of a weedy North American population of feral rye (*Secale cereale*).

race of *S. c. ancestrale* from northern Iran is referred to as a separate species in many publications (14,26).

Although cytogenetic differences exist among *Secale* species, few, if any, good morphological characters separate them. For instance, in a detailed morphometric study, Frederiksen and Petersen (20) found only perennial habit and spike fragility differentiating mountain rye (*S. strictum*, previously *S. montanum*) from cereal rye (*S. cereale*). We have observed that even spike fragility is variable in *S. strictum*, as shown in PI 272388 of the USDA (United States Department of Agriculture) germplasm accession. Similarly, the "daralgesii" race of mountain rye has a non-fragile spike. Because *Secale* species are at least partially interfertile, differences between taxa are even further reduced in zones of sympatry, presumably from occasional hybridization and introgression (20).

Species relationships have been deduced primarily through study of meiosis in F_1 hybrids between species. Crosses between *S. cereale* and *S. strictum* produce hybrids whose chromosomes do not properly pair during meiosis. Similarly, crosses between *S. sylvestre* and *S. cereale* result in poor seed set and low fertility of F_1 offspring. Reduced fertility in F_1 hybrids can be rapidly overcome by backcrossing to a parental population. Khush (26) performed reciprocal crosses between several races of *S. c. ancestrale* and *S. c. cereale* and found no incompatibilities between them, with the exception of the "vavilovii" race in which he found some evidence of chromosomal rearrangement. Similarly, *S. s. africanum* and *S. s. strictum* have no cytological incompatibility despite their complete geographic isolation from one another (*S. s. africanum* is endemic and restricted to southern Africa). Recent comparisons of internal transcribed spacer (ITS) sequences, simple sequence repeats (SSRs), and heterochromatic chromatin across *Secale* species and subspecies reconfirm that *S. strictum* is ancestral to other *Secale* and that the "vavilovii" race is distinct from other *S. cereale* subspecies (14,15). At least one collection of a "vavilovii" race from northern Iran showed greater similarity in chromosome arrangement to *S. strictum* than to *S. cereale*. This population is now considered to be a self-fertile form of S. *strictum* that had been previously misidentified (21,40). Other "vavilovii" populations may actually be hybrid *S. strictum* *S. cereale* derivatives.

Cereal rye differs from mountain rye by two reciprocal chromosomal translocations that provide a partial reproductive barrier between the two species (38,39). These translocations appear to have occurred between the short arms of chromosome 2 and chromosome 6 and the long arm of chromosome 7 (7,14,16). How these translocations became fixed is not entirely clear. But, assuming that translocants of *S. strictum* were conferred with a selective advantage along with partial reproductive isolation from their parental species that the chromosomal translocation imparted, they could form a new species. Stutz (43) proposed an alternative scenario in which *S. cereale cereale* originated from a complex of weedy races that were derived from "vavilovii" race *S. strictum* hybrids. The "vavilovii" type, in turn, originated from the self-compatible annual *S. silvestre* through a single translocation event. Self-incompatibility in cereal rye would thereby

TABLE 12.2
Old[a] and New[b] Taxonomy of *Secale* Species

Species[a]	Description	Current Status[b]
	Annual spp.	
Secale cereale L.	Cultigen with stiff rachis and either winter or spring annual habit.	*S. cereale cereale*
S. turkestanikum*	Self-fertile, semibrittle cultigen in central Asia and Transcaucasia.	*S. cereale cereale*
S. afghanicum (Vavil.)	Weed of Caucasus, western Asia, and India.	*S. cereale ancestrale*
Secale ancestrale (Zhuk.)	Brittle annual weed with long culms (300 cm), small seeds, and fragile rachis, restricted to sandy ditches south of Izmir, near Aydin in southern Turkey.	*S. cereale ancestrale*
S. segetale (Zhuk.)	Semibrittle polymorphic annual weed of grain fields occurring from the Black Sea east to Lake Balkhash.	*S. cereale ancestrale*
S. dighoricum (Vavil.)	Semibrittle annual weed (more so than "segetale"), restricted to grain fields of northern Ossetia, Caucasus.	*S. cereale ancestrale*
S. vavilovii Grossh.*	Low-growing, morphologically variable annual psammophyte with fragile rachis. Found on lower slopes of Mt. Ararat and Araxis River. Some populations are selfing. According to Stutz (personal communication), populations from northern Iran are misidentified *S. cereal* *S. strictum* hybrids.	*S. cereale ancestrale*
	Perennial spp.	
S. strictum (Presl.)	Polymorphic perennial species with brittle rachis widely distributed across dry mountainous regions from the middle Atlas Mountains of Morocco and the Sierra Nevada of Spain eastward to the Caspian Sea.	*S. strictum strictum*
S. anatolicum Boiss.	Highly polymorphic perennial weed with brittle rachis common throughout Turkey and western Iran and Iraq.	*S. strictum strictum*
S. ciliatoglume (Boiss.)	Isolated perennial weedy populations with pubescent culms, leaf blades, and sheaths, and brittle rachis. Occurs in orchards or vineyards near Mardin, Turkey (Kurdistan).	*S. strictum strictum* var. *ciliatoglume*
S. dalmaticum Vis.	Narrowly distributed blue-green, broad-leafed perennial with brittle rachis. Found in Montenegro at Johannes fortress above Kotor, Yugoslavia.	*S. strictum strictum*
S. daralgesii Tumanian	Weedy/semicultivated perennial race with non-fragile rachis narrowly distributed along roadcuts and ditchbanks of Armenia. It may be of hybrid *S. strictum* *S. cereale* origin.	*S. strictum strictum*
S. kuprijanovii Grossh.	Narrowly distributed broad-leafed perennial with robust culm. Found in mountain meadows of northern and western Caucasus Mountains.	*S. strictum strictum*
S. africanum Stapf	Self-compatible, fine-textured perennial endemic to southern Africa.	*S. strictum africanum*
Secale sylvestre Host	Low growing, selfing annual psammophyte with brittle rachis. Found from central Hungary eastward to sandy steppes of southern Russia. Distinct from other *Secale* in having awned glumes.	*S. sylvestre*

[a] Older designations according to Stutz (43), Zohary (52), and Frederiksen and Petersen (20).
[b] Current status according to Frederiksen and Petersen (21).
* "Vavilovii" populations are often still referred to as *S. vavilovii*.

have had to evolve from a selfing progenitor. Stutz's hypothesis is less parsimonious than that of reciprocal translocation without hybridization, but is a more reasonable explanation of how translocations could have become fixed through initial inbreeding in precursors of *S. cereale*.

12.2 HISTORY OF DOMESTICATION

Cereal rye was probably domesticated as late as 4000 to 5000 years BP (before the present) as a secondary crop from a weedy ancestor that grew in wheat and barley fields of eastern Turkey and neighboring areas (40,48,52). Uncertainty remains as to the exact time of domestication, in part because domestication has been a continuous process until recently (8), but also because preserved cultivated rye grains from archeological sites cannot be easily differentiated from wild material or even specifically from *S. strictum*. Its entry into central and northern Europe came with transport of wheat and barley northward (beginning around 2800 BP) by the Celts (40).

The immediate ancestor of cultivated rye is believed by many to be the "segetale" race of *S. c. ancestrale*, a semibrittle and persistent weed of cultivated fields that still occurs in eastern Europe and the Middle East (26,52). As mentioned above, Stutz (43) alternatively proposed that the ancestor of cereal rye was the "vavilovii" race, which produced the ancestral stock of cereal rye via natural hybridization with mountain rye (*S. strictum*). The immediate weedy ancestor is generally accepted to have evolved from mountain rye, a perennial species with shattering seed head that is still widely distributed across southern Europe and Asia Minor (18,26) (Figure 12.2).

Based on currently existing centers of *S. cereale* diversity and the presence of an accessory "B" chromosome in eastern Asian populations, there may have been two primary zones of rye domestication: one in the mountainous regions of Afghanistan, northeastern Iran, and Turkmenistan and one to the west of the Caspian Sea in Asia Minor (26). Linguistic evidence suggests its origin as a weed in other crops: the Persian, Arabic, and Turkish names for rye literally translate to mean "weed growing among wheat and barley" (26,40). As cereal crops and their associated weeds were carried northward, cereal rye itself ultimately became valued as a crop because of its superior cold and drought tolerance relative to other grain crops.

Today's cultivated varieties are bred to have an intact (*non-shattering*) seed head that holds seeds together until harvest. In contrast, naturally occurring and weedy races or subspecies of cereal

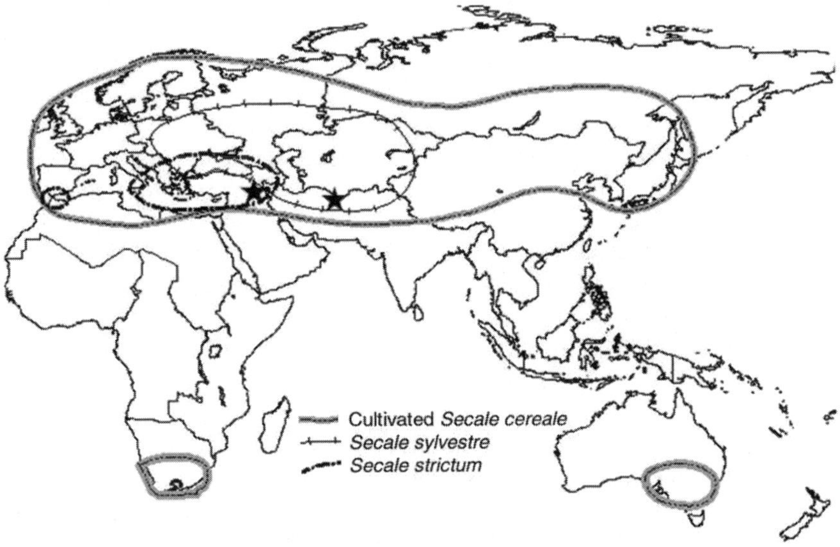

FIGURE 12.2 Eurasian distribution of *Secale*. Stars represent likely origins of cultivation. (Adapted from Sencer and Hawkes (40).)

rye have shattering seed heads (6,26). Although the brittle rachis phenotype in rye is controlled by a single dominant gene with non-shattering as recessive (27), various degrees of shattering have still been reported in wild and weedy plants (6,27,43). The variation in spike fragility found across races of annual and perennial *Secale* could be due either to environmental factors or to minor gene modifiers that have not, as yet, been identified (27). Khush (26) suggested that large grain size in *S. cereale cereale* evolved from competition with crop plants under cultivation (crop mimicry) and perhaps also from threshing practices. Erect growth habit and robust culms could have arisen from direct selection under domestication or from selection of correlated traits (26). The stiff non-shattering rachis typical of cultivated rye was directly or indirectly selected for by farming practices.

12.2.1 Distribution of Cultivated Rye in Eurasia

Cereal rye is cultivated across northern Europe, the Near East, and Asia, ranging northward to Norway and eastward to Japan (Figure 12.2). Its distribution as a crop has changed little in recent history, but its global economic importance has dropped precipitously across most of its range. The Russian Federation remains the primary producer and consumer of cereal rye, with 3.5 million ha harvested in 2003 (18a). Local landraces that are especially well adapted to regional environmental conditions, such as cold winters in Scandinavia and Russia, moist summers in north central Europe, or hot, dry summers of eastern Turkey are still grown across the wide geographic range of rye cultivation (36,37). Winter rye varieties from Bulgaria and the former Yugoslavia require only 15 to 25 days of vernalization to flower, whereas Siberian rye varieties require 35 to 40 days. Similar geographic variation exists for the long-day length requirement to initiate flowering: northern varieties require, on average, longer days or more intense light than southern varieties (36).

12.2.2 Wild and Weedy Rye in Eurasia

Secale cereale ancestrale has, by all accounts, been an agricultural weed since the advent of agriculture in Eurasia (47). Weedy populations occur outside cultivation across much of the southern end of the historic agricultural range of cereal rye (6,52). Populations of shattering weedy rye that persist outside of cultivation are more common toward the western end of the range of ancestral rye races, predominantly in Turkey (i.e., the "anatolicum," "vavilovii," and "ancestrale" races of *S. c. ancestrale*). Freely shattering populations of rye are absent from western and northern Europe (28,44). Kranz (28) has hypothesized that the moist, maritime climate of northern Europe limits natural dispersal by inhibiting seed heads from shattering.

In western and central Asia, non-brittle winter annual forms of rye (the "segetale" and "dighoricum" races of *S. c. ancestrale*) are weeds in grain fields, especially in wheat. In fact, they occur predominantly as non-brittle winter annual weeds of wheat (6). Vavilov (48) suggested that these populations are remnants of cultivated rye that had been admixed in cultivated wheat fields and had eventually supplanted wheat in the northern climates to which they were better adapted. Zohary (52), in contrast, suggested that farmers have allowed weedy forms to persist over centuries and that they, in turn, have converged in seed size and other characteristics to wheat crops. In an extensive survey of rye in central Asian agroecosystems, Balabajev (6) found that weedy rye was common in wheat fields and along field edges. He further found that its abundance as a weed in wheat fields increased with elevation, reaching a maximum density near the limits of winter wheat cultivation at 2450 m.

Mountain rye (*S. strictum*) occurs in mountain regions and dry plateaus from south central Europe to as far east as the Caucasus, with its center of distribution located in Turkey (Figure 12.2) (28). Isolated populations also occur in Sicily, the Sierra Nevadas of Spain, and northern Morocco (Figure 12.2). Kranz (28) even reported that some (probably introduced) *S. strictum* could be found in northern Germany in the Lueneburger Heide. Mountain rye is partially compatible with cereal rye, but can often be distinguished from the latter by its thinner leaves, perennial habit, and freely shattering seed heads. Weedy hybrid mountain cereal rye swarms occur in central Turkey and along the Anatolian Plateau. Morphologically, hybrids are indistinguishable from cereal rye with the exception of having a shattering seed head. According to Zohary (52) their hybrid status has been cytogenetically confirmed. Wild and weedy races of *S. cereale*, in contrast, extend well beyond the range of mountain rye and are mostly associated with disturbed habitats and agriculture. Zohary (52) suggested that the range expansion and adaptation of weedy *S. cereale* to diverse human-modified environments have been facilitated by introgression from mountain rye. We furthermore suggest that repeated backcrossing with cultivated rye can quickly obscure past hybridization. The evolutionary role of gene flow from cultivated cereal rye into weedy populations is probably large but entirely unknown.

Secale sylvestre is an annual inbreeding species that is distributed from eastern Europe (Hungary) to the Altai Mountains in central Asia (21). Although it is cytogenetically more similar to *S. strictum* than to *Secale cereale*, the two species are ecologically, reproductively, and, over most of their range, also geographically isolated (40). *Secale sylvestre* occurs primarily on sandy soil along shorelines, pastures, and sand dunes and is not considered weedy, despite having a shattering seed head.

Both cultivated cereal rye and its related wild and weedy forms are adapted to a winter annual habit along the higher latitudes and altitudes of their distribution, whereas they tend toward a spring habit in warmer, dryer climates of eastern Turkey and south central Asia (40). Collections of primitive rye specimens from northern Iran have, for instance, a spring habit.

Because weedy progenitor races of *S. cereale* still exist and occasionally co-occur with cereal rye near its center of origin, the origin of weedy races in agricultural systems remains unknown. Reversion of cultivated cereal rye to feral rye via endoferality may be possible but has never been documented. Endoferal rye, if it exists, cannot at present be differentiated from either weedy ancestral races or hybrid (exoferal) lineages between cereal rye and related species or races. High potential gene flow across varieties, races, and even species of *Secale* renders identification of clear origins difficult, if not impossible. Rye of every intermediate form, from annuals with non-disarticulating rachis, to those with a semibrittle rachis, to those with a completely shattering rachis, as well as perennials with both shattering and even non-shattering seed heads have been collected throughout Asia Minor eastward to northern Iraq and Iran (26,52). In this chapter, we will focus on the introduction of rye and the distribution and ecology of its weedy forms in North America, because of the difficulties in assigning evolutionary relationships to Old World populations.

12.2.3 Distribution of Rye and Weedy Rye in North America

12.2.3.1 Rye in North America

Cereal rye has been cultivated in North America for seed (primarily for bread flour and whiskey), forage, hay, green manure, and as a cover crop since Old World agriculture was introduced. Rye grows well in regions with little rainfall or on acidic or otherwise poor-quality soil. Up to the early 1900s, rye was primarily grown in the northeastern U.S. (29) (Figure 12.3A). By the 1950s, however, rye production had shifted to the central U.S. and extended westward into the Great Basin ecoregion (10) (Figure 12.3B). Distribution maps for rye based solely on grain harvests may, however, underestimate rye agriculture, because rye was historically and is still planted also as a cover crop. In 2003, for instance, 555,610 ha were planted with rye in the U.S., whereas only 137,188 ha were harvested (46a). Cereal rye remains a popular winter cover crop that produces an ample amount of leafy biomass, or green manure, which is tilled under in spring before planting. Today, Oklahoma, Georgia, and South Dakota are, in order of importance, the major rye producing states in the U.S., harvesting a total of 62,322 ha in 2003 (46a). The average yield of cereal rye in the U.S. was 1.71 tons/ha in 2003, notably lower than the average worldwide yield of 2.14 tons/ha (18a).

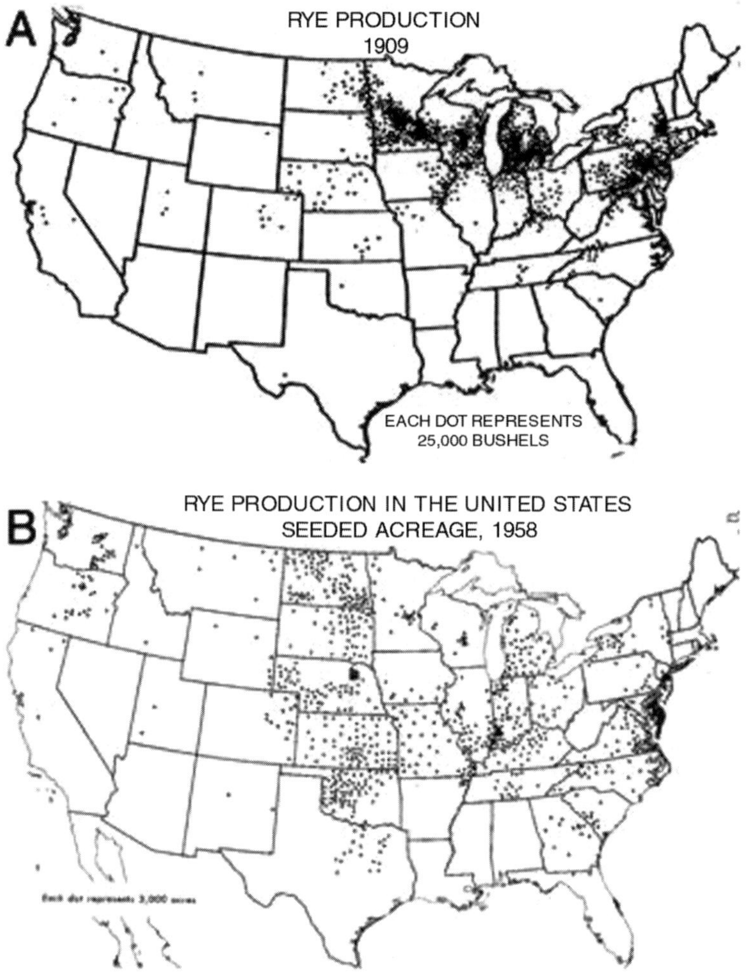

FIGURE 12.3 Historic distribution of cultivated rye in the U.S. A: Rye production in 1909. B: Rye production in 1958.

In the early 1900s, cereal rye and, to a lesser extent, its close perennial relative, mountain rye, began to be seeded for erosion control and as quick, dependable forage on overgrazed rangelands and cleared land. A short-lived perennial mountain rye cereal rye hybrid derivative dubbed "Michels grass" enjoyed brief fame as a forage and rangeland grass in the late 1930s and early 1940s (51). Ironically, those qualities that make cereal rye a problematic grain crop in some regions, that is, its propensity to volunteer and its profuse vegetative growth, made Michels grass and mountain rye attractive for revegetation projects. The USDA Soil Conservation Corps, U.S. Forest Service, Civil Conservation Corps, and various state transportation departments seeded cereal rye in eroded areas and road cuts to stabilize the soil (33). They also experimented with mountain rye, and to a lesser extent, with Michels grass. Both Michels grass and mountain rye had the additional advantage of, at least initially, being perennial and self-dispersing (i.e., having brittle seed heads with freely dispersing spikelets).

When used for erosion control or revegetation and range land forage, rye is typically allowed to reseed itself rather than being harvested. Cereal rye grown in this manner is often grazed by livestock, but is released completely from the selective pressure of cultivation. More recently, cereal rye has also been used effectively to control serious invasive weeds such as *Centauria maculata* and *C. diffusa* in Okanogan County, Washington (19). Similarly, mountain rye has been used experimentally for reclamation of weed-infested rangeland and coal mine spoils because of its great ability to bind soil, tolerate acidic soils, and suppress weeds (2,3,12).

The brief but varied history of rye in the U.S. is made more complex by the succession of different rye cultivars employed, the repeated use of locally grown seed, and the past localized introductions of hybrid Michels grass and mountain rye. The names of cultivars used by farmers and range land managers were unfortunately often not documented. Even today, commercially available rye is frequently given no more specific a name than *winter*, *spring*, or *common rye*. The term *common rye* is also used in records and refers to a commonly available, presumably locally adapted seed source from a particular region (30). Unlike wheat, rye has relatively few cultivars. Of these, fewer still have been introduced into North America. Records of introductions, popularity, and distribution of different cultivars have been kept in agricultural research experimental station reports and farmers' bulletins (10,30) and probably give an accurate representation of popular cultivars. We assume that most unnamed varieties originated from one or more of these documented cultivars.

Most early cultivars in the U.S. were of the winter type (30). Winter types are sown in the fall and are often cold-tolerant. In contrast, spring types are usually sown in the spring, are less cold-tolerant, but mature more quickly than winter types. Popularity of cultivars varies by region across the U.S. (Table 12.3). In the northeast, the 'Balbo,' 'Rosen,' and 'Tetra-Petkus' were historically popular. In the south, both 'Wrens Abruzzi' and 'Balbo' were common, though the former was used primarily for green manure. In the midwest, 'Rosen,' 'Minnesota No. 2,' and 'Prolific' were popular. In the western states, 'Rosen,' 'Petkus,' 'Prolific,' and, in California, 'Merced' were common, and 'Abruzzi' was often used as a cover crop. Overall, 'Rosen' rye was historically probably the most widely planted cultivar in the western states (41). More recently, a spring type cultivar ('Gazelle') from Canada has become popular. The decline in interest in rye farming has unfortunately left farmers with few commercially available cultivars to choose from.

TABLE 12.3
Most Common Rye Cultivars Historically Planted across the U.S.

Cultivar	Introduction/ Release	Country/ State of Origin	Habit	Region Planted
Abruzzi/Wrens Abruzzi	1900, 1904	Italy	Variable	Southeast
Balbo	1933	Italy	Winter	Midwest, East
Dakold (ND #959)	1902	North Dakota	Winter	Northern Midwest, Northwest
Gazelle	1974	Canada	Spring	Northwest
Imperial (White rye)	1929	Wisconsin	Winter	Northern Midwest
Kung (KingII)	1930s	Sweden	Winter	Northwest
Merced	1947		Spring	California
Petkus, Tetra-Petkus	1900	Germany	Winter	Northeast, Northwest, Wisconsin,
Prolific	Pre-1940	Canada (Germany)	Variable	Northwest, Northern Midwest
Rosen	1912	Russia	Winter	Northwest, Midwest, Northeast
Schlanstedt (Pedigree #2)	1900	Germany	Winter	Wisconsin, North Dakota
Swedish Rye (Minn. #2)	1895	Minnesota	Winter	Northern Midwest, Northwest
Vern	1905	Idaho	Spring	Northwest, California

Sources: Based on Briggle (10), Leighty (29), and Martin and Smith (30).

12.2.3.2 Weedy Rye in North America

Weedy rye occurs both as a volunteer from rye grain remaining after harvest and as a feral crop. Volunteer rye was already identified as a serious weed of wheat fields in the U.S. in the early 1900s (29). By the mid-1920s, before the Seed Purification Act was implemented, it had become a major grain contaminant (9). Farming agencies sent recommendations to farmers not to plant wheat after rye, as rye tended to reappear in years following harvest. Rye was and still is considered a costly weed primarily because rye grain cannot be separated from wheat after harvest. When the two are milled together, rye reduces the leavening properties of wheat flour enough to significantly reduce its value, though it produces more flavorful bread.

Weedy rye is still referred to as *volunteer rye* by many farmers and crop specialists, especially in the midwestern U.S. Crop volunteers are defined as originating directly from remnant seed of cultivated crops. Although we know that rye does volunteer, the mere fact that domesticated cereal rye is so rarely planted at present implies that the rye that persists in wheat fields is feral and not volunteer.

The northern Great Plains and Great Basin ecoregions of the U.S. continue to suffer heavy crop losses from feral rye (Figure 12.4). Interestingly, feral and volunteer rye does not appear to be a serious weed in the northeast and is of only minor importance in the southeast despite the fact that it has been grown there longer than in the west.

Feral rye in agricultural fields is reported to have low primary seed dormancy in contrast to several other serious weeds of wheat: most grains (99%) germinate when exposed to favorable conditions (4). If, however, seeds do not receive adequate moisture, they may persist in the soil seed bank for years (42).

12.2.4 Weedy Rye in the Western U.S.

Feral rye is an especially significant problem in the western wheat producing regions of the U.S. (Figure 12.4) (5). Rye is listed as a noxious weed in Washington and has been referred to in recent reference texts of important weeds of the western U.S. The Pacific Northwest Exotic Pest Council considers it invasive in Washington and Oregon (50) (http://www.nps.gov/plants/alien/).

Self-sustaining, naturalized populations of *S. cereale* have only been documented on uncultivated land since the 1960s (4,45). Whereas the total area of rye cultivation has decreased, feral rye, both in planted fields and out, appears to still be increasing in abundance and expanding in

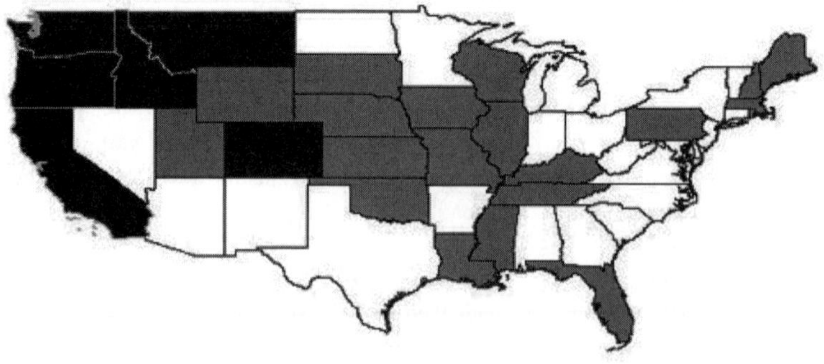

FIGURE 12.4 Distribution of feral and volunteer rye in the continental U.S. White: not persisting outside of cultivation. Gray: occasionally a serious weed of agricultural fields but not known to persist outside of cultivation. Black: a persistent naturalized weed outside cultivation and occasional serious weed within. Note: cereal rye is grown in all 50 continental states. (Source: Compiled from local floras from all 50 states and Internet weed databases.)

range. Naturalized populations of rye are generally assumed to be merely cultivated rye gone wild, but their superficial similarity to cereal rye may belie a more complex evolutionary origin.

As mentioned earlier, hybrids of *Secale cereale* and its perennial ancestor, *S. strictum*, were promoted and distributed for a limited time for erosion control and forage in the late 1930s and early 1940s (46,51). Feral rye populations persisting outside cultivation in the western U.S. share some traits of *S. strictum* in that they have shattering seed heads, more fragile culms, and smaller seeds than cultivated cereal rye (Burger, unpublished data). They share other traits with *S. cereale*, such as annual habit (25).

Naturalized populations of feral rye occur throughout the western U.S. (Figure 12.4). Large populations exist in northern California and in the dry canyons and roadsides of Oregon and eastern Washington. Smaller populations occur in southern California along roadsides and sandy riverbanks above approximately 600 m where no cereal rye is known to have been planted historically. Although rye has long been recognized as a volunteer weed by wheat farmers (see Section 12.2.3.2), naturalized populations were not observed in the northwestern U.S. until the mid-1950s (22,31,46). To our knowledge, large populations of feral rye do not occur in the major rye producing regions of the U.S. (e.g., Minnesota, the Dakotas, Maine, Oklahoma, Georgia, and Tennessee). The cause for the curious distribution pattern of feral rye remains a mystery, though both the regional climate and the history of *Secale* spp. introductions differ across the western U.S. and the remainder of North America.

12.2.5 Introductions of Mountain Rye and Hybrid Michels Grass

Mountain rye originating from eastern Europe, Turkey, and the Caucasus was introduced into the U.S. for research purposes as early as the 1920s. Universities and federal agencies such as the USDA and the Forest Service became interested in mountain rye for land reclamation primarily because of its perennial habit and soil binding properties. The mountain rye accession P4888 (originating from Russia) outperformed other accessions tested in both forage production and longevity (34). It has become the most commonly used perennial rye variety since its introduction before 1940 (2,34). No natural hybrids between P4888 and other strains of mountain rye or cereal rye have ever been documented in North America. Although natural hybridization is possible, the hybrids formed by this means would have had to overcome three major obstacles:

1. The ranges of experimental plantings of mountain rye generally do not overlap with those of cereal rye planting, because the former are grown in previously wooded disturbed land.
2. Mountain rye has a delayed flowering phenology relative to cereal rye.
3. F_1 hybrids would have an initial reduction in offspring fitness of approximately 50 to 80%. Reduced fitness of these hybrid offspring is due to improper chromosome pairing during meiosis and disappears with successive backcrossing.

Low fitness in hybrids can be easily overcome if early hybrid generations are artificially selected. Michels grass, an artificial mountain rye cereal rye hybrid derivative purportedly created by C.A. Michels at the University of Idaho in 1932, is one such case of a successful hybrid from two partially incompatible parents (32). Michels grass was initially mistakenly promoted as a Mosida (6N) wheat *Elymus condensatus* hybrid. Based on chromosome number and morphology it was later reidentified as a mountain cereal rye hybrid. Its mistaken identity was probably the result of hybrid cereal rye mountain rye from another experiment germinating in Michels test plots (Vogel, deceased, transcribed by Metzger, personal communication). Michels grass was promoted as a forage crop and ground cover and was in high demand for a short time (49,51). Early generations of the hybrid were apparently perennial, but perenniality degraded soon after its commercial release so that plants were primarily annual and biennial by 1940 (49). Traits such as fragile rachis, bronze glumes, and long awns apparently distinguished Michels grass from cereal rye cultivars (25).

Mountain rye does not persist in North America (23), but feral rye may be partly derived from the Michels grass mountain rye cereal rye hybrid. Introductions of mountain rye and Michels have been documented (13,34), but documentation does not include the extensive commercial sale of Michels grass from 1938 to 1940. Most introductions of Michels grass were into the arid western U.S. where feral rye is currently distributed (Figure 12.4). Mountain rye and more recently constructed mountain rye cereal rye hybrids continue to be studied for their usefulness in controlling weeds, erosion, and providing forage (1). Seed of non-shattering mountain rye has recently become commercially available in North America.

12.3 THE CASE OF NATURALIZED FERAL RYE IN THE WESTERN U.S.

Based on morphological and isozyme data from northern California populations, several studies have proposed that feral rye is of hybrid mountain rye cereal rye (Michels grass) origin (25,45,46). We suggest that the conclusion of hybrid origin is premature (though perhaps not false), because morphological intermediacy between parental types in feral rye is not reflected by isozyme marker intermediacy. We have evidence that the situation in northern California is not unique; feral rye throughout the western U.S. is morphologically indistinguishable from that in northern California and should share either a common origin or a common mechanism for its evolution. Unfortunately, hybrid origin is difficult to prove for several reasons: Michels grass is no longer available, the identity of parental seed stock used to make hybrids was never recorded, and introduced hybrids have probably been subjected to many generations of gene flow from cultivated cereal rye.

Suneson et al. (46) and, later, Jain (25) observed traits such as winter habit, brittle spikes, and bronze lemmas in feral rye and considered these to be associated with Michels grass. In a careful genetic study of northern California feral rye populations using isozyme markers, Sun and Corke (45) found high genetic diversity and little genetic differentiation among feral populations. Genetic diversity of feral populations was similar to that of both cultivar and mountain rye accessions. Despite the similarity of feral populations to one another, they were also able to detect local geographic structure across feral populations. The evidence provided for hybrid origin was, however, ambiguous. We can assume that with continued introgression from cultivated rye, traces of hybrid origin, if present, have only weakened in the approximately 14 years since their study. The ambiguous origin of feral rye populations in northern California raises the question of whether all feral populations in the western U.S. share a common ancestry. Indeed, naturalized populations of feral rye with the brittle rachis phenotype occur from eastern Washington south to southern California, along roadsides, sandy washes, pasture land, fallow land, and occasionally even in undisturbed otherwise native habitat (Burger, unpublished data).

Here we present the preliminary results of a genetic and morphological comparison of feral rye collected throughout the intermountain zones and mountain ranges of the western U.S. Our goal here is to identify the genetic origins of and relationships between feral rye populations across a much broader geographic range than previously investigated and to document any phenotypic divergence from cultivars and across environments.

12.3.1 GENETIC ANALYSIS

We compared 18 feral populations from 3 geographic locations with 6 historically grown cultivars and 2 mountain rye accessions using isozyme markers. Feral populations originated from 3 distinct regions along a latitudinal gradient — eastern Washington, northern California, and southern California. The cultivars tested were Abruzzi, Dakold, Gazelle, KingII, Merced, Petkus, and Rosen. The mountain rye accessions were PI 401400 and P4888. Plants (n = 20–40/population) were grown from collected seeds (1 seed/spike) to seedling stage and tissue was extracted according to Sun

TABLE 12.4
Genetic Diversity of Feral Rye Populations, Cultivars, and Mountain Rye

Origin	Type	Number of Populations	Number of Plants Sampled	H_e (SEM)
Eastern Washington	Weedy	6	237	0.190 (0.003)
Northern California	Weedy	5	206	0.208 (0.006)
Southern California	Weedy	7	257	0.190 (0.008)
Cultivar	Cultivated	7	240	0.167 (0.005)
Mountain rye	Sister species	2	55	0.105 (0.005)

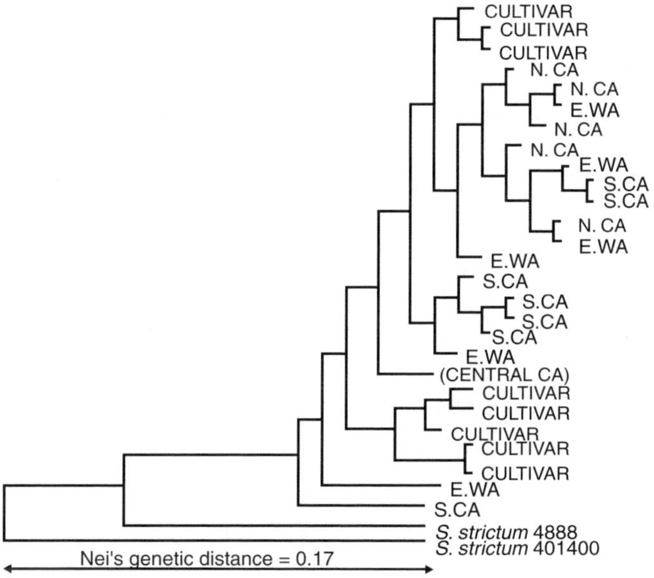

FIGURE 12.5 Genetic distances among feral, cultivated, and mountain rye populations. The unweighted pair group method with arithmetic mean (UPGMA) was used with isozyme marker data to estimate genetic distances among 18 feral, 7 cultivated, and 2 mountain rye populations. Feral populations were collected from eastern Washington, northern California, and southern California. See text for names of cultivars.

and Corke (45) immediately before electrophoresis. Three 12% starch gel and buffer systems (pH 8 morpholine-citrate, pH 6.3 lithium hydroxide-borate, and pH 7 *tris*-citrate-histidine were used to separate isozyme alleles (45). Fourteen loci belonging to 9 enzyme systems (AAT, ACO, IDH, MDH, PGD, PGI, PGM, SKDH, TPI) were resolved and scored. Ten of these were variable at greater than 5%. Most cultivars and mountain rye accessions were obtained from the USDA Genetic Resources Information Network.

In contrast to Sun and Corke (45), we found higher genetic diversity in feral populations relative to cultivars and mountain rye (Table 12.4). Contrary to expectations for direct hybrid origin, feral populations were not intermediate between cultivars and mountain rye accessions (Figure 12.5). Mountain rye formed a clear outgroup not only to cultivars, but also to feral populations. Cultivars formed two distinct clusters (clades) based on their multilocus genotype frequencies. Both clades included winter- and spring-type rye and were not morphologically distinct in any way. Most feral populations were intermediate to these clades. With two exceptions, alleles found in feral populations were shared by one or both clusters of cultivars (allele data not shown). Two feral populations contained unique population-specific alleles at low frequency (data not shown). The two mountain rye accessions also contained unique accession-specific IDH alleles at relatively high frequency. Consequently, the maximum genetic distance measured between any two populations was between the two *S. strictum* accessions and not between *S. strictum* and *S. cereale*. Weedy populations did not cluster with cultivars, nor did they cluster geographically.

We conclude that mountain cereal rye hybrid ancestry is either absent or obscured. Isozyme data instead suggest that the weedy populations originated from a mixture (hybridization) of two genetically distinct clades of cultivars. Interestingly, none of the 18 feral populations sampled clustered with one or the other of the two clades of cultivars. Whether initial hybridization between cultivars and later backcrossing with cultivars produces any fitness advantage or is the source of the higher genetic diversity in feral populations remains unknown. A mountain cereal rye origin, however, still cannot be excluded. Either we may not have sampled the correct mountain rye or cultivar ancestors or, more likely, years of gene flow from cultivated cereal rye have masked hybrid origin.

12.3.2 Ecological Comparison

Some level of genetic divergence exists between feral and cultivated rye, but genetic divergence may not translate into ecological divergence. We compared cultivars (Abruzzi-winter/spring mix, Gazelle-spring, and Rosen-winter), mountain rye (P4888), and three feral populations from each of the above-mentioned locations in a greenhouse environment with and without vernalization (cold treatment). Our goal was to quantify morphological and phenological differences between feral populations and cultivars and to estimate phenotypic divergence in spring and winter types in feral populations across a latitudinal gradient.

Forty seedlings of each of 13 populations (feral, cultivar, and mountain rye) were germinated and planted in 50-celled flats. Twenty seedlings of each population were vernalized for 7 weeks at 4°C while the remainder was left under greenhouse conditions. After vernalization, all plants were exposed to floral inductive long-day conditions. Flowering phenology was recorded.

All feral populations (with the exception of a single plant from northern California) and cultivars flowered regardless of the presence or absence of vernalization, indicating that winter habit is probably day-length and not cold-temperature dependent in the cultivars and populations studied. Northern California and eastern Washington populations flowered later than all three cultivars, independently of cold treatment (data not shown). In contrast, southern California populations flowered earlier and with only a few days delay from spring-type cultivars (data not shown). We suggest that some level of adaptation to local climate has occurred in weedy populations and that it has been facilitated by high levels of genetic variation. Variation in flowering phenology may, however, still be due to maternal effects from material collected in the wild. We will conduct a reciprocal transplant in the future to test our preliminary conclusions of divergence and local adaptation in feral populations.

12.4 SUMMARY AND DISCUSSION

Our literature review suggests that weedy rye has no single clear origin worldwide. In its native range in eastern Europe and southwest Asia, feral rye may have either directly originated from cultivated *S. cereale*, from its wild and weedy *S. cereale ancestrale* progenitor, or from hybrid *S. cereale* *S. strictum*. Because there are only weak reproductive barriers isolating *Secale* species and none isolating subspecies, we are left to conclude that feral rye in Eurasia is a polytypic admixture of several forms of *S. cereale* as well as of *S. strictum*.

Previous taxonomic definitions of *Secale* in Eurasia were based on ecological habits of rye, such as its ability to persist outside cultivation, the size of its seed, its spring or winter habit, and the brittleness of its seed head. Because both cultivated cereal rye and feral rye outcross, they are genetically diverse and highly adaptable. There are no good morphological or cytological character traits that clearly differentiate most traditionally classified subspecies and species of *Secale* (21). Uncultivated races of both spring and winter rye with shattering seed heads are now lumped together into the *Secale cereale ancestrale* subspecies. Whether they are really monophyletic remains doubtful. Hybridization with mountain rye occurs in its native range and is, in places, common (52). Weedy "ancestrale" races may have originated from these hybrid swarms or from more ancient divergence from mountain rye. Natural hybridization with *S. sylvestre*, a selfing relative, is undocumented, but should be investigated. The distribution of brittle races of rye follows approximately the geographic limits of the Mediterranean and temperate dry-summer climate in eastern Europe and the Near East.

Feral rye is abundant in wheat fields of central Asia and Siberia. It is probably of ancient origin, as both wheat and rye have been grown in these parts for at least 2000 years. These stands could represent remnants of ancient progenitor rye or an admixture of precultivated and cultivated rye. As yet, no genetic comparisons exist of these populations.

North American populations of volunteer and feral rye are rapidly nearing the complexity of weedy ryes in Eurasia. Because no native *Secale* species occur in North America and the introduction of rye is relatively recent, their origins should, however, be simpler. In North America, rye occurs as a crop, as a weed of crops, and as a persistent, naturalized weed of uncultivated land. Feral rye varies in shattering, seed color, seed size, straw color, awn length, phenology, and cold tolerance, to name a few traits. The diversity of forms suggests a complex origin, but also puts the traditional taxonomic definitions of *Secale* in its native habitat into question.

Gene flow is a fact of life with cereal rye as with any other outcrossing crop (17). Feral crops act as bridges for cultivar alleles to move from human-managed habitats to wild environments and vice versa. An understanding of their origins and divergence from crops is essential to predict the probability and rate of cultivar gene escape into the wild.

Our research demonstrates that naturalized feral rye populations in the western U.S. are polymorphic and genetically diverse relative to cultivars. We can exclude the possibility that feral populations originated from a single introduction of cereal rye or from a single introduction of hybrid Michels grass without subsequent genetic mixing with cultivars. If feral populations originated from hybrid Michels grass, then their hybrid origin has been obscured by gene flow from cultivars since introduction. If they originated solely from cultivars, then feral populations are the product of multiple introductions and possibly even a distinct hybridization event between two clades of cereal rye (Figure 12.5). Whatever its origin, feral rye in the western U.S. has diverged from cultivars, both genotypically and phenotypically. Furthermore, it is likely to remain a well-established component of our introduced weedy flora.

We suggest that the divergence of phenotypes across environments will increase as the cultivation of rye continues to dwindle. Wherever cereal rye is cultivated, feral rye, if present, will acquire alleles from cultivars, and, reciprocally, must transmit feral traits into cultivars. The effects of high rates of gene flow between cultivated and feral rye as well as the effects of decreasing cultivation of rye on feral populations remain as a fascinating topic for future study.

ACKNOWLEDGMENTS

The authors would like to thank A. Burger, D. Burger, J. Clegg, S. Hegde, J. Heraty, A. Lukaszewski, A. Montalvo, G. Waines, and, especially, S. Lee for technical assistance. J. Gressel and an anonymous reviewer made helpful suggestions on an earlier version of the chapter. This work was funded in part by Project No. CA-*-BPS-7194-CG awarded by NRI-CSREES-USDA to N.C. Ellstrand.

LITERATURE CITED

1. Acharya S, Mir Z, Moyer JR, Orshinsky BR, Thomas JE. 2003. Effect of row spacing and seeding rate on forage yield and quality of perennial cereal rye (*Secale cereale* L.). *Can. J. Plant Sci.* 83:363–369.
2. Aldon EF, Springfield HW, Garcia G. 1975. Can soil amendments aid revegetation of New Mexico coal mine spoils? *Rocky Mountain Forest and Range Experiment Station* 292:1–7.
3. Andersen MR, Depuit EJ, Abernethy RH, Kleinman LH. 1989. *Suppression of annual bromegrasses by mountain rye on semiarid mined lands.* Presented at Symposium on cheatgrass invasion, shrub die-off and other aspects of shrub biology and management, Las Vegas, NV.
4. Anderson RL. 1998. Ecological characteristics of three winter annual grasses. *Weed Technol.* 12:178–183.
5. Anderson RL. 2002. Conservation tillage fact sheet. USDA — ARS Central Great Plains Research Station.
6. Balabajev GA. 1926. On the distribution of weed rye *Secale cereale* L. in the mountains of Central Asia. *Bull. Appl. Bot. Plant-breed.* 16:101–131.
7. Benito C, Gallego FJ, Zaragoza C, Frade JM, Figueiras AM. 1991. Biochemical evidence of a translocation between 6RL/7RL chromosome arms in rye (*Secale cereale* L.): a genetic map of 6R chromosome. *Theor. Appl. Genet.* 82:27–32.
8. Bensin M. 1933. Autogamous Turkestan rye (*Secale turkestanicum*). *Bull. Torrey Bot. Club* 60:155–160.
9. Black RH, Haller CR. 1922. Foreign material in spring wheat. *Farmers' Bull.* 1287:3–21.
10. Briggle LW. 1959. Growing rye. *Farmers' Bull.* 2145:3–16.
11. Buman RA, Abernethy RH. 1988. Temperature requirements for mountain rye, Hycrest crested wheatgrass, and downy brome germination. *J. Range Manag.* 41:35–39.
12. Buman RA, Monsen SB, Abernethy RH. 1988. Seedling competition between mountain rye, 'Hycrest' crested wheatgrass, and downy brome. *J. Range Manag.* 41:30–39.
13. Crowder WA, Kelley CA. 1982. *Performance summary of mountain rye, Secale montanum Guss. at the Pullman Plant Materials Center,* USDA Soil Conservation Service, Pullman Plant Materials Center, Pullman, WA.
14. Cuadrado A, Jouve N. 2002. Evolutionary trends of different repetitive DNA sequences during speciation in the genus *Secale*. *J. Hered.* 93:339–345.
15. De Bustos A, Jouve, N. 2002. Phylogenetic relationships of the genus *Secale* based on the characterisation of rDNA ITS sequences. *Plant Syst. Evol.* 235:147–154.
16. Devos KM Atkinson MD, Chinoy CN, Francis HA, Harcourt RL, Koebner RMD, Liu CJ, Masojc P, Xie DX, Gale MD. 1993. Chromosomal rearrangements in the rye genome relative to that of wheat. *Theor. Appl. Genet.* 85:673–680.
17. Ellstrand NC. 2003. *Dangerous liaisons: when cultivated plants mate with their wild relatives.* Baltimore, MD: Johns Hopkins University Press. 244 pp.
18. Evans GM. 1976. Chapter 33. Rye: *Secale cereale* (Gramineae-Triticinae). In *Evolution of crop plants*, Simmonds NW, Ed., pp. 108–111. London: Longman.

18a. FAOSTAT data 2004.

19. Featherstone R. 2003. Rye redemption. *Wheat Life* 46:16–19.
20. Frederiksen S, Petersen G. 1997. Morphometrical analyses of *Secale* (Triticeae, Poaceae). *Nordic J. Bot.* 17:185–197.
21. Frederiksen S, Petersen G. 1998. A taxonomic revision of *Secale* (Triticeae, Poaceae). *Nordic J. Bot.* 18:399–420.
22. Gray W. 2004. Wheat farmer, retired. Pullman, WA. Personal communication.
23. Hitchcock AS. 1971. *Manual of the grasses of the United States.* New York: Dover.
24. Jain SK. 1960. Cytogenetics of rye (*Secale* sp.). *Biblio. Genet.* 19:1–86.
25. Jain SK. 1977. Genetic diversity of weedy rye populations in California. *Crop Sci.* 17:480–482.
26. Khush GS. 1963. Cytogenetic and evolutionary studies in *Secale III.* Cytogenetics of weedy ryes and origin of cultivated rye. *Econ. Bot.* 17:60–71.
27. Kranz AR. 1963. Die anatomischen, oekologischen und genetischen Grudlagen der Aehrenbruechigkeit des Roggens. *Beit. Biol. Pflanz.* 38:338–472.
28. Kranz AR. 1973. *Wildarten und Primitivformen des Roggens (Secale L.).* Berlin: Paul Parey. 60 pp.

29. Leighty CE. 1917. Rye growing in the southeastern states. *Farmers' Bull.* 894:3–14.
30. Martin JH, Smith RW. 1923. Growing rye in the western half of the United States. *Farmers' Bull.* 1358:3–18.
31. Metzger R. 2004. Wheat geneticist, retired, USDA. Corvallis, OR. Personal communication.
32. Michels CA. 1940. *Wheat-wild rye hybrid grass. Rep. Mimeo Leaflet 40,* University of Idaho, Moscow, ID.
33. Monsen SB. 2004. Botanist, retired, USDA Forest Service Shrub Lab. Provo, UT. Personal communication.
34. Monsen SB, Shaw NL. 1984. *Development of a select cultivar of Secale montanum Guss. for western ranges — progress and status of research by the Intermountain Forest and Range Experiment Station, January 1984. Rep. W190-4-26,* USDA Soil Conservation Service, West National Technical Center, Portland, OR.
35. Nalborczyk E, Sowa A. 2001. Physiology of rye. In *Rye: production, chemistry, and technology,* Bushuk W, Ed., pp. 53–68. St. Paul, MN: American Association of Cereal Chemists.
36. Nuttonson MY. 1958. *Rye-climate relationships and the use of phenology in ascertaining the thermal and photo-thermal requirements of rye.* Washington, D.C.: American Institute of Crop Ecology. 219 pp.
37. Persson K, Von Bothmer R. 2002. Genetic diversity amongst landraces of rye (*Secale cereale* L.) from northern Europe. *Hereditas* 136:29–38.
38. Riley R. 1955. The cytogenetics of the differences between some *Secale* species. *J. Agric.* 46:377–383.
39. Schiemann E, Nuernberg-Krueger U. 1952. Neue Untersuchungen an *Secale africanum* Stapf. II. *Secale africanum* und seine Bastarde mit *Secale montanum* und *Secale cereale*. *Naturwissenschaften* 39:136–137.
40. Sencer HA, Hawkes G. 1980. On the origin of cultivated rye. *Biol. J. Linn. Soc.* 13:299–313.
41. Spragg FA. 1918. The spread of Rosen rye. *J. Heredity* 9:42–44.
42. Stump WL, Westra, P. 2000. The seedbank dynamics of feral rye (*Secale cereale*). *Weed Tech.* 14:7–14.
43. Stutz HC. 1972. On the origin of cultivated rye. *Am. J. Bot.* 59:59–70.
44. Sukopp H, Sukopp U. 1993. Ecological long-term effects of cultigens becoming feral and of naturalization of non-native species. *Experientia* 49:210–218.
45. Sun M, Corke H. 1992. Population genetics and colonizing success of weedy rye in northern California. *Theor. Appl. Genet.* 83:321–329.
46. Suneson CA, Rachie KO, Khush, GS. 1969. A dynamic population of weedy rye. *Crop Sci.* 9:121–124.
46a. USDA NASS Crop Production 2003.
47. Vavilov NI. 1917. On the origin of cultivated rye. *Bull. Appl. Bot. Gen. Plant Breed.* 10:561–590.
48. Vavilov NI. 1926. Studies on the origin of cultivated plants. *Bull. Appl. Bot. Gen. Plant Breed.* 16:1–248.
49. Weihing RM, Robertson, EM. 1940. Claims regarding wheat-wild rye hybrid grass found to be without foundation. *Col. Agr. Exp. Farm Bull.* 2:3.
50. Whitson TD, Burrill LD, Dewey SA, Cudney DW, Nelson BE, Lee RD, Parker R. 2000. *Weeds of the West.* Newark, CA: Western Society of Weed Science. 628 pp.
51. Young VA. 1941. A promising new hybrid grass for certain burned-over forest lands. *J. Forest.* 39:930–934.
52. Zohary D. 1971. Origin of South-west Asiatic cereals: wheats, barley, oats, and rye. In *Plant Life of South-west Asia,* Davis P, Ed., pp. 235–263. Edinburgh.

QUESTIONS AND ANSWERS

Klaus Ammann: Why has cultivation of rye dwindled in California?

Answer: Rye was never an important crop in California, but it was initially the only grain crop that would grow well in the dry, cold climates of northern California. As irrigation became available to farmers, they switched to wheat. Weedy rye probably played a factor in the loss in popularity of cereal rye, especially because much of the cereal rye grown in northern California was reseeded locally and consequently was surely contaminated by weedy rye both directly from harvested weed plants and indirectly from weedy rye pollen.

Baorong Lu: Cytogenetic studies showed that there are two translocations between cultivated rye and mountain rye. How do you explain the rich diversity found in weedy types? This may indicate the hybrid origin of the weedy type rather than its dedomestication.

Answer: High genetic diversity in weedy populations often suggests hybrid origin, but at this stage in our analysis, molecular evidence does not. We are by no means convinced yet that hybrid origin has not been obscured by years of gene flow from cereal rye.

Rick Roush: Have you looked for the presence of translocations, especially of the two reciprocal translocations that distinguish cereal and mountain rye, in any of your samples?

Answer: As yet, we have not. But it is unlikely that translocation polymorphisms would still exist in weedy populations, because they would be strongly selected against.

Rick Roush: Could you try to regenerate hybrids between the ryes to see what sorts of progeny are generated?

Answer: We will be making crosses both between mountain and cereal rye and between the two clades of cereal rye in the future. Attempts to make interspecific hybrids failed this year due to differences in flowering phenology between species.

Jonathan Gressel: Would not the perennial populations have a selective advantage? Do they flower annually? Can you reconstruct the putative hybrid? Is rye a recent weed?

Answer: Disturbed environments with dry, hot summers should favor annual, not perennial habit. In southern California, especially, both the introduced weed flora and the native forb/grass flora are dominated by annuals. We do have some evidence of low frequency occurrence of perenniality in at least one northern California population.

We will try to reconstruct at least the F_1 of the mountain rye cereal rye hybrid, but its performance and appearance may be drastically different from later backcrossed generations. Additionally, we cannot be certain as to the exact identity of the original parental plant material of the hybrid Michels grass that we will be trying to recreate. Rye is a recent weed (since the 1950s) in the western U.S. and was, according to farmers, initially grown without posing much persistent trouble as a volunteer.

Duncan Vaughan: Could you comment on the relative usefulness of analysis and understanding diversity of weedy/feral populations using neutral markers as opposed to adaptive markers/traits?

Answer: We hesitate to use traits that we know are under selection when trying to resolve the genetic origins of feral rye. Morphological or life history traits may be under different selection pressure in different environments. Adaptive markers are useful, however, when defining the traits involved in making a crop a weed and, as you mention, in studying the actual diversity of forms in the field.

André Bervillé: Did you try to study your data with multilocus analysis programs such as Structure or Geneclass? I am not confident in the dendrogram based on Nei's genetic distances.

Answer: Good suggestion. We plan to conduct a more rigorous analysis of relationships between feral populations and putative ancestors using Structure once we have acquired more marker data. Until then, we hope that the reader will view the phenogram we provide as preliminary and only designed to resolve relative degrees of relatedness and not ancestor/descendant relationships.

13 Can Feral Radishes Become Weeds?

Allison A. Snow and Lesley G. Campbell

13.1 INTRODUCTION

The title of this chapter — Can Feral Radishes Become Weeds? — is phrased as a question for three reasons. First, the literature on this topic is incomplete and we expect that further investigations will help elucidate the answer to this question. Second, many feral populations of *Raphanus sativus* have resulted from hybrids between the crop and a closely related weed, *R. raphanistrum*, and it is not known whether this external "trigger" is essential for feral populations to become weedy. Third, the answer depends on one's definition of a weed. Inconsistent use of this term can lead to confusion about the weed status of feral radishes. Agricultural scientists define *weeds* as plants that reduce the productivity of crops, other managed plantations (e.g., vineyards, orchards, planted pine forests), or pastures and range land. A broader definition of weeds includes any plant that causes economic or environmental harm — for example, in agriculture, lawns, recreational areas, wetlands, and other natural areas. Weedy plants that displace native vegetation often are referred to as *invasives*, and economic losses due to invasive plants can be substantial (48). Another category of weeds, which we will not employ here, includes species that occupy disturbed, ruderal habitats without necessarily causing economic or environmental harm.

In this chapter, we ask whether feral *R. sativus* can occur as weedy or invasive populations, as defined above. In particular, we describe how feral radishes evolve and ask whether they interfere with the production of food crops anywhere in the world. Based on a review of the available literature, our answer to this question is "yes, in some cases," but this may require the help of genes from weedy *R. raphanistrum*. Likewise, *Raphanus raphanistrum* could become even more troublesome if it were to acquire fitness-enhancing genes from the crop. Here, we review what is known about ferality in radishes and we explore some human-induced evolutionary changes that could allow feral radishes and their sexually compatible relatives to become weedier in the future.

13.2 EARLY DOMESTICATION

Radish and its feral and wild relatives are insect-pollinated, self-incompatible plants with an annual or biennial life cycle. Radish is an ancient crop that appears to have multiple origins from several wild species (12,60). It is likely that radishes were independently domesticated in both Eurasia and eastern Asia. Herodotus (circa 484–424 BC) suggested that radish was already an important crop in Egypt nearly 5000 years ago, having been depicted on the 4000-year-old walls of the pyramids. Radish was also cultivated in eastern China more than 2000 years ago, before the establishment of the "Silk Road" that permitted extensive trade with central Asia. The present diversity of culinary and morphological types of radishes is greatest in Asia, particularly in China and Japan.

Deciphering the origins of cultivated radish is complicated by the crop's many forms and uses in different parts of the world, and by its long history of dispersal around the globe. Crisp (12) recognized several categories of radishes. "European" radishes have small swollen roots (actually part hypocotyl and part root) and are grown primarily in short-season, temperate regions to be eaten as fresh vegetables. Large-rooted daikon radishes are grown mostly in Asia in both temperate and tropical regions. Daikon radishes are eaten raw, as a cooked vegetable, canned, or pickled. Two less common types of radishes have been bred for their leaves (fodder radish) or seedpods, and the latter have been selected as either oil-seed crops or vegetables. "Rat-tail" radish has been selected for both its leaves and its edible immature seedpods, which are up to 80 cm long. In modern-day Pakistan, some traditional landraces of radish are grown for both their immature seedpods and their swollen roots (43). Thus, there is a huge variety of ways in which the fruits, seeds, sprouts, leaves, and roots are grown for traditional dishes around the world.

Pistrick (42), followed by Specht (52), divided cultivated radishes (all *Raphanus sativus*) into three main taxonomic groups:

1. Convar. *oleifera* — oilseed and fodder radishes
2. Convar. *caudatus* — "rat-tail" radish, also known as var. *mougri*, grown for their edible immature seedpods
3. Convar. *sativus* — all forms with edible roots, with many different varieties

Here, we focus mainly on the last group, convar. *sativus*, which is composed of widespread and economically important root crops. The swollen root is the primary commercial crop, but the seeds have economic value for seed supply sales and the consumption of germinated radish sprouts.

An elegant study by Yamagishi and Terachi (60) showed that the most likely ancestors of cultivated radish are *Raphanus raphanistrum*, *R. maritimus*, *R. landra*, or their earlier progenitors. These investigators compared configurations of two mitochondrial gene regions, *cox1* and *orfB*, among three wild species and cultivated radish from Europe, Asia, and Japan. Mitochondrial genes are useful for tracing phylogenetic histories because they are maternally inherited and they seldom recombine. Five mitochondrial haplotypes were found among accessions of *R. sativus* (Table 13.1). Cultivated varieties that share haplotypes with wild relatives probably descended from one or more of these wild species or their progenitors.

Reconstructing the phylogenetic history of a crop is complicated by the fact that wild accessions may include wild or feral relatives that have hybridized with the crop during the past millennia.

TABLE 13.1
Evidence for Multiple Origins of Cultivated Radish Based on Mitochondrial DNA Haplotypes in Radish and Its Putative Wild Ancestors (*R. raphanistrum*, *R. maritimus*, and *R. landra*)

	Haplotypes				
	Numbers of Accessions with Each Haplotype				
	1	2	3	4	5
Wild species					
R. raphanistrum	5	6	4	1	1
R. maritimus	4	1	1	1	0
R. landra	1	0	0	0	0
Cultivated varieties					
Japan	10	1	10	2	1
Asia	1	2	9	3	0
Europe	0	7	0	0	0

Source: Based on Yamagishi and Terachi (60). Used with permission.

Also, conclusions from this approach can be influenced by the diversity of wild and crop accessions that are examined. Bearing these caveats in mind, it appears that cultivated radishes have originated multiple times from wild taxa that were similar to *R. raphanistrum*, *R. maritimus*, and *R. landra*. This is consistent with earlier evidence for independent domestication in Eurasia and eastern Asia.

Raphanus raphanistrum, *R. maritimus*, and *R. landra* are able to hybridize with each other and with *R. sativus* (all have 2n = 18 chromosomes (34)), and most recently they have been classified as subspecies of *R. raphanistrum* (8,30). Because *R. sativus* sometimes hybridizes with *R. raphanistrum* in the field, several authors have suggested that these taxa also should be consolidated into a single species (e.g., 4,12,51). In fact, one study involving crop-wild hybrids showed perfect colinearity between genomes in the positions of 144 informative restriction fragment length polymorphism (RFLP) markers on radish's 9 chromosomes (4). Given the close taxonomic relationships among the putative wild ancestors of radish, we simplify our discussion below by assuming that the progenitors of cultivated radish were wild forms of radish that shared many attributes with weedy *R. raphanistrum* (*R. raphanistrum* ssp. *raphanistrum*).

Domestication of radish from its wild relatives involved selection for a larger and more flavorful root, often with a red or purple skin, along with high seed production for propagating the crop. Selection for a larger root probably resulted in delayed flowering and a tendency to be biennial. As with many other crops, domestication also selected for seeds that are easier to harvest. In weedy *R. raphanistrum*, tough, hard-to-crack fruits protect the seeds from bird predation and other types of damage. Fruits of *R. raphanistrum* are shed gradually as they mature on the maternal plant, and the fruit does not split open to release the seeds. Instead, it breaks into distinct sections, each of which encapsulates a single seed (Figure 13.1 and Figure 13.2). In a sense, the fruit wall of *R. raphanistrum* acts as a protective "outer seed coat" for the seeds, which lack the strong, impervious seed coat found in many other annual species. In domesticated *R. sativus*, however, seeds are contained within indehiscent fruits, but the fruits remain firmly attached to the plant after it has senesced and they are not divided into sections (Figure 13.2). In many modern varieties of radish, the spongy seedpods crush easily by hand, which allows the seeds to be extracted from the seedpod more efficiently. The conditions needed for dedomestication of *R. sativus* and its potential to become an agricultural weed are discussed in a later section.

13.3 MODERN RADISH VARIETIES WITH EDIBLE ROOTS

Before we evaluate the potential for radishes to revert to a feral condition, perhaps aided by genes from *R. raphanistrum*, it is useful to consider salient features of modern-day radishes. As an open-pollinated crop that is self-incompatible, radishes have been bred for root color, taste, texture, size, and shape, as well as agronomic performance, day-length requirements to prevent premature bolting, seed yield, and other traits (2,12). As noted above, small-rooted, spherical, red radishes are widely consumed in Europe, North America, and other temperate regions of the world, but the large-rooted daikon radishes have greater commercial importance in Asia. In some mountainside villages in Japan, several hundred landraces of daikon radishes are still grown for traditional dishes, thereby preserving a great deal of genetic and phenotypic diversity (61).

Radishes are fast-growing plants — *raphanos* comes from Greek for "quick appearing." Varieties grown in temperate climates germinate in the early spring and are tolerant of cool temperatures. The spatial scale at which radishes are grown ranges from small kitchen gardens to large, industrial-scale farms. Like other members of the mustard family, the plants produce glucosinolates that provide a peppery flavor and may aid in herbivore defense and allelopathy (1,29). Asiatic radishes have been selected for resistance to various pathogens, such as *Fusarium*, *Albugo candida*, *Peronospora parasitica*, and viruses (12). Historically, disease resistance was not a high priority for the small-rooted European varieties, because they were grown in the spring and the roots were harvested before diseases become prevalent. More recently, European varieties have been selected for multiple plantings per season, involving both annual and biennial life spans, and breeders have

FIGURE 13.1 The habit and reproductive structures of weedy *Raphanus raphanistrum* L. Note the woody, segmented fruits (siliques) with constrictions between individual seeds: 1. Habit, 2. Seedling, 3. Flower, 4. Open flower, 5. Fruit, 6. Seed. (From Holm et al. (26) used with permission.)

FIGURE 13.2 Morphological differences among fruits of cultivated, feral, hybrid, and weedy *Raphanus* taxa. From left to right: one fruit from cultivated *R. sativus* (Red Silk variety), two from F_2 feral radish plants, two from F_2 wild-crop hybrids, and one from *R. raphanistrum*, which typically has constrictions between individual seeds. Note that two fruits remain attached to the pedicel. All fruits except the Red Silk variety were collected from experimental field populations in Michigan in 2003 (see text); digital photographs were converted to high-contrast images; for scale, the wild fruit is 4 cm long.

introduced various types of resistance to fungal disease pathogens (e.g., *Fusarium, Rhizoctonia* spp.). Chinese and Japanese radishes sometimes have been used as sources of resistance genes (58).

The yield and overall performance of radishes have been enhanced by the development of F_1 hybrids. Cytoplasmic male sterility (CMS) was found in many Japanese and Chinese cultivars (40), and it has been used to produce hybrid seed in large-rooted cultivars (12). More recently, the Ogura type of CMS has been used to produce F_1 hybrids of small-rooted European varieties, but open-pollinated seed production without CMS is still common. A great deal of genetic diversity is maintained in open-pollinated radish varieties, relative to wild *Raphanus* populations (15). In regions where landraces are grown (e.g., Japan, Pakistan), genetic diversity is especially high because plants that are grown for seed often cross-pollinate with other cultivated and feral varieties (43,61). The genetic diversity that is maintained in cultivated radish could facilitate the establishment of feral populations, as we discuss further below.

Conventional breeding in radish is expected to become more sophisticated with the use of genomics-guided strategies. The first genetic map of the *Raphanus sativus* genome was published in 2003 (4), paving the way for the use of marker-assisted breeding and comparative mapping with *Arabidopsis* and *Brassica* species. Radish has been transformed using a floral dip method (13), but transgenic radish has not been an attractive target for seed companies to date because of a relatively small market. Radish is rarely an important staple crop in the west, and future breeding may be focused on the aesthetic value of its color patterns and shapes, for sale in both novel and traditional food markets (12). Breeders have also focused on transferring a late-flowering trait to commercially important varieties, but conventional breeding techniques have been unsuccessful, producing a hybrid of low quality (33). Curtis et al. (14) obtained late-flowering radishes by expressing an antisense *gigantea* gene fragment from *Arabidopsis*. This trait is a possible candidate for the development of transgenic radishes.

13.4 DEDOMESTICATION AND WEED EVOLUTION IN RADISHES

13.4.1 Characteristics of Weedy *Raphanus raphanistrum*

It is useful to compare *R. raphanistrum* with feral radish to understand which characteristics may prevent feral radish from becoming a serious agricultural weed. Known as jointed charlock or wild radish, *R. raphanistrum* has entered the ranks of the 180 worst weeds worldwide (5,26,35,39,57). This species occurs in agricultural fields, especially where small grains and alfalfa are cultivated, in waste places, and along sheltered beaches in many cool and temperate areas of the world. It is found in North America, South America, Australia, Africa, and Eurasia, but has only recently colonized eastern Asia (28). Worldwide, *R. raphanistrum* has been reported as a weed problem in more than 45 crop species in at least 65 countries (26). In North America, *R. raphanistrum* is a common agricultural weed in the northeastern and central U.S., eastern and western coastal regions of Canada, and the Pacific Northwest (36). *Raphanus raphanistrum* is especially troublesome in wheat, oat, and barley (11,20,57), and it is considered to be one of the most serious agricultural dicot weeds in Australia (9). In Australia and South Africa, *R. raphanistrum* has evolved resistance to several acetolactate synthase (ALS)-inhibiting herbicides (23,49,62). Its current prominence in Australian wheat fields is related to the fact that it has become resistant to multiple modes of herbicide action, including ALS-inhibitors, an auxin analog (2,4D), photosystem II-inhibitors (atrazine, metribuzin), and a phytoene desaturase-inhibiting herbicide (diflufenican) (55). A thorough description of the ecology and weed status of *R. raphanistrum* is provided by Warwick and Francis (56).

Raphanus raphanistrum has three essential features of successful agricultural weeds:

1. The ability to disperse its seeds widely
2. The ability to produce a large population of dormant seeds in farmers' fields
3. The ability to compete with crop plants

For example, the tough, woody fruits fall from the plant as soon as they mature, and they later break apart into small segments that are difficult to clean from grain crop seed supplies (26). This allows the seeds to spread wherever contaminated grain seed has been shipped. In the early 1800s, *R. raphanistrum* was cited as one of the worst weeds of arable fields in England (26), and this weed undoubtedly dispersed to other countries along with grain shipments. In the early 1980s, *R. raphanistrum* was a major contaminant of grain seed that was transported throughout Australia during a four-year drought (53). Even with modern methods of cleaning certified grain seed, there are many other routes by which the seeds can disperse. The silique segments pass through livestock intact and can be dispersed with manure and hay. In Iran, for example, *R. raphanistrum* seeds are dispersed in sheep manure, as well as by contamination of farmer-saved seed (26). Once the weed becomes established, it can build up a prodigious seed bank that can last for up to 15 to 20 years (7,26,45,46,47). With its long-lived seed bank, early emergence after tilling, and rapid life cycle, *R. raphanistrum* is often a difficult-to-manage weed. Without proper attention to the problem, the plants can produce thousands of seeds per square meter and can drastically reduce wheat yields (5,26, Powles, personal communication).

13.4.2 Weedy Traits and Endoferality

13.4.2.1 General Considerations

As discussed in Chapter 1, endoferality refers to dedomestication of the crop without the aid of gene flow from wild or weedy populations. New mutations and inherent variation within the crop gene pool, including crop-derived off-types, provide the raw material on which selection can act. For cultivated radishes to become feral and evolve into agricultural weeds, strong selection is needed to remove deleterious crop traits and to increase the frequencies of weedy traits. We are not aware of any published studies of endoferality in radishes, although our ongoing research in Michigan will begin to fill this apparent gap (see Section 13.5.4). In this section, we speculate about which traits are likely to be most important in the dedomestication of radish. The next section focuses on exoferality, which involves progeny from hybrids between the crop and its weedy relatives and is well documented in radishes.

In some cases, what looks like endoferality may actually involve genes from other taxa. It is possible that weed-to-crop gene flow has occurred in seed production fields, allowing low levels of weed genes to enter seed supplies. Although this may be rare in industrialized countries, especially those with strictly enforced standards for certified seeds, it is probably quite common where landraces co-occur with *R. raphanistrum* (or feral *R. sativus*). Thus, it is difficult to be sure of the source of dedomestication genes in the crop. Nonetheless, the concept of endoferality still can be useful. Crop seeds are shipped and planted repeatedly over vast geographic areas, and it is important to know whether the crop has an inherent tendency to produce volunteer and feral populations. Some crops are able to spawn new weed populations *de novo*, as discussed elsewhere in this book. Others will do so only when aided by genes from a wild or weedy relative, resulting in exoferality.

To understand genetic changes that can lead to endoferality, it is helpful to know whether key domestication traits are controlled by a few genes or many genes, and whether these genes are recessive or dominant in the crop. The genetic basis for specific domestication traits in radishes is known in some cases, and a more thorough survey of agricultural publications, especially in Europe and Japan, would probably uncover relevant information. Here, we examine some of the traits that occur in *R. raphanistrum* and are needed by feral *R. sativus*, building on an excellent earlier summary by Panetsos and Baker (41). We also discuss initial results from our field experiments in Michigan, where we established four artificial populations of volunteer plants from a common cultivated variety (Red Silk), beginning in 2002 (see Section 13.5.4.3 for further details about this experiment).

13.4.2.2 Earlier Flowering and a Less Swollen Root

The evolution of early flowering may be one of the first traits needed for the establishment of volunteer and feral radish populations. Early flowering is advantageous in arable fields because plants can often complete their life cycle before being harvested with crop plants or killed by cold or drought. This is possible even when germination is induced later in the growing season. Also, because most fields are tilled in early spring, the combination of rapid germination and a short life cycle can allow the plants to avoid herbivores, diseases, and other weeds that appear later in the season (50).

Early flowering and a less swollen root are positively correlated in segregating populations of crop-wild hybrids, perhaps because of resource allocation trade-offs that occur during early root growth (41, Campbell and Snow, unpublished data). Radish is considered to be a long-day species that delays flowering until the next year, although this varies among varieties (2). Earliness in days to bolting and to flowering appears to be a highly heritable, dominant trait that is controlled by a few major genes (63). Bolting and induction of flowering are also influenced by two environmental cues — low temperatures and long day lengths (14,33). In North America, when small-rooted radishes are planted in the spring and then remain unharvested, they frequently bolt and produce copious amounts of seeds by late summer. Volunteers of large-rooted radishes are more likely to have a biennial life cycle than small-rooted ones (2,43,61). Because of heritable variability within and among cultivated varieties, we hypothesize that radishes are able to evolve to have wild-type roots and earlier flowering, allowing them to become short-lived annuals. This transition probably occurs more easily in small-rooted "European" varieties than in large-rooted daikon radishes.

13.4.2.3 Early Abscission of Mature Fruits

In contrast with *R. raphanistrum*, fruits of cultivated and feral *R. sativus* are retained on the parent plant long after the plant has senesced. This makes the fruits more susceptible to predation and limits the temporal and spatial scales of seed dispersal. Lack of fruit abscission in feral plants could be a major deterrent to the buildup of a seed bank in the soil. We have observed that fruit abscission is not essential for maintenance of feral population at our experimental sites, but it could contribute to their relatively small population sizes. Whether the lack of early abscission persists over the long term in feral populations is unknown.

13.4.2.4 Thicker and Woodier Fruits

One obstacle to the establishment of feral populations may be seed predation, especially by birds, because the crop has thin, easily punctured fruit capsules (41). Tougher fruits also aid in the persistence of seeds in the soil. In weedy *R. raphanistrum*, the presence of a thick, impervious fruit capsule prevents seeds from germinating in the fall, thereby enforcing a certain degree of dormancy, and allowing deeply buried seeds to remain viable for several years (7,26,45,46,47). Thus, the evolution of tougher fruit capsules is expected to promote ferality. The extent to which this occurs spontaneously is not known. In our experimental populations in Michigan, the fruits of feral radishes have thinner walls than *R. raphanistrum* and their seeds often germinate within the fruit and die in late summer. Nonetheless, two of our original four populations have persisted over the course of two years in local field conditions.

13.4.2.5 Segmented Fruit Capsules that Break into Single-Seeded Sections

Weedy *R. raphanistrum* fruits typically have constrictions between the seeds (Figure 13.1 and Figure 13.2), although many weedy populations also have individuals that lack this characteristic (Snow and Campbell, personal observation). Single-seeded fruit segments are more difficult to remove from seed supplies of small grains than whole fruits, and as such, they may represent an example of crop seed "mimicry" (3). This trait could also aid in local seed dispersal and avoidance of predators. However, intact, non-segmented, corky fruits may stay buoyant in water for a longer period of time, which could be important for coastal populations of *R. raphanistrum*. In any case, we have not seen segmented fruits in cultivated radishes or their early generation descendants (Figure 13.2). We hypothesize that genes conferring segmented fruits are lacking in the crop.

13.4.2.6 Staggered Seed Germination

Crop plants are strongly selected for rapid and synchronous seed germination, but weedy species often exhibit great variability in germination dates, making them more difficult to eradicate. Cultivated and volunteer radish populations lack two characteristics that promote staggered germination dates. First, they lack the tough, segmented fruit capsules that must crack and begin to decompose before the seed can germinate (Figure 13.2). Second, we have observed that, in the greenhouse, extracted seeds from weedy *R. raphanistrum* germinate over a wider period (about 1 to 3 weeks) than seeds from early-generation feral radishes (less than 1 week) (Campbell and Snow, unpublished data). If feral plants lack staggered seed germination, they could be more susceptible to being killed by extreme weather, weed control methods, and other types of disturbances.

13.4.2.7 Resistance to Insect Herbivores and Pathogens

The complex effects of insect herbivores and diseases on weedy populations of *R. raphanistrum* are not well known, but these factors may limit population growth rates in some years and locations (1). It is conceivable that domestication and selection for flavor and appearance have reduced the ability of volunteer and feral plants to withstand these pressures. However, some cultivars have been bred for resistance to fungal pathogens (*Fusarium, Rhizoctonia, Plasmodiophora, Thelaviopsis* spp.) (10), and many have probably acquired some tolerance of abiotic stress from centuries of selective breeding. Herbivore and disease pressures may be relatively unimportant in the evolution of endoferality, but this assumption should be investigated further, especially if transgenes for insect and pathogen resistance are introduced into cultivated radishes. Hypothetically, resistance to fungal pathogens, if expressed in the fruit or seed coat, might lead to greater seed longevity, perhaps allowing feral populations to become established in agricultural fields. Mature fungal-resistant plants might also prosper in fields where disease pressure is high.

13.4.2.8 Greater Genetic Diversity

The amount of genetic diversity that is maintained in weed populations can enhance their ability to evolve in step with changing environmental conditions, including various methods of weed management such as tilling and herbicide applications. In some crops, especially those that are vegetatively propagated or are self-pollinated, breeders have removed much of the genetic diversity that exists in landraces and wild relatives (15). However, radishes have always been bred as open-pollinated, self-incompatible crops (but see 24), and cultivars have retained nearly as much genetic diversity as *R. raphanistrum* (15,31). Therefore, it is unlikely that inbreeding and other manifestations of low genetic diversity prevent feral radish populations from becoming established. Moreover, seed-mediated gene flow is likely to distribute genetic diversity among disjunct populations, and recurring gene flow from the crop could add new alleles to feral radish populations.

13.5 EXOFERALITY VIA CROP-WEED HYBRIDIZATION

13.5.1 General Considerations

Spontaneous hybridization between crops and related weeds has the potential to lead to the evolution of more troublesome weeds in some crop-weed systems (6,16,17,18). Therefore, it is useful to examine the dynamics of hybridization in radishes and to ask whether this process could result in exoferality and enhanced weediness. For convenience, we have drawn a distinction between endoferality and exoferality, recognizing that a whole spectrum of situations exists in radishes. Feral radishes have acquired genes from *R. raphanistrum*, and *R. raphanistrum* populations have acquired genes from *R. sativus*. Further research involving molecular markers is needed to uncover the varying degrees of cryptic introgression that have occurred in these taxa.

Several lines of evidence suggest that genes from *R. raphanistrum* have introgressed into feral populations of *R. sativus* (41), and that genes from the crop have introgressed into weedy populations of *R. raphanistrum* (51). Cytological data, flower color polymorphisms, and fruit characteristics have been used as evidence for hybridization and introgression in both directions (i.e., into feral *R. sativus* and into weedy *R. raphanistrum*). As discussed below, it is likely that gene flow from *R. raphanistrum* into *R. sativus* has facilitated the establishment of feral populations in California and perhaps Brazil. This has probably occurred repeatedly in other parts of the world where *R. raphanistrum* and the crop co-exist. Thus, radishes appear to show a weak tendency to evolve endoferality and a strong tendency for both exoferality and the introgression of crop alleles into populations of *R. raphanistrum*. The occurrence of bilateral gene flow means that the distinction between feral *R. sativus* and weedy *R. raphanistrum* can become quite blurred. Nonetheless, these arbitrary taxonomic categories are useful for identifying populations that are more similar to the crop vs. those that have had little or no influx of crop genes.

Feral radish populations have been reported in Europe (54), North America (19,37,41), and South America (21). In Japan and Korea, wild *R. sativus* var. *hortensis* f. *raphanistroides*, a biennial plant, is common along beaches, cliffs, abandoned fields, and other ruderal areas, and sometimes occurs adjacent to cultivated daikon radishes (27,28,61). However, a recent study by Yamagishi (60) suggests that these populations are derived mainly from *R. raphanistrum* rather than *R. sativus*. In any case, volunteer and feral populations of daikon radish landraces (*R. sativus* var. *hortensis*) have been observed as ruderal species in mountainside villages in Kyushu, Japan (61), providing clear evidence of ferality. Below, we describe specific case studies of feral *R. sativus* in the Brazil and the U.S.

13.5.2 Herbicide-Resistant Feral *Raphanus Sativus* in Southern Brazil

Raphanus sativus has become weedy in temperate South America and has evolved resistance to ALS-inhibiting herbicides in southern Brazil (21,25). Herbicide resistance was reported in Heap (25) by Theisen (Fundacep Fecotrigo, Cruz Alta, Brazil), who kindly provided us with the following details. In southern Brazil, the main agricultural system has been no-till since about 1980, with soybean and corn grown during the summer and a variety of other crops in the winter, including wheat, barley, oats (*Avena sativa*), rye, canola, and flax. Before the adoption of no-till methods, *R. raphanistrum* was a common weed in wheat and other winter crops, as has been reported elsewhere (26). After no-till agriculture was adopted, a type of *Raphanus sativus* called "turnip" or "forrageiro" was planted widely as one of several popular cover crops. This radish cover crop was probably *R. sativus* convar. *oleiferus*, and it started to occur as a volunteer weed in summer crops, where it was treated with various herbicides. Because *R. raphanistrum* was also present in several winter and summer crops, it is possible that crossing occurred between this species and cultivated or volunteer *R. sativus*. By 2001, feral *R. sativus* acquired resistance to several ALS-inhibiting herbicides in the southern state of Rio Grande do Sul. Based on photographs of the flowers and fruits of these plants (25), and information provided by Theisen, feral *R. sativus* does not appear to have the constricted fruits or yellow flower pigmentation that are common in *R. raphanistrum*. Thus, it is considered to be *R. sativus* even though it may have hybridized with *R. raphanistrum*.

Theisen noted that feral *R. sativus* is not as aggressive as many other weeds, but it is common in winter crops (mainly wheat, barley, and oat) as well as summer ones (soy, corn, and bean). In this case, it is likely that the evolution of ferality was facilitated by planting *R. sativus* as a cover crop in regions where it could hybridize with *R. raphanistrum*. It is interesting to note that *R. raphanistrum* has also evolved resistance to ALS-inhibiting herbicides in Australia, but this has not been reported Brazilian populations of *R. raphanistrum*.

13.5.3 Crop-Weed Hybridization in California

Raphanus sativus and *R. raphanistrum* first appeared in California in the 19th century (41). These taxa hybridized to such an extent that distinct populations of *R. raphanistrum* have largely disappeared, but *R. sativus* bearing occasional traits of the weed (e.g., yellow petal color) is common (41,51). California populations, which have the common name of wild radish, can occur as an agricultural weed, especially in cole crops, such as broccoli, cauliflower, and cabbage (59). However, they are most common in ruderal areas (37,41; Fennimore, University of California at Davis, personal communication to Snow; Snow personal observation). The Nature Conservancy includes feral *R. sativus* on a state list of invasive plants (44) because it is a non-native species that occurs in disturbed natural areas, including coastal dune habitats.

Panetsos and Baker (41) concluded that hybridization between feral *R. sativus* and weedy *R. raphanistrum* "appears to have been a major factor in converting the erstwhile crop plant into a highly successful weed." Citing Panetsos and Baker (41), Ellstrand and Schierenbeck (16) went a step further to argue that hybridization was a stimulus for the evolution of invasiveness in *R. sativus*. However, their criterion for "invasiveness" was that the hybrid derivative "must replace at least one of its parent taxa or invade a habitat in which neither parent is present." In this case, feral *R. sativus* may have displaced weedy *R. raphanistrum*, but it is not clear whether this process caused new problems in natural or agricultural areas. At most, hybridization may have made it easier for *R. sativus* to establish feral populations that sometimes occur as weeds. In terms of agricultural areas, we found some evidence that feral populations of *R. sativus* are regarded as problematic weeds in California (59), but it is not known whether they are weedier or more invasive than their wild parent, *R. raphanistrum*. In any case, wild radish populations have the potential to become more abundant due to continued gene flow from the crop. Klinger and Ellstrand (32) demonstrated this in a fitness experiment. Their findings suggest that feral *R. sativus* can benefit from further episodes of hybridization with the crop, perhaps due to heterosis, because F_1 plants produced more seeds per plant than local genotypes of wild *R. sativus*.

13.5.4 Crop-Weed Hybridization in Experimental Populations in Michigan

13.5.4.1 Overview of Experimental Populations

Our current research is designed to test the hypothesis that hybridizing populations derived from *R. sativus* and *R. raphanistrum* can evolve into more successful weeds than either of these parental taxa. To address this question, we established 2 long-term studies at the University of Michigan Biological Station in Pellston, Michigan. The first study, which started in 1996, consists of 4 artificial hybrid populations in which 3 crop-specific genes are being monitored from year to year to evaluate their ability to persist (51, Snow et al. in preparation). The second study, which started in 2002, is a replicated experiment that is more relevant to the evolution of ferality. In this experiment, we established 4 populations of F_1 *R. sativus* (small-rooted Red Silk variety), 5 populations of F_1 weedy *R. raphanistrum* obtained from a nearby farm, and 5 populations of F_1 weed-crop hybrids derived from these 2 parental taxa. All 18 populations in both experiments are spatially isolated from local populations of *R. raphanistrum* and from each other to prevent gene flow among them. Results from these 2 studies are useful for understanding the process of introgression in radishes, as well as the evolution of weediness.

13.5.4.2 Fitness of F_1 Weed-Crop Hybrids and Persistence of Crop Alleles

The evolutionary outcome of hybridization is often influenced by how chromosomes and linkage groups of the parental taxa interact. In *R. sativus* and *R. raphanistrum*, the rate at which introgression occurs is probably affected by an interesting chromosomal translocation between the crop and the weed. Many investigators found that F_1 hybrids between *R. sativus* and *R. raphanistrum* are heterozygous for a reciprocal translocation (41). A few studies did not detect this heterozygosity, perhaps due to past episodes of gene flow (41), but this seems to be unusual. In any case, plants that are heterozygous for the translocation have problems with chromosome pairing: quadrivalents are formed instead of bivalents between homologous pairs. This reduces pollen fertility to approximately 60% and seed set per fruit to approximately 50% relative to levels of approximately 95% for both traits in either parent (41,51). Thus, F_1 hybrids have lower fitness than *R. raphanistrum*, which is a locally abundant weed in Michigan. Despite this partial barrier to gene flow, three crop-specific markers remain fairly common in all of the populations that we established in 1996 and have monitored in each subsequent year (51; Snow et al., in preparation).

Panetsos and Baker (41) found that the reciprocal translocation was linked to genes that influence root structure and flowering time, although their sample sizes were small. If true, the presence of this linkage group has two important implications. Linkage among these genes may have made it easier to maintain important domestication traits during centuries of crop breeding. Conversely, linkage could facilitate the loss of deleterious crop genes following crop-wild hybridization. Because heterozygosity for the translocation reduces plant fitness, hybridizing populations should become fixed for one form or the other, in a positive, frequency-dependent fashion. Thus, when crop-to-weed hybridization is infrequent, *R. raphanistrum* populations should quickly eliminate the fertility-reducing translocation derived from the crop, as well as linked, fitness-reducing genes for swollen roots and delayed flowering. This is precisely what we have observed within a period of 3 to 4 generations in experimental populations that initially consisted of 25% crop alleles and 75% alleles from *R. raphanistrum* (Figure 13.3; Snow et al., in preparation).

Also, in a separate selection experiment that was conducted in a greenhouse, we found that flowering time is highly heritable in F_2 crop-weed hybrids, as expected (Campbell and Snow, in preparation). This heritable variation can easily allow the hybrids to evolve earlier flowering times. In California, where *R. sativus* appears to have "displaced" *R. raphanistrum*, and where heterozygosity for the translocation subsequently remained common (41), it is not known how these

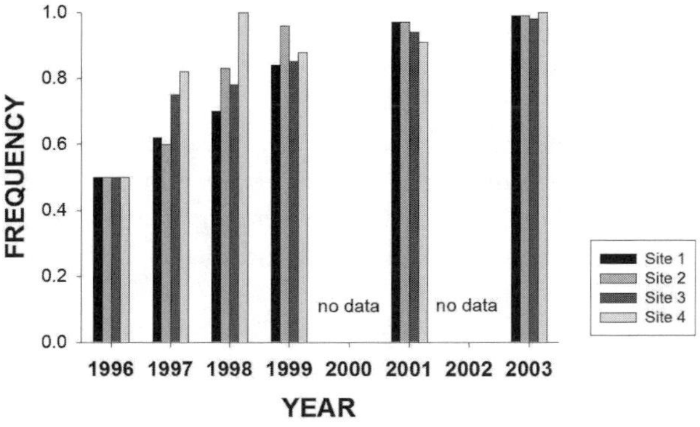

FIGURE 13.3 Rapid restoration of male fertility in hybrid radish populations. Changes in the frequencies of plants with normal pollen (greater than 70% viable) in four experimental populations in Michigan are shown (see text). Plants that are heterozygous for a reciprocal translocation are likely to have less than 70% viable pollen. In 1996, half of the plants were *R. raphanistrum* and half were F_1 wild-crop hybrids that were heterozygous for the translocation. No data were collected in 2000 or 2002.

dynamics have played out. California populations have non-swollen roots and early flowering, similar to *R. raphanistrum*. Further research involving mapped molecular markers (4) would help elucidate patterns of introgression involving various crop-specific and weed-specific genes and their linkage groups.

13.5.4.3 Persistence of Feral Populations

Our efforts to establish feral experimental populations in Michigan have been partially successful, in that two out of four field populations that were started in 2002 persisted to the summer of 2004 (this study is ongoing). At the same time, it has been much easier to establish vigorous experimental populations of crop-weed hybrids as opposed to volunteer plants from the crop, as predicted by Panetsos and Baker (41). Many plants in our second-generation feral radish populations had swollen roots and failed to flower and set seed before winter, and others flowered only 3 to 16 days later than the wild populations that serve as controls. In a milder climate, such biennial individuals would be able to tolerate winter conditions and persist (it is possible that mild winters could facilitate the evolution of ferality in radishes). To summarize our initial findings to date, this experiment suggests that endoferality is possible in radishes. We also expect that progeny from the hybridizing populations will have greater fecundity and greater population growth rates than those from control populations of *R. raphanistrum*, but this remains to be seen.

13.6 CONCLUSIONS

Radishes are capable of establishing volunteer and feral populations, as either annuals or biennials, but they do not appear to be serious agricultural weeds in most areas. Rather, non-weedy feral radish populations may be regarded as "incipient" weeds — we expect that some populations could evolve into worse weeds given the right combination of selection pressures and sufficient genetic diversity, as occurred in no-till rotations in Brazil (25). New genes from crop breeding or from weedy *R. raphanistrum* have the potential to accelerate this process, especially if there is a great deal of gene flow into feral populations.

Weed scientists are familiar with the potential of weed populations to shrink or expand rapidly in response to tilling, herbicide use, biological control, crop rotations, cover crops, annual variations in the weather, and other factors, such as the use of clean, certified crop seeds. For example, intermittent low-till practices often favor increases in the abundances of biennial weeds. Therefore, it is worth recognizing that weeds that are relatively harmless at present, such as feral *R. sativus*, could be propelled into greater dominance in the future by specific types of changes. Also, a more thorough search of the literature on feral *R. sativus* may reveal other cases in which this has already occurred.

The economic and environmental harm that weeds cause also depends on where they occur. For example, weedy *R. raphanistrum* is not a great concern in potato, alfalfa, or corn in north central Michigan (Snow and Campbell, personal observation) or along the New England coast, where it occurs in low frequencies above the storm tide level on sheltered, cobble beaches. In contrast, this plant is one of the worst weeds in Australian wheat fields; it is a nemesis of grain crops in general (26), and it has recently spread to eastern Asia. In the future, it is conceivable that feral populations of *R. sativus* could also disperse over great distances and encounter favorable conditions for invasion.

Because *R. raphanistrum* is already an economically important weed, great care will be needed to decide which types of novel transgenic traits will be introduced into cultivated radishes, if any (32,51). Single-gene traits for enhanced resistance to herbicides, diseases, herbivores, and abiotic stress are likely to spread to weedy populations by means of pollen flow and seed dispersal, and this has the potential to lead to worse weed problems. Traits that do not offer fitness-related benefits are unlikely to pose risks, nor is the intentional use of fitness-reducing transgenes as a means of biological confinement (22; Gressel and Al-Ahmad, Chapter 22 of this volume). However, no method of biological confinement is likely to be 100% effective (38). In our view, this method of reining in questionable transgenes should be used cautiously and only when possible risks are offset by significant economic or environmental benefits.

In any case, it does not seem likely that new varieties of transgenic radishes will be developed in the near future. The crop is not essential for food security, it is not a widely traded commodity, which would make it attractive to large seed companies, and it is already quite well adapted for commercial production. In the future, weedy *R. raphanistrum* and feral populations of the crop will continue to evolve in response to natural- and human-induced factors. The immediate threat that they pose for crop productivity is probably the recent and increasingly common evolution of herbicide resistance in both *R. raphanistrum* and feral *R. sativus*.

ACKNOWLEDGMENTS

We sincerely thank Norman Ellstrand, Steven Fennimore, Jonathan Gressel, Subray Hegde, Susan Mazer, Steven Powles, Barry Rice, Neal Stewart, Sharon Strauss, Giovani Theisen, and Suzanne Warwick for helpful comments on the topics presented in this chapter.

LITERATURE CITED

1. Agrawal A. 1999. Induced responses to herbivory in wild radish: effects on several herbivores and plant fitness. *Ecology* 80:713–723.
2. Banga O. 1976. Radish. In *Evolution of crop plants*, Simmonds NW, Ed., pp. 60–62. New York: Longman.
3. Barrett SCH. 1983. Crop mimicry in weeds. *Econ. Bot.* 37:255–282.
4. Bett KE, Lydiate DJ. 2003. Genetic analysis and genome mapping in *Raphanus*. *Genome* 46:423–430.
5. Blackshaw RE, Lemerle D, Young KR. 2002. Influence of wild radish on yield and quality of canola. *Weed Sci.* 50:344–349.
6. Boudry P, Morchen M, Saumitou-Laprade P, Vernet P, Van Dijk H. 1992. The origin and evolution of weed beets: consequences for the breeding and release of herbicide-resistant sugar beets. *Theor. Appl. Genet.* 87:471–478.
7. Chancellor, RJ. 1986. Decline of arable weed seeds during 20 years in soil under grass and the periodicity of seedling emergence after cultivation. *J. Appl. Ecol.* 23:631–637.
8. Chater, AO. 1993. *Raphanus* L. In *Flora Europea*, Tutin TG et al., Eds., 2nd ed., Vol. 1. p. 417. Cambridge: Cambridge University Press.
9. Cheam AH, Code GR. 1995. The biology of Australian weeds. 24. *Raphanus raphanistrum* L. *Plant Prot. Q.* 10:2–13.
10. Chen CY, Seguin-Swartz G. 1999. Reaction of wild crucifers to *Leptosphaeria maculans*, the causal agent of blackleg of crucifers. *Can. J. Plant Pathol.* 21:361–367.
11. Cousens R, Warringa J, Cameron J, Hoy V. 2001. Early growth and development of wild radish (*Raphanus raphanistrum*) in relation to wheat. *Aust. J. Agr. Res.* 52:755–769.
12. Crisp P. 1995. Radish, *Raphanus sativus* (Cruciferae). In *Evolution of crop plants*, Smartt J, Simmonds NW, Eds., 2nd ed., pp. 86–89. Harlow, UK: Longman Scientific & Technical.
13. Curtis IS, Nam HG. 2001. Transgenic radish (*Raphanus sativus* L. *longipinnatus* Bailey) by floral-dip method — plant development and surfactant are important in optimizing transformation efficiency. *Transgenic Res.* 10:363–371.
14. Curtis IS, Nam HG, Yun JY, Seo KH. 2002. Expression of an antisense GIGANTEA (GI) gene fragment in transgenic radish causes delayed bolting and flowering. *Transgenic Res.*11:249–256.
15. Ellstrand NC, Marshall DL. 1985. The impact on domestication on distribution of allozyme variation within and among cultivars of radish, *Raphanus sativus* L. *Theor. Appl. Genet.* 69:393–398.
16. Ellstrand NC, Schierenbeck KA. 2000. Hybridization as a stimulus for the evolution of invasiveness in plants? *Proc. Natl. Acad. Sci. USA* 97:7043–7050.

17. Ellstrand NC, Prentice HC, Hancock JF. 1999. Gene flow and introgression from domesticated plants into their wild relatives. *Annu. Rev. Ecol. Syst.* 30:539–563.
18. Ellstrand NC. 2003. *Dangerous liaisons? When cultivated plants mate with their wild relatives.* Baltimore, MD: Johns Hopkins University Press. 244 pp.
19. Fernald ML. 1950. *Gray's manual of botany.* 8th ed. New York: Van Nostrand Reinhold Co., 1632 pp.
20. Fischer D, Harvey R, Oplinger E, Maloney T. 1999. Response of oat (*Avena sativa*) varieties and wild radish (*Raphanus raphanistrum*) to thifensulfuron plus tribenuron. *Weed Technol.* 13:144–150.
21. Gómez-Campo C, Prakash S. 1999. Origin and domestication. In *Biology of Brassica coenospecies*, Gómez-Campo C, Ed., pp. 33–58. Amsterdam: Elsevier.
22. Gressel, J. 1999. Tandem constructs: preventing the rise of superweeds. *Trends Biotechnol.* 17:361–366.
23. Hashem A, Bowran D, Piper T, Dhammu H. 2001. Resistance of wild radish (*Raphanus raphanistrum*) to acetolactate synthase-inhibiting herbicides in the western Australian wheat belt. *Weed Technol.* 15:68–74.
24. Hawlader MSH, Mian MAK. 1997. Self-incompatibility studies in local cultivars of radish (*Raphanus sativus* L.) grown in Bangladesh. *Euphytica* 96:311–315.
25. Heap, IM. 2004. International survey of herbicide resistant weeds. Available at www.weedscience.com. Accessed May 11, 2004.
26. Holm LG, Doll J, Holm E, Pancho J, Herberger J. 1997. *World weeds: natural histories and distributions.* New York: John Wiley and Sons, 1129 pp.
27. Huh MK, Ohnishi O. 2001. Allozyme diversity and population structure of Japan and Korean populations of wild radish, *Raphanus sativus* var. *hortensis* f. *raphanistroides* (Brassicaceae). *Genes Genet. Syst.* 76:15–23.
28. Huh MK, Ohnishi O. 2002. Genetic diversity and genetic relationships of East Asian natural populations of wild radish revealed by AFLP. *Breed. Sci.* 52:79–88.
29. Irwin RE, Strauss SY, Storz S, Emerson A, Guibert G. 2003.The role of herbivores in the maintenance of a flower color polymorphism in wild radish. *Ecology* 84:1733–1743.
30. Jalas, J., Suominen, J, and Lampinen, R. (Eds.) 1996. *Raphanus.* Atlas Florae Europaeae — Distribution of vascular plants in Europe. Vol. 11:290–293. Helsinki: Helsinki University Printing House.
31. Kercher S, Conner JK. 1996. Patterns of genetic variability within and among populations of wild radish, *Raphanus raphanistrum* (Brassicaceae). *Am. J. Bot.* 83:1416–1421.
32. Klinger T, Ellstrand NC. 1994. Engineered genes in wild populations: fitness of weed-crop hybrids of *Raphanus sativus*. *Ecol. Appl.* 4:117–120.
33. Lee SS. 1987. Bolting in radish. In *Improved vegetable production in Asia*, No. 36, Asian and Pacific Council, pp. 60–70. Taipei, Taiwan: Food and Fertilizer Technology Center.
34. Lewis-Jones LJ, Thorpe JP, Wallis GP. 1982. Genetic divergence in four species of the genus *Raphanus*: implications for the ancestry of the domestic radish *R. sativus*. *Biol. J. Linn. Soc.* 18:35–48.
35. Monjardino M, Pannell DJ, Powles SB. 2003. Multispecies resistance and integrated management: a bioeconomic model for integrated management of rigid ryegrass (*Lolium rigidum*) and wild radish (*Raphanus raphanistrum*). *Weed Sci.* 51:798–809.
36. Mekenian M, Willemsen R. 1975. Germination characteristics of *Raphanus raphanistrum*. I. Laboratory studies. *Bull. Torrey Bot. Soc.* 102:243–252.
37. Munz, PA. 1973. *A California Flora.* 3rd ed. Berkeley: University of California Press. 1681 pp.
38. NRC (U.S. National Research Council) 2004. *Biological confinement of genetically engineered organisms.* Washington, DC: National Academies Press. 255 pp.
39. Norsworthy JK. 2003. Allelopathic potential of wild radish (*Raphanus raphanistrum*). *Weed Technol.* 17:307–313.
40. Ogura H. 1968. Studies on the new male sterility in Japanese radish with special reference to utilization of this sterility toward the practical raising of hybrid seeds. *Mem. Agric. Kagoshima Univ.* 6:39–78.
41. Panetsos CA, Baker HG. 1967. The origin of variation in "wild" *Raphanus sativus* (Cruciferae) in California. *Genetica* 38:243–274.
42. Pistrick K. 1987. Untersuchungen zur Systematik der Gattung *Raphanus* L. *Kulturpflanze* 35:225–321.
43. Rabbani MA, Murakami Y, Kuginuki Y, Takayanagi K. 1998. Genetic variation in radish (*Raphanus sativus* L.) germplasm from Pakistan using morphological traits and RAPDs. *Genet. Resources Crop Evol.* 45:307–316.

44. The Nature Conservancy. Wildlands Invasive Species Team. http://tncweeds.ucdavis.edu/esadocs/raphsati.html. Accessed May 20, 2004.
45. Reeves T, Code G, Piggin C. 1981. Seed production and longevity, seasonal emergence, and phenology of wild radish, *Raphanus raphanistrum. Austral. J. Exp. Agric. Anim. Husb.* 21:524–530.
46. Roberts HA, Boddrell JE. 1983. Seed survival and periodicity of seedling emergence in eight species of Cruciferae. *Ann. Appl. Biol.* 103:301–309.
47. Roberts HA, Neilson JE. 1981. Changes in the soil seed bank of four long-term crop/herbicide experiments. *J. Appl. Ecol.* 18:661–668.
48. Sakai AK, Allendorf FW, Holt JS, Lodge DM, Molofsky J, With KA, Baughman S, Cabin RJ, Cohen JE, Ellstrand NC, McCauley DE, O'Neil P, Parker IM, Thompson JN, Weller SG. 2001. The population biology of invasive species. *Annu. Rev. Ecol. Syst.* 32:305–332.
49. Smit J, Cairns A. 2001. Resistance of *Raphanus raphanistrum* to chlorsulfuron in the Republic of South Africa. *Weed Res.* 41:41–47.
50. Snow AA, Stanton ML. 1988. Aphids limit fecundity of a weedy annual radish (*Raphanus sativus*). *Am. J. Bot.* 75:589–593.
51. Snow AA, Uthus KL, Culley TM. 2001. Fitness of hybrids between weedy and cultivated radish: implications for weed evolution. *Ecol. Appl.* 11:934–943.
52. Specht CE. 2001. *Raphanus*. In *Mansfeld's Encyclopedia of Agricultural and Horticultural Crops*. Hanelt P, Ed., Vol. 3:1476–1481. Heidelberg: Springer-Verlag, 6 vols.
53. Thomas AG, Gill AM, Moore PHR, Forcella F. 1984. Drought feeding and the dispersal of weeds. *J. Austral. Inst. Agri. Sci.* 50:103–107.
54. Trouard-Riolle Y. 1914. Recherches morphologiques et biologiques sur le radis cultives. *Ann. Sci. Agron. Franc. Etrang.* 31:295–322.
55. Walsh MJ, Powles SB, Beard BR, Porter SA. 2004. Multiple-herbicide resistance across four modes of action in wild radish (*Raphanus raphanistrum*). *Weed Sci.* 52:8–13.
56. Warwick SI, Francis A. The biology of Canadian weeds. *Raphanus raphanistrum.* L. *Can. J. Plant Sci.,* in review.
57. Weaver S, Ivany J. 1998. Economic thresholds for wild radish, wild oat, hemp-nettle, and corn spurry in spring barley. *Can. J. Plant Sci.* 78:357–361.
58. Williams PH, Pound GS. 1963. Nature and inheritance of resistance to *Albugo candida* in radish. *Phytopathology.* 53:1150–1154.
59. Witson TD, Ed. 1996. *Weeds of the West.* 5th ed., Newark, CA: The Western Society of Weed Science.
60. Yamagishi H, Terachi T. 2003. Multiple origins of cultivated radishes as evidenced by a comparison of the structural variations in mitochondrial DNA of *Raphanus. Genome* 46:89–94.
61. Yamaguchi H, Okamoto M. 1997. Traditional seed production in landraces of daikon (*Raphanus sativus*) in Kyushu, Japan. *Euphytica* 95:141–147.
62. Yu Q, Zhang XQ, Hashem A, Walsh MJ, Powles SB. 2003. ALS proline (197) mutations confer ALS herbicide resistance in eight separated wild radish (*Raphanus raphanistrum*) populations. *Weed Sci.* 51:831–838.
63. Yoon HM, Pyo HK. 1977. Studies on the inheritance of bolting, flowering time and other traits in radish (*Raphanus sativus* L.). *Korean J. Breed.* 9:45–57.

14 Ferality — Risks of Gene Flow between Sunflower and Other *Helianthus* Species

André Bervillé, Marie-Hélène Muller, Bernard Poinso, and Hervé Serieys

14.1 INTRODUCTION

14.1.1 Botany and Economic Importance of *Helianthus* Species

Sunflower (*H. annuus* L.) and Jerusalem artichoke (*H. tuberosus* L.) belong to the genus *Helianthus* in the Heliantheae, a tribe of the Asteraceae, the most abundant family of plants. None of the Heliantheae, which include about 200 genera, had naturally spread out from the American continents before the colonization by Europeans. *Helianthus* spp. were first introduced to Europe as decorative plants, then later as foodstuffs, either for oil consumption (sunflower) or as a vegetable (Jerusalem artichoke). These species are pollinated by bees or flies and are self-incompatible. Natural crosses between *Helianthus* and other Asteraceae are not probable and have not been observed in Europe (15) or elsewhere.

Sunflower for seed oil production is of major economic importance in Europe and it is the fourth most abundant oil crop in the world after soybean, palm, and rapeseed. Sunflower oil is the most consumed oil in Europe. It is used mostly after refining and is appreciated by consumers for salad and frying and, moreover, it is relatively inexpensive. However, the high 55 to 60% linoleic acid content of classic sunflower oil raises dietary concerns relating to prevention of heart disease. Recently bred varieties released as NewSun cultivars have a 20% linoleic acid and 60% oleic acid content (OAC), which mimics olive oil. Sunflower oil is sold as classic, as high oleic acid content (HOAC), or as a mixture of normal with HOAC oil in appropriate proportions.

The underground tubers of Jerusalem artichoke are used as vegetables in different countries either boiled or baked. Jerusalem artichoke tubers have been used to feed animals, for alcohol production and to prepare biofuel. However, in Europe, this crop is negatively associated with the food restrictions of the last world war, and it is not widely used. There is recent interest in its well-documented ability to prevent digestive diseases (namely, colon cancer) (8,34).

14.1.2 Domestication and Breeding Sunflower and Jerusalem Artichoke

Sunflower has been domesticated by Amerindians who needed fat sources (19). Seeds were smashed or ground and cooked on a heated stone to make biscuits or pancakes. Such sunflower probably led to confectionery sunflower, which is relatively low fat (25 to 30% of oil in the seed) and after grilling is hulled by machinery for confectionery use or with fingers for consumption. It is also called edible sunflower. The native American sunflower varieties and landraces form a genetically cohesive group as seen by random amplified polymorphic DNA (RAPD) evidence (7). Early remains of *Helianthus annuus* discovered at the San Andres site in the Gulf Coast region of Tabasco, Mexico, are the earliest record of domesticated sunflower. Age determinations of one large domesticated seed and an achene produced dates of about 4000 years ago (24). Those discoveries demonstrated that sunflower was domesticated in eastern North America. Moreover, Schwarzbach and Rieseberg (37) recognize several domestication events based on cytoplasmic and nuclear DNAs. Jerusalem artichoke tubers have probably been used since the same period.

Sunflower and Jerusalem artichoke were first introduced in Europe as ornamental plants early after the discovery of America, by the 16th century.

Other annual species (*H. bolanderi* or serpentine sunflower, *H. debilis* or cucumber leaf sunflower, *H. argophyllus* or silver leaf sunflower) and some perennial *Helianthus* spp. (*H. strumosus* or strumose sunflower, *H. rigidus* or stiff sunflower, *H. giganteus* or giant sunflower) were probably introduced at the same time. Such decorative plants have been scattered in private gardens, some botanical gardens, and famous parks. They did not spread too much and stayed confined.

Confectionery sunflower was introduced into Russia by the middle of the 18th century and was liked by people for its fat (25 to 30% oil), especially during lent, as its consumption was not prohibited by the Orthodox Church. By the 1930s, various Russian institutes started breeding programs aimed at enhancing the oil content of sunflower seeds. This has been performed in Russia first at the Oil Crop Institute (VNIIMK) at Krasnodar, but also later at Armavir, Rostov, and Dnipropetrovs'k (49). This crop with 40% oil in seeds was called "oilseed sunflower" in contrast to the "confectionery sunflower." Further breeding developed lines that better exploited hybrid vigor, and the oil content was enhanced up to 55% in recent cultivars. Selfing in sunflower populations causes tremendous inbreeding depression and some progeny do not survive. However, the best lines obtained are checked for combining ability to restore vigor. The discovery of cytoplasmic male sterility (CMS) by Leclercq (23) and efficient male fertility restoration genes (50) enhanced hybrid seed production, and in a few years, sunflower became the fourth oil crop worldwide, with about 10% of world oil production (9,14,48).

14.1.3 Where Did *Helianthus* Establish in the Wild Outside Its Native Area?

14.1.3.1 Europe

Sunflower became a widely spread crop by the 1970s, when some hybrid cultivars were produced on a large scale. It became common to see volunteer sunflowers in fields of different crops, soybeans, peas, and less in cereals and sugar beets (probably due to herbicide treatment at pre-sowing stages). It is now common to see sunflower volunteers in different countries of Europe and Russia; they are often visible in autumn and frequently are able to produce progeny. The fate of seeds remaining after harvest in the soil has not been studied in depth and some questions arise about the longevity of such seed banks (4,12) and about the continuity of volunteer and feral populations that seed banks may generate (41).

In addition to these volunteers, several feral and weedy populations of sunflower have been observed in Spain and Italy, where they interfere with cropped sunflower. The mechanisms by which they appeared, established, and became weedy, have to be investigated.

14.1.3.2 Other Continents

Several annual *Helianthus* species spread to different continents where they were not native. In Argentina, wild sunflower including *H. petiolaris* or prairie sunflower (11,33) escaped from gardens and formed permanent populations (11,32). In Mozambique (38), Australia, and China, *H. argophyllus* escaped and spread as feral populations (15).

Jerusalem artichoke also escaped from gardens in Australia and formed feral populations in the region of Adelaide. Recently, we heard about wild sunflower spreading in Russia north to the Caspian Sea toward Kazakhstan. We know of no work on these populations.

14.1.4 Why Do We Care about These Volunteer or Feral Populations?

14.1.4.1 Gene Flow from Volunteers or Feral May Modify Sunflower Oil Composition

Recently, the Pervenets mutant discovered and obtained by chemical mutagenesis in Krasnodar (Russia) now enables the cultivation of a sunflower with an HOAC of 80 to 90% (43). This new crop is now spreading in Europe, Argentina, and North America due to its advantages for human diet and several industrial applications (lubricants, cosmetics, biofuel). Because of the risk of pollen flow between HOAC and classic varieties, the crops are supposed to be cropped in separate areas. The regulations require a separation by at least 400 m (15). These requirements are due to the dominance of the Pervenets mutation and to the fact that seed oil composition is determined by the embryo genotype. Because both parental lines are homozygous for the Pervenets mutation, all the pollen grains on the F_1 plants that are usually grown for commercial sunflower should carry the Pervenets mutation. Once these pollen grains pollinate a classic sunflower crop, the resulting embryo will produce HOAC oil.

Industry prefers oil with oleic acid content level either classic (20 to 25%) or high (at least 83%). Conversely, there is a suppressor of the Pervenets mutation that can modify the oleic acid composition of Pervenets cultivars when there is gene flow toward Pervenets varieties (22). This decreases the income for the sunflower crop growers.

Other types of sunflower oils exist with different fatty acid compositions or varying tocopherol composition including a high α-tocopherol with high vitamin E activity and with high antioxidant activity through β, γ, δ tocopherols, and conversely with low antioxidant and low vitamin E activities, respectively. Each of these crops requires production with the same isolation distances to avoid gene flow between them and to prevent adulteration (15).

14.1.4.2 Impact of Crop Alleles in the Wild

Gene flow and the fate of crop alleles in wild sunflower have been studied by the Rieseberg group in the U.S. (10,16,25,31,51). The genetic basis of commercial hybrid cultivars is narrow and all the hybrid cultivars are made based upon a unique CMS source (20) and restoration fertility system (48). Simple sequence repeat (SSR) alleles of cultivated and wild sunflowers are frequently different in size (30,46). Thus, it is easy to differentiate the two forms with molecular markers, but surprisingly no allele is specific to the crop. There are only differences in frequency between crop and wild sunflowers, except for the Pervenets allele in HOAC sunflower, which is crop-specific and whose impact and fate in wild sunflower have never been studied. No transgenes have been developed for commercial sunflowers in Europe, but release of some cultivars with disease (*Sclerotinia*) resistance, parasitic plant (*Orobanche*) resistance, drought tolerance, and herbicide resistance are in preparation and several trials are being performed under containment in the U.S. (http://www.nbiap.vt.edu/cfdocs/fieldtests1.cfm.) and elsewhere, but not in Europe.

14.1.4.3 How Did These Introduced *Helianthus* Species Evolve from the Native Forms?

Sunflower, Jerusalem artichoke, and their relatives, although native to America, are now spread over Europe either as crops or have escaped from cultivation. Sunflower volunteers, escaped and permanent wild sunflowers, and Jerusalem artichoke populations represent the dominant feral *Helianthus* species. No true wild sunflower exists in Europe; all escaped and permanent populations may correspond to more or less introgressed forms according to the main domestication traits (15). Wild-type sunflowers in Europe probably derived from true wild types imported with seeds from the U.S. and further crossed and backcrossed with the crop. However, no studies have been undertaken on these populations and it is of interest to understand how they evolved from the original wild form, and how much backcrosses with the crops contributed to their evolution.

14.1.4.4 Feral and Volunteers May Constitute Gene Reservoirs for Crop Alleles and Become Invasive

For the last century, Jerusalem artichoke, other decorative *Helianthus* species, and sunflower have all escaped from cultivation and colonized many places in Europe — wild, ruderal, and agroecosystems. Thus, gene flow may occur between these escaped populations and sunflower crops. Such gene flow is possible from the escaped populations of these species or from some other introduced decorative relatives (*Verbesina*, *Tithonia*, *Simsia*) which share common genomes with *Helianthus* (44). Consequently, those escaped and permanent *Helianthus* forms may represent gene reservoirs for cultivated alleles. Because some have the tendency to become invasive, the mechanism of such establishment outside agroecosystems is a concern worth studying to understand the invasive processes. Our aims have been to determine the origins of the different kinds of sub-spontaneous populations of *Helianthus* (volunteers, feral, or weedy) in Europe, to study potential gene flow to or from sunflower crops, and study the population dynamics of such populations. Such studies are important to understand whether or not these spontaneous populations may represent a reservoir hiding escaped crop alleles, and if so, to assess the risk of gene flow back to the crop. Moreover, we would like to understand how these forms may become established outside agriculture to control their dissemination.

14.2 STUDIES ON *HELIANTHUS ANNUUS*

Based on history, two types of volunteer or feral sunflowers are expected to further escape and become permanently established in Europe:

1. Common volunteers escaped from commercial hybrids, probably corresponding to advanced-hybrid generations
2. Feral or weedy populations of wild sunflowers, more or less crossed with commercial hybrids

We used the term feral for wild sunflower outside of agro- and ruderal ecosystems, whereas weedy is used for the same dedomesticated sunflower in crops. In both cases we are talking of populations with a phenotype having some similarities to the wild *H. annuus* in North America. We sometimes called them "wild sunflower populations" to underline this wild phenotype, even if they are not completely similar to the wild populations of the native area. Volunteers and wild sunflower populations represent different entities that raise different questions. For permanent (naturalized) wild sunflower populations, the questions deal with their origin, their invasiveness, and gene exchange with cultivated forms. Questions on volunteers deal with their origin, their ability to establish, and their eventual gene flow with the crop. Moreover, population studies of these two compartments require different approaches and some specific tools.

14.2.1 Escaped and Permanent Wild Sunflower Populations in Italy and Spain

14.2.1.1 The Problem

Wild *Helianthus annuus* plants have colonized several thousand hectares in Italy and Spain (Baldini, University of Udine, personal communication and Fernandez-Martinez, CSIC Cordoba, personal communication). This invasion started by the 1970s and cultivating the crop has been abandoned in areas too infested with wild sunflowers. These populations are permanent and are amenable to population dynamics studies. This is also a situation where transgenic herbicide-resistant sunflowers could be used to control the wild sunflowers, and the consequences of gene flow on population dynamics need to be studied with and without transgenic failsafe mechanisms (26).

14.2.1.2 Hypotheses on Their Origins

Olivieri (University of Udine, personal communication) presumes that the origin of such invasive sunflower plants was probably as contaminants from the first cultivated hybrids imported from the U.S. (5). Indeed, the first imported hybrids in Europe were grown in Texas, where commercial seeds were probably contaminated with those of wild plants at harvest. Gene flow may also have naturally occurred there, from wild to cultivated female plants in the seed producing fields (10,28). These contaminated seeds (crop-wild hybrids or true wild sunflower) were exported to Europe and germinated together with the crops, which probably led to the present escaped and permanent weedy populations. Hybrid seed production now takes place in other countries, particularly in Turkey, to avoid this problem.

Wild sunflowers from the U.S. produce tiny seeds (thousand seed weight fluctuates from 3 to 10 g) in comparison to cultivars (thousand seed weight varies between 30 and 80 g). Wild seeds create long-lived seed banks with dormant seeds (41). Some seeds from feral sunflowers originating from Italy have been grown in Montpellier for observation. The plants look like wild *Helianthus annuus*; they are multiheaded, highly vigorous, and often with colored (anthocyanin-containing) stems. These characters are quite different from the crop, supporting the hypothesis of seed contamination and rendering the hypothesis of dedomestication of hybrid volunteers less probable. Moreover, thousand seed weights fluctuate from 16 to 25 g. These are intermediate between wild (3 to 10 g) and cultivated and could be from dedomestication or from hybrids. Head and seed sizes vary from wild to intermediate types, which suggests that these populations may have resulted from crosses with cultivars. Genuine American wild sunflower populations, which we regularly grow in Mauguio, do not have this kind of intermediate type.

These feral and weedy populations are observed every year in Italy. They will be studied in further detail in collaboration with Italian teams at Udine, Padova, and Perugia.

14.2.1.3 Planned Work

We will investigate the origin of these feral populations using SSR markers, in comparison with about 350 American wild sunflower populations and other annual subspecies kept in the *Helianthus* collection at INRA (*Institut National de la Recherche Agronomique*) Montpellier. Hybridization levels with cultivated forms and estimation of recurrent gene flow will be determined with molecular markers and analyzed using population genetics programs.

Linkage disequilibrium at loci implicated to encode domestication traits (branching, height, stem color, seed size, and oil content) and molecular loci will be examined to determine whether wild alleles have a tendency to cluster. Indeed, whether cultivars are fixed for such alleles and display specific combinations in comparison to wild sunflower.

14.2.2 A Case Study — Model of Wild Sunflower Establishment Near Montpellier

14.2.2.1 Unwanted Sunflowers Become an Experimental Plot

At our INRA station in France, several genotypes including 25 wild *H. annuus* (sunflower) accessions and 20 wild other annual *Helianthus* species were grown in a nursery field at our domain (Mauguio) substation near Montpellier in 1985 (Table 14.1). Other crops have been grown in this field successively through 2001 (sunflower, wheat, maize, rapeseed, and sorghum), but we still observed the proliferation of wild *Helianthus* (probably wild sunflower) (Figure 14.1).

TABLE 14.1
Wild Annual *Helianthus* Species Grown Since 1985 in a Weedy Field that Became a Sunflower Nursery

Helianthus Annual Species	Subspecies	Number of Accessions
H. annuus wild sunflower	—	25
H. agrestis		1
H. anomalus		3
H. argophyllus		1
H. bolanderi		1
H debilis		2
H. debilis	*debilis*	1
H. debilis	*sylvestris*	1
H. deserticola		2
H. exilis		1
H. niveus	*tephrodes*	1
H. petiolaris		1
H. petiolaris	*fallax*	1
H. petiolaris	*petiolaris*	1
H. praecox	*hirtus*	1
H. praecox	*praecox*	1
H. praecox	*runyonii*	1

FIGURE 14.1 Wild sunflower population established in a sorghum crop at Mauguio near Montpellier by September 2001.

We then decided to use this "polluted" field as an experiment to investigate the evolution of ferality, the consequences of different cultural practices (mainly rotations) on the evolution of weedy plants, and the consequences of gene flow from cultivated sunflower. Interestingly, we still have most of the primary accessions that were cultivated in 1985. Thus, we might determine the sources of invading forms and ascertain which eventual mix of genotypes has contributed by hybridization events to the feral population.

The field was divided into 140 plots of 5 10 m in 2001. Several seed samples were harvested from these plants taking into account the density of wild *Helianthus* (Table 14.2, Figure 14.2). Seed size was highly variable (thousand seed weight ranging from 2.8 to 18 g), suggesting genetic exchanges with the previously cultivated sunflower crops (Figure 14.3).

The field at Mauguio was further split in 2002: half was sown with wheat and half with cultivated sunflower. Wheat did not receive any herbicide treatment. Large populations of wild sunflowers were

TABLE 14.2
Sampling Method: Number of Samples Chosen according to Density to Evaluate Seed Size of Sunflowers in the Experimental Field (2001–2003)

Populating the 5 m 10 m Plot	Sampled Plant Number
1	1
≤10	3
>10 ≤25	5
>25 ≤50	10
>50	15

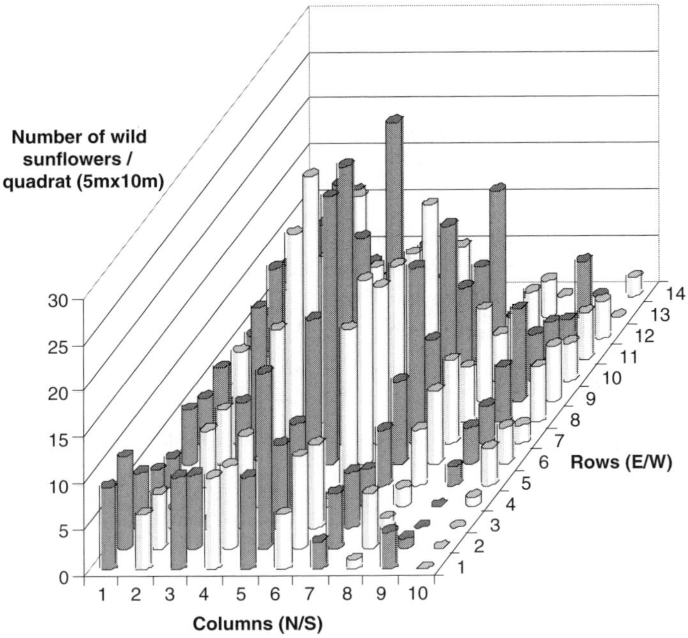

FIGURE 14.2 Plant density of wild sunflowers in a sorghum crop at Mauguio near Montpellier by September 2001. Gray and white bars correspond to alternate rows.

observed at harvest, both in wheat and sunflower subplots. Cultural practices and combine harvesting strongly limited wild plant populations in the wheat crop, where only few mature seeds were produced, although in the sunflower crop most of wild sunflowers reached maturity and generated seeds.

We kept the same crop rotations in 2003. No wild *Helianthus* appeared in summer in the half with wheat, but some appeared at the edge. In autumn, only a few wild sunflowers appeared. Wild sunflower proliferation increased drastically in the plot cropped with commercial sunflower (Figure 14.4A and Figure 14.4B), and their density was noted (Figure 14.5). Wild sunflowers increased more than 30-fold in two generations.

The flowering of the cultivated sunflower crop and some earliest germinating wild *Helianthus* coincided, so that gene flow could have occurred between the crop and the wild forms. Wild sunflower density was recorded in each of the 70 plots of 50 m^2. Leaf disks were harvested on 3 plants per plot, 1 resembling wild, 1 intermediate, and 1 resembling cultivated sunflower, a total of 210 samples for further genotyping. We recorded morphologic traits such as branching, anthocyanin

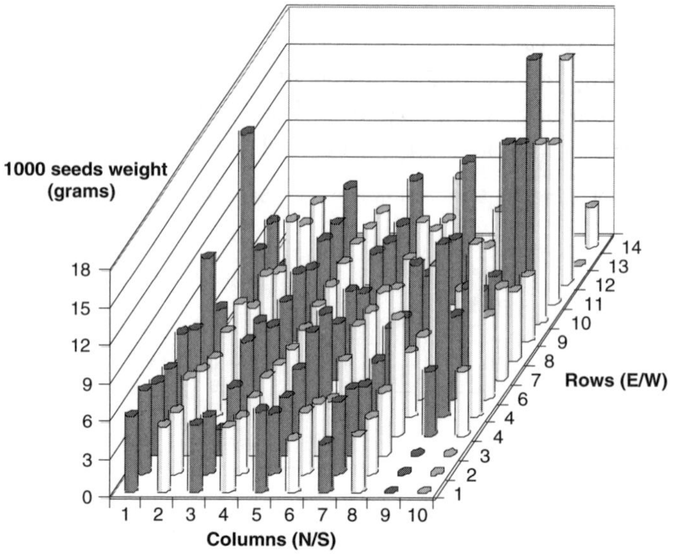

FIGURE 14.3 Spatial distribution of wild sunflower seed size in a sorghum crop at Mauguio near Montpellier September 2001. Gray and white colors correspond to alternate rows.

pigments, and head and stem diameters. The crop was also sampled as a control. In the wild sunflower population sampled in 2003, 17% of the plants did not show any branching or anthocyanin pigments: this shift of phenotypes from wild toward cultivated habit between 2001 and 2003 suggests substantial gene flow from the crop.

14.2.2.2 Planned Work

Evolution will be estimated by measuring changes of SSR allele frequencies between 2001 and 2004. Gene flow with cultivars will be estimated. To measure gene flow between crop and feral wild sunflower in both directions, we will introduce in the spring of 2004 the Pervenets allele by growing an HOAC hybrid.

The different agronomic practices will allow us to determine those practices that may be favorable for controlling weedy sunflower. Other trials are being performed at INRA Dijon under different agronomic practices (plowed or scarified) either abandoning sunflower cropping (fallow land) or cropping another species (winter peas).

14.2.3 Existing Volunteer Populations — Record, Localization, Fate, and Establishment

Volunteer populations may have their origin in the commercial hybrid or in impurities of the hybrid. They may exchange genes with successive sunflower crops and they may have evolved under selection pressure depending on which crop allele they inherited, as the cultivars are heterozygous at several loci: branching, male fertility restoration, and downy mildew resistance. Thus, their fate may depend on gene exchange with the successive sunflower crops that may bring new alleles for disease resistance to volunteers.

Volunteer populations in Europe are expected to carry the PET1 cytoplasm (23) giving the CMS trait to female lines, and they are expected to segregate for the fertility restoration (*Rf*) genes, as the hybrid plants are heterozygous at these loci. The male lines of commercial hybrids have several heads (branched plant) to ensure longer pollen release. The branching trait is controlled by one locus that came from a wild American sunflower: the alleles are coded *Br*/*br* and branching allele *br* is recessive. This means that if you have volunteers from hybrids, they will segregate and

A

B

FIGURE 14.4 Wild sunflowers population established in a sunflower crop at Mauguio near Montpellier. A: July 2003. B: August 2003.

some will appear feral (branching). Some recent restorer lines do not carry the *br* allele, which prevents diagnosis of the volunteer generation without knowing the genetic origin of the cultivated hybrid. Evaluation of advanced generation levels will be pursued by studying linkage disequilibria of SSR loci and by using markers linked to *Rf* or *Br* loci.

Moreover, downy mildew (due to *Plasmopara halstedii*) is a prevalent disease in Europe and cultivars must be resistant to pathotypes, as fungicides have insufficient persistence to ensure adequate protection and consequently adequate yield (47). In commercial hybrids, only the male line carries a resistant allele (*Pl* gene), and in a few cases, female and male lines may carry complementary alleles at one or two loci. Consequently, volunteers will segregate for these alleles and the permanent presence of downy mildew in fields should lead to selective pressures that may affect the frequency of *pl*/*Pl* alleles. No studies have been carried out to determine the set of pathotypes present in one sunflower field. New pathotypes appear regularly in France, and we will examine whether volunteers influence their evolution. Moreover, it is obvious that volunteers carrying downy mildew resistance alleles will be highly favored to produce progeny. It is then clear that mildew will exert a strong pressure on the fate of volunteer populations, as Snow et al. (40) have documented the effect of pest resistance genes in wild sunflower populations.

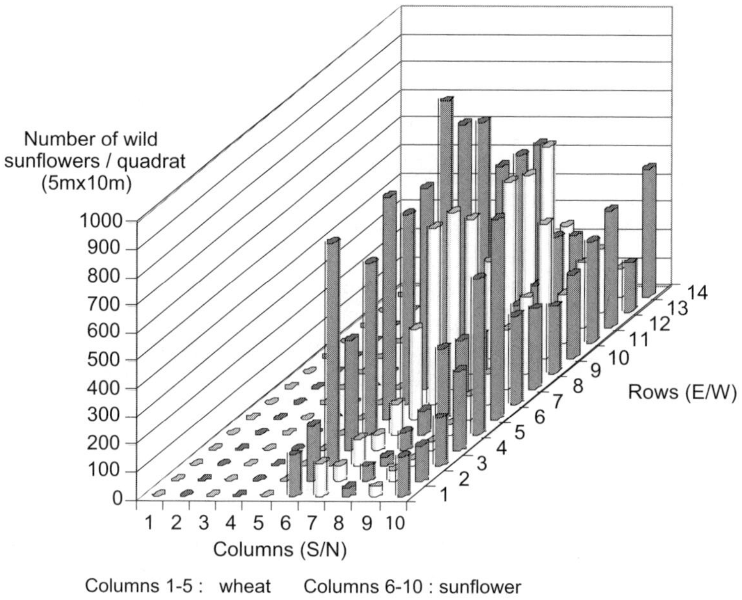

FIGURE 14.5 Wild sunflowers in wheat and cultivated sunflower rotations. Plant density per plot at Mauguio, July 2003. The field was split into two sections: one cultivated with wheat (Column 1 to Column 5) and the other cropped with sunflower (Column 6 to Column 10).

We aim to mark downy mildew pathotypes (virulence genes) that will be identified on these samples to determine their spectra and their relationships with the prevalent strains in culture. Gobbin et al. (17) studying *Vitis/Plamopara viticola* have shown that dot blots of infection by the pathogen on the couple *Vitis/Plamopara viticola* in the field correspond to a few major strains, although the diversity of the fungus is wider. Stark-Urnau et al. (45) have also shown a reduction in genotypic diversity between primary and secondary downy mildew infections suggesting strong environmentally directed selection among them. Different parameters that may influence competition between sunflower downy mildew strains will be assessed: volunteer density, population size, and duration since the establishment of populations. The spread of virulence genes in volunteer populations should indicate the age of these populations, because we know the chronology of introduction for commercial hybrids and therefore the chronology for *Pl* alleles. The co-evolution of mildew on volunteers could lead to new pathotypes. Understanding this situation may well assist in forecasting the appearance of new mildew pathotypes. Conversely, downy mildew may suppress the establishment of volunteers in feral populations, depending on the generation of new pathotypes.

Sunflower transgenes carrying resistance to downy mildew could escape and establish in volunteer or escaped and permanent populations or cross with such populations. A new pathotype could evolve in the diverse wild populations, overcoming the resistance. Engineered *Sclerotinia*-resistant sunflower cultivars may also introgress resistance genes into weedy or feral sunflower populations causing serious problems, as this disease is prevalent in western Europe.

14.3 STUDIES ON JERUSALEM ARTICHOKE

14.3.1 Hybridization Rate Efficiency with Sunflower

Annual *Helianthus* species are classified in sect. Helianthus or sect. Agrestes whereas perennial *Helianthus* species belong to sect. Atrorubentes or sect. Ciliares. Each section can be differentiated from each other by specific RAPD markers enabling differentiation and recognition of annual or perennial species and also allowing estimation of gene flow (44). Jerusalem artichoke and sunflower

TABLE 14.3
Hybridization Rates between Sunflower and Jerusalem Artichoke, in Both Directions

	H. tuberosus Accessions	
Pollen and Seed Production Characteristics	**325**	**570**
Pollen fertility (%)	84.6	(n d)
1000 seeds weight (g)	8.3	5.1
Percent of Fertilized Achenes (%)		
*H. tuberosus** selfed	0.8	1.4
*H. tuberosus** polycross (insects)	12.7	15.6
*CMS H. annuus** *H. tuberosus* (manual cross)	0.8	6.6
*CMS H. annuus** *H. tuberosus* (insects)	0.5	0.6
*H. tuberosus** *H. annuus* (manual cross)	10.6	22.3
*H. tuberosus** *H. annuus* (insects)	2.6	2.2

(*) Harvested parent.
(n d) Not determined.
% fertilized achenes = mean value of crosses with HA89 and 92B6 inbred lines.

can hybridize despite the chromosome numbers being 102 and 34, respectively, with hybrids bearing 68 chromosomes. The interspecific hybrids are poorly male and female fertile and they bear just a few seeds. Sunflowers have been used to improve several agronomic traits, thus some Jerusalem artichoke clones may have derived from hybrids between the two species. Jerusalem artichoke cultivars are maintained by tubers; if propagated by seeds, the cultivar is lost. Wild Jerusalem artichoke may be propagated either as clones (each plant will lead to a clone) or by seeds.

We have used hybridization to determine the crossing efficiency between the two species. Crosses were handmade or performed by bees on fertile or female CMS sunflower inbred lines. We used early Jerusalem artichoke accessions, flowering in June and July: one from a cultivated clone (INRA 325), the other from a native U.S. wild population (INRA 570).

The rates of self- and intercrosses between the two species are displayed in Table 14.3. It appears that the level of self-fertility of Jerusalem artichoke is low (less than 1.5% fertilized achenes), but intercrossing produced significantly higher seed set (approximately 14%). When *H. tuberosus* was pollinated with sunflower, seed set varied between 2 and 22% fertilized achenes, but due to *H. tuberosus* crossing with each other, the true interspecific hybridization level is probably lower. To accurately define the hybridization ratio, the genetic status (6x or 4x) of the seeds remains to be established. Reciprocal crosses performed on male sterile (CMS) sunflower provide more realistic hybridization level, which ranged between 0.6 and 6.6% for the 325 and 570 accessions, respectively. In spite of their relatively low crossing levels, the risk of gene flow between these two species is likely in natural conditions. Studies on the fitness (survival status, male and female fertility) and the fate of hybrids and derived progeny should provide additional information on the ability to establish in new environments and to exchange genes with sympatric compatible species. Since 1978, we have already produced and maintained in collection, hybrid plants that resemble Jerusalem artichoke and consequently may mask or hide sunflower genes escaped from crop or volunteers if such natural crosses occurred.

14.3.2 Feral Jerusalem Artichoke Populations

Jerusalem artichoke has been widely cultivated throughout Europe. By 1955, the crop reached 150,000 ha in France (5), due to its ability to grow on poor soils and its drought tolerance. It became the symbol of disliked vegetables from the austerity period during and after World War II, and most people abandoned Jerusalem artichoke, turning to other more attractive vegetables. Its culture was then abandoned, but the tubers remaining in soils proliferated and erosion has led to their migration along the valleys and colonized not only stream banks in the Cévennes Mountains, but also river banks in the plains. In France, it colonized the badlands represented by "les Causses" (700 to 1000 m elevation), where it had been previously cropped. Such Jerusalem artichoke populations have escaped and become established for several decades and are still expanding in the valleys and on the plateaus. They also appear near sunflower seed production fields leading to exoferal offspring. Consequently, gene flow between sunflower and Jerusalem artichoke needs to be studied. Interspecific crosses between these two species are naturally possible and we will determine their efficiency in natural conditions.

Feral Jerusalem artichoke populations have escaped from cultivation and become permanent in the Orb and Hérault River valleys 50 to 60 km north from Montpellier. There are frequent and abundant fixed populations on sandy and shaly river banks. They move down with repeated floods and colonize along the banks. Some Jerusalem artichoke crops were also abandoned around cities and proliferate in ruderal waste grounds in the suburbs.

14.3.3 Studies on Jerusalem Artichoke Populations

The main questions arising about feral Jerusalem artichoke are whether they may introgress genes from sunflower; and if so, whether crop genes may modify their evolution; and whether gene flow back to sunflower is possible. Probably, pests may have some influence on these populations; they remain to be identified. Such gene flow is hard to visually detect by looking for hybrid plants in the progeny of a crop, unless a marked phenotypic effect result, such as bolting in sugar beet (36). Otherwise, gene flow will remain difficult to quantify.

For each patch of feral Jerusalem artichoke plants, we will record homogeneity; description of flowers (abundance of flowers, male and female fertility); height and color of stems; and form, color, and smoothness or roughness of tubers.

We will check the transfer of SSR markers between annual to perennial *Helianthus* spp., as there is a common genome between *Helianthus* sections based on previous results (44). We will try to define a set of markers that can be used to differentiate between different annual and perennial *Helianthus* spp. and to reveal hybrid plants between sunflower and Jerusalem artichoke. Then we will use these molecular markers to further characterize each clump of plants to determine whether it is a clone or it comprises several plants. Relationships between patches will be established using clustering methods, along with multiloci genetic analysis methods (29).

Karyotype analyses also allow identification of sunflower Jerusalem artichoke hybrids. The technique is routine in our laboratories, but is more time-consuming than using molecular markers, if there are appropriate markers.

Some Jerusalem artichoke clones have poor fertility, but others produce copious amounts of seeds and we cannot eliminate the possibility that single plants originate from seeds. We have a collection of 140 cultivated accessions for Jerusalem artichoke and can compare tuber molecular profiles to those of tubers in our collection. Referenced clones will delineate the origin of a plant from a tuber. If we do not find any entry in our collection, it will suggest an origin from a seed. Another possible origin is an imported form escaped from a nursery. We have not yet collected such plants, but probably such ornamental plants will display a recognizable phenotype. Referenced clones with intermediate chromosomal composition ($2n = 4x = 68$) probably exist in our collection and they will enable differentiation between eventual new hybrid individuals that resulted from gene flow between escaped Jerusalem artichoke and domesticated sunflower crop and *vice versa.*

TABLE 14.4
Situations, Hypotheses, and Consequences of Gene Flow from Crop to Wild Populations

Donor Crop Population	Recipient Wild Population	Concepts and Designation in Population Genetics	Expectation
Crop allele fitness lower	Wild allele fitness higher	Neutral	No effect Crop and wild coexist
Crop allele fitness higher	Wild allele fitness lower	Genetic assimilation	Crop allele replaces wild alleles Wild will look like crop
Hybrids less fertile than parents	Small populations	Demographic swamping	Endangered wild population
	Habitat disruption	May aggravate genetic assimilation	Probably more frequent cases
	Immigration		Short-term effect
	Important populations	Population meltdown	Drastic reduction of wild population
	Recurrent gene flow		Long-term effect
Hybrids more fertile than parent	Recurrent gene flow	Hybrid expansion	Hybrid may be invasive Not obvious since gene flow decreases hybrid frequency

14.4 MODELING THE IMPACT OF GENE FLOW AND FATE OF WILD RELATIVES

Gene flow from cultivated toward wild sunflower raises questions on the fate of cultivated alleles once transferred in wild plants and on the fate of wild populations receiving cultivated alleles. Studies on gene flow between crop and volunteers or feral populations that enable an understanding of the fate of the feral population are incomplete or do not exist (13). Thus, some researchers have tried to model the impact of gene flow to predict the fate of crop-alleles in wild populations. By extension, they consider that their results could be valid for volunteers or feral populations and the models can be used to examine the possible consequences on the dynamics of the feral population.

14.4.1 Contact

Gene flow may occur in many contact zones where crop and wild relatives coexist (14). We strongly suspect that such crosses occurred for sunflower in Italy and at Mauguio, where wild sunflower populations have escaped and are settled. We have seen fertile hybrids between even genetically distant species within the genes. As fertile hybrids between cultivated and wild forms may establish gene flow, one must consider the consequences of such recurrent gene flow. The consequences on wild populations depend on several factors such as hybrid fertility, crop allele fitness, wild population size, and recurrent flow. Most of the time, this occurs without consequence for the wild forms, but that may lead to more invasive forms in certain environments. Several concepts of population genetics have been used to model these different effects. The main results are summarized (Table 14.4) and the conclusions are discussed.

14.4.2 Modeling

Models simulating the fate of a wild population recurrently receiving pollen from a crop fixed at one or two independent loci attempt to predict whether replacement can occur, what its consequence of population dynamics for the wild population is, and how different factors affect that replacement. We summarize here the main conclusions of interest to us (18):

- The conditions for genetic assimilation are not stringent. Recurrent gene flow can fix the crop allele in the wild, even if it is disfavored. Of course, if the crop allele is favorable (herbicide resistance, for instance), the fixation can occur and then prevent elimination of such plants with herbicide treatment (26). Computation of generation numbers to reach fixation is not accurate because parameters do not include genetic drift that may have a major effect over time.
- If the disfavored crop allele is dominant and if the wild population is small, genetic assimilation can involve threshold and hysteresis. It means that a small increase in immigration may lead to the fixation of the transgene in the wild population.
- The fixation of a disfavored allele may lead to a drastic shrinkage of the wild population. If the population is small, demographic swamping may occur more quickly. Small populations are thus more vulnerable to recurrent gene flow from crop.

These models may apply to crop with volunteers and wild sunflowers that are frequently adjacent in the U.S. However, the wild sunflower populations are not small, endangered populations. In European countries, we lack accurate data to evaluate crop and sunflower volunteer gene flow. Recent observations of downy mildew on volunteers suggest that *Pl* genes may play a major role in volunteer spreading. New imidazolinone-resistant sunflower cultivars will probably be used in areas infested by feral populations or wild sunflowers and we can predict that the resistant allele will be introgressed in these forms. However, the present models do not enable prediction of the fate of these feral or wild populations, nor can they predict whether introgression will be more rapid than the evolution of resistance from within the wild sunflower genome.

The timing of replacement of wild alleles by crop alleles and relationship between frequency of crop gene and the wild population size have been modeled to depend on the immigration rate, selection pressure on hybrid and wild plants, dominance level (crop-wild alleles), and population size. Whatever the other restrictive hypotheses done by Huxel (21) and Wolf et al. (52), Ronce and Kirkpatrick (35) have also shown the possibility of migrational meltdown in smaller populations.

In agroecosystems, this means that two consequences may occur for surrounding wild forms: they may fix more strongly disfavored alleles, but drastic shrinkage of the wild population may also occur. Volunteers may also contribute to gene exchange by gathering alleles present in successive sunflower cultivars as *Pl* genes used against downy mildew. New allelic combinations with unpredictable fate may occur and disperse.

14.4.3 Sunflower

Can these models (18,21,35,52) apply to the situation of sunflower in Europe? We can consider both the case of volunteers and of weedy or feral forms. Sunflower crop alleles once moved to wild relatives may modify their evolution. For several loci, frequency differences between crop and wild alleles are important; and in a few cases, for linked loci, a cluster of crop alleles may be transferred to wild relatives.

Whether volunteers are the first step in the evolution of escaped plants toward ferality is not yet documented in sunflower. Previous results suggest that Italian wild sunflowers did not derive from volunteers, but from wild contaminant seeds. However, no studies have dealt with volunteers so the question remains about the consequences of further gene flow from crops to volunteers. In Spain and Italy, cultivated sunflowers and feral or weedy populations of sunflowers often grow in close proximity to each other. Sunflower is allogamous and some studies have shown that cultivated alleles can persist for several generations in wild populations (6). Similarly, our data on European weedy populations suggest that they have been subjected to crop gene flow.

Then, to more precisely assess the consequence of this gene flow and the suitability of the model, one must evaluate parameters such as the fitness of crop-wild hybrid or the size of the populations. Wild sunflowers are invasive in some places in Europe, although they have probably been subjected to recurrent gene flow from cultivated sunflower. Models did not predict invasiveness of wild forms except in extreme conditions (18). Reasons for misfit with modeled conclusions are not understood, but suggest that the model may be invalid. Factors that promote such invasive behavior remain to be elucidated.

14.4.4 Transgene Containment

One method to prevent transgene persistence in wild forms has been called transgenic mitigation (TM) (2). It consists of one construct with a favorable gene for the crop (e.g., herbicide resistance) and a couple of genes disfavored for the wild plants, but neutral or favorable for the crop (e.g., dwarf genes). With this construct, the hemizygous plants for the transgenic construction appeared to be weak competitors with the wild type in greenhouse and without herbicide treatment (2). The results demonstrate that crop-wild hybrid plants competing with wild plants were disadvantaged when the selector (herbicide) was not used. These tandem constructs may then help keep the transgene at a low frequency in wild populations. Such sequences that are advantaged in crop sunflower and disadvantaged in wild sunflower correspond to domestication traits that are being mapped by several sunflower teams. Thus, it will soon be possible to introduce these regions in tandem constructs as suggested by Al-Ahmad et al. (2) to verify whether they can prevent transgene diffusion from crop to feral or wild sunflower populations.

However, extrapolation on the fate of transgene based on further generations as suggested by Al-Ahmad et al. (2), may be modified through modeling considerations on recurrent gene flow especially the prediction of Haygood et al. (18) described above but including genetic drift. They predict that invasiveness can occur if gene flow is recurrent and sufficiently important. However, with indicating threshold effect and hysteresis, they mean that invasiveness may happen, but the genetic drift not considered in the model may considerably change the prediction.

Further experiments are required to estimate the effects of gene flow that may transfer the tandem construct in the wild by dint of recurrent crosses.

14.5 WEEDINESS, FERALITY, AND INVASIVENESS IN *HELIANTHUS*

Once feral populations are permanent, the risk of seeing some crop alleles or transgenes escaping to these populations is inevitable. The impact and consequences of such transgenes on volunteers and feral populations are considered with examples dealing with downy mildew resistance and other crop specific traits. The fate of feral populations under the pressure of such gene flow is discussed below.

14.5.1 Major Risks of Crop-Allele Gene Flow in Europe

14.5.1.1 Risk of Crop-Allele Escape

After 2 years of sympatry with cultivated sunflower, wild sunflower displayed additional domesticated traits in our experimental plot in Mauguio. It is then obvious that we cannot prevent gene flow from sunflower crop to volunteers, which are numerous and prevalent, nor to other *Helianthus* species that have escaped and are permanently established in Europe. Thus, the problem is now to contain cultivated alleles to cultivation. Several possibilities to contain transgenes in crop are being studied, such as chloroplast insertion and transgene carrying sequences disadvantaged in the wild population. Probably a combination of strategies should be advocated to prevent spreading of transgenes in volunteers (2).

Gene escape via pollen can be restrained by full CMS and no restoration allele. Such permanent CMS does exist in our collection (39) and could be developed for engineered sunflower crop producing interesting compounds before seed formation (e.g., functional foods, industrial compounds such as latex). However, seed escape can still occur and the volunteers can cross with the CMS crop giving the same nuclear hybrid. It is more difficult to prevent seed escape and to consider solutions to prevent F_2 seed from being produced, to germinate, or to systematically kill F_2 plants. Several transgenic solutions are possible (2) and not discussed here. A natural way could be to maintain sunflower lines constructed with introgressed fragments from other annual *Helianthus* species and differing by translocations. Such lines should be self-fertile but will lead to deep male and female sterility with other species if carrying three or more translocations.

14.5.1.2 The Risk of Volunteers Turning into Weedy Populations

There are no studies in Europe, and we do not know whether volunteers may have evolved toward weedy characters. This remains to be studied. We plan also to look for off-types in imported seed from Turkey, South Africa, and Chile. Gene flow toward Jerusalem artichoke is possible and it probably has occurred in hybrid seed production areas where sunflower adjoins escaped and in permanent populations of Jerusalem artichoke.

14.5.1.3 The Risk of Seed Volunteers Turning into Feral Populations

The spread of volunteers seems controlled by farmers in most fields by crop rotation, but in some areas (fallow land, waste areas), volunteers may reproduce for several generations, and consequently they may have evolved potential invasive properties. Invasive sunflower looks like wild sunflower with branching plants and small seeds, but this phenotype is probably due to recombination events between some crop-allele loci. Another explanation could be mutation of crop-alleles back to wild alleles. Here, we have documented that wild seed contamination is more plausible than back mutation from crop to wild allele. Back mutation is unlikely because all the alleles required to make a wild sunflower are already present, but dispersed in different crop sunflower genomes. It is simpler in terms of frequency to hypothesize recombination between domestication loci carrying wild allele traits than several back mutations from crop-to-wild alleles. Moreover, the wild sunflower types appeared only at places where wild sunflower probably had been imported as documented for Italy and Spain. Furthermore, risks due to wild sunflower seeds contaminating commercial seeds have been overcome because hybrid seed production takes place in Turkey, South Africa, or South America. Presently there are no formal studies on volunteers although volunteers are now present and expanding.

14.5.1.4 The Risk of Gene Flow Back from Feral Populations to Crop

It is too early to draw any conclusions, except that Jerusalem artichoke populations may represent a reservoir for crop-alleles and in the future, this could include transgenes. Possible gene flow back to sunflower also has to be studied in Jerusalem artichoke.

The invasiveness of wild sunflower has to be unraveled in the two experimental studies on volunteers and in wild-escaped and permanent sunflowers. Each experiment brings complementary levels of knowledge to Italy where it escaped and became naturally permanent and in Mauguio where it is due to cultivation of a *Helianthus* collection. We will learn about their origins and gene flow with sunflower crops.

14.5.1.5 Evolution of Weedy and Feral Populations toward Wild Sunflower

An olive tree that has been abandoned looks like an oleaster. The change is only phenotypic, without any genetic change. The progeny of this tree will be feral oleasters (see Chapter 15). In most cases, it is obvious that weedy and feral sunflower populations look like more or less wild sunflower (Figure 14.4 and Figure 14.5). In several descriptions of such weedy or feral populations of rice, *Setaria*, rapeseed, and *Raphanus* (see other chapters), the term of mimicry is used because there are some differences between the wild populations. To evolve from a crop shape to a wild shape some genetic changes have to take place. If they are located at the same loci as those implicated in the domestication process, we have to call the process dedomestication. But, if different loci are involved in such a resemblance, this is typically a convergence from an evolutionary point of view and not only a mimicry that is purely descriptive. Until now, to our knowledge, there is no evidence for dedomestication or convergence in any cases, except possibly Tibetan wheat (see Chapter 11, Section 11.5).

14.5.2 Major Risk of *Helianthus* Gene Flow at Center of Origin

The sunflower crop in the U.S. is almost always surrounded by wild sunflowers (10). Sunflower is in contact with its wild progenitor throughout the range of sunflower cultivation. Moreover, the phenology of the crop and of the wild form overlaps most of the time leading to possible crop-wild hybridization. Thus, wild sunflower is a weed in sunflower crops in the U.S. This situation enabled Alexander et al. (3) to examine the seed size of wild sunflower plants receiving gene flow from cultivated sunflower. They showed that the average seed size was twice the size of wild controls. However, the fate of the seeds depends on their size and larger seeds are preferably eaten by predators. This reduces hybrid fitness and would presumably slow down the spread of transgenes into wild populations. Moreover, Snow et al. (42) have shown that wild plants can benefit from the insect-resistant *Bt* transgene after being backcrossed to the crop carrying such a transgene. The transgene was not associated with a fitness cost, as the wild plants produced more seeds due to reduced herbivory damage. Therefore, by enhancing resistance to herbivory in the crop may quickly increase the population size of wild sunflower in the crop. Gene flow between sunflower and Jerusalem artichoke, in both directions, has not been studied in field conditions and could also lead to surprises.

14.5.3 Adaptation to Drought and Poor Soils

Sunflower and Jerusalem artichoke are among the most naturally drought-resistant crops. Escaped populations have a tendency to establish in poor dry lands abandoned from other crops, where competition with other natural species is low and human management is non-existent. Consequently, the risk of ferality is much higher than the risk of weediness in Europe where agriculture is intensive. In contrast, the risk to see *Helianthus* species establishing is high.

14.6 CONCLUSION

Sunflower and Jerusalem artichoke became and are still becoming feral or weedy in Europe where *Helianthus* species are not native. The small amount of data we presently have on feral or weedy sunflower populations suggests that they probably derived from wild sunflower seeds that contaminated the first sunflower hybrid varieties imported from the U.S. It is too early for us to judge the impact of volunteers and whether they represent the source of potential invasive forms.

In Europe, there is reticence to advocate engineered sunflower as a crop. More studies are required to manage sunflower volunteers and to eliminate escaped and permanent wild sunflower populations. They represent a real danger to generate new downy mildew pathotypes and to be a reservoir for crop-alleles. Our concerns with Jerusalem artichoke are to determine the extent of bidirectional gene flow between sunflower and perennial escaped and permanent populations, and to analyze the genetic diversity in these species propagated both by seeds and tubers in escaped populations and vegetatively in the crop.

14.7 SYNOPSIS OF MATERIALS AND METHODS

14.7.1 Interspecific Hybridization

Interspecific hybridization between Jerusalem artichoke and cultivated sunflower was performed under controlled conditions (tunnels) to discard any illicit pollination, in 2001 and 2002. We used two early flowering Jerusalem artichoke accessions — one represented a cultivated clone (INRA 325), the second a native wild population from the U.S. (INRA 570) — and two cultivated inbred lines HA89, 92B6.

14.7.1.1 Pollination under Isolation Cages with Bees

Ten plants of each Jerusalem artichoke accession were grown under separate isolation cages (80 m^2), covered with an insect proof net, together with cultivated sunflowers. To ensure concurrent flowering concomitance with Jerusalem artichoke, 4 sowing dates were performed for cultivated inbreds and the pollination was ensured by honeybees (hive with 10,000 to 20,000 insects). In the first cage, we realized pollination of Jerusalem artichoke with fertile sunflower inbred lines (HA89 or 92B6, respectively, in 2001 and 2002). Seed set measured on 15 heads per Jerusalem artichoke plant provided a global estimation of both interspecific hybridization and Jerusalem artichoke intercrosses. In the second cage, we performed pollination of cultivated sunflower with Jerusalem artichoke, and we used CMS isogenic inbreds (CMS HA89 and CMS 92B6, respectively, in 2001 and 2002) to directly estimate the interspecific hybridization ratio. An average of 15 heads was sampled at each of the 4 sowing dates. The number of initiated achenes was determined by counting the total number of developed ovaries on heads for Jerusalem artichoke accessions, HA89 and 92B6 inbreds. The potential achene numbers were 112.7, 93.5, 1320, and 1699, respectively, for JA 325, JA 570, HA89, and 92B6. The percentage of fertilized achenes was expressed as the ratio of seed set to the potential achene number.

14.7.1.2 Pollination with Manual Assistance

A similar crossing scheme was performed on the same material, using manual pollination on bagged plants.

14.7.2 Molecular Tools Available for Sunflower

SSR markers or microsatellites have recently been developed in sunflower in the CARTISOL program (a consortium between companies, INRA, and Clermont-Ferrand University to develop molecular markers and to map some agronomic traits in sunflower). This enabled the construction of several genetic maps and a diversity analysis (16,46). Moreover, the SSRs are the most informative markers in population genetic studies due to co-dominance. We therefore chose different sets of SSRs that have been mapped and that differ by their location in the genome and the distances between loci in some linkage groups (53). In previous diversity analysis studies, several authors have already observed that the diversity in sunflower is narrow in comparison with wild sunflowers (46).

Different programs and modeling tools are available (admixture programs and models). They could be used to determine whether volunteers have their origin in the crop or in impurities of the crop and whether they correspond to advanced-hybrid generations and, if so, which generation. Recombination rates between known loci are known, and by also knowing the phase (the alleles linked in each parent) of alleles in hybrid cultivars, and following volunteer populations, it is possible to estimate the number of generations that have occurred after the initial hybridization event. Multiloci analyses based on Bayesian and admixture programs should lead to interpretation of the framework of population genetics models.

ACKNOWLEDGMENTS

We are indebted to Professor Angelo Olivieri (Udine), Dr. Mario Baldini (Udine), and Professor Jose Fernandez-Martinez (Cordoba) for their help in providing information and seeds from weedy or feral sunflower populations in Italy and Spain, respectively.

LITERATURE CITED

1. Al Khatib K, Miller J. 2000. Registration of four genetic stocks of sunflower resistant to imidazolinone herbicide. *Crop Sci.* 40:869–870.
2. Al-Ahmad H, Galili S, Gressel J. 2004. Tandem constructs to mitigate transgene persistence: tobacco as a model. *Mol. Ecol.* 13:697–710.
3. Alexander HM, Cummings CL, Khan L, Snow AA. 2001. Seed size variation and predation of seeds produced by wild and crop-wild sunflowers. *Am. J. Bot.* 88:623–627.
4. Alexander HM, Schrag AM. 2003. Role of soil seed banks and newly dispersed seeds in population dynamics of the annual sunflower, *Helianthus annuus*. *J. Ecol.* 91:987–998.
5. Anonymous 2002. Larousse Agricole. Paris: Larousse.
6. Arias DM, Rieseberg LH. 1994. Gene flow between cultivated and wild sunflowers. *Theor. Appl. Genet.* 89:655–660.
7. Arias DM, Rieseberg LH. 1995. Genetic-relationships among domesticated and wild sunflowers (*Helianthus annuus*, Asteraceae). *Econ. Bot.* 49:239–248.
8. Bach Knudsen K, Hessov I. 1995. Recovery of inulin from Jerusalem artichoke (*Helianthus tuberosus* L.) in the small intestine of man. *Br. J. Nutr.* 74:101–113.
9. Burke JM, Gardner KA, Rieseberg LH. 2002. The potential for gene flow between cultivated and wild sunflower (*Helianthus annuus* L) in the United States. *Am. J. Bot.* 89:1550–1552.
10. Cantamutto MA, Poverene MM. 2003. Los recursos genéticos del girasol silvestre. *IDIA* 3:152–157.
11. Chojnowski M, Corbineau F, Come D. 1997. Physiological and biochemical changes induced in sunflower seeds by osmopriming and subsequent drying, storage and aging. *Seed Sci. Res.* 7:323–731.
12. Ellstrand N. 2003. *Dangerous liaisons? When cultivated plants mate with their relatives.* Baltimore, MD: Johns Hopkins University Press.
13. Ellstrand NC. 2003. Current knowledge of gene flow in plants: implications for transgene flow. *Phil. Trans R. Soc. London Ser. B Biol. Sci.* 358:1163–1170.
14. FAOSTAT. 2003. http://apps.fao.org.
15. Faure N, Serieys H, Bervillé A. 2002. Potential gene flow from cultivated sunflower to volunteer, wild *Helianthus* species in Europe. *Agric. Ecosys. Environ.* 89:183–190.
16. Gentzbittel L, Mestries E, Mouzeyar S, Mazeyrat F, Badaoui S, et al. 1999. A composite map of expressed sequences and phenotypic traits of the sunflower (*Helianthus annuus* L.) genome. *Theor. Appl. Genet.* 99:218–234.
17. Gobbin D, Pertot I, Gessler C. 2003. Genetic structure of a *Plasmopara viticola* population in an isolated Italian mountain vineyard. *J. Phytopathol.* 151:636–646.
18. Haygood R, Ives AR, Andow DA. 2003. Consequences of recurrent gene flow from crops to wild relatives. *Proc. R. Soc. London Ser. B Biol. Sci.* 270:1879–1886.
19. Heiser CB. 1951. The sunflower among the North American Indians. *Proc. Am. Philos. Soc.* 95:432–448.
20. Horn R, Hustedt JE, Horstmeyer A, Hahnen J, Zetsche K, et al. 1996. The CMS-associated 16 kDa protein encoded by orfH522 in the Pet1 cytoplasm is also present in other male-sterile cytoplasms of sunflower. *Plant Mol. Biol.* 30:523–538.
21. Huxel G. 1999. Rapid displacement of native species by invasive species: effect of hybridization. *Biol. Conserv.* 89:143–152.
22. Lacombe S, Kaan F, Léger S, Bervillé A. 2001. An oleate desaturase and a suppressor loci direct high oleic acid content of sunflower (*Helianthus annuus* L.) oil in the Pervenets mutant. *C.R. Acad. Sci. III -Life Sci.* 324:839–845.

23. Leclercq P. 1969. Une stérilité mâle cytoplasmique chez le tournesol. *Ann. Amélior. Plantes* 19:99–106.
24. Lentz DL, M. Pohl ED, Pope KO, Wyatt AR. 2001. Prehistoric sunflower (*Helianthus annuus* L.) domestication in Mexico. *Econ. Bot.* 55:370–376.
25. Linder CR, Taha I, Seiler GJ, Snow AA, Rieseberg LH. 1998. Long-term introgression of crop genes into wild sunflower populations. *Theor. Appl. Genet.* 96:339–347.
26. Marshall MW, Al-Khatib K, Loughin T. 2001. Gene flow, growth, and competitiveness of imazethapyr-resistant common sunflower. *Weed Sci.* 49:14–21.
27. Massinga RA, Al-Khatib K, St. Amand P, Miller JF. 2003. Gene flow from imidazolinone-resistant domesticated sunflower to wild relatives. *Weed Sci.* 51:854–862.
28. Moreau E, Bervillé A. 1991. A dot blot assay for the determination in routine of maintainer contaminants in cytoplasmic male sterile seed stocks of sunflower (*Helianthus annuus* L). *Plant Var. Seeds* 5:19–26.
29. Paetkau D, Slade R, Burden MS, Estoup A. 2004. Genetic assignment methods for the direct, real time estimation and migration rate: a simulation-based exploration of accuracy and power. *Mol. Ecol.* 13:55–65.
30. Paniego N, Echaide M, Fernández L, Muñoz M, Torales S, et al. 2002, Microsatellite isolation and characterization in sunflower *(Helianthus annuus L.). Genome* 45:34–43.
31. Pilson D, Snow AA, Rieseberg LH, Alexander HM. 2002. Fitness and population effects of gene flow from transgenic sunflower to wild *Helianthus annuus*. Columbus, OH: Workshop on the Ecological Effects of Transgenic Crops.
32. Poverene M, Ureta S, Cantamutto M. 2003. Wild *Helianthus* species and wild-sunflower hybridization in Argentina. Sevilla, Spain: SUNBIO Conference.
33. Poverene MM, Cantamutto MA, Carrera AD, Ureta MS, Salaberry MT, et al. 2002. El girasol silvestre (*Helianthus* spp.) en la Argentina: caracterización para la liberación de cultivares transgénicos. *Rev. Investig. Agropec. (Arg.)* 31:97–116.
34. Ritsema T, Smeekens S. 2003. Fructans: beneficial for plants and humans. *Curr. Opin. Plant Biol.* 6:223–230.
35. Ronce O, Kirkpatrick M. 2001. When sources become sinks: migrational meltdown in heterogeneous habitat. *Evolution* 55:1520–1531.
36. Santoni S, Bervillé A. 1992. Evidence for gene exchanges between sugar beet (*B. vulgaris* L.) and wild beets: consequence for transgenic sugar beets. *Plant Mol. Biol.* 20:578–580.
37. Schwarzbach AE, Rieseberg LH. 2002. Likely multiple origins of a diploid hybrid sunflower species. *Mol. Ecol.* 11:1703–1715.
38. Scotti I, Di Bernardo N, Dellacasa S, Vischi M, Olivieri AM. 2003. Hybridisation between wild sunflowers (*Helianthus argophyllus* and *H. debilis*) in southern Africa — morphological and molecular characterisation. 11th New Phytologist Symposium, Antigonish, Nova Scotia: St. Francis Xavier University.
39. Serieys H. 2002. Identification, Study and Utilization in Breeding Programs of new CMS sources. Progress report of the FAO working group, Montpellier, France.
40. Snow AA, Pilson AD, Rieseberg LH, Alexander HM. 2002. Ecological effects of pest resistance genes that disperse into weed populations. Beijing: 7th International Symposium on The Biosafety of Genetically Modified Organisms.
41. Snow AA, Pilson AD, Rieseberg LH, Wszelaki A, Seiler GJ. 1998. Fecundity, phenology, and seed dormancy of F1 wild-crop hybrids in sunflower (*Helianthus annuus*, Asteraceae). *Am. J. Bot.* 85:794–801.
42. Snow AA, Pilson AD, Rieseberg LH, Paulsen M, Pleskac N, et al. 2003. A Bt transgene reduces herbivory and enhances fecundity in wild sunflowers. *Ecol. Appl.* 13:279–286.
43. Soldatov KI. 1976. Chemical mutagenesis in sunflower breeding. Krasnodar (Russia): VIIth Int. Sunflower Conf.
44. Sossey-Alaoui K, Serieys H, Tersac M, Lambert P, Schilling E, et al. 1998. Evidence for several genomes in *Helianthus*. *Theor. Appl. Genet.* 97:422–430.
45. Stark-Urnau M, Seidel M, Kast WK, Gemmrich AR. 2000. Studies on the genetic diversity of primary and secondary infections of *Plasmopara viticola* using RAPD/PCR. *Vitis* 39:163–166.
46. Tang S, Knapp S 2003. Microsatellites uncover extraordinary diversity in native American landraces and wild populations of cultivated sunflower. *Theor. Appl. Genet.* 106:990–1003.

47. Tourvieille J, Meliala C. 2000. Présent sur les cinq continents, le mildiou a suivi le tournesol. *Le mildiou du tournesol*. Tourvieille de Labrouhe D, Nicolas P, Vear F, Eds., Paris: CETIOM et INRA Editions: 17–21.
48. Vear F. 1992. *Le tournesol. Amélioration des espèces végétales cultivées*. Bannerot H, Gallais A, Eds., Paris: INRA:146–160.
49. Vrânceanu AV. 2000. *Floarea-soarelui hibrida*. Bucuresti, Romania: Ceres.
50. Vrânceanu AV, Stoenescu F. 1978. Genes for pollen fertility restoration in sunflower. *Euphytica* 27:447–454.
51. Whitton J, Wolf DE, Arias DM, Snow AA, Rieseberg LH. 1997. The persistence of cultivar alleles in wild populations of sunflowers five generations after hybridization. *Theor. Appl. Genet.* 95:33–40.
52. Wolf DE, Takebayashi N, Rieseberg LH. 2001. Predicting the risk of extinction through hybridization. *Conserv. Biol.* 15:1039–1053.
53. Yu JST, Slabaugh MB, Heesacker A, Cole G, Herring M, et al. 2003. Towards a saturated molecular genetic linkage map for cultivated sunflower. *Crop Sci.* 43:367–387.

QUESTIONS AND ANSWERS

Ulrich Sukopp: Why did you make the difference between dedomestication and convergence for the weedy or feral populations that look like (mimic?) wild sunflowers? Why did you not say dedomestication only?

Answer: For me, mimicry means that the plants look like each other without any explanation on the mechanisms. Dedomestication means that the same loci that have led to domestication about 4000 years ago for sunflower had new allele mutations. The events at these loci have still to be described. Convergence means that some other loci than those involved in domestication were modified leading to a resembling wild genotype. Both processes are difficult for experimentation, but we have to keep in mind that they may exist.

Henri Darmency: There is considerable concern about the fitness of pathogen resistance genes in feral hybrid and wild relatives. Could your results about the conventional resistance genes to mildew in volunteer derivatives provide direct field estimate of fitness value and impact upon populations?

Answer: No, it is too early to answer your question. Our observations are recent and we did not quantify the plants carrying alleles for resistance. However, it is clear that we have to pay attention to this situation. To the best of my knowledge, it is the first report on this situation of mildew that may limit volunteer populations and Bernard Poinso is starting to study it.

BaoRong Lu: Gene flow will impact the crop allele. Could you explain what type of impact to the crop allele by gene flow, allele frequency, or something else?

Answer: Sunflower produces oil in the seed. Oil composition depends upon both genes from female and male alleles as usually the crop is a F_1 hybrid. When two cultivars with different oil composition, namely, classic and high oleic acid oil, as an example, are grown in adjacent fields, gene flow affects oil composition in both fields and the quality is not valued.

BaoRong Lu: Gene flow from wild to crop: What will be the impact or consequences of the flowing back of genes from wild?

Answer: In North America, sunflower crop is always surrounded by wild sunflower and may be infested by weed sunflower (mixed with the crop). Consequently, crop alleles may escape and, depending on their effect, they may induce deep changes in wild populations. Herbicide-resistant alleles and transgenes will escape rapidly and they may modify the dynamic of the wild populations. They may contribute to spread crop alleles back in the weedy population annihilating the advantage of the crop on the weed. Thus, herbicide resistance and disease resistance alleles may spread from crop to weed via feral populations.

Suzanne Warwick: The feral Jerusalem artichoke plants on river banks looked like seedlings (i.e., small). When you dig up the plants, can you tell if they establish from a seed or a rhizome?

Answer: I do not think so, the pictures are from this spring and the shoots correspond to a patch. Because of the position of the patch in the riverbed, I think that floods have dispersed rhizomes along the banks. When the plant was lifted out of the ground, it grew from a tuber and they are not seedlings.

Allison Snow comments: In the U.S., we see off-type cultivated plants that are even more likely to establish volunteer and feral populations than the seeds from normal cultivated plants (see Reagan and Snow, in preparation).

15 Issues of Ferality or Potential for Ferality in Oats, Olives, the *Vigna* Group, Ryegrass Species, Safflower, and Sugarcane

André Bervillé, Catherine Breton, Ken Cunliffe, Henri Darmency, Allen G. Good, Jonathan Gressel, Linda M. Hall, Marc A. McPherson, Frédéric Médail, Christian Pinatel, Duncan A. Vaughan, and Suzanne I. Warwick

15.1 INTRODUCTION

The other chapters in this book demonstrate the vast array of systems that plants utilize to evolve endoferality and exoferality, and mixtures between the two. This chapter presents short case histories of other crop species that extend this biodiverse array and accentuates other issues and implications that may derive from ferality. Clearly, other cases could be added to this, as many as there are additional other crops, but the idea is not to be exhaustive. Our intent is to supply a feeling of further issues and to provide parameters that must be discerned when analyzing the potential problems, if any, that may arise when a given crop becomes a persistent volunteer or introgresses genes from a wild relative.

15.2 OATS*

15.2.1 Cytogenetic Interrelationships and Crossability

Oats (*Avena* species) are major temperate cereals that occur at three ploidy levels on an x = 7 basis. Cultivated forms of the wild diploid *A. strigosa* Schreb. (2n = 14, genome AA) evolved in Spain and are still cultivated locally throughout Europe on a small scale. The tetraploid *A. abyssinica* Hochst. (2n = 28, genomes AABB) evolved from the weedy tetraploid *A. barbata* Pott and is restricted to Ethiopia where it is tolerated in and harvested with barley. The two allohexaploid cultivated oat species *A. sativa* L. and *A. byzantina* C. Koch (2n = 42, 21 bivalents at meiosis, genomes AACCDD) were domesticated in Europe 2000 to 3000 years ago from the wild-weedy hexaploid *A. sterilis* L. brought as a weed with wheat and barley from the Fertile Crescent (35). Molecular data (isozyme, RAPD — random amplified polymorphic DNA, and chromosomal translocation) suggest a separate path of domestication for each species (175). *Avena sterilis* occurs wild in herbaceous plant communities and as a weed in the Middle East and throughout the

* Prepared by Henri Darmency.

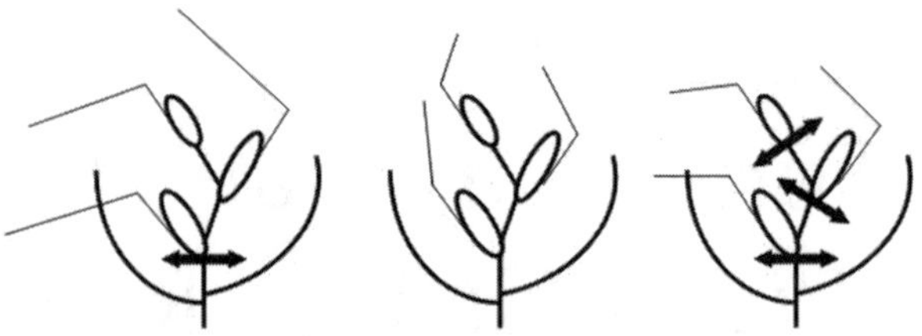

FIGURE 15.1 Simplified representation of the spikelet type of hexaploid oats (glumes, peduncles, seeds, and awns). Arrows indicate the presence of abscission cell layers resulting in a sucker mouth scar and whole spikelet shedding in *A. sterilis* and single seed shedding in *A. fatua*. Threshing is necessary to fracture the peduncle in *A. sativa* and *A. byzantina*.

Mediterranean Basin, but is progressively replaced by another hexaploid weedy species, *A. fatua* L., which is progressing northward in Europe. *Avena fatua* is confined to crops and could be considered as a specialized ecotype of *A. sterilis* or a derivative of the crop. Volunteers of domesticated oats are frequent, but hardly persist as feral populations because they lack primary seed dormancy, spontaneous germination is a dominant trait (99), and because they cannot compete with wild oat populations.

The main distinction between the two weed species *A. fatua* and *A. sterilis* is the type of seed shedding. The primary floret is separated from the rachilla by an oval disarticulation surface (sucker mouth) at the base of the seed in both species, but the upper florets all disarticulate similarly in *A. fatua,* whereas they remain attached to the rachilla in *A. sterilis*. The dispersal unit is therefore a single seed in *A. fatua,* but it is the whole spikelet (minus glumes) in *A. sterilis*, and the whole panicle for both crop species (Figure 15.1). Otherwise, all four hexaploid species are rather morphologically similar with respect to the other ploidy levels and completely interfertile so that they can be considered as the same biological species. Unlike in most of the cases discussed in this book and diploid and tetraploid oats (131), in all hexaploid *Avena* crosses, nonshedding is dominant. Florell (56) proposed a two gene model: B... is the non-disarticulating crop type; bbS. is the *sterilis* type with basal floret disarticulation only; bbss is the *fatua* type with all florets disarticulating. Other seed traits such lemma hairiness and presence of twisted awns are linked to floret separation, which reinforces the easy visual distinction between the weedy and crop species (98).

The spontaneous occurrence of "fatuoids" and "steriloids" in the progeny of cultivated oats has long been controversial. Such off types occur at rates up to 1% (13) without apparent dependence to the lineage of the cultivar (22). Fatuoids and steriloids differ from the corresponding cultivar only with respect to the complex trait comprising floret base, basal hairiness, and awn of *A. fatua* and *A. sterilis*, respectively. The genes that produce the fatuoid characters were found to be allelic with and recessive to those encoding for *sterilis* type in fatuoid   *A. sterilis* cross (119), but deviations from this one-gene scenario have also been observed (78). A simplified organization of the sequence of genes involved in the epistatic interaction to the fatuoid syndrome is suggested in Nishiyama and Yabuno (132), but the mode of gene interaction remains to be clarified.

15.2.2 Origin of Fatuoids

There are two possibilities of the origin of fatuoids: hybridization (1) and mutation in a broad sense (130). Although oats are considered to be self-pollinating species, outcrossing is frequently observed: up to 6% among cultivated varieties (98) and up to 12% between *A. sativa* and *A. sterilis* or *A. fatua* (128,151). This offers large opportunities for gene exchange and evolving exoferal populations that can retain seed shedding as an important weedy trait. However, this also produces considerable variability in segregating progeny, which is not the case for most of the fatuoids that generally resemble the cultivar from which they are issued (13,23). In contrast, residual variability within cultivar and intravarietal crosses, therefore not detectable upon examination of cultivar characteristics, could cause genetic instability resulting in the loss of epistatic control of the fatuoid syndrome (13,23)

The mutant nature of fatuoids was demonstrated by Nilsson-Ehle (130) who discussed its similarity with the speltoid mutation in wheat and postulated a loss of chromosome blocks including several loci. Huskins (84) studied the cytogenetics of fatuoid forms and classified them into three series according to deficiency on the long arm of a chromosome of genome D, later identified in monosomic line MK9 (132). Fatuoids α are attributed to a small mutation of the genes complex: they produce a 1:2:1 normal:heterozygous fatuoid:homozygous fatuoid segregation ratio when selfed. Fatuoids β are monosomic with $2n = 41$: they mainly produce hetero- and homozygous fatuoids and are dwarf and rather sterile because of other genes lost on the short arm of the chromosome. Fatuoids γ are intermediate between the two previous classes with $2n = 41 +$ a telocentric or altered chromosome. Considerable chromosomal variation is known in hexaploid oats and could generate such genome instability (115,172).

The endoferal origin of *A. fatua* through the spontaneous release of fatuoids α from *A. sativa* is therefore possible, which could explain its initial distribution restricted to north Europe and its quasi-unique adaptation to arable fields, exactly as the domesticated oat did. The phenotypic dominance of the nonshedding crop characteristics, in contrast to what occurs in most other crop/ancestor relationships, and the adaptive advantage of single seed dispersal unit in arable fields, could explain the apparition by deletion of the novel seed shedding *A. fatua* type, which was not present in the ancestor species *A. sterilis*. *A. fatua* is now one of the world worst weeds, distributed worldwide everywhere cereals are grown (81,149), and not just a problem of domesticated oats. It has been successful in displaying high variability and adaptation to various farming systems, from Europe to the Americas and Australia, including herbicide resistances (71). However, new cytogenetics tools could imbalance this hypothesis (172).

Both hybridization and fatuoid mutation are of concern for the escape of transgenes from genetically modified oat cultivars. Attempts have been made to release such cultivars (e.g., virus resistance), but the phenomenon presented above and its high frequency must be sufficient to ban transgenes conferring high adaptive value to the weedy partner of the wild-crop-weed oat gene pool. Such a new trait would be immediately transmitted to wild oat populations and make them more troublesome and difficult to control in cereals.

15.3 OLIVES*

Olive (cultivated) and oleaster (wild form) belong to *Olea europaea* ssp. *europaea* and are in sympatry all around the Mediterranean Basin. Oleaster diversity is deeply structured according to geography (19). In the east of the Mediterranean Basin, oleasters display only one chlorotype and one mitotype, whereas in the west, oleasters display three cytotypes (three chlorotypes and three mitotypes) (18). The nuclear diversity based upon RAPD, amplified fragment length polymorphism (AFLP), and simple sequence repeat (SSR) polymorphisms has been also found higher in the west than in the east (27). Probably oleaster differentiated in the west of the Mediterranean Basin.

The genetic diversity in oleasters is much higher than in olive. Therefore, we can observe the bottleneck between oleaster and cultivar forms (27). Genetic relationships between olive and oleaster were recently studied and several studies converge to sustain several domestication events of olive in oleasters (17,27,158). Olive was domesticated from oleasters in the east by 5800 BP (177) and in several other places, probably concomitantly, in the west (158). Domestication in olive

* Prepared by C. Breton, F. Medail, C. Pinatel, and A. Bervillé.

was probably based upon a few traits including drupe oil content, ability to be propagated by cuttings, fruit size, and fruit abscission at maturity. Phoenician, Greek, and Roman civilization probably spread olive cultivation by 2500 to 2000 before present (BP) from the eastern origin that is prevalent in cultivars. Olive breeding had never been conscientious by crossing cultivars, but people picked up advantageous wild or feral forms to displace them and to constitute orchards. Here, feral tree means that the tree belongs to a progeny of a cultivar. Further vegetative propagation of one tree by cuttings or grafting spread the genotypes that may become a major cultivar. Because of the long while between nuts and progenies, a history of olive cultivars does not exist. Tales and legends about the origin of cultivars probably compensated for this ignorance.

Because of deep freezes and the appearance of less expensive seed oils, olive declined for several decades since 1850. However, due to longevity of olive trees, most of those that were abandoned from cultivation developed looking like oleaster although they keep the same genotypes. When they reproduced, they were feral olives. Once all traces of cultivation have disappeared, in wild lands, abandoned olive trees, feral trees, and genuine oleasters cannot be differentiated based on any morphological traits. Clustering statistical methods based on molecular markers leading to dendrograms enabled in most cases *a posteriori* differentiation of the three trees status.

Recently assigning methods based on multilocus analyses have been developed and applied to oleaster and olive (Breton et al., submitted). Without any *a priori* knowledge about the cultivated, abandoned, feral, or genuine status of a tree, it becomes possible to classify most trees in the four categories according to the molecular data. Only analyses of Corsican oleasters have revealed feral olive (28); none has been detected elsewhere. Surprisingly, when classical clustering methods were unable to reveal feral olive, in contrast assigning methods, and admixture analyses (137) using the STRUCTURE package pointed out unambiguously that one-third of the same sample of trees (62/166) considered as oleasters were cryptic hybrid between olive and oleasters. Moreover, we showed with the same methods that some cultivars are genuine oleasters displaced in orchards.

In the Mediterranean Basin, recurrent gene flow from cultivars to oleasters has therefore more consequences than thought before our studies. Oleaster genetic stocks may be so introgressed in cultivars that we may have concern with their genetic conservation.

Olive cultivars have been introduced in the Americas and Australia, where they escaped and feral populations established with dramatic effects on local flora and fauna. In California, olive cultivars have spread from cultivation in relatively small areas (155). Reconstitution of olive history, introduction, and cultivation could enable us to understand how trees escaped. In Australia, the situation is more dramatic (66). First, several *Olea* ssp. may have escaped (*cuspidata*, *europaea*, R. Carter); thus the trees may be advanced-hybrid generations between *O. e. europaea* and *O. e. cuspidata*. We continue with Sedgley (124) (Adelaide University) to identify DNA samples provided to us of these feral trees by molecular analyses to verify which subspecies is the origin in the diversity of *O. europaea* (17,61).

Olive is an excellent model to study both the domestication (by 5800 BP) process, because some trees more than 2000 years old are still living, and the dedomestication process, because once-abandoned olive trees began to look like oleasters without any genetic changes. Abandoned olive trees have led to feral populations even in southern France. Some of these populations are under study to understand whether they correspond to endoferality or exoferality.

Reverse oleaster gene flow, otherwise known as olive cultivar pollination by oleasters, is frequent because several widespread cultivars are male sterile or display a cytoplasmic male sterility (CMS) (20). When olive orchards are surrounded by oleasters, as in Algeria, France, Libya, Morocco, and Spain, olive may be pollinated by oleasters, but this has no consequence on the oil composition of fruits, which is determined by the maternal tissue of the fruits. However, such hybrid fruits may lead again to feral trees. Once set up in an orchard, such progeny could display valuable improvement for several agronomic traits.

15.4 *VIGNA* IN ASIA*

The genus *Vigna* consists of 11 cultivated or domesticated species, of which mungbean (*V. radiata*) and cowpea (*V. unguiculata*) are the most important. The presumed wild progenitors of cultivated *Vigna* and weedy intermediates between wild and cultivated *Vigna* grow in similar habitats throughout Asia, generally in herbaceous or shrubby communities on road, rail, or field embankments, along riverbanks and irrigation channels or recently fallow land. Humans or natural events, such as flooding, disturb all these habitats to varying extents. Weedy forms are generally not serious weeds in cultivated fields and may even be harvested from ruderal habitats. The genus includes species, such as azuki bean, for which transformation systems are well developed (102).

15.4.1 COWPEA

Cowpea [*Vigna unguiculata* (L.) Walp.] was domesticated in Africa, but it is a common crop in Asia. It is believed that the natural distribution of the wild relatives and progenitors of cowpea is in Africa. However, in Thailand and Japan, many populations of *V. unguiculata* grow in ruderal habitats. Wild populations in Thailand tend to have small seeds and plants have a scrambling habit compared to the generally bushy habit of domesticated cowpea. An AFLP study that compared one population growing in the wild with cultivated cowpea from Thailand revealed that they were genetically well diverged (152). This suggests that this wild population is not a recent escape from cultivation. Grain (cv. *gr. unguiculata*) and vegetable (cv. *gr. sesquipedalis*) types of cowpea are grown in Thailand; therefore repeated introductions of cowpea from Africa to Asia may have resulted in seeds of the wild or weedy relatives of cowpea being introduced.

Wild cowpea populations in Japan have an erect bushy habit with rather large black seeds. They can commonly be found in southern parts of Japan growing in disturbed habitats. RAPD analysis of cultivated and wild cowpea revealed that there has been little genetic differentiation between these two types (Jinno, Hokkaido Experiment Stations, Japan, unpublished data). This suggests that in Japan, cowpeas growing in ruderal habitats are feral escapes from cultivation.

15.4.2 RICE BEAN

Rice bean [*Vigna umbellata* (Thunb.) Ohwi and Ohashi] is cultivated in Thailand mainly in the north of the country. Rice bean in Thailand was a component of shifting, slash and burn cultivation agricultural systems but is now cultivated in non-shifting mixed cropping systems, intercropped with maize.

However, in Myanmar, rice bean is still cultivated in shifting cultivation fields. Wild rice bean growing in ruderal habitats has small black seeds and cultivated rice bean has larger usually red seeds. However, occasionally wild populations are found with intermediate and small seed size and seed color not typical of usual wild rice bean. AFLP analysis results suggest that these represent populations with gene introgression from the cultigen. Sometimes farmers' cultivated seed lots have a few seeds with atypical color that farmers usually discard. These atypical colored seeds may be the result of gene flow between wild and cultivated rice bean (152).

15.4.3 AZUKI BEAN

Azuki bean [*Vigna angularis* (L.) Ohwi and Ohashi] and its presumed wild progenitor [*V. angularis* var. *nipponensis* (Ohwi) Ohwi and Ohashi] is found across East Asia as far west as the Himalayan foothills of Nepal, Bhutan, and India (178). *V. angularis* var. *nipponensis* grows in herbaceous communities. This crop complex, including wild and weedy forms, has been studied in detail in

* Prepared by D. Vaughn.

Japan (170,171). Gene flow from cultivated azuki bean to wild azuki bean has recently been demonstrated using SSR (microsatellite) markers (164). Further studies using morphological and SSR marker analyses of populations across Japan have revealed that there is a low level of outcrossing from cultivated azuki bean into wild populations of azuki bean (Miranda-Jonson, Kaga, National Institute of Agrobiological Sciences, Japan, in preparation). These studies show that azuki populations in ruderal habitats have a wide range in the number of introgressed alleles from the cultigen in the area of Japan (central Honshu) where the azuki bean complex is most diverse. It appears that proximity to intensively cultivated areas, that includes azuki bean fields, is one factor that helps explain population genetic structure, but microlevel factors are also important. In addition, the results reveal that adaptive potential and spread of seeds are also important in explaining the high level of population genetic complexity in the azuki bean complex in Japan.

15.4.4 Considerations for Transgenic Biosafety for *Vigna* Crops

Azuki bean has a well-developed system for routine genetic transformation that been used in the development of breeding lines having bruchid resistance conferred by an α–amylase inhibitor gene (102). Transformation of cowpea for resistance to various insect pests has been proposed (47). Based on gene flow studies in Japan, introgression occurs where cultivated and wild azuki are sympatric. AFLP analysis has shown that gene flow between wild and domesticated cowpea in Africa has resulted in a large crop weed complex (41). Given the prominent colored flowers of all *Vigna* species, insect facilitated gene flow would be expected wherever cross-compatible wild and cultivated *Vigna* occur. There is a need to accurately identify *Vigna* crop relatives, and to determine those that are cross-compatible with cultivated *Vigna* and their distribution. This would help establish appropriate measures to prevent unwanted gene escape from cultivated *Vigna*. In addition, some cultivated *Vigna* species, for example, cowpea (*V. unguiculata*) and rice bean (*V. umbellata*), are able to establish viable populations outside cultivated fields. Hence, the potential of transgenic *Vigna* to become feral needs also to be recognized.

15.5 THE RYEGRASS COMPLEX*

Perennial ryegrass (*Lolium perenne*) is the most widely grown temperate forage grass globally (67). Advances in the technology to manipulate genes provide opportunities to improve plant adaptation to various environmental constraints beyond the realm of the species genome. There is enhanced potential to manipulate metabolic pathways, improve the feed value, or radically change agronomic characteristics of the species. Transgenic pasture grasses, including *Lolium perenne*, are currently being developed and released for field evaluation (156).

Perennial ryegrass is an outcrossing, wind-pollinated (anemophilous) species. There are no biological barriers and frequently no physical barriers to gene exchange between cultivated and wild, weedy, or ruderal populations of *Lolium perenne* (173). *Lolium perenne* also frequently grows in close proximity to a range of other closely related grass species of the *Lolium* and *Festuca* genera. A close and sexually compatible relative of perennial ryegrass, annual ryegrass (*Lolium rigidum*) exists as one of the most troublesome weeds in temperate cropping systems (140).

Interfertility between species of the *Lolium–Festuca* (fescue) complex suggests that gene flow may also pose a risk of introgression from novel perennial ryegrass into a number of related species (90) with possible development of feral weed problems. The main difference between cultivars and weedy biotypes would be that the cultivars are genetically much narrower than their weedy relatives due to the domestication process. This means that some genotypes from diverse weed populations would be more likely to survive strong selection pressures.

---

* Prepared by K. Cunliffe.

15.5.1 Taxonomy of the *Lolium–Festuca* Complex

Hybrids between *Lolium* species and intergeneric hybrids with *Festuca* exist naturally (32). Due to the close similarity between these genera, introgression from ancestral hybrids appears to have played a major role in their evolutionary development (90). The ability to hybridize with related species signifies a risk of introgression of novel genes from transgenics across the entire species complex.

15.5.1.1 *Lolium–Festuca* Complex — A Taxonomic Problem

Early taxonomy in the *Lolium–Festuca* species complex relied on morphological differences to distinguish between species (160). Taxonomic division is made difficult by the inter-gradation in morphology within and among species. Jauhar (90) contend that the genera *Lolium*, *Festuca,* and *Vulpia* (at one time included in the *Festuca* genus) are in fact so similar that they should be amalgamated into a single genus. Spicate (*Lolium*) vs. paniculate (*Festuca*) inflorescence type was the most important distinguishing characteristic that separated the genera of *Lolium* and *Festuca.* This difference could be ascribed to a single spontaneous gene mutation and apart from this, these genera are almost inseparable (90).

Morphological and electrophoretic data have been used to describe the relationships between the different species and genera of the family Poaceae subfamily Festucoideae (30); principal component analysis of morphological data clearly separated the three genera, although the trends were less distinct between *Festuca* and *Vulpia*. Cluster positions also varied depending on the type of morphological characteristic used in the analysis. A close association between *Lolium* and *Festuca* section Bovinae was evident. Protein electrophoretic data confirmed a close association between the outcrossing *Lolium* species and the *Festuca* sect. Bovinae, but indicated that a separate genus for the inbreeding *Lolium* species might be appropriate. An analysis of morphological and isozyme variation among accessions of *L. perenne*, *L. rigidum*, *L multiflorum*, *L. remotum,* and *L. temulentum* showed close similarity between the outcrossing *Lolium* species and some genetic distance between the outcrossing *Lolium* species and the inbreeding *L. temulentum* and *L. remotum* (16).

A species by classical definition is "a genetically distinctive, reproductively isolated, natural population." The complete interfertility and intergradation between some members of the *Lolium* genus and partial affinity for members of the *Festuca* genus have created difficulties for taxonomists in delineating species (159). The merging of species implies that there will be a gradation of fertility and tendency to gene flow between *Lolium perenne* and its sexually compatible relatives, which increases with increasing relatedness (Figure 15.2).

15.5.2 Ecogeographical Adaptation — Spatial Separation of the Species

The *Lolium–Festuca* complex is indigenous to temperate or mountain areas of Europe, Asia, and northern Africa (166). Widespread adventive proliferation of both genera through the temperate world has accompanied human settlement and agriculture with the genus *Festuca* having a much wider range of adventive distribution than *Lolium* (166,167). Bennett et al. (16) hypothesized that changing patterns of agriculture in the main regions of distribution of these species have resulted in an increase in the frequency of interspecific hybrids. This could have important implications for the introgression of transgenes across the species complex.

15.5.2.1 Phylogeography within the Genus *Lolium*

The genus *Lolium* consists of just eight predominantly mesophytic species adapted to moist and well-drained habitats (167). The close association of *Lolium perenne* (as sown pasture) and *Lolium rigidum* (as a weed) with agriculture in Europe was confirmed by assessing maternally inherited chloroplast DNA (9). These two species particularly are disposed to gene flow and the development of feral ecotypes.

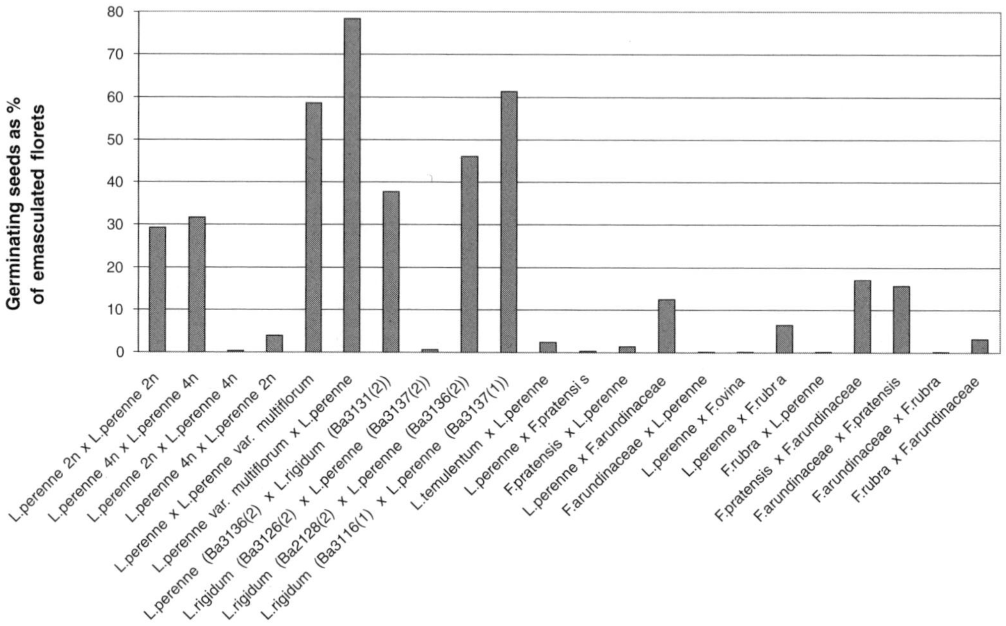

FIGURE 15.2 Success of interbreeding among species of the *Lolium–Festuca* complex. The percentage of germinating seeds produced by hand-pollinated florets was measured. Caution must be exercised in comparing these data compiled from research of different authors using different methods.

15.5.2.2 Phylogeography of *Lolium* vs. *Festuca*

In contrast to the 8 species of *Lolium*, which are all diploid, there are some 360 or more *Festuca* (fescue) species of variable ploidy. The distribution of *Festuca* species is inclusive and beyond that of the genus *Lolium* (167). Fescues occupy a much wider ecological range — from helophytic (marshy areas) through mesophytic to xerophytic (dry) areas (166). Polyploidy has had the effect of extending the ecological range over which the members of this genus are distributed (93).

Adaptation of individual species to specific ecological niches has physically separated some species within the complex. Irrespective of reproductive compatibility, ecogeographical separation presents a significant barrier to hybrid formation and gene flow. *Lolium perenne* and *Festuca ovina*, for example, only occur in close proximity where there is an abrupt change in habitat (93). Spatial isolation, including separation from other sexually compatible species (crops, wild, ruderal, or weeds), has long been employed in the seed industry to maintain genetic purity within acceptable limits. Spatial isolation is likely to be one of the mainstay strategies for the containment of transgenes.

15.5.3 Interfertility

Jenkin (94) studied some 75,000 *Lolium perenne* specimens and reported phenological and morphological differences between and within populations from different habitats. Although plants from different habitats conformed to broadly distinct morphological types, they remained fully interfertile. *Lolium perenne* should be considered to have no sub-speciation.

15.5.3.1 Self-Incompatibility and Self-Fertility

The outcrossing *Lolium* and *Festuca* species, including *Lolium perenne*, have a self-incompatibility system controlled by multiple alleles at two independently inherited (S and Z) loci (38,55). Self-compatibility in perennial ryegrass ranged from 0 to 6.1% in controlled pollination experiments

(15). Self-pollinated progeny of *Lolium perenne* have been shown to exhibit significantly reduced vigor from inbreeding depression (14). Both environment and genotype influence self-compatibility (92). High temperature (34°C) during anthesis has been shown to increase seed set due to selfing from 2 to 31% in clones of perennial ryegrass (168).

In populations with narrow genetic diversity, the large amount of incompatible pollen produced by neighboring plants due to common S and Z alleles will render these plants more susceptible to crossing with compatible pollen from more distant sources.

Other *Lolium* species (*L. subulatum, L. temulentum, L. remotum,* and *L. persicum)* are largely self-pollinated and therefore show a low propensity for gene flow from elsewhere. Annual *Lolium* species show a greater tendency for self-fertility than perennial *Lolium* species (62).

15.5.3.2 Karyotype Fertility Barrier

The entire genus *Lolium* is diploid in its natural form (160). Autotetraploid populations that have been developed by breeders are reproductively isolated from their diploid progenitor (25). Meiotic irregularity associated with trivalent and quadrivalent formation frequently results in the development of aneuploid progeny that are usually less fertile than euploids (48). Where aneuploid plants do survive, they are likely to succumb to competition from more vigorous euploids in the grass sward.

Successful pollination of diploid *Lolium multiflorum* by the diploid pollen from tetraploid forms is negligible (64). In subsequent work on *Lolium perenne,* no fertilization barrier between diploid and tetraploid karyotypes was reported (65). Ovules of both diploid and tetraploid karyotypes were fertilized to a comparable extent by n and 2n pollen. However, poor endosperm development resulted in greatly reduced germination of triploid progeny. Interploid pollination of adjacent diploid and tetraploid perennial ryegrass crops results in reduced germination capacity because of the formation of non-viable seeds.

Use of uncommon artificial karyotypes, such as tetraploid perennial ryegrass, may provide a useful barrier to gene flow between transgenic pastures and surrounding diploid populations. Furthermore, differences in karyotype could be employed to reproductively isolate transgenic perennial ryegrass from other sexually compatible species. Diploid *Lolium perenne* crossed with tetraploid *Festuca pratensis* was able to yield triploid hybrids that were highly vigorous (31). However, the progeny are likely to be completely sterile.

15.5.3.3 Interfertility between the Outcrossing *Lolium* Species

Of the eight species of *Lolium* that were recognized by Terrell (160), three — *L. perenne, L. multiflorum,* and *L. rigidum* — were considered to be outcrossing and self-incompatible. *L. canariense* is also partly outcrossing (32).

The phenomenon of interfertility persists with crosses between *Lolium perenne* and the other outcrossing *Lolium* species (91,96). Minor meiotic irregularities were observed in some crosses, but most authorities consider that hybrids between outcrossing *Lolium* species are fully fertile (25,32,91,97).

In some instances, for example, seed germination was lower than expected in crosses between *Lolium perenne* and *Lolium rigidum* (96). This indicated a degree of incompatibility or abnormal caryopsis development. However, fully fertile hybrids were obtained. Charmet et al. (33) suggested that the three outcrossing *Lolium* species should be regarded as a "single biological species."

Gene flow between *Lolium perenne* and all other outcrossing *Lolium* species must be regarded as highly probable where the species grow in close proximity.

15.5.3.4 Interfertility between *Lolium Perenne* and the Inbreeding *Lolium* Species

Hybrids between *Lolium perenne* and the self-pollinating *Lolium* species (*L. temulentum, L. remotum, L. persicum, L. subulatum,* and to some extent *L. canariense*) may form at low frequency

(159). Generally, these hybrids will be male sterile, but backcrossing is likely to increase fertility (90). Gene flow from transgenic perennial ryegrass to the self-pollinating *Lolium* species is likely to present a lower risk than within the outcrossing group of *Lolium* species.

15.5.3.5 *Lolium–Festuca* — Interfertility beyond the Genus *Lolium*

Both the genera *Lolium* and *Festuca* are thought to have a common ancestor with a base chromosome number x = 7 (30). Diploid fescue species preceded *Lolium* species and polyploid fescues. The outcrossing *Lolium* species gave rise to the inbreeding *Lolium* species.

The outbreeding *Lolium* species are particularly closely related to the *Festuca* section Bovinae (broad-leaved *Festuca* species) (25). *Lolium perenne* *Festuca pratensis* hybrids are reported to occur frequently in nature (32). These hybrids are completely sterile despite quite regular meiosis (33). Hybrids with other fescue species are considered rare (90).

Intergeneric reciprocal crosses between *Lolium* and *Festuca* behave differently with respect to caryopsis development (93). Hybrids between the genera *Lolium* and *Festuca* are generally more successful when *Lolium* is the maternal parent (76). This has implications for gene flow from novel perennial ryegrass, in that introgression is more likely to occur in the reciprocal direction than from *Lolium* to *Festuca*.

Hexaploid *Festuca arundinacea* appears to be derived from diploid *F. pratensis* and most probably has a genetic contribution from the genus *Lolium* (25). Interfertile diploid *Lolium* species hybridize with diploid *F. pratensis* and hexaploid *F. arundinacea* to a similar extent as the *Fescue* species (*F. pratensis* and *F. arundinacea*) do with one another; although in both instances, differences in ploidy confer a major reproductive barrier (42). Embryo rescue was necessary to generate hybrid seedlings of these intergeneric crosses, suggesting that there was poor caryopsis development. The F_1 generation was entirely male sterile and meiotic abnormality hindered backcrossing (33). The low level of cross-compatibility between *Lolium perenne* and *Festuca arundinacea* varies with genotype (29).

Where F_1 hybrids develop and survive from intergeneric crosses, they are often more vigorous than either parent. This was shown for hybrids of *L. multiflorum* *F. arundinacea* (83). However, these hybrids are usually completely sterile.

With great difficulty, it was possible to successfully retrieve one viable hybrid from *in vivo* crosses between tetraploid *Lolium multiflorum* cultivar "Malmi" and *Dactylis glomerata* (133). In this case, treatment with auxin was necessary to overcome strong postzygotic barriers. The hybrid produced non-dehiscing anthers and showed a low propensity to backcross with its *Lolium multiflorum* parent (as male). *Lolium perenne* when crossed with *Dactylis glomerata* was able to stimulate the ovaries and yielded a low level of poorly developed embryos. Reciprocal crosses failed to produce any embryos at all.

Different groups at different times, attempted to produce different experimental intergeneric crosses (65,91,93,95,133). These authors used a range of genotypes and procedures, so a direct comparison does not apply in all cases (Figure 15.2). The intergeneric fertility barrier is greatly increased in crosses between *Lolium perenne* and non-Bovinae section *Festuca* species.

The risk of gene flow from *Lolium perenne* to the closely related *Festuca* section Bovinae species may be regarded as slight, but a low incidence of backcrossing may lead to some gene flow. Wide natural intergeneric hybridization within and beyond the *Lolium–Festuca* complex is improbable.

Interrelationships that exist within the *Lolium–Festuca* complex indicate that gene flow from perennial ryegrass to even its more remote relatives in a stepwise manner is possible through successive generations of interspecific hybridization and backcrossing.

15.5.4 Pollen Flow

15.5.4.1 Amount and Timing

Pollen in chasmogamous grasses is liberated in large quantities for relatively short periods of the day during favorable atmospheric conditions (85). It is estimated that a single perennial ryegrass spike releases approximately 2.3 million pollen grains (153). Each spike produces approximately 200 florets and in a fertile and free pollinating situation, approximately 50% of florets may go on to produce viable seed (165). The high pollen to seed ratio (approximately 23,000:1) is typical of wind-borne pollination systems. Much of the pollen produced falls from the air and does not reach receptive stigmas. Pollen may be lost through asynchronous stigma presentation and pollen shed, excessive pollen landing on stigmas, non-viability, and sexual incompatibility. The reduction in airborne pollen concentration is an important factor accounting for the decline in gene flow over distance.

A range of temporal differences in flowering is observed across the species complex, which may result in isolation. Heterogeneous populations have a wider range in time of pollen shed than populations with a narrow genetic background. In general, the usefulness of diurnal differences in flowering time as an isolation tool to prevent gene flow increases with genetic distance between relevant species.

15.5.4.2 Distance of Pollen Flow

Separation by distance is the primary isolation method used to prevent genetic contamination of seed crops and, more recently, to prevent gene flow from transgenic crops. Although much research and experiential data exist for seed crop isolation, this is not true of transgenic crops, which made their commercial debut as recently as 1994 (143).

15.5.4.3 Distribution of Pollen and Gene Flow over Distance

Studies of pollen dispersal with regard to the maintenance of seed purity have provided a basis for predicting the rate of gene flow from perennial ryegrass populations (63). The rate of pollen flow has been measured and modeled by various authors. A strongly leptokurtic (exponential decay) distribution of pollen from its source is consistently reported for wind-pollinating plant species and is confirmed by studies of gene flow (43).

Some models attempt to explain the influence of wind speed and direction on pollen and gene flow from wind-pollinating grass species (59,142). Significant differences in the rate of pollen dispersal from perennial ryegrass for different directions of wind have been reported (60). This research concerned pollen deposition over short periods and may bear little resemblance to the pattern of gene flow.

15.5.4.4 Pollen Viability and the Outer Limits of Gene Flow

Little work has been done on pollen viability in perennial ryegrass; however, *in vitro* tests for viability of *Festuca arundinacea* pollen indicate a gradual decline in viability from the time of anther dehiscence to about 48 hours, followed by a rapid decline thereafter (134). Most pollen that has not deposited by day precipitates during the night (54). Thus, pollen would usually only be effective on the day of release.

Pollen sometimes travels in rising turbulent air masses, which could be hundreds or thousands of meters across (68). Rising air masses may develop as:

- Thermal lift on warm to hot calm days
- Ridge lift, where the wind is forced upward over a slope, creating turbulent eddies on the windward and leeward side
- Wave lift, where atmospheric waves cause turbulence and lift closer to the ground
- Frontal lift, where an air stream displaces an air mass of a different temperature, including sea breezes
- Orthographic lift, where one air stream rises over another
- Anabatic lift, where warm air rises up a hillside

The pollen load of these rising air cells could then be released some distance from its source. If viable pollen is able to travel in rising air cells, contamination over distance may not always be so rare or random and this may explain some aspects of atypical pollen and gene flow. Quantification of this neglected aspect of pollen and gene flow is merited. An outside distance of 50 km has been suggested for rare fertility events due to surviving pollen transportation over distance for most wind-pollinated crops (54).

15.5.5 The Control of Gene Flow Using Non-paternally Inherited Transformation or Male Sterility

There are several possible technologies whereby transgenes may not be mobile through pollen dispersal:

- Transgenic plant endophytic fungi of the species *Acremonium* are being developed to deliver pest control compounds to perennial ryegrass (161). Endophytic fungi are not transmitted through pollen.
- Cytoplasmic DNA, such as that from chloroplasts or mitochondria, is mainly inherited maternally and not through pollen (45).
- Apomictic or otherwise asexually multiplied transgenic cultivars in combination with methods for controlling male fertility would permit the production of seed, but avoid production or release of viable pollen and therefore gene flow through it (70).

15.5.6 Concluding Remarks

Lolium perenne is an important pasture species worldwide. It is a member of the rather loosely defined and intergraded *Lolium–Festuca* complex. These species have a wide distribution covering most of the temperate regions of the world and beyond. A considerable range in interfertility exists between the members of this complex. Close relatives, like different ecotypes of the same species or all of the outcrossing *Lolium* species, are almost completely interfertile, whereas those further apart show little or no sexual affinity. The phenomenon of interfertility in the species complex indicates that over successive generations, introgression of novel genes from perennial ryegrass could occur even into its most distant relatives. They frequently grow in close enough as cultivated, wild, ruderal, or weed populations to permit cross-pollination. This could lead to problems with new feral weeds.

Many fitness related traits of organisms can be explained by natural selection. Weedy genotypes have already survived the rigors of natural selection over a diverse range of microhabitats, often too marginal for crop survival. These weedy genotypes may then be perpetuated in a free pollinating environment, continually expanding the genetic diversity through generations of mutation, introgression, and selection. Conversely, the process of domestication introduces a host of maladaptive traits (36). In pasture species, these would include palatability to grazing animals and a prolonged vegetative stage. Typically, *Lolium perenne* cultivars are derived from a polycross comprising a gene pool limited to about 200 individual genotypes from a single or just a few similar or closely related source populations. The narrow genetic base confers the best chance of a genetically stable and uniform cultivar that is agronomically adapted to the confines of a sown pasture situation.

At the interface of a crop and sexually compatible weed populations, gene flow may occur in either direction. The weed population would already harbor many adaptive traits. Any introgressed trait conferring added fitness would lead to a fitter weed population.

Innovative transformation involving maternally inherited and other non-genomic DNA and inclusion of male sterility systems may substantially reduce the risk of gene flow from perennial ryegrass through pollen dispersal.

15.6 SAFFLOWER — FERALITY IN A PLANT-MADE PHARMACEUTICAL PLATFORM*

Cultivated safflower (*Carthamus tinctorius* L.) is being evaluated as a biological platform for the field production of high value plant-made pharmaceuticals (PMPs). The need to differentiate crops

* Prepared by M.A. McPherson, A.G. Good, and L.M. Hall.

containing PMPs from food and feed crops has increased the interest of the pharmaceutical/biotechnology sector in using minor crops as a platform in Canada, including tobacco (*Nicotiana tabacum* L.), alfalfa (*Medicago* sp. L.), white clover (*Trifolium angustifolium* L.), Ethiopian mustard (*Brassica carinata* A. Braun), white mustard (*Sinapis alba* L.), flax (*Linum usitatissimum* L.), and safflower (*C. tinctorius* L.) (5).

Biosafety concerns for PMPs are similar to traditional transgenic crops, including outcrossing to other non-transgenic varieties or wild and weedy relatives, the need to maintain isolation distances from cross-compatible species, segregation of PMP from food or feeds, and the feral nature of the crop. Because PMP products may have biological activity, unlike crops with input traits such as herbicide resistance, the impact of inadvertent environmental release and contamination of the food or feed system may be greater (120).

Safflower (*Carthamus tinctorius*) has been grown extensively in the Mediterranean and is thought to have originated in the Euphrates Basin (109,154). Hybridization of safflower with sympatric (spatially) wild relatives may have played a significant role in the evolution of *Carthamus* L. and cultivated safflower in the Mediterranean (7,145). Hybridization between safflower and weedy relatives could facilitate a transfer of PMP or feral genes. The resulting hybrids could facilitate the introgression of a transgene into wild or weedy populations or the formation of feral safflower populations. Safflower, originally grown for dye production (carthamine), is currently grown for the extraction of oil for human consumption and for birdseed. Safflower, introduced in the New World in 1899, was commercially grown in the U.S. during the early 1950s (106,109). In recent years, approximately 80,000 ha per annum of safflower has been harvested in the U.S. In Canada, less than 1000 ha have been harvested annually (3,4). Feral populations, volunteer safflowers that have become established in the agroecosystem, have been reported in the U.S. (58,74,75,122,144,150), but have not been reported in Canada. Thus, this crop may have the potential to dedomesticate under certain production conditions. In this review, we will discuss the implications of growing transgenic safflower containing PMP and the potential for feral safflower populations to increase in both the U.S. and Canada.

15.6.1 Platform for Plant-Made Pharmaceuticals

"Centennial," a cultivated variety of safflower, has been genetically engineered to express two nuclear encoded gene cassettes. The first is a selective marker, for glufosinate-ammonium resistance, and the second is a trait generated by the fusion of a gene encoding a PMP with the oil seed transmembrane protein oleosin (114). The oleosin fusion protein enables the production and subsequent isolation of the PMP from oilseed crops (126,136). This has previously been demonstrated in safflower (114), where the isolation of the PMP from mature, oil rich safflower seeds (approximately 30 to 40% of the seed by dry weight) (154) has economic potential.

The selective marker, glufosinate resistance, is unlikely to confer a selective advantage to volunteers or feral populations because it is used only on a few crops and is not used in ruderal areas such as roadsides and waste areas. The fitness consequences of the PMP genes have not yet been experimentally determined.

15.6.2 Safflower, Systematics, Biology, Biogeography

15.6.2.1 Systematics and Hybridization Potential between Cultivated Safflower and Its Wild Relatives

Cultivated safflower is a member of the Asteraceae (Compositae), tribe Cardueae (thistles), and subtribe Centaureinae (113,162). The genus *Carthamus* L. has 16 recognized species *sensu* (118). It is a member of the *Carthamus–Carduncellus* complex that consists of several closely related genera including *Carduncellus* Adans., *Femeniasia* Susanna, *Lamottea* Pomel, and *Phonus* Hill (162). The cross-compatibilities among most of the members of *Carthamus* and some closely related

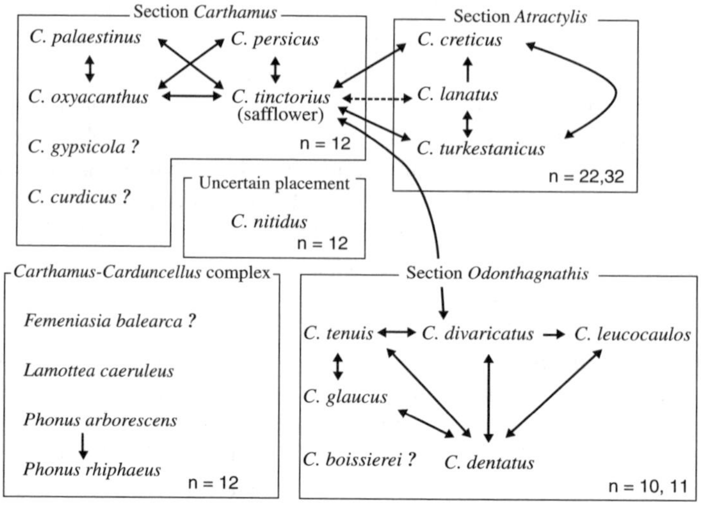

FIGURE 15.3 Artificial interspecific crosses that resulted in fertile progeny (see text for references). Solid lines indicate F_1 fertile hybrids with viable seed production. Dotted lines indicate hybridization occurred, but the F_1 hybrids could not be obtained without embryo rescue or treatments with colchicine. Arrows indicate the direction of the cross (male → female). Species in bold are naturalized in North America. Boxes indicate the taxa assigned to each section of *Carthamus sensu* (118) except the species that have since been reassigned to different genera within the complex. *Carthamus glaucus* contains several subspecies, which are all interfertile (111). However, interspecific crosses with *C. glaucus* ssp. *glaucus* from different geographical regions are sterile. Crosses between *C. divaricatus* and safflower produced self-incompatible hybrids, which had low fertility when backcrossed with safflower. *?* indicates taxa where crosses have not been reported. Documented locations of introduced species to the New World: *C. oxyacanthus,* Florida and Oregon; *C. creticus,* California; *C. lanatus,* California, Florida, Oklahoma, Oregon, and Texas; *C. leucocaulos,* California; safflower, California, Florida, Oklahoma, Oregon, and Texas. No wild or weedy relatives of safflower have been reported in Canada.

species from the *Carthamus–Carduncellus* complex have been evaluated (6–8,34,46,50–52,72, 86–88,101,104,105,107,108,111,113,145). These papers have been reviewed (123) and their findings summarized in Figure 15.3.

Cultivated safflower and the other members of the section *Carthamus* have a chromosome number of n = 12 (Figure 15.3). The ease of crossing and obtaining viable offspring among the members of this section of the genus has led several authors to consider these species (safflower, *Carthamus persicus* Willd., *Carthamus palaestinus* Eig, and *Carthamus oxyacanthus* Bieb.) to be a biological race rather than separate species (6,86). Genomic formulas for some species in the genus *Carthamus* have been determined (52,114). These formulas could be used to infer the potential for introgression of a nuclear encoded transgene from PMP safflower to a wild relative.

Vigorous hybrids from natural crosses of safflower (genomic formula BB) and *C. oxyacanthus* (BB) have been documented in both greenhouse studies when grown together and in fields in Pakistan and India, where the two species were sympatric (46,107,110) (Figure 15.3). Hybrids of safflower (BB) and *C. palaestinus* (B_1B_1) have been found in Israel where these species were sympatric (107,110) (Figure 15.3). Artificial crosses of safflower with *C. persicus* (B_1B_1) have produced fertile F_1 and F_2 progeny (86) (Figure 15.3). Hybridization experiments with *Carthamus gypsicola* Ilj., *Carthamus curdicus* Hanelt, and the newly discovered *Femeniasia balearica* Susanna have not yet been conducted. Numerous attempts to cross *C. nitidus* Boiss. (n = 12) with other members of the section *Carthamus* have not produced fertile hybrids (109,111) (Figure 15.3). Safflower has also been crossed with four species outside of the section *Carthamus* to produce viable hybrids. A single cross of *C. divaricatus* Beguinot and Vacc. (n = 11; Section *Odonthagnathis*) with safflower was obtained but the offspring from backcrosses with these hybrids and safflower had low fertility (51) (Figure 15.3). Crosses between safflower and *C. lanatus* L. ($A_1A_1B_1B_1$; n = 22; Section *Atractylis*) did not produce viable offspring without treatments with colchicine and embryo rescue (Figure 15.3). Thus, the probability of a hybridization event between these two species producing viable offspring in nature is relatively low. Crosses of safflower with two members of Section *Atractylis*, *C. creticus* L. ($A_1A_1B_1B_1$ A_2A_2; n = 32) and *C. turkestanicus* Popov ($A_1A_1B_1B_1$ A_3A_3; n = 32), produced viable offspring (Figure 15.3). The three recognized members of *Atractylis* are cross-compatible (Figure 15.3). Thus, a transgene could introgress from PMP safflower to a weedy relative and spread to other species via hybridization.

15.6.2.2 Carthamus — Biology and Biogeography

Carthamus oxyacanthus and *C. persicus* are weeds of wheat fields and roadsides of cooler climates (6,86,110,154). *Carthamus palaestinus* is a wild relative found in desert regions of the Mediterranean (southern Israel). *Carthamus persicus* is found in Iraq, Lebanon, and Syria. *Carthamus oxyacanthus* is endemic to Turkey, subtropical regions of western Iraq, Iran, northwest India, throughout Kazakhastan, Turkmenistan, and Uzbekistan (110,154).

Data obtained from artificial crosses of safflower with other species can be used as an initial indicator to predict the potential for hybridization and introgression of a transgene into a weedy population. Hybridization is only the first step to introgression. For a hybridization event to occur, two species must be sympatric both spatially and temporally (i.e., flowering synchronously). Four wild safflower relatives have been introduced in some areas of the New World (75) (Figure 15.3). Geographical locations (spatial) of these species have been documented in the U.S. (Figure 15.3). *Carthamus oxyacanthus* and *C. lanatus* is naturalized in Oregon and Florida (75). *Carthamus lanatus* has been a weed in California since 1891 (57). Kessler (103) documented *C. lanatus* as a serious weed growing in an isolated area of Oklahoma. *Carthamus lanatus* and *Carthamus leucocaulos* Sibth. *et* Sm. have been documented in Texas and collected from Argentina and Chile (39,75,121). *Carthamus creticus* and *C. lanatus* have been reported in California (75,82,127). Little information regarding the flowering times (temporal) in these localities has been reported.

Potential recipients of nuclear encoded transgenes from a cultivated PMP with safflower in the New World include safflower escapes from cultivation and four naturalized wild relatives (*C. creticus*, *C. lanatus*, *C. leucocaulos,* and *C. oxyacanthus*). Of these wild species, only *C. oxyacanthus* and *C. creticus* have been shown to produce viable hybrid offspring when crossed with cultivated safflower (123) (Figure 15.3).

15.6.3 Domestication Traits in Safflower

Safflower is considered one of humanities' oldest crops (100,110,154). Domestication of this thistle species has resulted in a suite of traits that increased seed recovery at harvest, reduced seed dormancy, and increased yield and quality parameters, previously defined as the "domestication syndrome" (69). Domestication traits have been compared between safflower and weedy relatives and their patterns of inheritance in hybrids used to infer the number of genes or loci involved and their dominant or recessive nature (112,176).

Shattering resistance is a well-established domestication trait that reduces seed loss and increases harvestability. The wild and weedy species *C. oxyacanthus, C. pericus,* and *C. palaestinus* have a shattering capitulum (head) (6,86,110,154). Nonshattering cultivated safflower lines were homozygous recessive (*sh*) for a single locus responsible for shattering, whereas *C. persicus* and *C. palaestinus* are homozygous dominant for this locus (*Sh*) (6,86).

The presence of a pappus (seed appendage for dispersal via water, wind, and adherence to animal fur) alters seed dispersal in the Asteraceae. Most of the achenes (seeds) of cultivated safflower lack a pappus and, when it is present, it is reduced (less than the length of the achenes). The allele controlling the presence of a pappus in *Carthamus persicus* is dominant (*P_*) and safflower is homozygous recessive for this locus (*pp*).

A single dominant gene (*Ro*) reduces the duration of the rosette stage in cultivated safflower. The presence of this gene reduces time to maturity. *Cathamus persicus* has a much prolonged rosette stage (*ro*) (86). The prolonged rosette stage of both *C. persicus* and *C. oxyacanthus* ensures that seeds are dispersed in the field after harvest of the cereal crops in which they commonly occur (86).

Reduced seed dormancy is a domestication trait of many crops that ensures synchronous germination and maturity and reduces volunteers in subsequent crops. However, complete loss of seed dormancy can lead to germination of mature seeds in the flower, prior to harvest, when conditions are humid (53). Cultivated safflower has a high germination rate suggesting that it exhibits low seed dormancy. Germination of seeds in mature heads prior to harvest after a heavy dew or rain has been identified as a production limitation for this crop (176). Seed dormancy in safflower is controlled by many loci, the genes are not additive, and heritability was high (112). The expected increase of dormancy under selection was approximately 10 to 25%.

It has been suggested that there is an association between lack of seed dormancy and some morphological traits. *Carthamus oxyacanthus* and *C. palaestinus* have pigmented (*w*) and mottled (*m*) seeds, which possess long- and short-term seed dormancy (112).

The weed *Carthamus lanatus* has a high level of secondary seed dormancy (approximately 90%) and seeds persist in the soil for up to 10 years (138,169). Dormancy of these seeds can be broken by a combination of leaching and exposure to red light (600 to 680 nm), conditions which are often present when the soil is prepared for planting (169). Seeds of *C. oxyacanthus* have seed dormancy (10% or more are dormant), but scarification can sometimes improve germination, although cold treatments do not (11,12). Scarification of white seeds reduced germination, whereas, it improved germination of seeds with pigmentation (10), suggesting a genetic association of seed dormancy with seed pigments.

Safflower seeds with striped hulls (*ww mm*) have more seed dormancy than those with the typical white seed coloration (*W_M_*). Safflower typically has cotyledons with a green midvein, whereas wild relatives such as *C. palaestinus* have purple midveins. Safflower cultivars with cotyledons expressing green midveins have more seed dormancy than those with purple midveins.

Domestication traits such as reduced shattering, lack of pappus, and short duration of the rosette stage ensure that the majority of safflower seeds are harvested. Reduced seed dormancy ensures the seeds germinate when planted and are less likely to persist in the seed bank. Other traits associated with domesticated safflower include a restriction of branching to the upper portion of the stem (rather than flowering side shoots), a reduction of hairs on stamen filaments, and a reduction in the spine length on the leaves (86). The genetic control of these traits and their functionality has not yet been studied in detail.

If a transgenic safflower were to hybridize with weedy relatives, recessive domestication traits may not be expressed. Hybrids may be more feral or invasive than cultivated safflower and could become problem weeds. The herbicide resistance trait may also confer an enhanced ability to survive in the agroecosystem. To reduce the risk of gene flow, transgenic safflower should be grown in areas free of wild relatives know to be cross-compatible.

15.6.4 Ferality of Safflower

Anecdotal reports suggest that safflower does not become established outside of agroecosystems (154). However, volunteer safflower has been documented in the U.S. from California (75), Iowa (144), Illinois (74), Kansas (58,144), New Mexico (122), Ohio (163), and Utah (150). It is not known how long these populations persist and whether they have become dedomesticated. Therefore, at least in some locations, PMP safflower has the potential to become established within the agroecosystem.

If PMP safflower is grown near existing feral populations, the PMP traits may move via pollen flow. Cross-pollination of safflower plants is facilitated predominantly by insects, but wind can also move pollen short distances (up to approximately 122 cm) between plants growing close together (8,34). Bees and other flying insects contribute to gene flow among safflower and its wild relatives over relatively large distances (34,100). The frequency of outcrossing was partially under genetic control, ranging from 0 to 100% among experimental cultivars (34). Thus, each cultivated variety of safflower considered for the production of PMP should to be evaluated to determine the potential outcrossing rate.

15.6.5 Information Required to Improve Our Prediction for PMP Safflower Ferality

Prior to the release or growing of broad-scale PMP safflower under confined release conditions, considerable information is required to assess environmental biosafety impacts, including:

- The likelihood of volunteers surviving and perpetuating in the natural environment
- A quantification of gene flow from PMP safflower between feral populations and conventional varieties
- The risk of introgression of the PMP genes to wild/weedy populations
- The potential for persistence in the environment of PMP safflower/weed hybrids

15.6.6 Conclusions

Biosafety concerns associated with the production of PMP in safflower need to be addressed prior to release, including the potential for ferality. Domestication traits have been selected in safflower and include decreased shattering, seed dormancy, pappus length, and duration of the rosette stage. Most alleles controlling domestication traits are recessive. In the New World, two safflower relatives, *C. oxyacanthus* and *C. creticus*, are known to be cross-compatible with cultivated safflower. Hybrids between safflower and these wild relatives could serve as sinks for PMP traits and sources of feral traits. Thus, hybrids could facilitate introgression of PMP genes into conventional safflower or weedy relatives. Alternatively, hybrids could transfer feral traits to PMP safflower, which may enhance ferality. For this reason, PMP safflower should not be grown in the Mediterranean area where many cross-compatible species are known and where extensive cropping of safflower for human and animal consumption occurs. Furthermore, regions in North America where wild relatives or conventional safflower are grown should be avoided or appropriate isolation distances determined and maintained. Production of PMP has great economic potential (114,126,136,154) and minor crops are platforms that may provide reduced risk to food and feed products. However, like all transgenic crops, risks and benefits must be evaluated on a case-by-case basis.

15.7 SUGARCANE*

Sugarcane is the most important sucrose-producing crop in the tropics and subtropics (141). Several large perennial grass species comprise sugarcane. Noble cane *Saccharum officinarum* accounts for most of the world's production of cane sugar and is vegetatively propagated by stem cuttings (49,141). Originating from New Guinea, *S. officinarum* is only found in cultivated conditions and not in the wild (141). There is no published information on its problem as being a volunteer weed.

* Prepared by S.I. Warwick.

Sugarcanes are derivatives of the "*Saccharum* complex," which is composed of the interbreeding genera *Saccharum*, *Erianthus*, *Scleropstachya*, *Narenga,* and *Miscanthus* (141). Various molecular marker studies have confirmed the close genetic relationships within *Saccharum* and related genera, including restriction fragment length polymorphisms (RFLPs) (89), RAPDs (73,129), AFLPs (117), microsatellites (37,146), sequence data from the ITS region of the nuclear ribosomal DNA (21,77,135), plastid *trnL* DNA (77), *rbcL* gene (174), and homologous centromeric repetitive sequences (157). Originally of tropical origin, it is believed that interspecific hybridization, principally with the wild species *S. spontaneum*, has extended the geographical range of economic sugarcane production. Its congeneric major weedy counterpart is the rhizomatous C_4 species *S. spontaneum.* Considered to be the most primitive *Saccharum* species and of Indian origin, it is adapted to diverse environments. A weed in 33 countries, it is a serious problem in tropical Asia, but also occurs as a sporadic weed in Africa, Europe, and the tropical New World (80). *S. spontaneum* is cross-compatible with the crop, and both crop and weed species are polyploid. There is evidence for ample introgression and convolution of complex *Saccharum* genomes. *S. officinarum* has chromosome numbers of $2n = 80$, and *S. spontaneum* ranges from $2n = 40$ to 128 (89,141). *S. spontaneum* is believed to have spontaneously crossed with *Miscanthus floridulus* to form *S. robustum*, which is one of the parents of noble sugarcane (141). *Saccharum barberi* ($2n = 81$–124) and *S. sinense* ($2n = 115$–120) are old cultivated species of India and China, respectively, that were also suggested to be natural hybrids of *S. officinarum* and *S. spontaneum* (141). This fact has been recently confirmed by molecular cytogenetic genomic *in situ* hybridization studies that detected species-specific repeated sequences from both parental taxa (44). Chromosomal and morphological evidence support the suggestion that *Erianthus maximus* is derived from a natural cross of the cultivated *S. officinarum* with the wild genus *Miscanthus*, as reviewed in (49).

Sugarcanes are wind-pollinated and show inbreeding depression (141). The crop has a narrow genetic base, with modern hybrids founded on only 20 noble and fewer than 10 *spontaneum* or *spontaneum* derivatives (141). Various types of molecular markers have confirmed the low levels of genetic diversity in *S. officinarum* compared to other *Saccharum* species (2,40,89,129). As with other crops, important agronomic characters (yield, sugar content) seem to be inherited quantitatively (79). Quantitative trait loci (QTLs) for sugar yield, stalk weight, and fiber content have been recently mapped (125). *S. spontaneum* has several traits of use for sugarcane breeding (80), including variation in sucrose content. It also has traits that could potentially increase the weediness of the crop, such as disease resistance, cold temperature tolerance, high tillering ability, vigorous growth capacity, and deep root system. Low temperature tolerance and disease resistance traits can also be introgressed into sugarcane from *Erianthus* (139).

An efficient system is now in place for improving diverse sugarcane cultivars by genetic transformation (24). The method has been used to introduce genes for resistance to several major diseases, insect pests, and herbicides. Research is underway to identify other genes for increased environmental stress resistance, agronomic efficiency, and yield of sucrose or other valuable products. Field tests in South Africa on genetically transformed glufosinate-resistant sugarcane have indicated morphological and agronomic characters, such as stalk height, diameter, fiber content, disease resistance, and yield, were not significantly different in the transgenic line (116). Sugarcane has also been transformed for various types of insect resistance. Transgenic lines expressing the snowdrop lectin (*Galanthus nivalis* agglutinin) were developed against the Mexican rice borer (*Eoreuma loftini*), the primary pest of Texas sugarcane (147,148). Transgenic sugarcane lines expressing the *Bacillus thuringinensis* gene *cry1*Ab against the sugarcane borer (*Diatraea saccharalis*) have demonstrated resistance under field conditions with no apparent loss in plant quality and yield (26). The perenniality of the crop and its ability to reproduce vegetatively may foster the persistence and spread of fitness-enhancing transgenic hybrids. The impact of such transgenes moving to *S. spontaneum* is of obvious concern, as it is already a serious weed problem in many areas.

LITERATURE CITED

1. Aamodt OS, Johnson LPV, Manson JM. 1934. Natural and artificial hybridization of *Avena fatua* and its relation to the origin of the fatuoids. *Can. J. Res.* 11:701–727.
2. Al-Janabi SM, McClelland M, Petersen C, Sobral BWS. 1994. Phylogenetic analysis of organellar DNA sequences in the Andropogoneae: Saccharinae. *Theor. Appl. Genet.* 88:933–944.
3. Anonymous. 2002. Field crops grown, 2001. *Alberta Agriculture, Food and Rural Development*: http://www1.agric.gov.ab.ca/$department/deptdocs.nsf/all/sdd5726/$FILE/5.4.pdf. Accessed June 24, 2004.
4. Anonymous. 2003. Crop production. *United States Department of Agriculture*: http://usda.mannlib.cornell.edu/reports/nassr/field/pcp-bb/2003/. Accessed June 22, 2004.
5. Anonymous. 2004. Response of the federal departments and agencies to the petition filed September 25, 2003, by Greenpeace Canada under the auditor general act. *Canadian Food Inspection Agency:* http://www.inspection.gc.ca/english/sci/biotech/tech/pharmcae.shtml.
6. Ashri A, Efron Y. 1964. Inheritance studies with fertile interspecific hybrids of three *Carthamus* L. species. *Crop Sci.* 4:510–514.
7. Ashri A, Knowles PF. 1960. Cytogenetics of safflower (*Carthamus* L.) species and their hybrids. *Agron. J.* 52:11–17.
8. Ashri A, Rudich J. 1965. Unequal reciprocal natural hybridization rates between two *Carthamus* L. *Crop Sci.* 5:190–191.
9. Balfourier F, Imbert C, Charmet G. 2000. Evidence for phylogeographic structure in *Lolium* species related to the spread of agriculture in Europe. A cpDNA study. *Theor. Appl. Genet.* 101:131–138.
10. Bassiri A, Kheradnam M. 1976. Relationships between seed color and viability, germination and seedling growth of wild safflower ecotypes. *Can. J. Plant Sci.* 56:911–917.
11. Bassiri A, Rouhani I. 1976. Effect of seed treatment on germination of wild safflower. *Weed Sci.* 24:233–234.
12. Bassiri A, Rouhani I, Ghorashy SR. 1975. Effect of temperature and scarification on germination and emergence of wild safflower, *Carthamus oxyacantha* Bieb. *J. Agric. Sci. Camb.* 84:239–243.
13. Baum BR. 1969. The use of lodicule type in assessing the origin of *Avena* fatuoids. *Can. J. Bot.* 47:931–944.
14. Bean EW, Yok-Hwa C. 1972. An analysis of growth of inbred progeny of *Lolium*. *J. Agric. Sci.* 79:147–153.
15. Beddows AR. 1973. Biological flora of the British Isles: *Lolium multiflorum* L. *J. Ecol.* 61:587–600.
16. Bennett SJ, Hayward MD, Marshall DF. 2002. Electrophoretic variation as a measure of species differentiation between four species of the genus *Lolium*. *Genet. Resour. Crop Evol.* 49:59–66.
17. Besnard G, Baradat P, Breton C, Khadari B, Berville A. 2001. Olive domestication from structure of oleasters and cultivars using nuclear RAPDs and mitochondrial RFLPs. *Genet. Select. Evol.* 33:S251–S268.
18. Besnard G, Berville A. 2000. Multiple origins for Mediterranean olive (*Olea europaea* L-ssp *europaea*) based upon mitochondrial DNA polymorphisms. *C. R. Acad. Sci.-Life Sci.* 323:173–181.
19. Besnard G, Khadari B, Baradat P, Berville A. 2002. *Olea europaea* (Oleaceae) phylogeography based on chloroplast DNA polymorphism. *Theor. Appl. Genet.* 104:1353–1361.
20. Besnard G, Khadari B, Villemur P, Berville A. 2000. Cytoplasmic male sterility in the olive (*Olea europaea* L.). *Theor. Appl. Genet.* 100:1018–1024.
21. Besse P, McIntyre CL, Berding N. 1996. Ribosomal DNA variations in *Erianthus*, a wild sugarcane relative (Andropogoneae-Saccharinae). *Theor. Appl. Genet.* 92:733–743.
22. Bickelmann U. 1989. The pedigree of German oat cultivars (*Avena sativa* L.) and the occurrence of fatuoids. *Plant Breed.* 103:163–170.
23. Bickelmann U. 1993. On fatuoids and hybrids of *Avena* spp. and their significance for seed production. III. Description by prolamin electrophoretic characters. *Plant Var. Seeds* 6:65–73.
24. Birch RG. 1996. New gene technologies and their potential value for sugarcane. *Outlook Agric.* 25:219–226.
25. Borrill M. 1976. Temperate grasses. In *Evolution of crop plants*, Simmonds NW, Ed., pp. 137–142. London and New York: Longman.

26. Braga DPV, Arrigoni EDB, Silva-Filho MC, Ulian EC. 2003. Expression of the Cry1Ab protein in genetically modified sugarcane for the control of *Diatraea saccharalis*. *J. New Seeds* 5:209–221.
27. Breton C, Besnard G, Berville A. 2003. Using multiple types of molecular markers to understand olive phylogeography. In *Documenting domestication: new genetic and archaeological paradigms*. Zeder MA, Decker-Walters D, Bradley D, Smith B, Eds., Washington, DC: Smithsonian Press.
28. Bronzini de Caraffa V, Maury J, Gambotti C, Breton C, Bervillé A, Giannettini J. 2002. Mitochondrial DNA variation and RAPD mark oleasters, olive and feral olive from western and eastern Mediterranean. *Theor. Appl. Genet.* 104:1209–1216.
29. Buckner RC. 1960. Cross-compatibility of annual and perennial ryegrass with tall fescue. *Agron. J.* 52:409.
30. Bulinska-Radomska Z, Lester RN. 1988. Intergeneric relationships of *Lolium, Festuca,* and *Vulpia* (Poaceae) and their phylogeny. *Plant Syst. Evol.* 159:217–227.
31. Carnahan HL, Hill HD. 1955. *Lolium perenne* L. x tetraploid *Festuca elatior* L. triploid hybrids and colchicine treatments for inducing autoallohexaploids. *Agron. J.* 47:258–262.
32. Charmet G, Balfourier F. 1994. Isozyme variation and species relationships in the genus *Lolium* L (ryegrasses, Graminaceae). *Theor. Appl. Genet.* 87:641–649.
33. Charmet G, Balfourier F, Chatard V. 1996. Taxonomic relationships and interspecific hybridization in the genus *Lolium* (grasses). *Genet. Resour. Crop Evol.* 43:319–327.
34. Claassen CE. 1950. Natural and controlled crossing in safflower, *Carthamus tinctorius* L. *Agron. J.* 42:381–384.
35. Coffman FA. 1946. Origin of cultivated oats. *J. Am. Soc. Agron.* 38:983–992.
36. Cooper M, Byth DE. 2004. Understanding plant adaptation to achieve systematic applied crop improvement — a fundamental challenge. In *Plant adaptation and crop improvement*, Cooper M, Hammer GL, Eds., pp. 5–23. Oxford, UK: CAB International.
37. Cordeiro GM, Pan Y-B, Henry RJ. 2003. Sugarcane microsatellites for the assessment of genetic diversity in sugarcane germplasm. *Plant Sci.* 165:181–189.
38. Cornish MA, Hayward MD, Lawrence MJ. 1979. Genetic control in ryegrass. 1. Genetic control in diploid *Lolium perenne* L. *Heredity* 43:95–106.
39. Correll DS, Johnston MC. 1970. *Manual of the vascular plants of Texas*. Renner: Texas Research Foundation. 1881 pp.
40. Coto O, Cornide MT, Calvo D, Canales E, D'Hont A, de Prada F. 2002. Genetic diversity among wild sugarcane germplasm from Laos revealed with markers. *Euphytica* 123:121–130.
41. Coulibaly S, Pasquet RS, Papa R, Gepts P. 2002. AFLP analysis of the phenetic organization and genetic diversity of *Vigna unguiculata* L. Walp. reveals extensive gene flow between wild and domesticated types. *Theor. Appl. Genet.* 104:358–366.
42. Crowder LV. 1953. Interspecific and intergeneric hybrids of *Festuca* and *Lolium*. *J. Heredity* 44:195–203.
43. Cunliffe KV, Vecchies AC, Jones ES, Kearney GA, Forster JW, et al. 2004. Assessment of gene flow using tetraploid genotypes of perennial ryegrass (*Lolium perenne* L.). *Aust. J. Agric. Res.* 55:389–396.
44. D'Hont A, Paulet F, Glaszmann JC. 2002. Oligoclonal interspecific origin of 'North Indian' and 'Chinese' sugarcanes. *Chromosome Res.* 10:253–262.
45. Darnell J, Lodish H, Baltimore D. 1990. *Molecular cell biology.* New York: Scientific American Books Inc.
46. Deshpande RB. 1952. Wild safflower (*Carthamus oxyacantha* Bieb.) — a possible oilseed crop for the desert and arid regions. *Indian J. Genet. Plant Breed.* 12:10–14.
47. deVries J, Toenniessen G. 2001. *Securing the harvest: biotechnology and seed systems for African crops*. Wallingford, UK: CABI Publishing. 208 pp.
48. Easton HS. 1973. Performance of aneuploids in an autotetraploid ryegrass population. *N. Z. J. Agric. Res.* 16:35–37.
49. Ellstrand NC. 2003. *Dangerous liaisons? When cultivated plants mate with their wild relatives.* Baltimore, MD: Johns Hopkins University Press. 244 pp.
50. Estilai A. 1977. Interspecific hybrids between *Carthamus tinctorius* and *C. alexandrinus*. *Crop Sci.* 17:800–802.
51. Estilai A, Knowles PF. 1976. Cytogenetic studies of *Carthamus divaricatus* with eleven pairs of chromosomes and its relationship to other *Carthamus* species (Compositae). *Am. J. Bot.* 63:771–782.

52. Estilai A, Knowles PF. 1978. Relationship of *Carthamus leucocaulos* to other *Carthamus* species (Compositae). *Can. J. Genet. Cytol.* 20:221–233.
53. Evans LT. 1993. *Crop evolution, adaptation and yield.* Cambridge, UK: Cambridge University Press. pp. 62–113.
54. Faegri K, van der Pijl L. 1979. Abiotic pollination. In *The principles of pollination ecology*, pp. 34–41. Oxford, UK: Pergamon Press.
55. Fearon CH, Hayward MD, Lawrence MJ. 1984. Self-incompatibility in ryegrass. 8. The mode of action of S-alleles and Z-alleles in the pollen of autotetraploids of *Lolium perenne* L. *Heredity* 53:415–422.
56. Florell VH. 1931. Inheritance of type of floret separation and other characters in interspecific crosses in oats. *J. Agric. Res.* 43:365–386.
57. Fuller TC. 1979. Ecology of some California weeds that also occur in Australia. *Proceedings 7th Asian-Pacific Weed Soc. Conf.* pp. 391–393.
58. Gates FC. 1940. *Flora of Kansas*. Manhattan: Agricultural experiment station, Kansas State College of Agriculture and Applied Science. pp. 99–251.
59. Giddings G. 2000. Modelling the spread of pollen from *Lolium perenne*. The implications for the release of wind-pollinated transgenics. *Theor. Appl. Genet.* 100:971–974.
60. Giddings GD, Hamilton NRS, Hayward MD. 1997. The release of genetically modified grasses. 2. The influence of wind direction on pollen dispersal. *Theor. Appl. Genet.* 94:1007–1014.
61. Green PS. 2002. A revision of *Olea* L. (Oleaceae). *Kew Bull.* 57:91–140.
62. Gregor JW. 1928. Pollination and seed production in the rye-grasses (*Lolium perenne* and *Lolium italicum*). *Trans. R. Soc. Edinburgh — Earth Sci.* 55:773–794.
63. Griffiths DJ. 1950. The liability of seed crops of perennial ryegrass (*Lolium perenne*) to contamination by wind-borne pollen. *J. Agric. Sci.* 40:19–38.
64. Griffiths DJ, Pegler RAD. 1966. Cross fertility between diploid and tetraploid ryegrass. *J. Br. Grassland Soc.* 21:162–166.
65. Griffiths DJ, Pegler RAD, Thnguthaisri T. 1971. Cross compatibility between diploid and tetraploid perennial ryegrass (*Lolium perenne* L.). *Euphytica* 20:102–112.
66. Guerin J. www.agwine.adelaide.edu.au/people.
67. Hampton JG. 1991. Temperate herbage seed production. An overview. *J. Appl. Seed Product.* Supplement:2–13.
68. Hardy KR, Otterstein H. 1968. Two scales of convention in the clear atmosphere. In *Proceedings of the International Conference on Cloud Physics*, pp. 534–538. Toronto.
69. Harlan JR. 1992. *Crops and man.* Madison, WI: Amer. Soc. Agron. 284 pp.
70. Hayward MD. 2001. The future of molecular breeding of forage crops. In *Molecular breeding of forage crops*, Spangenberg G, Ed., pp. 325–337. Victoria, Australia: Lorne and Hamilton.
71. Heap IM. 2004. International survey of herbicide-resistant weeds. http://www.weedscience.org.
72. Heaton TC, Klisiewicz JM. 1981. A disease-resistant safflower alloploid from *Carthamus tinctorius* L. X *C. lanatus* L. *Can. J. Plant. Sci.* 61:219–224.
73. Hemaprabha G, Rangasamy SRS. 2001. Genetic similarity among five species of *Saccharum* based on isozyme and RAPD markers. *Indian J. Genet. Plant Breed.* 61:341–347.
74. Henry RD. 1992. Some distributional records and floristic notes for the Illinois vascular flora. *Trans. Ill. State Acad. Sci.* 85:9–15.
75. Hickman JC. 1993. *The Jepson manual: higher plants of California*. Los Angeles: University of California Press. pp. 177, 220–227.
76. Hill HD, Carnahan HL. 1962. *Lolium perenne* L. x induced tetraploid *Festuca elatior* L. and hybrids with *Festuca arundinacea* Schreb. *Crop Sci.* 2:245–248.
77. Hodkinson TR, Chase MW, Lledo MD, Salamin N, Renvoize SA. 2002. Phylogenetics of *Miscanthus, Saccharum* and related genera (Saccharinae, Andropogoneae, Poaceae) based on DNA sequences from ITS nuclear ribosomal DNA and plastid trnL intron and trnL-F intergeneric spacers. *J. Plant Res.* 115:381–392.
78. Hoekstra GJ, Burrows VD, Mather DE. 2003. Inheritance and expression of the naked-grained and fatuoid characters in oat. *Crop Sci.* 43:57–62.
79. Hogarth DM. 1987. Genetics of sugar cane. In *Sugar cane improvement through breeding*, Heinz D, Ed., 603 pp. Amsterdam: Elsevier.

80. Holm L, Doll J, Holm E, Pancho J, Herberger J. 1997. *World weeds. Natural histories and distribution.* New York: John Wiley & Sons, Inc. 1152 pp.
81. Holm LG, Plucknett D, L, Pancho JC, Herberger JP. 1977. *The world's worst weeds. Distribution and biology.* Honolulu: Hawaii University Press.
82. Hoover RF. 1970. *The vascular plants of San Luis Obispo County, California.* Los Angeles: University of California Press. 350 pp.
83. Hovin AW, Buckner RC. 1966. Interspecific and intergeneric hybridization in breeding of *Festuca arundinaceae*. In *Proceedings of the 10th International Grassland Congress*, pp. 686–688.
84. Huskins CL. 1946. Fatuoid, speltoid and related mutations of oats and wheat. *Bot. Rev.* 12:457–514.
85. Hyde HA, Williams DA. 1945. Studies in atmospheric pollen. II Diurnal variation in the incidence of grass pollen. *New Phytol.* 44:83–94.
86. Imrie BC, Knowles PF. 1970. Inheritance studies in interspecific hybrids between *Cartharnus flavescens and C. tinctorius. Crop. Sci.* 10:349–352.
87. Imrie BC, Knowles PF. 1971. Genetic studies of self-incompatibility in *Carthamus flavescens* Spreng. *Crop Sci.* 11:6–9.
88. Jambhale ND. 1994. An attempt of hybridization between *Carthamus tinctorius* and *Helianthus annuus* aided by grafting. *Indian J. Genet.* 54:235–241.
89. Jannoo N, Grivet L, Seguin M, Paulet F, Domaingue R, et al. 1999. Molecular investigation of the genetic base of sugarcane cultivars. *Theor. Appl. Genet.* 99:171–184.
90. Jauhar PP. 1993. Cytogenetics of the *Festuca-Lolium* complex: relevance to breeding. Berlin: Springer-Verlag.
91. Jenkin TJ. 1931. The inter-fertility of *Lolium perenne* and *L. perenne* var. *multiflorum. Bull. Welsh Plant Breed. Stn.*12:121–125.
92. Jenkin TJ. 1931. Self-fertility in perennial rye-grass (*Lolium perenne* L.). *Bull. Welsh Plant Breed. Stn.* 12:100–119.
93. Jenkin TJ. 1933. Interspecific and intergeneric hybrids in herbage grasses. Initial crosses. *J. Genet.* 28:205–264.
94. Jenkin TJ. 1936. Natural selection in relation to the grasses. *Proc. R. Soc. Bot.* 121:52–56.
95. Jenkin TJ. 1954. Interspecific and intergeneric hybrids in herbage grasses V *Lolium rigidum* sens. ampl. with other *Lolium* species. *J. Genet.* 52:252–281.
96. Jenkin TJ. 1954. Interspecific and intergeneric hybrids in herbage grasses VII *Lolium perenne* L. with other *Lolium* species. *J. Genet.* 52:300–317.
97. Jenkin TJ. 1955. Interspecific and intergeneric hybrids in herbage grasses. XVII Further crosses involving *Lolium perenne*. *J. Genet.* 53:442–446.
98. Jensen NF. 1961. Genetics and inheritance in oats: inheritance of morphological and other characters. In *Oats and oats improvement*, Coffman FA, Ed., pp. 125–136. Madison, WI: Amer. Soc. Agron.
99. Johnson LPV. 1935. The inheritance of delayed germination in hybrids of *Avena fatua* and *A. sativa. Can. J. Res. (Sec. C)* 13:367–387.
100. Johnston AM, Tanaka DL, Miller PR, Brandt SA, Nielsen DC, et al. 2002. Oilseed crops for semiarid cropping systems in the northern Great Plains. *Agron. J.* 94:231–240.
101. Kadam BS, Patankar VK. 1942. Natural cross-pollination in safflower. *Indian J. Genet. Plant Breed.* 2:69–70.
102. Kaga A, Vaughan DA, Tomooka N. 2004. Molecular markers in plant breding and crop improvement of *Vigna*. In *Biotechnology in agriculture and forestry*, Lorz H, Wenzel G, Eds., Heidelberg, Germany: Springer-Verlag.
103. Kessler E. 1987. *Carthamus lanatus* L. (Asteraceae: Cynareae) — a potential serious plant pest in Oklahoma. *Proc. Okla. Acad. Sci.* 67:39–43.
104. Khidir MO, Knowles PF. 1970. Cytogenetic studies of *Carthamus* species (Compositae) with 32 pairs of chromosomes. II. Intersectional hybridization. *Can. J. Genet. Cytol.* 12:90–99.
105. Khidir MO, Knowles PF. 1970a. Cytogenetic studies of *Carthamus* species (Compositae) with 32 pairs of chromosomes I. Intersectional hybridization. *Am. J. Bot.* 57:123–129.
106. Knowles PF. 1958. Safflower. *Adv. Agron.* 10:289–323.
107. Knowles PF. 1969. Centers of plant diversity and conservation of crop germplasm: Safflower. *Econ. Bot.* 23:324–349.

108. Knowles PF. 1980. Safflower. In *Hybridization of crop plants*, Fehr WR, Hadley HH, Eds., pp. 535–547. Madison, WI: Am. Soc. Agron.
109. Knowles PF. 1989. Safflower. In *Oil crops of the world: their breeding and utilization*, Robbelen G, Downey RK, Ashri A, Eds., pp. 363–374. New York: McGraw Hill.
110. Knowles PF, Ashri A. 1995. Safflower: *Carthamus tinctorius* (Compositae). In *Evolution of crop plants*, Smartt J, Simmonds NW, Eds., pp. 47–50. Harlow, UK: Longman.
111. Knowles PF, Schank SC. 1964. Artificial hybrids of *Carthamus nitidus* Boiss. and *C. tinctorius* L. (Compositae). *Crop Sci.* 4:596–599.
112. Kotecha A, Zimmerman LH. 1978. Genetics of seed dormancy and its association with other traits in safflower. *Crop Sci.* 18:1003–1007.
113. Kumar H. 1991. Cytogenetics of safflower. In *Chromosome engineering in plants: genetics, breeding, evolution, Part B, Developments in plant genetics and breeding*, Tsuchiya T, Gupta PK, Eds., pp. 251–277. Amsterdam: Elsevier.
114. Lacey DJ, Wellner N, Beaudoin F, Napier JA, Shewry PR. 1998. Secondary structure of oleosins in oil bodies isolated from seeds of safflower (*Carthamus tinctorius* L.) and sunflower (*Helianthus annuus* L.). *Biochem. J.* 334:469–477.
115. Ladizinsky G. 1970. Chromosomal rearrangements in the hexaploid oats. *Heredity* 25:457–461.
116. Leibbrandt NB, Snyman SJ. 2003. Stability of gene expression and agronomic performance of a transgenic herbicide-resistant sugarcane line in South Africa. *Crop Sci.* 43:671–677.
117. Lima MLA, Garcia AAF, Oliveira KM, Matsuoka S, Arizono H, et al. 2002. Analysis of genetic similarity detected by AFLP and coefficient of parentage among genotypes of sugar cane (*Saccharum* spp.). *Theor. Appl. Genet.* 104:30–38.
118. López-González G. 1989. Acerca de la clasificación natural del género *Carthamus* L., s.l. *Anal. Jardín Bot. Madrid* 47:11–34.
119. Lyrene PM, Shands HL. 1974. Groat protein percentage in *Avena sativa* L. fatuoids and in a fatuoid x *A. sterilis* L. cross. *Crop Sci.* 14:765–767.
120. Ma JKC, Pascal MW, Drake MW, Christou P. 2003. The production of recombinant pharmaceutical proteins in plants. *Nature Rev. Genet.* 4:794–805.
121. Marticorena C, Quezada M. 1985. *Catálog de la flora vascular de Chile. Gayana; Botanica. Universidad de Concepcion Chile.* 42:1–2. Concepcion, Chile: Universidad de Concepcion. pp. 1–2.
122. Martin WC, Hutchins Cr. 1981. A flora of New Mexico, Monticello, NY: Lubrecht & Cramer.
123. McPherson MA, Good AG, Topinka AKC, Hall LM. 2004. Theoretical hybridization potential of transgenic safflower (*Carthamus tinctorius* L.) with weedy relatives in the New World. *Can J. Plant Sci.* 84:923–934.
124. Mekuria GT, Collins G, Sedgley M. 2002. Genetic diversity within an isolated olive (*Olea europaea* L.) population in relation to feral spread. *Sci. Hort.* 94:91–105.
125. Ming R, Wang YW, Draye X, Moore PH, Irvine JE, Paterson AH. 2002. Molecular dissection of complex traits in autopolyploids: mapping QTLs affecting sugar yield and related traits in sugarcane. *Theor. Appl. Genet.* 105:332–345.
126. Moloney MM. 2000. Seeds as repositories of recombinant proteins in molecular farming. *Korean J. Plant Tissue Cult.* 27:283–297.
127. Munz PA, Keck DD. 1968. *A California flora,* pp. 1273. Berkeley: University of California Press.
128. Murray BG, Morrison IN, Friesen LF. 2002. Pollen-mediated gene flow in wild oat. *Weed Sci.* 50:321–325.
129. Nair NV, Nair S, Sreenivasan TV, Mohan M. 1999. Analysis of genetic diversity and phylogeny in *Saccharum* and related genera using RAPD markers. *Genet. Res. Crop Evol.* 46:73–79.
130. Nilsson-Ehle H. 1921. Fortgesetzte untersuchungen über fatuoidmutationen bein hafer. *Hereditas* 2:401–409.
131. Nishiyama I. 1987. Progressive differentiation of spikelet types in relation to ploidy in the genus *Avena. Plant Breed.* 99:98–106.
132. Nishiyama I, Yabuno T. 1978. Basic genetic behaviors of eu- and aneuploid fatuoid oats. *Jpn. J. Genet.* 53:353–360.
133. Oertel C, Fuchs J, Matzk F. 1996. Successful hybridization between *Lolium* and *Dactylis*. *Plant Breed.* 115:101–105.

134. Pacini E, Franchi GG, Lisci M, Nepi M. 1997. Pollen viability related to type of pollination in six angiosperm species. *Ann. Bot.* 80:83–87.
135. Pan YB, Burner DM, Legendre BL. 2000. An assessment of the phylogenetic relationship among sugarcane and related taxa based on the nucleotide sequence of 5S rRNA intergeneric spacers. *Genetica* 108:285–295.
136. Parmenter DL, Boothe JG, vanRooijen GJH, Yeung EC, Moloney MM. 1995. Production of biologically active hirudin in plant seeds using oleosin partitioning. *Plant Mol. Biol.* 29:1167–1180.
137. Pritchard JK, Stephens M, Donnelly P. 2000. Inference of population structure using multilocus genotype data. *Genetics* 155:945–959.
138. Quinlivan BJ, Pierce JR. 1969. Dormancy of life span of saffron thistle seeds. *West Aust. J. Agric.* 10:177.
139. Ram B, Sreenivasan TV, Sahi BK, Singh N. 2001. Introgression of low temperature tolerance and red rot resistance from *Erianthus* in sugarcane. *Euphytica* 122:145–153.
140. Rieger MA, Lamond M, Preston C, Powles SB, Roush RT. 2002. Pollen-mediated movement of herbicide resistance between commercial canola fields. *Science* 296:2386–2388.
141. Roach BT. 1995. Sugar canes. In *Evolution of crop plants*, Smartt J, Simmonds NW, Eds., pp. 160–166. Harlow, UK: Longman.
142. Rognli OA, Nilsson NO, Nurminiemi M. 2000. Effects of distance and pollen competition on gene flow in the wind-pollinated grass *Festuca pratensis* Huds. *Heredity* 85:550–560.
143. Rognli OA, Tomiuk J, Tufto J, Nurminiemi M, Jørgensen RB. 1999. Gene flow from transgenic crops to wild populations. *Nordtest Rep.* 409:1–42.
144. Rydberg PA. 1971. *Flora of the prairies and plains of central North America.* New York: Dover. 884 pp.
145. Schank SC, Knowles PF. 1964. Cytogenetics of hybrids of *Carthamus* species (Compositae) with ten pairs of chromosomes. *Am. J. Bot.* 1:1093–1102.
146. Selvi A, Nair NV, Balasundaram N, Mohapatra T. 2003. Evaluation of maize microsatellites markers for genetic diversity analysis and fingerprinting in sugarcane. *Genome* 46:394–403.
147. Setamou M, Bernal JS, Legaspi JC, Mirkov TE. 2002. Effects of snowdrop lectin (*Galanthus rivalis* agglutinin) expressed in transgenic sugarcane on fitness of *Cotesia flavipes* (Hymenoptera Braconidae), a parasitoid of the non-target pest *Diatraea saccharalis* (Lepidoptera: Crambidae). *Ann. Entomol. Soc. Am.* 95:75–83.
148. Setamou M, Bernal JS, Legaspi JC, Mirkov TE, Legaspi BC. 2002. Evaluation of lectin-expressing transgenic sugarcane against stalkborers (Lepidoptera : Pyralidae): effects on life history parameters. *J. Econ. Ent.* 95:469–477.
149. Sharma MP, Vanden Born WH. 1978. The biology of Canadian weeds: 27. *Avena fatua* L. *Can. J. Plant Sci.* 58:141–157.
150. Shaw RJ. 1989. *Vascular plants of northern Utah: an identification manual.* Logan, UT: State University Press. 412 pp.
151. Shorter R, Gibson P, Frey KJ. 1978. Outcrossing rates in oat species crosses (*Avena sativa* L. *A. sterilis* L.). *Crop Sci.* 18:877–878.
152. Seehalak W, Tomooka N, Waranyuwat A, Thipyapong P, Laosuwan P, et al. Genetic diversity of *Vigna* germplasm from Thailand and neighboring regions revealed by AFLP analysis. *Genet. Resour. Crop Evol.* Submitted.
153. Smart IJ, Tuddenham WG, Knox RB. 1979. Aerobiology of grass pollen in city atmosphere of Melbourne: effects of weather parameters and pollen sources. *Aust. J. Bot.* 27:333–342.
154. Smith JR. 1996. *Safflower.* Champaign, IL: AOCS Press. pp. 1–66, 142–184.
155. Soleri D. www.es.ucsb.edu/faculty/soleri.php.
156. Spangenberg G, Kalla R, Lidgett A, Sawbridge T, Ong EK, John U. 2001. Breeding forage plants in the genomic era. In *Molecular breeding of forage crops*, Spangenberg G, Ed., pp. 1–39. Dordrecht, Netherlands: Kluwer.
157. Takahashi C, Shibata F, Hizume M. 2002. FISH detection of sugar cane centromeric repetitive sequences in the chromosomes of two *Miscanthus* species. *Chromosome Sci.* 6:1–5.
158. Terral JF, Alonso N, Buxo R, Capdevila I, Chatti N, et al. 2004. Historical biogeography of olive domestication (*Olea europaea* L.) as revealed by geometrical morphometry applied to biological and archaeological material. *J. Biogeogr.* 31:63–77.
159. Terrell EE. 1966. Taxonomic implications of genetics in ryegrass (*Lolium*). *Bot. Rev.* 32:138–163.

160. Terrell EE. 1968. A taxonomic review of the genus *Lolium*. In *Technical Bulletin, 1392*, pp. 1–65. Washington, DC: USDA.
161. Van Heeswijk R, Hutchinson J, Kaul V, McDonald G, Woodward J. 1994. The role of biotechnology in perennial ryegrass improvement for temperate pastures. *N. Z. J. Agric. Res.* 37:427–438.
162. Vilatersana R, Susanna A, Garcia-Jacas N, Garnatje T. 2000. Generic delimitation and phylogeny of the Carduncellus-Carthamus complex (Asteraceae) based on ITS sequences. *Plant Syst. Evol.* 221:89–105.
163. Vincent MA, Cusick AW. 1998. New records of alien species in the Ohio vascular flora. *Ohio J. Sci.* 98:10–17.
164. Wang XW, Kaga A, Tomooka N, Vaughan DA. 2004. The development of SSR markers by a new method in plants and their application to gene flow studies in azuki bean (*Vigna angularis* (Willd) Ohwi and Ohashi. *Theor. Appl. Genet.* 109:352–360.
165. Warringa JW. 1997. *Physiological constraints to seed growth in perennial ryegrass*. The Hague, Netherlands: Wageningen: CIP-Gegevins Koninklijke Bibliotheek.
166. Watson L, Dallwitz MJ. 1999. Grass genera of the world — *Festuca* L. http://biodiversity.uno.edu/delta/grass/www/festuca.htm.
167. Watson L, Dallwitz MJ. 1999. Grass genera of the world — *Lolium* L. http://biodiversity.uno.edu/delta/grass/www/lolium.htm.
168. Wilkins PW, Thorogood D. 1992. Breakdown of self-incompatibility in perennial ryegrass at high temperature and its uses in breeding. *Euphytica* 64:65–69.
169. Wright GC, McWilliam JR, Whalley RDB. 1980. Effects of light and leaching on germination of saffron thistle (*Carthamus lanatus* L.). *Aust. J. Plant Physiol.* 7:587–594.
170. Xu RQ, Tomooka N, Vaughan DA. 2000. AFLP markers for characterizing the azuki bean complex. *Crop Sci.* 40:808–815.
171. Xu RQ, Tomooka N, Vaughan DA, Doi K. 2000. The *Vigna angularis* complex: genetic variation and relationships revealed by RAPD analysis, and their implications for in situ conservation and domestication. *Genet. Resour. Crop Evol.* 47:123–134.
172. Yang Q, Hanson L, Bennet MD, Leitch IJ. 1999. Genome structure and evolution in the allohexaploid weed *Avena fatua* L. (Poaceae). *Genome* 42:512–518.
173. Young S, Humphreys MO, Abberton MT, Robbins MP, Webb KJ. 1999. *The risks associated with the introduction of GM forage grasses and forage legumes*, Report to Ministry of Agriculture Forestry and Fisheries.
174. Zhang YW, Long HS, Fang YH, Yao YG, Cai Q, Zhang YP. 2002. Sequence variation of *rbcL* gene and evolution of *Saccharum* and related species. *Acta Bot. Yunnanica* 24:29–36.
175. Zhou X, Jellen EN, Murphy JP. 1999. Progenitor germplasm of domesticated hexaploid oat. *Crop Sci.* 39:1208–1214.
176. Zimmerman LH. 1972. Variation and selection for preharvest seed dormancy in safflower. *Crop Sci.* 12:33–34.
177. Zohary D, Hopf M. 1993. *Domestication of plants in the old world*. Oxford, UK: Oxford University Press.
178. Zong XX, Kaga A, Tomooka N, Wang XW, Han OK, Vaughan D. 2003. The genetic diversity of the *Vigna angularis* complex in Asia. *Genome* 46:647–658.

16 Asian Rice and Weedy Rice — Evolutionary Perspectives

Duncan A. Vaughan, Paulino L. Sanchez, Jun Ushiki, Akito Kaga, and Norihiko Tomooka

16.1 INTRODUCTION

Rice feeds more of humanity than any other crop and its importance is clearly reflected with the UN declaring 2004 as the International Year of Rice. Any factor that threatens the genetic base of rice or its productive potential has implications for global food security. Weedy rice and gene introgression from cultivated rice to wild rice populations are factors that threaten the genetic base and productivity of rice.

The rice crop complex, which is the AA genome gene pool of *Oryza*, includes the cultigen and its wild and weedy relatives. Although AA genome *Oryza* species are found worldwide, the focus is specifically on the Asian region in this chapter. The use of the word *rice* refers only to Asian rice, *Oryza sativa*, and not African rice, *O. glaberrima*.

The weedy rice members of the rice crop complex are either pre- or post-rice domestication ecotypes. In the context of this chapter, weedy rice is defined as rice unwanted by humans for which most spikelets are not captured during harvesting and that is specifically adapted to human-disturbed habitats. Weedy rice affects both yield and quality of harvested rice. Volunteer rice, which is not discussed, is distinguished from weedy rice by its non-shattering nature and as a problem can be solved by continuous planting of pure seeds or seedlings. Volunteer rice is usually unwanted because it affects the quality of harvested rice. Therefore, understanding weedy rice requires an understanding of rice evolution, domestication, and diversification, because weedy rice evolved in parallel with rice. These issues are discussed first before focusing on weedy rice.

16.1.1 The Genus *Oryza* in Relation to Rice

16.1.1.1 The Antiquity of *Oryza*

The most recent information suggests that grasses evolved between 55 and 70 million years ago (MYA) (31), well after the date that Gondwanaland broke up at 130 MYA (10). Kellogg (31) further suggested that grasses probably diversified 55 to 32 MYA. Thus, it seems unreasonable to assume that ancestral *Oryza* emerged during this period.

The estimated dates for diversification of the different sections of *Oryza* also vary (63), but most likely occurred 10 to 20 MYA (54,62).

Oryza is the only major cereal that belongs in the Bambusoideae, the most primitive of extant grasses (15). Other major cereals belong to either the Pooideae (barley, wheat, and oats) or Panicoideae (sorghum and maize). Therefore, rice has characteristics that set it apart from other cereals. Among these characteristics is that rice has a small genome, which is the best sequenced among cereals and is a basic genomic standard for comparison with the genomes of other cereals (3).

16.1.1.2 *Oryza* Genome and Species Diversity as It Relates to Rice

The genus *Oryza* consists of 23 species, divided into 4 taxonomic sections and 9 identified genomes. Most *Oryza* species can be grouped into 4 species complexes (Table 16.1). Section *Sativa* is split into 2 complexes — the *O. sativa* complex and *O. officinalis* complex. The *O. sativa* complex contains about 7 morphological species; all are diploid and share the AA genome. Despite many reproductive barriers among AA genome taxa (43), hybrids between these species can be produced without resorting to embryo rescue, as they all have some degree of fertility. There are reports of natural hybrids between many of these species, such as *O. sativa* and *O. longistaminata* (45). Thus, all AA genome *Oryza* belong to a single biological species.

The *Oryza officinalis* complex has species with four genomes (BB, CC, DD, EE), with both diploid and allotetraploid species. Although the global pan-tropical distribution of this species complex and the *O. sativa* complex are similar, genome diversification has only been a feature of the *O. officinalis* complex. An example of genome diversification within the *O. officinalis* complex is *O. australiensis.* This species has the largest genome among *Oryza* species, primarily due to a large number of copies of part of a retrotransposon (*RIRE1*) (58).

It is not clear why genetic divergence of morphological species in the *Oryza sativa* complex has not resulted in strong reproductive barriers. There is no present explanation for why the *O. sativa* species complex has not undergone comparable genomic diversification as the *O. officinalis* complex, even though it is thought to have evolved over a similar period.

16.1.1.3 The Perennial–Annual Axis of Diversity in the Primary Gene Pool of Rice AA Genome (Axis 1)

The primary evolutionary forces acting on populations of the *Oryza sativa* complex leading to diversification are related to local variation in fluctuations of water level in their habitats. The effect of hydrological regime can be illustrated by variation in AA genome of wild rice in Papua New Guinea. In the Sepik River basin of northern Papua New Guinea, where the river level does not fluctuate greatly, *O. rufipogon* is a strongly vegetative, perennial, and produces few seeds. In Lake Murray, southern Papua New Guinea, the lake level rises and falls greatly during the year leaving shore side populations of *O. rufipogon* above the water line, but in wet soils, during the dry season. This perennial ecotype produces abundant seeds.

Wild AA genome *Oryza* species that grow in permanently deep water have perennial characteristics, whereas populations growing in seasonally dry ponds have evolved annual characteristics. Variations in hydrological regimes in mainland Asia resulted in the evolution of a perennial–annual continuum that is characterized by changes in the breeding system from perennial outcrossing to annual inbreeding.

Thus, cultivated rice arose from a gene pool of aquatic grasses that primarily diversified in relation to the hydrological regime and this resulted in variation in breeding systems where they grow. Although rice reflects the life cycle and breeding system of an annual wild rice, it is remarkably diverse in response to hydrological regime, with varieties growing in fields that flood to 1 to 3.5 m (deep water rice) and varieties in *unbunded* (field lacking surroundings to retain water) as well as in dry land fields (upland rice) (20).

16.1.1.4 Distribution of the Wild AA Genome *Oryza* Species

Even though AA genome *Oryza* species have a pan-tropical distribution, there are many areas in the tropics where they cannot be found. In the Philippines, there is only one known population of wild AA genome *Oryza*, *O. rufipogon.* This population is in a small volcanic crater lake on the southern island of Mindanao. Most of Tamil Nadu, India, has no AA genome wild rice although across the Jaffna Straight in Sri Lanka it is common. Thus, the role of wild *Oryza* species in the evolution of weedy rice needs to be viewed in relation to their distribution and that of the cultigen

TABLE 16.1
***Oryza* Species: Taxonomic Sections and Complexes, Genome Group, Those Reported to Be Weeds in Rice Fields, Distribution**

Complex Species	Genome Group	Weed Species in Rice[1]	Distribution
		Section Oryza	
Oryza sativa complex			
Oryza sativa L.	AA	Yes	Cultigen, worldwide
O. rufipogon sensu lacto (syn: *O. nivara* for the annual form *O. rufipogon* sensu stricto for the perennial form)	AA	Yes	Annual forms only on continental Asia, perennial forms continental and insular Asia to New Guinea and northern Australia Probably also introduced into Latin America
O. glaberrima Steud.	AA	Yes	West Africa
O. barthii A. Chev.	AA	Yes	Sub-Saharan Africa
O. longistaminata Chev. et Roehr.	AA	Yes	Sub-Saharan Africa
O. meridionalis Ng[2]	AA		Australia, New Guinea
O. glumaepatula Steud.[2]	AA		Latin America
O. officinalis complex			
O. officinalis Wall ex Watt	CC	One report	Asia to New Guinea and northern Australia
O. minuta J.S.Presl ex C.B.Presl.	BBCC		Philippines and New Guinea
O. rhizomatis Vaughan	CC		Sri Lanka
O. eichingeri Peter	CC		West, East Africa and Sri Lanka
O. malapuzhaensis Krishnaswamy and Chandrasakaran	BBCC		Southern India
O. punctata Kotschy ex Steud.	BB, BBCC	Yes	Sub-Saharan Africa
O. latifolia Desv.	CCDD	Yes	Latin America
O. alta Swallen	CCDD		Latin America
O. grandiglumis (Doell.) Prod.	CCDD		South America
O. australiensis Domin	EE		Australia
		Section Ridleyanae Tateoka	
O. schlechteri Pilger	Unknown		New Guinea
O. ridleyi complex			
O. ridleyi Hook.	HHJJ		Southeast Asia to New Guinea
O. longiglumis Jansen	HHJJ		New Guinea
		Section Granulata Roschev.	
O. granulata complex			
O. granulata Nees et Arn ex Watt	GG		Asia (mainly continental)
O. meyeriana (Zoll. et Mor.ex Steud.)Baill.	GG		Asia (mainly insular) and New Caledonia
		Section Brachyantha B.R. Lu	
O. brachyantha Chev. et Roehr.	FF		Sub-Saharan Africa

[1] From Oka (45), Second (53,55), and Vaughan et al. (61). The species is unknown as a weed of rice, unless otherwise stated.
[2] These species could be a weed in rice fields, but the authors are not aware that they in fact are.

Source: Modified from Vaughan et al. (64).

(Table 16.1). In addition, the flowering time of sympatric wild and cultivated rice species is an important factor determining whether gene flow can occur among them. Around Alor Setar, Malaysia, wild *O. rufipogon* is common, but it flowers much later than cultivated rice. Hence, *O. rufipogon* is not implicated in the emergence of weedy rice in the region. Conversely, on the other side of Malaysian Peninsula in Kelantan, introgression between wild and cultivated rice does occur.

16.2 RICE DOMESTICATION

16.2.1 Domestication Processes

16.2.1.1 The Wild Rice–Domesticated Rice Evolutionary Axis (Axis 2)

Crop domestication is evolution under a selection regime of harvesting and of planting what is harvested. Domestication involves ecological adaptation to human-made habitats and a genetic architecture that makes plant populations dependent on human interference in such habitats. Even though domestication is the result, in areas where crops exist in a complex with wild and weedy forms, all intermediary steps toward domestication may be expected (25). In contrast, wild relatives of crops have a genetic architecture that allows them to thrive outside the sphere of human activity. Weedy relatives of crops have a genetic architecture that allows them to thrive independently of humans for propagation, but are largely dependent on human-disturbed habitats to grow.

An increase in seed number per panicle has been more important than an increase in seed size during the domestication of rice. The general planting pattern of rice that involved direct seeding on the soil surface or direct seeding in seedbeds does not provide a selection pressure for increased seed size.

The domestication process in rice can have the following steps:

- Gathering wild rice — In parts of southern Asia, wild rice has religious significance and is harvested today for religious purposes. Wild rice has also been used in times of famine in the recent past in some countries such as Cambodia.
- Cultivation — In parts of India, semi-domesticated rice is grown. Primitive cultivars found in western India appear similar to typical annual wild rice but are more homogeneous, flower earlier, and have a more erect habit and have shorter awns than wild populations (56). Cultivation results in a number of changes to wild plants. The most important consequence is the amount of food obtained compared to gathering (17). Semi-domesticated rice in western India has a more synchronous maturity than wild rice, which enhances yield.
- Repeated harvesting and planting over successive generations provides the selection pressures that lead to domestication. This quickly favors non-shattering types, more uniform ripening, and non-dormant types or types that lose dormancy between harvesting and subsequent planting.

16.2.1.2 Indica–Japonica Evolutionary Axis (Axis 3)

The domestication of rice has characteristics not found in the other major cereals. Principal among these is that rice evolved from the wild rice gene pool into two eco-genetic varietal types called indica and japonica. The vast majority of rice varieties can be readily divided into these two eco-genetic types. Traditional indica and japonica varieties have distinct differences in morphological and adaptive traits. In addition, a series of diagnostic characters enable the two groups to be reliably distinguished (45) (Table 16.2). The majority of varieties in the hot tropics are indica varieties whereas japonica varieties are generally grown in temperate regions or at higher altitude in the tropics.

Indica and japonica varieties have differences in chloroplast types (11,16,28) and wild rice has differences, such as isozyme variation, related to indica–japonica differentiation (38,52). Several studies have suggested that perennial type of wild rice is similar to japonica varieties and annual wild rice is similar to indica varieties (12,38,74). Thus, it is now generally considered that indica and japonica varietal groups derive from different domestication lineages in wild rice, rather than within cultivated rice.

TABLE 16.2
Trait Differences that Help to Differentiate Indica from Japonica Varieties and Their Characteristics

Trait	Indica [allele]	Japonica [allele]	Gene[a]	Linkage Group	Discrete/ Continuous	Diagnostic/ Adaptive	Ref.
			Biochemical (Isozymes), Chemical Response				
Acid phosphatase 1	[1(-4)]	[2(+9)]	*Acp*-1	12	d	d	49
Catalase 1	[1]	[2]	*Cat*-1	6	d	d	49
Esterase 2	[1,2]	[0]	*Est*-2	6	d	d	49
Peroxidase 2	[1(4C)]	[0]	*Pox*-2	12	d	d	49
Phosphoglucose isomerase 1	[1]	[2]	*Pgi*-1	3	d	d	49
Phosophoglucose isomerase 2	[2]	[1]	*Pgi*-2	6	d	d	49
Amino peptidase 2	[2]	[1]	*Amp*-2	8	d	d	50
Phenol reaction (see also black hull)	Positive	Negative	*Bhc* (Ph)	4	d	d	50
$KClO_3$ susceptibility	Susceptible	Resistant			d	d	50
			Molecular				
p-SINE1 loci *r55,r54,r60, r210*	Present	Absent			d	d	44
p-SINE1 loci *r505,r506,r507*	Absent	Present			d	d	44
ORF100 delection region (chloroplast DNA)	Deletion	Nondeletion type			d	d	11
RAPD primer CMN-A32 (nuclear DNA)	Negative	Positive			d	d	57
			Morphological				
Apiculus hair length	Short	Long	*Aph*	6	d/c	d	39
Red pericarp	Occasional	Uncommon	*Rc*	7	d	?	39
Black hull	Sometimes	None	*Bha, Bhb, Bhc*		d	?	50
Spikelet length/width ratio	Larger	Smaller	*wgl*(t)	5	c	a	45
Awnedness	Rare	Sometimes	Many	Many	d/c	a	50
Mesocotyl length	Shorter	Longer			c	a	45
Hull pubescence	Dense	Not dense	*Hg*	3		?	13
Leaf pubescence	Dense	Not dense	*Hla* (6), *Hlb*, *gl*1 (5), *gl*2	6,5 plus		a	13
			Quality Related				
Digestability of endosperm in KOH	Lower	Higher			c	a	45
Hardening of endosperm time	Faster	Slower			c	a	45
			Environment Related				
Low temperature sensitivity (seedling stage)	Sensitive	Insensitive	*Cts*1, *Cts*2(t)	4	d	a	39
Drought resistance (seedlings)	High	Low			d	a	39
Tillering (reduced tiller number)	Many	Fewer	*rcn*	4,6		a	8
Lodging resistance	Low	High				a	45
Nitrogen response to grain yield	Low	High				a	45
Competitive ability	High	Low				a	45

TABLE 16.2 (continued)
Trait Differences that Help to Differentiate Indica from Japonica Varieties and Their Characteristics

Trait	Indica [allele]	Japonica [allele]	Gene[a]	Linkage Group	Discrete/ Continuous	Diagnostic/ Adaptive	Ref.
Temp. response of seed germination and growth rate	High	Low				a	45
Seed longevity in storage	Long	Short				a	45
Ease of shattering	Easier	Harder	Many	Many	d/c	a	8

[a] Information from the Oryzabase Web site of the National Institute of Genetics, Japan (http://www.shigen.nig.ac.jp/rice/oryzabase/genes/).

16.2.2 Places of Domestication

The progenitors of rice are and continue to be widespread in Asia. Whenever a progenitor is widespread, there is always the possibility of multiple domestication events especially of a wild plant so productive that it is still harvested long after it was first domesticated (25). Current opinion now suggests that rice was domesticated more than once in Asia and probably in widely separated geographic locations (12,21,37,51). In particular, this important contention is supported by the diverse non-nuclear (cytoplasmic) genomes among *indica* landraces and between *japonica* and *indica* varietal groups (51). Independent domestication of indica and japonica varietal groups would help explain the rather discontinuous nature of variation between these two groups. The earliest evidence of domesticated rice, dated to about 8000 before present (BP), is from an archaeological site in the middle Yangtze Valley (Figure 16.1). This area is about 500 km north of where most wild rice populations in China grow today. However, archeo-historical data reveal that wild rice grew well north of the Yangtze River at the time when rice was first being domesticated (9) (Figure 16.1). The early domestication of rice helps explain why the center of rice diversity in Asia and

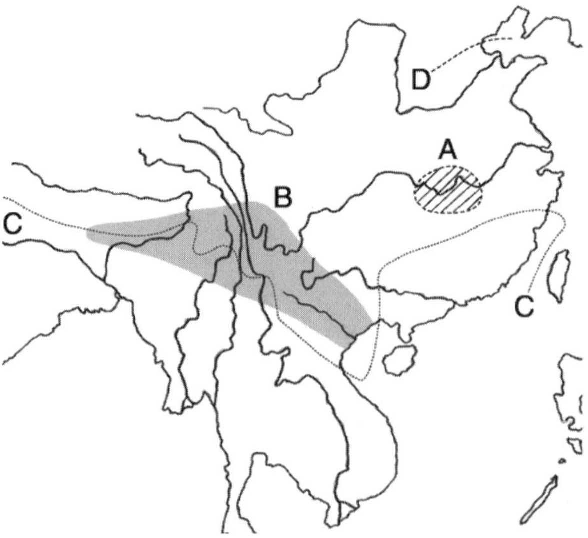

FIGURE 16.1 The domestication of rice: A. Place of first domestication; B. Center of rice diversity; C. present day (····); D. past (---) northern limit of wild rice distribution. (Modified from Vaughan et al. (65).)

the area currently believed to be the area of first rice domestication do not coincide (65). Rice was also domesticated in southern Asia and possibly also mainland southeastern Asia, probably a few millennia later.

16.2.3 Genetics of Domestication

16.2.3.1 Neutral Alleles and Domestication

Domestication results in a loss of diversity compared to that of the wild progenitor gene pool. The loss of diversity for neutral alleles or non-domestication related traits is about 30% for rice and most other cereals (5). It is presumed and seems logical, that this loss involved mass selection within large wild populations (5). Lack of strong selection in large populations during domestication would help explain why useful genes for rice productivity escaped selection. An abundance of yield-related traits present in wild rice populations that can be exploited by rice breeders have recently been found (72).

16.2.3.2 Domestication Related Traits and Domestication

Genomic regions associated with domestication traits have a lower level of variation in the cultigen than in their wild relatives. For example, the promoter for *teosinte branched 1*, the gene responsible for the conversion of teosinte from a multiple tillering plant to a single tillering maize, has 61-fold less diversity than in its closest wild relative (69). However, the considerable loss of genetic diversity associated with particular genes in maize has little effect on neighboring genes (14).

16.2.3.3 Gene Clusters (Closely Linked Genes)

Traits related to domestication of rice and other crops often appear to be clustered at specific locations on the genome (7,18). For example, quantitative trait loci (QTLs) for domestication-related shattering, dormancy, awn length, and seed fertility map to the distal end of rice chromosome 8 (7) (Figure 16.2). This genomic region also has a series of QTLs associated with indica–japonica differentiation, potassium chlorate resistance, apiculus hair length, germination rate, low temperature resistance, and distance from panicle base to lowest branch (7) (Figure 16.2). These clusters of domestication-related traits on the genome provide the genetic basis for the "domestication syndrome" concept (23).

16.2.3.4 Gene Association (Unlinked Genes or Cryptic Linkage)

Although gene clusters resulting from close linkage have been reported in rice, there are also genes associated with evolutionary trends that are not linked in the rice crop complex (40). Genes related to indica and japonica varieties are found on many rice chromosomes (Table 16.2). Their association can be explained by selection of co-adapted gene sets. Losses of shattering and secondary dormancy are the most important trait changes in the domestication of cereals. There have been several mapping studies that have analyzed these two traits in different crosses involving rice (4,6,7,33,68,73). These studies reveal that both traits are controlled by many loci. Loci related to shattering have been reported on 8 and for dormancy on 10 of the 12 chromosomes of *Oryza* AA genome species (Figure 16.3). Five loci associated with shattering and dormancy appear to be closely linked to one another (Figure 16.3). However, because shattering and dormancy segregate from each other in F_2 populations of crosses between wild and cultivated rice, the loci for these traits that have been selected in cultivated rice are not closely linked. This suggests that these characters and possibly others associated with domestication results from natural selection of co-adapted traits (6).

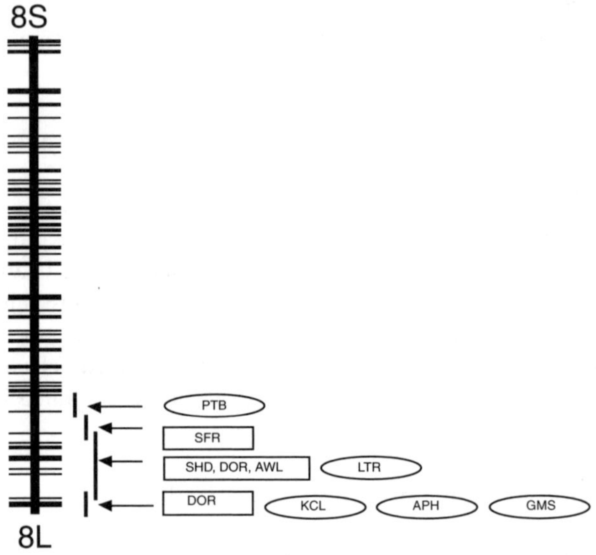

FIGURE 16.2 Clustering of domestication [▭] and indica/japonica [⬭] related traits at the distal end of chromosome 8 in rice APH — apiculus hair length; AWL — awn length; DOR — dormancy; GMS — germination speed; $KClO_3$ potassium chlorate resistance, LTR — low temperature resistance; PTB — panicle base to lowest branch; SFR — seed fertility; SHD — seed shedding. (Based on data in Cai and Morishima (7).)

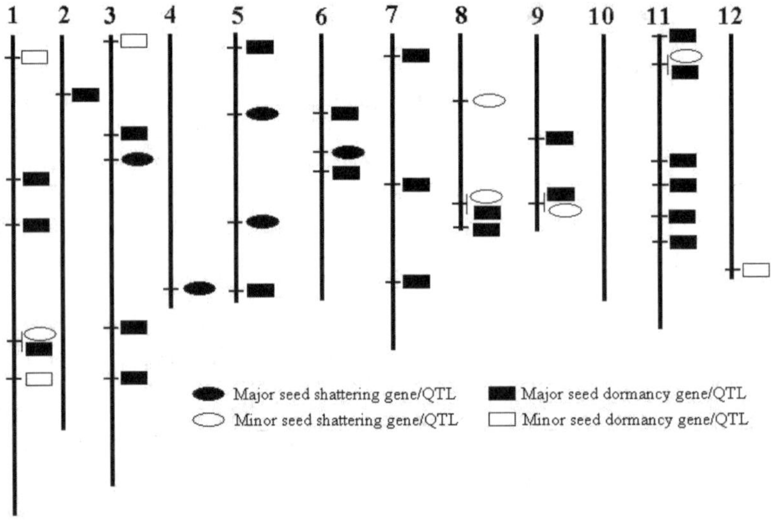

FIGURE 16.3 Location of major and minor loci (genes or QTL) for shattering and dormancy in AA genome *Oryza* species. Major loci are considered those highly expressed generally having a log likelihood (LOD), score of >5% and phenotypic variance of >15%. (Based on Cai and Morishima (6,7), Lin et al. (33), Miura et al. (35), Nagai et al. (42), Sanchez et al. (48), Wan et al. (68), and Xiong et al. (73).)

16.3 DIVERSIFICATION OF RICE

The number of rice accessions in gene banks exceeds that of any other crop and reflects the extraordinary diversity of the cultivated rice gene pool. The analysis of this diversity points to a broad zone from Nepal to Vietnam, where cultivated traditional landraces of rice are most diverse or have been in the recent past (Figure 16.1).

TABLE 16.3
Rice Varietal Diversity in Two Villages in Northwest Vietnam

Ethnic Group: Thai				Ethnic Group: Humong			
Village: Muong Pon, Altitude 450 m				Village: Sothan, Altitude 1175 m			
Variety Name	Culture Type	Variety Group	Amylose %	Variety Name	Culture Type	Variety Group	Amylose %
Tu Vong	Upland	Japonica	2	Blau Ta	Upland	Japonica	21
Khao Pit	Upland	Japonica	2	Blau La	Rainfed	Indica	26
Khao Lech	Upland	Japonica	3	Ble Blau	Rainfed	Indica	3
Khao lep Trong	Upland	Japonica	2	Blau Tan	Upland	Japonica	3
Khao Coi Noi	Upland	Japonica	2	Blau Blau	Rainfed	Japonica	1
Pu Loah	Upland	Indica	2	Talenoi	Rainfed	Indica	17
Khao Coi Loi	Upland	Japonica	3	Blau Bla	Rainfed	Not determined	1
Khao Pen Tien	Upland	Japonica	4				
Khao Tu Rong	Rainfed	Japonica	2				
Khao Ten Lai	Rainfed	Japonica	1				

Two interacting factors can explain why rice is more diverse than other crops. The first is the eco-geographic diversity of the center of rice diversity, which reflects natural selection. Yunnan province, China, and neighboring areas are at the center of this zone. Here rivers tumble from the Tibetan plateau and then flow north, east, south, and west. These rivers have cut steep valleys that have long been used for cultivating rice. The eco-geographic diversity of this region exceeds any other where a major crop diversified.

The second factor is a high diversity of ethnic groups. Each ethnic group has its own preference for eating quality of rice and each has applied selection pressure to the rice gene pool adapting it to the different cultural requirements and environments where these ethnic groups live.

The landrace diversity in two nearby villages in Lai Chau province, Vietnam is illustrative of how this combination of natural and human selection has affected rice diversity (Table 16.3). Landraces from the ethnic Thai village of Muong Pon (21:30ŸN, 103ŸE, altitude 450 m) and the ethnic Humong village of Sothan (22:30ŸN, 103:40ŸE, altitude 1175 m) were collected in 1989. The Thai ethnic group prefers sticky rice; hence, all varieties from that village have low amylose content. The Humong ethnic group consumes both sticky and non-sticky rice. Because each village across Asia often has (or had) 8 to 10 traditional rice varieties, it quickly becomes apparent why the number of local rice landraces in Asia is so great. Traditional varieties are far less common now than a few decades ago. Landraces are primarily confined to adverse rice growing environments where modern varieties are not suitable. For example, Sri Lanka, a country famous for its traditional rice diversity, in 2003, grew landrace cultivars on less than 1% of the rice land (Rajapakse, Plant Genetic Resources Center, Perideniya, Sri Lanka personal communication, 2004) compared to about 28% of 20 years ago (22).

16.3.1 Evolution of Weedy Rice

The process of evolution can be summarized as the differential survival of genetic variation by selection in a diploid plant such as rice. Genetic variation is the result of mutation that spreads and fluctuates through populations as a result of hybridization and gene flow. Selection is the result of human or natural (non-human) selection pressures. The evolutionary process is common to cultivated, wild, and weedy plants, but the relative contribution of human or natural selection pressures on the different components of the rice crop complex varies. The unit of evolution is the population, but the population has no meaning except as the summed activities of its individuals and their

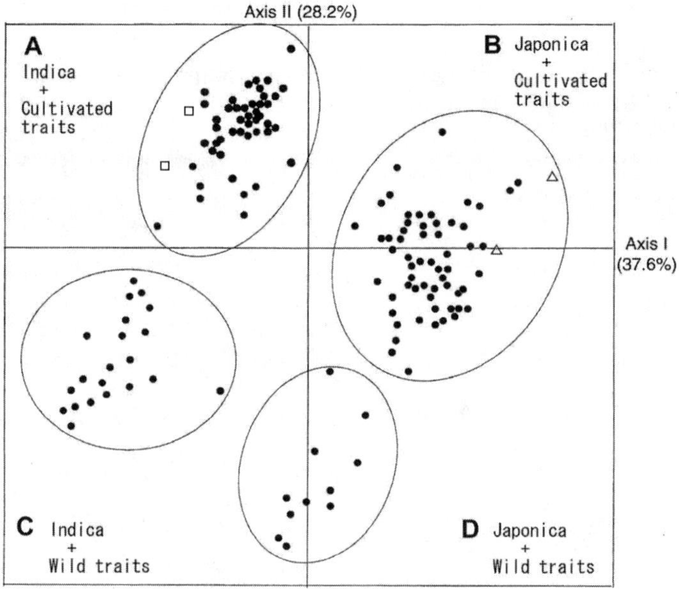

FIGURE 16.4 Diversity in a global collection of weedy *Oryza sativa* (●) compared with two indica (□) and two japonica (△) cultivars. The principal component analysis is based on domestication related traits (100-grain weight, degree of seed dormancy, degree of shattering), the indica/japonica differentiating traits (susceptibility to potassium phosphate, hull hair length, phenol reaction), and 14 isozyme loci. (Modified from Suh et al. (57) with permission.)

interactions (26). The limits of the population may be blurred between the crop complex components in the context of crop complexes.

Broad-based studies of morphological traits and isozymes in weedy rice reveal two major evolutionary trends in rice. One is the divergence between weedy rice strains having either wild or cultivated traits. The other reflects divergence of weedy rice in relation to indica or japonica characteristics (57) (Figure 16.4). These two major trends result in four broad groups of weedy rice:

1. (A) Indica- or (B) japonica-type plants with some cultivated characteristics — These two groups flourish where wild rice does not occur and may reflect old or recent cultivars that have become weedy either as volunteer plants (non-shattering spikelets) or shattering rice (see case studies below). Weedy rice with cultivated characteristics can occur anywhere rice is grown from the high altitude of Bhutan to rice fields of Uruguay.
2. (C) Indica- or (D) japonica-type plants with wild characteristics such as highly shattering spikelets and dormancy — These two groups tend to be found in areas where wild rice occurs today or used to be in the past. This group usually consists of the derivatives of hybridization between cultivated and wild plants. In a few cases, weedy rice in Group C may represent domestication intermediates between wild and domesticated rice.

These four groups reflect basic evolutionary trends along two axes of diversity of wild-cultivated and indica–japonica groups. Rice is mainly grown in Asia where rice was domesticated, diversified, and its wild relatives occur, so the detailed story is much more complex. Genetic analyses of areas outside the homeland of rice also reveal great variation in weedy rice (19,66; Chapter 19, Chapter 20). However, caution is necessary when interpreting weedy rice diversity based on neutral molecular markers, as they may not be associated with key characters for ecogenetic differentiation among or within populations (7,49).

Weedy rice derived from rice does not escape from the agricultural setting and become established in natural habitats; it remains confined to rice fields and their immediate ruderal surroundings.

However, genes from cultivated rice can become established in wild populations because of hybridization with sympatric wild rice (45), which is discussed below.

Apart from AA genome *Oryza* species, only three other *Oryza* species, all in the *O. officinalis* complex, have been reported as weeds in rice fields: *O. latifolia* in Central America, *O. officinalis* in the Philippines, and *O. punctata* in Africa (Table 16.1). Of these, *O. punctata* is a particularly serious weed of rice in southern Africa.

16.4 CASE STUDIES

The two major evolutionary trends described above in rice reflect basic ways weedy rice can evolve. The first is associated with gene flow between wild and cultivated rice. The other is associated with diversification and selection processes within cultivated rice itself. Case studies below have been chosen to illustrate these two ways weedy rice can evolve.

16.4.1 Gene Flow in the Rice Crop Complex

16.4.1.1 Introgression

Wild and cultivated rice are sympatric across large parts of India. Oka and Chang (46) undertook an analysis of populations of wild and cultivated rice in the vicinity of Raipur (21:10YN, 81:40YE) in central India. They analyzed four populations:

1. An annual type wild population growing in pools in grassland
2. An annual type wild population growing in a disturbed area between rice fields
3. Weedy rice growing in a rice field
4. The cultivar in the rice field from which the weedy rice was collected (Figure 16.5A)

A discrimination function was calculated for individuals from each population using six morphological traits distinguishing between wild and cultivated plants. Plotting this function against heading date revealed a wild-to-cultivated continuum in the samples (Figure 16.5B).

16.4.1.2 Gene Dispersal and Hybrid Swarms

Wild rice populations do not have a glutinous endosperm except when growing near glutinous rice varieties. In Laos, north and northeastern Thailand, and the Shan state of Myanmar, glutinous rice is mostly preferred for daily consumption. Hence, this region has been called the glutinous rice zone (70). The glutinous trait enables rapid assessment of gene flow from rice to wild populations. Analysis of populations from a wide area of north and northeast Thailand in the early 1960s revealed the two-way gene flow between wild and cultivated rice enabling estimates of outcrossing (Table 16.4) (47). Outcrossing in one population was particularly high (44%) and this population, representing a hybrid swarm, was studied further. A discriminant function, based on the combined measurement of spikelet number per panicle, panicle length, and spikelet width was determined for 106 plants in the hybrid swarm population that was situated in a stream between fields of glutinous rice. It maximized the differences between plants that were homozygous for wild-type genes and homozygous for the waxy gene (*wxwx*) (Figure 16.6A). Plotting this function against seed shedding revealed the continuous and high level of diversity in the population (Figure 16.6B). Hybrid swarms are considered to be rather transient (24) and are not commonly found in the rice crop complex of Asia. Hybrid swarms generate great genetic diversity when they do occur. Evolutionary radiation is possible from hybrid swarms in natural and cultivated environments.

These two classic studies of population genetic variation in India and Thailand, where wild and cultivated rice are sympatric, reveal trends related to the wild-cultivated axis of differentiation. They also reveal the continuous and local nature of genetic variation in the rice crop complex.

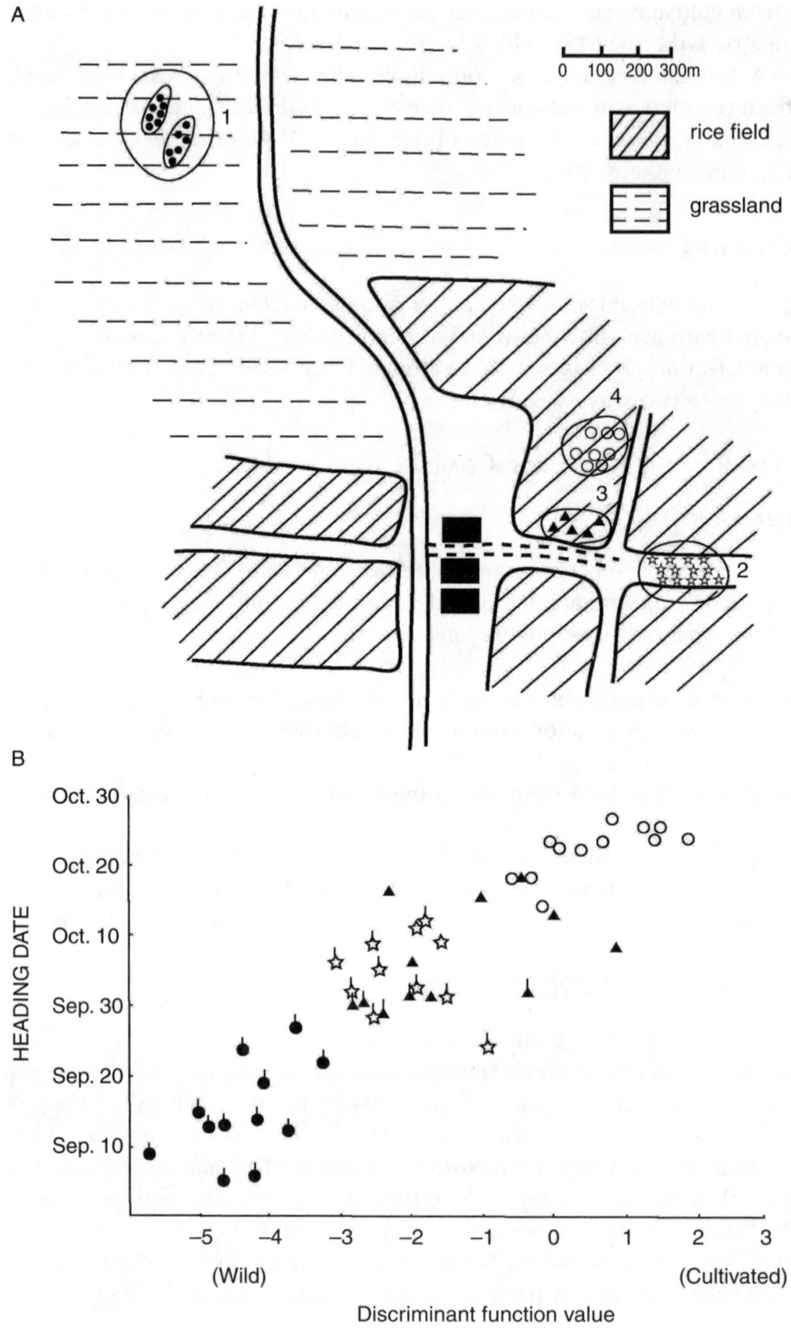

FIGURE 16.5 Spatial relationship and morphogenetic analysis of sympatric wild and cultivated rice in a central Indian village, India. (A) Sketch map showing the location of populations analyzed south of Raipur, Madhya Pradesh, India. (B) Scatterplot of discriminant score for wild/cultivated characters determined from 130 lines grown from the samples taken in the four populations plotted against heading date. Population 1 (●), population 2 (☆), population 3 (▲), population 4(○). Presence of spicule indicates sample seeds were awned. (Modified from Oka and Chang (46).)

TABLE 16.4
Frequency of the Glutinous Gene (*wxwx*) in Wild and Cultivated Rice from Different Parts of North and Northeast Thailand

Accession Number	Plant Type	Site	No. of Plants ++	+*wx*	*wxwx*	Freq. of *wx* Gene	% Outcrossing
W217	Perennial	Chiengmai	73	1	1	2.0	
W219	Perennial	Chiengmai	59	1	0	0.8	
W225	Perennial	Kon Kaen	54	0	0	0.0	
W229	Perennial	Kon Kaen	42	0	0	0.0	
W223	Perennial	Nong Kai	28	0	0	0.0	
W218	Intermediate	Chiengmai	55	42	9	28.3	43.98
W220	Intermediate	Chiengmai	48	6	0	5.6	
W222	Intermediate	Chiengmai	115	2	0	0.9	
W228	Intermediate	Kon Kaen	31	5	0	6.9	
W235	Intermediate	Nong Kai	11	3	3	26.4	
W226-7	Annual	Kon Kaen	67	0	0	0.0	
W230	Annual	Kon Kaen	18	0	0	0.0	
W236	Annual	Udon	28	0	0	0.0	
W240	Cultivated	Chiengmai	13	2	30	68.9	0.21
W241	Cultivated	Chiengmai	6	0	22	78.6	2.06
W242	Cultivated	Chiengmai	2	0	78	97.5	0.00
W243	Cultivated	Kon Kaen	4	0	20	83.3	3.58
W244	Cultivated	Kon Kaen	4	1	7	62.5	3.41
W246	Cultivated	Nong Kai	6	0	57	90.5	2.53
W248	Cultivated	Nong Kai	5	0	68	93.2	1.25

Source: From Oka and Chang (47).

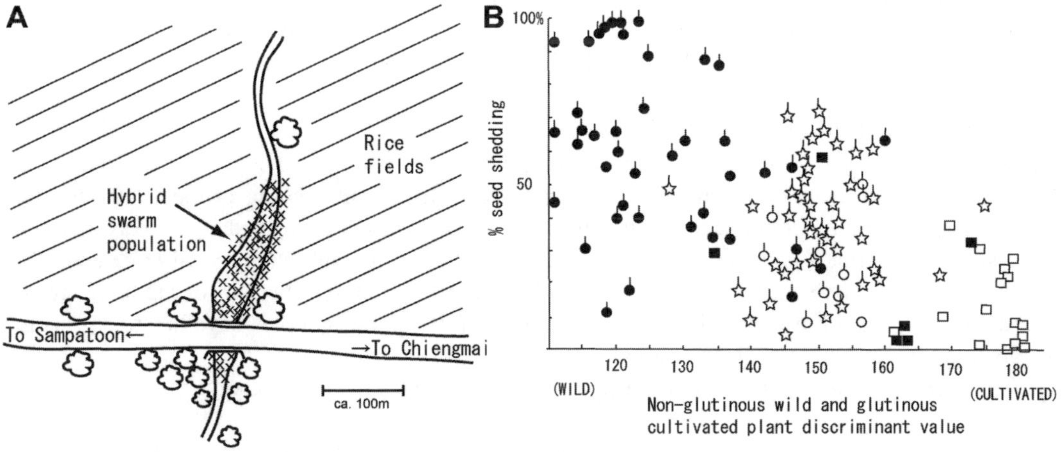

FIGURE 16.6 Location of a hybrid swarm population between fields of rice at Sampatoon, Thailand and its morphogenetic analysis. (A) Sketch map of the Sampatoon hybrid swarm population. (B) Scatterplot of discriminant score for wild non-glutinous/cultivated glutinous traits in 106 plants from the Sampatoon hybrid swarm population plotted against % seed shedding. Wild plant genotypes ● wild type ++, ☆ heterozygote +*wx*, ○ glutinous *wxwx*. Cultivated rice genotype ■ wild type ++, □ glutinous rice *wxwx*. Presence of spicule indicates the samples seeds were awned. (Modified from Oka and Chang (47).)

16.4.2 Evolution of Weedy Rice in Rice Fields

16.4.2.1 Weedy Rice in Japan

Weedy rice has been an occasional problem throughout history in Japan. Currently weedy rice is restricted to just two areas of Japan — Nagano and Okayama prefectures. Although their origin can only be inferred, the differences between weedy rice in these two prefectures provide a clear example of the concurrent local evolution of weedy rice.

16.4.2.1.1 Nagano Weedy Rice

Weedy rice in Nagano has a red pericarp. Currently, it is more widely distributed than it was 30 years ago. The morphogenetic characteristics of weedy rice have changed over the last 30 years. Samples of weedy rice collected in 1970 were later maturing and taller in stature compared to samples collected in the last few years. Weedy rice now matures before the most widely grown japonica cultivar, Koshihikari, and is about the same medium height. The decline in stature of Nagano weedy rice means it is much more difficult for farmers to rogue it out. Isozyme analyses revealed that it belongs to the japonica group (59). Although there are a few specialty rice varieties with red pericarp grown in Nagano primarily associated with shrines or organic farming, these varieties are non-shattering and their limited production does not suggest they are a source of the scattered populations of weedy rice. Red rice was commonly grown in Japan in the 18th century (1) and weedy remnants of earlier commonly grown rice varieties seem to be the most likely source of weedy red rice in Nagano. Why then has weedy rice only become a problem in Nagano when direct seeding is more widely practiced in other prefectures such as Yamagata, Aichi, and Fukushima and they have no weedy rice problem? Possible explanations are:

- Some farmers in Nagano, but not other prefectures, raise seedlings in seedbeds located in paddy fields rather than the general practice of raising seedlings outside the field (36). Thus red rice may have emerged in the seedbed and been distributed across the fields during transplanting. Red rice seed banks may have become established and red rice became a problem when farmers changed planting practices to direct seeding.
- Many paddy fields in Japan including Nagano have been redeveloped to improve field size and shape. Consequently, soil has been moved from location to location. In Nagano, this may have resulted in the spread of weedy rice seeds.

16.4.2.1.2 Okayama Weed Rice Complex

Okayama weedy rice is generally white or pale brown in contrast to Nagano red weedy rice. Okayama is also the area of Japan that first introduced dry direct seeding in the 1940s and now has the largest area of direct seeding in Japan. Weedy rice has been recorded in Okayama since the 1940s (29). Farmers can control weedy rice by reverting to transplanting. Weedy rice genotypes in Okayama have alleles associated with either indica (restricted distribution) or japonica (widely distributed) varietal groups based on isozyme analysis (59). Some samples have heterozygous loci suggesting they are indica japonica hybrids.

There are several hypotheses concerning the origin of weedy rice in Okayama:

- The majority of weedy rice varieties are japonica and are morphogenetically similar to major japonica varieties grown in Okayama such as Akebono and Asahi (29,59). Unlike many parts of Japan, Okayama continues to grow old varieties, such as Akebono released in 1949 and Asahi bred even earlier. Japonica weedy rice has probably evolved from these old varieties that have been grown over many decades in Okayama.
- The confined location of weedy rice with indica alleles suggests that their origin may have been derived either from recently introduced indica varieties, such as Tetep and Moletsu that are currently used for cattle forage, or derived from old indica varieties.

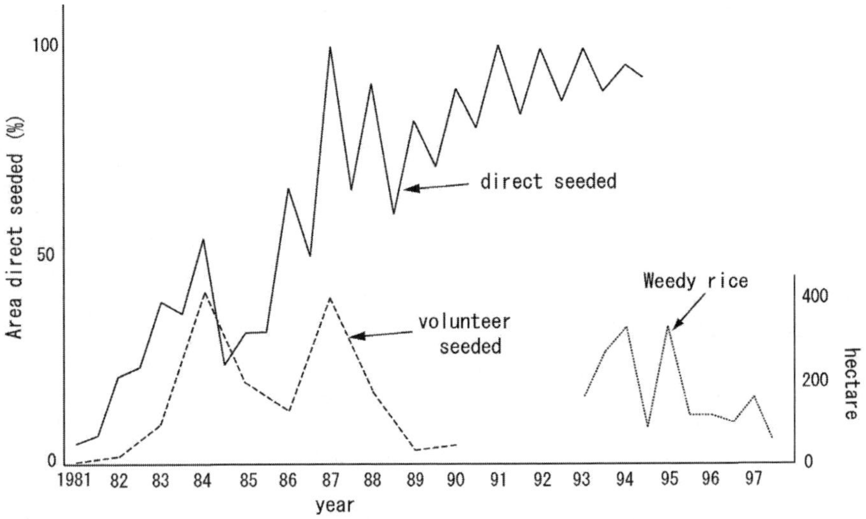

FIGURE 16.7 Changes in rice seeding practices and the emergence of weedy rice in Muda irrigation scheme, Malaysia from 1981–1997. Direct-seeded % (—) and volunteer-seeded % (----) (left scale) and observed area (ha) of weedy rice (····) (right scale). (Based on data in Azmi et al. (2), Ho (27), and Watanabe et al. (71).)

Analysis of isozyme profiles of weedy rice, old, and more recently grown varieties in Japan suggests weedy rice shares alleles with recently introduced indica varieties (30,60).

- The weedy rice that appears to be indica–japonica hybrids may reflect hybridization between japonica and indica varieties in the past. Currently grown cattle forage varieties, such as Hoshiyutaka and Kusahonami that are indica–japonica hybrids and grown in Okayama, are morphologically different from this type of weedy rice. Consequently, these cattle forage varieties may not be the source of this type of weedy rice.

16.4.2.2 Weedy Rice in Malaysia

The emergence of weedy rice ("padi angina") in the Muda irrigation scheme (approximately 100,000 ha of rice land) in the northwest of west Malaysia followed the rapid adoption of direct seeding in the 1980s. Direct seeding in this area was characterized by the practice of volunteer seeding that was adopted in years when there was a drought and insufficient irrigation water. Volunteer seeding is the incorporation of shattered seeds from combine harvesting into the soil by dry rotovation with tractors (27). Due to drought in 1987, 93% of the Muda area was dry-plowed and 98% of the rice was dry-seeded (58%) or volunteer-seeded (40%) (Figure 16.7). The drought with the high proportion of volunteer seeding seems to have been the trigger for the emergence of weedy rice in the Muda irrigation scheme. Within 3 years, weedy rice was observed (67); by 1993, the problem became obvious with more than 100 ha infested; and by 1994, the invested area had increased to 325 ha.

Rice production in west Malaysia depends heavily on machinery. Harvesting with large cooperative combine harvesters is common and these travel from area to area, as rice ripens. Therefore, it is easy for seeds to be rapidly dispersed from one area to another.

16.5 CONCLUDING COMMENTS

The genomes of rice have been sequenced and comparative genomics have revealed many similarities to other cereal genomes, yet rice has many characteristics that set it apart from other cereals. Broad generalizations across crops can be misleading when considering rice evolution in relation to weedy rice. For example, rice is a diploid and not considered an ancient (e.g., maize) or more recent (e.g., wheat) polyploid species. Most rice production is in Asia where rice was domesticated and major production centers of rice are sympatric with its immediate progenitors.

The three major evolutionary axes in the Asian rice crop complex — (Axis 1) perennial (outcrossing)-annual (inbreeding), (Axis 2) wild-domesticated, (Axis 3) indica–japonica forms have been introduced. Although Axis 1 and Axis 2 exhibit an evolutionary continuum (see Figure 16.5 and Figure 16.6), Axis 3 is rather discontinuous although intermediate forms occur (13). The genetic basis for these trends can best be explained in the context of divergence of multi-locus systems that exhibit different patterns of associations among character states (40). At the genome level, some of these characters are represented by alleles that are closely linked in gene clusters (Figure 16.2). Other characters are inherited as unlinked gene associations (e.g., genes for apiculus hair length and phenol reaction both closely associated with indica–japonica differentiation are on different chromosomes). For associated characters that are not linked, selection forces have resulted in particular coadapted sets of genes being preserved in the gene pool.

Seed shattering is the single characteristic of all weedy rice in rice fields. Most of their spikelets fall from their panicles so they are not captured during harvesting of the rice crop. Weedy rice can directly evolve from rice in a few seasons when there is sufficient genetic variation and strong selection for shattering. In many areas, weedy rice is the most serious weed of rice, as it can appear similar to rice in nearly all respects except shattering. Weedy rice has a competitive advantage in the highly disturbed environment in and around rice fields. Weedy rice evolves in specific localities, but its seeds can easily become widely dispersed. Unlike some crops where escaped shattering forms of the cultigen can become established as viable populations in wild habitats where no wild relative of the crop exists, this has not happened in rice.

Hybrid derivatives from the association between cultivated and wild rice are adapted to a wide spectrum of habitats from the wild to cultivated fields. Gene flow from cultivated rice into wild rice (the usual direction) frequently occurs. The locations where potential gene escape into wild populations can occur can be mapped to a high degree of resolution and monitored so that effective measures can be taken to prevent undesirable genes from the cultivated rice getting into wild populations.

Traditional elimination of weedy rice in rice fields has relied on careful attention to cultural practices that will lower the amount of weedy rice in the soil seed bank. Water control and careful tilling are prerequisites to do this. Weedy rice in California was eliminated by precise field preparation, excellent water control, and use of certified seed (26A). Rural labor costs in Asia are increasing because of economic development and human migration to urban areas. There is also increasing competition between urban and rural areas for water (34). Consequently, the application of the necessary rice cultural practices to control weedy rice and other weeds can be problematic.

Weedy rice is not a problem in transplanted fields with good water control, provided seeds sown in the seedling nursery are pure. Hence, current trends from transplanting to broadcasting rice may need to be reversed. Such transitions from broadcasting to transplanting have occurred in the past (e.g., Sri Lanka in the early 1900s). However, weedy rice is only one part of rice field weed flora. New concepts for control of weedy rice require an integrated approach that considers the entire rice field weed flora.

Gene flow between rice and wild rice has been occurring over millennia in many rice-growing areas. Weedy types of rice have been suggested as a more usable source of genes for rice improvement than wild rice because weedy rice is more similar to cultivated rice than wild rice (63). One of the most often quoted examples of "wild" rice uses in rice breeding is grassy stunt virus resistant *Oryza nivara* (IRRI acc. 101508). However, because this wild rice has straw hulled spikelets, a characteristic of cultivated rice, it is probable that this accession has introgressed genes from cultivated rice. In addition, the most important source of cytoplasmic male sterility used for hybrid rice production originated in a population of wild rice that was considered to be a hybrid between wild and cultivated rice (32).

Weeds are the most important pests of rice in Asia. Among them, weedy rice is becoming an increasingly serious problem in some areas that requires a response commensurate with their importance.

ACKNOWLEDGMENTS

The authors would like to thank Professor Hiroko Morishima for providing valuable comments on a draft of this chapter and inspired research leadership on rice and *Oryza* evolution. The authors also thank Professor Hak Soo Suh, Yeungnam University, Korea, for permission to use a modified version of his figure that appears as Figure 16.4 in this chapter. Some of the ideas on which this chapter are based were formulated while the first author was a recipient of an OECD fellowship award in 2003, for which he is thankful.

LITERATURE CITED

1. Arashi K. 1974. *Consideration on red rice in Japan.* Tokyo: Yusankaku (in Japanese). 295 pp.
2. Azmi M, Abdullah MZ, Mislamah B, Baki BB. 2000. Management of weedy rice (*Oryza sativa* L.): the Malaysian experience. In *Wild and weedy rice in rice ecosystems in Asia — a review.* Limited Proceedings No. 2. Baki BB, Chin DV, Mortimer M, Eds., pp. 91–96, Los Banos, Philippines: International Rice Research Institute.
3. Bennetzen J. 2002. Opening the door to comparative plant biology. *Science* 296:60–63.
4. Bres-Patry C, Lorieux M, Clément G, Bangratz M, Ghesquiere A. 2001. Heredity and genetic mapping of domestication-related traits in a temperate japonica weedy rice. *Theor. Appl. Genet.* 102:118–126.
5. Buckler IV ES, Thornsberry JM, Kresovich S. 2001. Molecular diversity and domestication of grasses. *Genet. Res.* 77:213–218.
6. Cai HW, Morishima H. 2000. Genomic regions affecting seed shattering and seed dormancy in rice. *Theor. Appl. Genet.* 100:840–846.
7. Cai HW, Morishima H. 2002. QTL clusters reflect character associations in wild and cultivated rice. *Theor. Appl. Genet.* 104:1217–1228.
8. Chang TT. 1964. *Present knowledge of rice genetics and cytogenetics.* Technical Bulletin 1. Los Banos, Philippines: International Rice Research Institute. 96 pp.
9. Chang TT. 1983. The origin and early cultures of the cereal grains and food legumes. In *The origins of Chinese civilization*, Keightley DN, Ed., pp 65–94, Berkeley, CA: University of California Press.
10. Chang TT. 2003. Origin, domestication and diversification. In *Rice: evolution, history, production and technology*, Wayne Smith C, Dilday RH, Eds., Hoboken, NJ: Wiley and Sons, pp. 27–65.
11. Chen WB, Nakamura I, Sato YI, Nakai H. 1993. Distribution of deletion type in cpDNA of cultivated and wild rice. *Jpn. J. Genet.* 68:597–603.
12. Cheng C, Motohashi R, Tsuchimoto S, Fukuta Y, Ohtsubo H, Ohtsubo E. 2002. Polyphyletic origin of cultivated rice: based on the interspersion pattern of SINEs. *Mol. Biol. Evol.* 20:67–75.
13. Cheng KS. 1985. A statistical evaluation of the classification of rice cultivars into hsien and keng subspecies. *Rice Genet. Newsl.* 2:46–48 (http://143.48.220.116/newsletters/rice_genetics/rgn2/v2II4html).
14. Clark RM, Linton E, Messing J, Doebley JF. 2004. Pattern of diversity in the genomic region near the maize domestication gene *tb1*. *Proc. Natl. Acad. Sci. USA* 101:700–707.
15. Clayton WD, Renvoize SA. 1986. Genera Graminum, London HMSO publications, 387 pp.
16. Dally AM, Second G. 1990. Chloroplast DNA diversity in wild and cultivated species of rice (genus *Oryza* section *Oryza*): cladistic mutation and genetic distance analysis. *Theor. Appl. Genet.* 80:209–222.
17. de Wet JMJ. 1992. The three phases of cereal domestication. In *Grass evolution and domestication*, Chapman GP, Ed., Cambridge, UK: Cambridge University Press.
18. Epimaki M, Koinange K, Singh SP, Gepts P. 1996. Genetic control of the domestication syndrome in common bean. *Crop Sci.* 36:1037–1045.

19. Federici MT, Vaughan DA, Tomooka N, Kaga A, Wang XW, Doi K. 2001. Analysis of Uruguayan weedy rice genetic diversity using AFLP molecular markers. *Electron. J. Biotechnol.* 4(3) December 15, 2001 (http://ejb.ucv.cl/content /vol4/issue3/abstract/3/index.html).
20. Garrity DP.1984. Rice environmental classification: a comparative review. In *Terminology for rice growing environments*, p. 11–26, Los Banos, Philippines: International Rice Research Institute.
21. Glaszmann JC. 1988. Geographic pattern of variation among Asian native rice cultivars (*Oryza sativa* L.) based on fifteen isozyme loci. *Genome* 30:782–792.
22. Gunawardena, I. 1983. Sri Lanka. In *1983 Rice germplasm conservation workshop*, pp. 38–39, Los Banos, Philippines: International Rice Research Institute and International Board for Plant Genetic Resources.
23. Hammer K. 1984. The domestication syndrome. *Kulturpflanze* 32:11–34 (In German).
24. Harlan JR. 1965. The possible role of weed races in the evolution of cultivated plants. *Euphytica* 14:173–176.
25. Harlan JR. 1992. Origin and processes of domestication. In *Grass evolution and domestication*, Chapman GP, Ed., Cambridge, UK: Cambridge University Press, pp. 159–175.
26. Harper JL. 1977. Population biology of plants. London: Academic Press, 892 pp.
26A. Hill JE, Smith RJ, Bayer DE. 1994. Rice weed control: current technology and emerging issues in temperate rice. *Aust. J. Exp. Agric.* 34:1021–1029.
27. Ho NK. 1991. Comparative ecological studies of weed flora in irrigated rice fields in the Muda area. Alor Setar, Malaysa: Muda Agricultural Development Authority, 97 pp.
28. Ishii T, Terauchi T, Tsunewaki T. 1988. Restriction endonuclease analysis of chloroplast DNA from A-genome diploid species of rice. *Jpn. J. Genet.* 63:523–536.
29. Ishii T. 2001 Occurrences of weedy rice in dry-seeded rice fields in Okayama Prefecture, *J. Agric. Sci. (Nogyou-gijyutsu)*. 56:257–261 (in Japanese).
30. Ishikawa R, Nakamura I, Nishihara T, Kikuchi M, Senda M, Akada S, Harada T, Niizeki M. 2002. Origin of cytoplasmic substituted red cultivars found in Japan. *Theor. Appl. Genet.* 105:608–613.
31. Kellogg EA. 2000. The Grasses: a case study in macroevolution. *Annu. Rev. Ecol. Syst.* 31:217–238.
32. Li JM, Yuan LP. 2000. Hybrid rice: genetics, breeding and seed production. *Plant Breed. Rev.* 17:15–158.
33. Lin SY, Sasaki T, Yano M. 1998. Mapping quantitative trait loci controlling seed dormancy and heading date in rice, *Oryza sativa* L., using backcross inbred lines. *Theor. Appl. Genet.* 96:997–1003.
34. Mew TW, Leung H, Savary S, Vera Cruz CM, Leach JE. 2004. Looking ahead in disease research and management. *Crit. Rev. Plant Sci.* 23:103–127.
35. Miura, K, Lin SY, Yano M, Nagamine T. 2002. Mapping quantitative trait loci controlling seed longevity in rice (*Oryza sativa* L.). *Theor. Appl.Genet.* 104:981–986.
36. Miyajima Y, Takahashi N. 1974. Occurrences of red rice "Toukon" in Nagano Prefecture, *J. Agric. Sci. (Nogyou-gijyutsu)* 29:453–455 (in Japanese).
37. Morishima H. 2001. Evolution and domestication of rice. In *Rice genetics IV*, Khush GS, Brar DS, Hardy B, Eds., pp. 63–77 Enfield, New Hampshire: Science Publishers Inc.
38. Morishima H, Gadrinab LU. 1987. Are the Asian common wild-rices differentiated into the indica and japonica types? In *Crop exploration and utilization of genetic resources*, Hsieh SC, Ed., pp. 11–20. Changhua, Taiwan: Taichung Distr. Agric. Improv. Sta.
39. Morishima H, Oka HI. 1981. Phylogenetic differentiation of cultivated rice, XXII. Numerical evaluation of the indica-japonica differentiation. *Jpn. J. Breed.* 31:402–413.
40. Morishima H, Sano Y, Oka HI. 1992. Evolutionary studies in cultivated rice and its relatives. *Oxford Surv. Evol. Biol.* 8:135–184.
41. Miura K, Lin SY, Yano M, Nagamine T. 2002. Mapping quantitative trait loci controlling seed longevity in rice (*Oryza sativa* L.). *Theor. Appl. Genet.* 104:981–986.
42. Nagai YS, Sobrizal, Sanchez PL, Kurakazu T, Doi K, Yoshimura A. 2002. *Sh3*, a gene for seed shattering, commonly found in wild rices. *Rice Genet. Newsl.* 19:74–75 (http://143.48.220.116/newsletters/rice_genetics/rgn19/b31.html).
43. Ogawa T. 2003. Genome research in genus *Oryza*. In *Monograph on genus* Oryza, Nanda JS, Sharma SD, Eds., pp. 171–212. New Dehli: Oxford & IBH Publishing Co. Pvt. Ltd.
44. Ohtsubo H, Cheng C, Ohsawa I, Tsuchimoto S, Ohtsubo E. 2004. Rice retrotransposon *p-SINE1* and origin of cultivated rice. *Breed. Sci.* 54:1–11.

45. Oka HI. 1988. *Origin of cultivated rice*. Tokyo: Japan Scientific Societies Press, 254 pp.
46. Oka HI, Chang WT. 1959. The impact of cultivation on populations of wild rice, *Oryza sativa* f. *spontanea*. *Phyton* 13:105–117.
47. Oka HI, Chang WT. 1961. Hybrid swarms between wild and cultivated rice species *Oryza perennis* and *O. sativa*. *Evolution* 15:418–430.
48. Sanchez PL, Kurakazu T, Hirata C, Sobrizal, Yoshimura A. 2002. Identification and mapping of seed shattering genes using introgression lines from wild rice species. *Rice Genet. Newsl.* 19:78–79 (http://www.shigen.nig.ac.jp/rice/oryzabase/rgn/newsletter.jsp).
49. Sano R, Morishima H. 1992. Indica-japonica differentiation of rice cultivars viewed from variations in key characters and isozymes, with special reference to landraces from the Himalayan hilly areas. *Theor. Appl. Genet.* 84:266–274.
50. Sato YI, Ishikawa R, Morishima H. 1990. Non-random association of genes and characters found in indica x japonica hybrids of rice. *Heredity* 65:75–79.
51. Sato YI. 2000. Origin and evolution of wild, weedy and cultivated rice. In *Wild and weedy rice in rice ecosystems in Asia — a review,* Baki BB, Chin DV, Mortimer M, Eds., pp. 7–15, Los Banos, Philippines: International Rice Research Institute.
52. Second, G. 1985. Evolutionary relationships in the *sativa* group of *Oryza* based on isozyme data. *Genet. Sel. Evol.* 17:89–114.
53. Second G. 1989. Additional observations and collection of the *Oryza latifolia* Desv. species complex. *Rice Genet. Newsl.* 6:73–76 (http://143.48.220.116/newsletters/rice_genetics/rgn6/v6p73.html).
54. Second G. 1991. Molecular markers in rice systematics and the evaluation of genetic resources. In *Biotechnology for agriculture and forestry vol. 14 Rice*, Bajaj YPS, Ed., pp. 468–494, Berlin: Springer-Verlag.
55. Second G. 1991. *Trip to San Jose area, Occidental Mindoro.* Mimeographed. Los Banos, Philippines: International Rice Research Institute Library.
56. Second G, Ghesquerre A. 1984. Collection and observation of wild rice (*Oryza*) species in western India. Report of a survey from October 1–28, 1984. Mimeographed. Los Banos, Philippines: International Rice Research Institute Library.
57. Suh SH, Sato YI, Morishima H. 1997. Genetic characterization of weedy rice (*Oryza sativa* L.) based on morpho-physiology, isozymes and RAPD markers. *Theor. Appl. Genet.* 94:316–321.
58. Uozu S, Ikehashi H, Ohmido N, Ohtsubo H, Ohtsubo E, Fukui K.1997. Repetitive sequences; cause of variation in genome size and chromosome morphology in the genus *Oryza*. *Plant. Mol. Biol.* 35:791–799.
59. Ushiki J, Ishikawa R, Ishii T, Tezuka M, Sakai N. 2004. Genetic polymorphism and speculation of origin of weedy rice strains in Japan. *Weed Sci. Soc. Am. Abstracts* 44:14.
60. Ushiki J, Ishikawa R, Ishii T. 2004. Genetic polymorphism of japonica and indica weedy rice strains found in Okayama prefecture. *Breed. Res.* 6186 (in Japanese).
61. Vaughan DA, Watanabe H, HilleRisLambers D, Zain MA, Tomooka N. 1999. Weedy rice complexes in direct seeding rice cultures. In *Proc. Int. Symp. World Food Security,* Horie T, Geng S, Amano T, Inamura T, Shiraiwa T, Eds., pp. 277–280. Kyoto: Crop Sci. Soc. Japan.
62. Vaughan DA, Tomooka N, Kaga A, Shcherban A, Xu RQ, Jinno Y, Doi K, Kashiwaba K. 2000. Population analyses using different molecular techniques to determine strategies for in-situ conservation — examples from the genera *Glycine*, *Oryza* and *Vigna*. In *Integration of biodiversity and genome technology for crop improvement*, Oono K, Komatsuda T, Kadowaki K, Vaughan DA, Eds., pp. 69–72. Tsukuba, Japan: National Institute of Agrobiological Resources.
63. Vaughan DA, Morishima H. 2003. The biosystematics of the genus *Oryza*. In *Rice: Evolution, history, production and technology,* Wayne Smith C, Dilday RH, Eds., pp. 27–65. Hoboken, NJ: Wiley and Sons.
64. Vaughan DA, Morishima H, Kadowaki K. 2003. *Oryza* diversity research. *Curr. Opin. Plant Biol.* 6:139–146.
65. Vaughan DA, Miyazaki S, Miyashita K. 2004. Rice gene pool and human migrations. In *Biological resources and migration*, Wessel G, Ed., pp. 1–13. Berlin: Springer-Verlag.
66. Vaughan LK, Ottis BV, Prazak-Havey AM, Bormans CA, Sneller C, Chandler JM, Park WD. 2001. Is all red rice found in commercial rice really *Oryza sativa*? *Weed Sci.* 49:468–476.

67. Wahab AH, Suhaimi O. 1991. Padi angin adverse effects and methods of its eradication (in Malay, English abstract). *Teknol. Padi, MARDI* 7:27–31.
68. Wan J, Nakazaki T, Kawaura K, Ikehashi H. 1997. Identification of marker loci for seed dormancy in rice (*Oryza sativa* L.). *Crop Sci.* 37:1759–1763.
69. Wang RL, Stec A, Hey J, Lukens L, Doebley J. 1999. The limits of selection during maize domestication. *Nature* 398:236–239.
70. Watabe, T. 1967. Glutinous rice in northern Thailand. Reports on Research in Southeast Asia Natural Science Series N-2. Kyoto: CSEAS, Kyoto University, 160 pp.
71. Watanabe H, Man A, Ismail MZ. 1996. Ecology of major weeds and their control in direct seeding rice culture of Malaysia. MARDI/MADA/JIRCAS collaborative study. Tsukuba, Japan: Japan Int. Res. Center Agricultural Sci., 202 pp.
72. Xiao J, Li J, Grandillo S, Ahn SN, Yuan LP, Tanksley SD, McCouch SR. 1998. Identification of trait-improving quantitative trait loci alleles from wild rice relative, *Oryza rufipogon*. *Genetics* 150:899–909.
73. Xiong LZ, Liu KD, Dai XK, Xu CG, Zhang Q. 1999. Identification of genetic factors controlling domestication-related traits of rice using an F_2 population of a cross between *Oryza sativa* and *O. rufipogon*. *Theor. Appl. Genet.* 98:243–251.
74. Yamanaka S, Nakamura I, Nakai H, Sato YI. 2003. Dual origin of the cultivated rice based on molecular markers of newly collected annual and perennial strains of wild rice species, *Oryza nivara* and *O. rufipogon*. *Genet. Res. Crop Evol.* 50:529–538.
75. Yamasaki M. 1929. On the variation of rice varieties in the resistance to the toxic action of potassium chlorate, and its practial significance. *National Institute of Crop Science, Koonosu, Saitama, Ministry of Agriculture and Forestry, Jpn. Annu. Rep.*1:1–27 (in Japanese with English summary).

QUESTIONS AND ANSWERS

Henri Darmency: Is there any segregation for shattering, even low rate of shattering in the F_2, when crossing varieties from different lineages?

Answer: Yes, shattering can appear when both parents are non-shattering. In rice, there is a wide diversity for degree of shattering (or threshability). Tropical japonicas of Indonesia are particularly hard to thresh whereas indica varieties tend to be easily threshed. Environmental factors can also influence shattering. These differences are manifest in the different methods used by farmers to thresh their harvested rice.

Jonathan Gressel: What are the genetics of green revolution dwarf types?

Answer: Some key domestication traits are covered in the chapter. For the semi-dwarf character that triggered the 1960s green revolution in rice this is controlled by a single recessive gene *sd1*. For an up-to-date database on known genes in rice, please refer to http://www.shigen.nig.ac.jp/rice/oryzabase/.

Baorong Lu: Do you think interspecific hybrids in rice fields (weeds) could live in natural habitats?

Answer: Interspecific hybrid populations in rice fields will evolve traits that have a selective advantage in rice fields. Interspecific hybrid populations in natural (wild) habitats will evolve traits that have a selective advantage in wild habitats. Should interspecific hybrids get from rice fields into natural habitats, their persistence will depend on their population size, genetic variation, and selection pressures. It is worth mentioning that in some places the boundary of inside and outside the rice field can be blurred and this is particularly true in deep-water rice cultures.

Baorong Lu: What are the origins of weedy rice — de-domestication or from hybridization with wild species?

Answer: Weedy rice can be derived from cultivated rice, wild rice, and inter- or intra-specific hybrids among components of the rice crop complex. Their evolution will depend on available variation and selection pressures.

Jonathan Gressel: What is the basis for the potassium chlorate test to discriminate indica and japonica varieties?

Answer: The original report (75) related to rice varietal differences in relation to application of potassium chlorate suggested that variation was due to permeability of root cells. I am not aware of subsequent research into this issue. In various tests conducted by H.I. Oka at the National Institute of Genetics, Japan, the potassium chlorate test was found to be one of the most reliable of those he used to distinguish indica and japonica varieties.

17 The Damage by Weedy Rice — Can Feral Rice Remain Undetected?

Bernal E. Valverde

17.1 INTRODUCTION

Weeds are the most important biological constraint to rice production worldwide (106). Perhaps the most difficult to control weed (actually a weed complex) in rice is weedy rice (*Oryza* spp.). This chapter describes general characteristics and the negative agronomic importance of weedy rice. The variability of weedy rice and the genetic relationships between weedy and cultivated rice are discussed in terms of the difficulty to identify weedy rice and the likelihood that it will remain undetected until infestations are sufficiently high to cause major losses in crop productivity. Practical approaches to field management of weedy rice, including chemical control and use of herbicide-resistant rice varieties, are also discussed. Finally, consideration is given to the spread of weedy rice and farmer attitudes that affect management of this weed complex.

17.2 DISTRIBUTION AND DIVERSITY OF WEEDY RICE

17.2.1 Weedy Rice Species and Their Distribution

Weedy rice is present in practically all rice growing regions and agroecosystems in more than 50 countries (52), causing substantial yield losses and sometimes the abandonment of otherwise productive land. Its similarity to cultivated rice in the vegetative stage makes it difficult to remove weedy rice by hand. Weedy rice species share characteristics resembling wild rice as well as the ability to exist in agroecosystems that make them qualify as undesirable plants in the rice crop. Some of the weedy rice species are interfertile in crosses with the rice crop, yet hybridization is one of the factors involved in the evolution of many weedy types. Oka (84) characterized weedy rice as intermediate between wild and domesticated forms. It occurs in direct-seeded rice fields, irrigation ditches, and dikes, but does not survive in natural habitats.

Vaughan et al. (Chapter 16) describes the taxonomic status of *Oryza* species, some of which are found as weeds in rice fields at particular locations. The most widely documented weedy rice species is *O. sativa* itself. Wild rice species, particularly *O. rufipogon*, the immediate ancestor of cultivated Asian rice (*O. sativa*), is also weedy especially in Asia (52,102, Chapter 16). *Oryza rufipogon*-like populations that invade rice fields are generally forms with introgressed genes from cultivated rice that intimately associate with the crop. These weedy forms have been documented in the Americas (91,97,108,110,111, Chapter 19; Valverde, Skov, and Andersen, unpublished) and in Australia (108), quite far from their mostly Asian centers of origin. The widespread nature of weedy rice in the Americas, as described in Chapter 19 and Chapter 20, is also increasing in importance in Europe (37,46,76, Chapter 21). Unless stated otherwise, this chapter deals primarily with weedy *O. sativa* and the introgressed *O. rufipogon* types.

In parts of West Africa, where *O. glaberrima* is the cultivated rice species, its ancestor *O. barthii* becomes a major weedy rice form, particularly in direct-seeded rice (23,57). Weedy intermediate forms also are abundant, but it is unknown whether they represent secondary products of introgression or a transient domestication state between the wild and cultivated rice (108). Other species reported as weedy in rice fields in Africa are *O. longistaminata* (59), a species known to hybridize with *O. sativa* (108), and the diploid form of *O. punctata* (52,58,108). *O. officinalis* has been reported as a weed in rice in the Philippines (9). In Central America, the native tetraploid species, *O. latifolia* is weedy in rice fields. *O. glumaepatula*, also native to this area, has the potential to become weedy in rice fields (74), with some accessions apparently being introgressed types (3).

17.2.2 Genetic Relationship and Diversity of Weedy and Cultivated Rice

The origin and variability of weedy rice is presented in detail by others (see Chapter 16, Chapter 19, and Chapter 20). Briefly, weedy rice may have originated by hybridization between cultivated *indica* or *japonica* rice and wild types or from cultivated varieties reverting to feral forms (12,103,104,108,109). In tropical rice-growing areas it is difficult to find truly wild populations without introgression of genes from cultivated rice (74,104). Comparative genome mapping studies also suggest that some weedy populations could have derived from *indica* *japonica* hybrids (11). Although originally defined as intersterile groups, *indica* and *japonica* types may form hybrids having widely varying fertility from one cross to another (50). Mixtures of primitive *japonica* and *indica* cultivars are sometimes planted in northern Laos. Hybrids from these fields have been characterized based on isozymes and types of glutinous endosperm (54). The morphological variability seen in these intermediates resembles that of weedy rice types elsewhere. Mating between weedy rice and weedy rice cultivated rice hybrids also occur (44).

The genetic relationships and diversity of Latin American weedy rice, particularly from Costa Rica and Colombia, is discussed in detail in Chapter 19. Weedy rice in this region is quite diverse with some accessions resembling cultivars and others more closely related to wild relatives. We also found major groupings of weedy and cultivated accessions from five Latin American countries (Colombia, Costa Rica, Nicaragua, Panama, and Venezuela) in a study conducted to determine their relatedness and diversity using microsatellite primer pairs (Skov, Valverde, Wellendorf, and Andersen, unpublished):

- Of 57 *O. sativa* accessions representing the most ruderal types with highest within accession diversity, formed a cluster according to their genetic similarity.
- A cluster was composed of 21 weedy and 7 cultivated accessions closely related with comparable within accession diversity.
- A cluster grouping had only cultivars.
- There were 6 accessions that did not belong to any cluster.

Similar trends in the grouping of weedy rice were observed in Uruguay (32). It is likely that weedy rice in Latin America initially arrived with and has been primarily disseminated by contaminated rice-crop seed, but is continually undergoing introgressive as well as intraspecific evolution to feral forms.

17.3 AGRONOMIC AND MARKET IMPACT OF WEEDY RICE

17.3.1 Direct Competitive Effects of Weedy Rice

The negative impact of weedy rice can be so severe that farmers abandon planting rice on their infested land, as weedy rice species are competitive with the crop. Season-long interference of 20 plants m^{-2} of weedy *O. sativa* reduced rice-crop grain yield by up to 86% under U.S. experimental field conditions (61). At a lower density of 5 plants m^{-2}, yield losses of 22% were observed in the

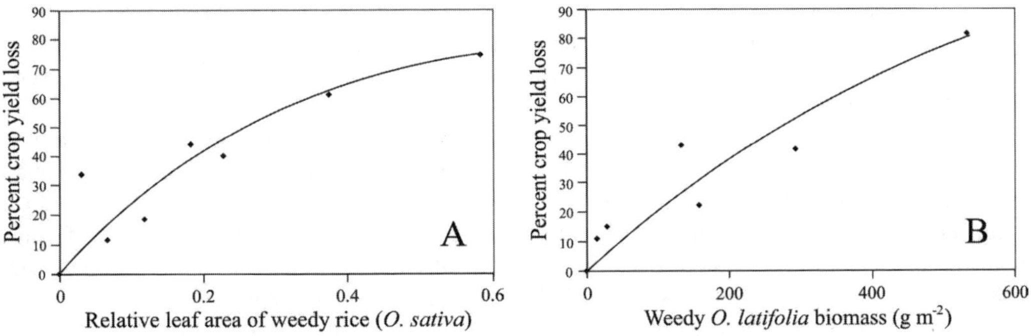

FIGURE 17.1 Suppression of rice yield by weedy *O. sativa* (A) and *O. latifolia* (B) on rice yield. Observations obtained at commercial rice fields in Costa Rica. Valverde and Madsen, unpublished.

U.S. (28). Similarly, 40 weedy rice plants m^{-2} in Greece reduced rice-grain yield 46 to 58%, depending on the cultivar (29). The reduction of Korean rice yield directly caused by weedy rice is calculated to be 5 to 10% of the total rice production every year (18). An estimated 16 to 20 g of rice grain is lost for each weedy rice panicle m^{-2} in Latin America (6,26,73). Rice grain losses of 85% in fields heavily infested by *O. longistaminata* have been documented in field trials in Mali (58). The impact of increasing infestation of two weedy rice species on cultivated rice in Costa Rica is shown in Figure 17.1, where the loss can be near total at high weed densities. Data were gathered from commercial rice fields at two locations: Parrita, Puntarenas Province for *O. latifolia* and Bagaces, Guanacaste Province for weedy *O. sativa*.

17.3.2 Indirect and Market Effects of Weedy Rice

Besides directly reducing yields by competition, weedy rice is also an alternate host of pathogens including rice blast, as well as insect pests. In Central America, early maturing weedy rice is often affected by grain sucking insects, especially *Oebalus insularis*, which later move to the maturing crop. Nymphs and adults of this bug feed on maturing rice panicles producing empty and shriveled grains. They also leave dark spots at feeding sites on the grain, affecting grain quality and providing an opportunity for fungal infection.

Another indirect effect is on the quality and value of the rice grain. The presence of red grains in commercial rice usually results in lower prices being paid to farmers. Detrimental effects on the quality of milled (polished) grain by the presence of pigmented weedy rice grain are greater when grains of both types are similar in size and shape (86). Grain lots contaminated with weedy rice grain require additional polishing at the rice mill to eliminate the red-colored pericarp. In this process, regular crop grain is easily broken. There is a close relationship between the degree of contamination and crop-grain quality after milling (70). At 4% contamination of the crop grain, losses due to grain breaking were 60%; at "normal" levels of 0.5%, losses reached 10% in broken grain (17). More intense rice milling affects the quality and nutritional value of rice: proteins and minerals are more abundant in the outer than in the inner portion of the kernel (55).

17.4 FIELD MANAGEMENT OF WEEDY RICE

17.4.1 Agronomic Practices and Weedy Rice Management

No single control tactic is sufficiently efficient to provide reliable weedy rice control. The best strategy is integrated management, which better farmers in most parts of the world observe to some extent. A key practice to prevent the introduction and spread of weedy rice is the use of clean crop seed. The uses of certified seed, water seeding, and continuous flooding have been key components of weedy rice management under monoculture in California, which has been weedy rice free for the last 4 decades, until recently (93). Unfortunately, in almost all growing areas in the world, rice farmers commonly save grain to be used as seed in the following season, a practice that deserves special attention in relation to the spread of weedy rice.

The cropping system and the agricultural practices associated with modern rice production can have an important impact on weedy rice infestations. Weedy rice in Malaysia, locally known as *padi angin*, was first detected in 1988 in the Tanjung Karan area and in 1990, sporadically in the Muda, the main rice production area (68,116, Chapter 16). By 2002, 91% of the rice farms in peninsular Malaysia were infested by weedy rice with 88% having at least a 10% infestation. Thus, the weedy rice spread widely in a short period becoming economically important.

Direct seeding and volunteer seedling culture are implicated in the rapid spread of weedy rice (68). Additionally, a large proportion of the rice area is planted with grain saved by farmers (113).

In a 3-year field trial in Korea, the highest proportion of weedy rice panicles in a rice field was observed under continuous direct seeding (37%) and wet seeding (31%), whereas under water seeding and machine transplanting it was only 15 and 1%, respectively. Where weedy rice can be easily distinguished from cultivated rice at the transplant stage, a change of cultivation method from dry seeding to machine transplanting can reduce weedy rice infestation by 99% (60), but the implications of the added cost were not assessed. Up to 35% of rice fields in Korea are infested with weedy rice, which is expected to spread, as direct seeding areas are estimated to triple by 2005 from the 70,000 ha in 1999 (60).

Weedy rice had not been a problem in the major irrigated-rice production areas of Brazil, until the late 1960s and early 1970s when long-season rice varieties were introduced into the states of Rio Grande do Sul, Santa Catarina, Parana, and Sao Paulo (40). The new varieties (such as IRGA 409 and IRGA 410) allowed weedy rice to shatter its seed before commercial rice harvesting. Over the years, populations increased to an average infestation of 95% of the area planted in Rio Grande do Sul. About 50% of this area is now under no-till rice planting in an effort to cope with heavy weedy rice infestations. About 145,000 ha are planted with pregerminated rice seed in Santa Catarina as part of their weedy rice management tactics (Foloni 2004, Universidad Estadual de Campinas, Brazil, personal communication). These two tactics also influence the weedy rice soil seed bank. The size of the seed bank at a 10-cm depth after 3 years of conventional planting was 2000 seeds m^{-2}, whereas under no-till planting and broadcasting of pregerminated seed, there were 600 and 90 seeds m^{-2}, respectively (7).

Rice can sometimes be rendered more competitive with weedy rice by altering agronomic practices. Procedures such as increasing the seeding rate, narrowing the row spacing, and selecting appropriate or competitive cultivars based on plant height and crop cycle duration help, but are not without detrimental effects (69). Some of these procedures significantly increase the cost to the farmer, and tall varieties are more prone to lodging, especially under high nitrogen fertilization. Increasing the seeding rate of cultivated rice from 50 kg ha^{-1} up to 150 kg ha^{-1} had only a minor effect on weedy rice seed production in field tests in Arkansas (30). High seeding rates of about 180 kg ha^{-1} are common in Central America. Modern, short varieties are usually at disadvantage with weedy rice, which is usually more competitive than the crop (39). Weedy rices from Colombia outcompeted the modern variety Oryzica-1 when planted at a 1:1 ratio under experimental field conditions (71).

Crop rotation and fallow also play an important role in weedy rice management. Herbicides are used in rotation crops, especially soybean and sorghum, to eliminate weedy rice in the U.S. Farmers typically rotate out of rice and into soybeans for one growing season to limit weedy rice infestations in the subsequent rice crop (5). Field trials were established in an area heavily infested (about 50% based on panicle counts) with weedy rice in Colombia for 3 years (6 cropping cycles) to determine the effect of crop rotation on rice yield and weedy rice infestation (73). Rotation of rice with soybeans every other cropping cycle or with 3 successive soybean cycles drastically reduced weedy rice infestation (to about 10%) and increased rice yield from 4 to 6 ton ha^{-1} at the end of the experiment. Under the same schemes, sorghum or *Crotolaria* (as a cover crop) better suppressed weedy rice and further improved rice yields.

Weedy rice infestations are not severe in Uruguay where, in addition to widespread use of certified seed, there is a long history of crop-pasture rotations (43). Crop rotation with sorghum, soybeans, or maize in Brazil also effectively reduced the weedy rice soil seed bank (4,65). Viability of seed buried at 12 cm decreased from 98% at the onset of the experiment to 69, 46, 11, and 1% after 1, 2, 3, and 4 years, respectively (41). A fallow period after rice harvest also promotes weedy rice seed decay. Weedy rice seed on the soil surface lost its viability after 12 months in Brazil, and the size of the seed bank (up to 15-cm depth) was reduced to 10% during the same period (67).

The intensive preparation of the soil under water-saturated conditions known as puddling was effective in Spain in reducing number of weedy rice panicles at the end of the cropping season (16). Puddling by itself, however, is not totally effective in controlling weedy rice. Most of the rice area in Venezuela is puddled, yet weedy rice remains a serious problem. Puddling also adds to the production costs, is hard on farm machinery, and is fuel expensive.

Manual weedy rice control is practiced by small farmers around the globe, but also at larger operations, especially at maturity, when weedy rice is more easily distinguished from the crop and rogued out or its panicles slashed. This can only be successfully performed when infestation levels are rather low and works best if the weedy plants are completely removed from the field. It does not increase rice yield, but does lower the level of weedy rice seed in rice. It is imperative that this be done in the part of the field to be harvested for saved seed for planting. When infestations are high, farmers often slash the panicles of tall, early maturing weedy rice plants, which provides only a temporary cosmetic improvement. This practice stimulates flowering of secondary tillers that will result in weedy rice seed maturing along with that of the crop.

17.4.2 Chemical Control and Herbicide-Resistant Rice

Weedy rice control with herbicides in the rice crop relies mostly on non-selective products that can be used only if contact with the crop can be avoided. There are only a few products with marginal selectivity, which requires accurate timing to prevent crop toxicity. Stale seedbed preparation is a useful practice involving herbicides. In this technique, rice planting is delayed to stimulate and allow weedy rice emergence, which is then eliminated with a broad-spectrum herbicide such as glyphosate. Stale seedbed preparation is common in Colombia and other Latin American countries and it is also practiced in other rice-growing regions (34,38). Stale seedbed has also proven to be effective for controlling weedy *O. longistaminata* in Tanzania (Riches 2004, Natural Resources Institute, U.K., personal communication). A drawback of this technique is the shortening of the cropping season and, in rain-fed production areas, there is a risk of not having suitable conditions for mechanical sowing once the rainy season begins. Glyphosate is typically used in no-till systems prior to mechanical planting or immediately after to control weedy and volunteer rice emerging after irrigation or with residual soil moisture.

Preplant application of dimethenamid plus pretilachlor to flooded soil prevented weedy rice emergence by about 90% under experimental conditions in Italy, but the herbicide mixture is not commercially used (33). Molinate, thiobencarb, and oxadiazon are typically used for weedy rice control as preplant or preemergence treatments in several parts of the world (34,95). Dalapon at high doses has been used in Italy for weedy rice control prior to crop planting (37). Molinate incorporated to the soil prior to planting controlled weedy rice in water- and drill-seeded rice by 79 and 49%, respectively (61). Oxadiargyl has also been applied to standing water to eliminate weedy rice (107). Time must be allowed for dissipation of residual activity by this herbicide, shortening the season for rice. Atrazine has been used to control *O. latifolia* several weeks before planting in Costa Rica.

Some treatments can be used with the standing crop during the cropping season. Growth regulators such as maleic hydrazide and herbicides that suppress weedy rice seed production can be used with short-cycle rice varieties. Applications are made when the crop grain is already mature but weedy rice panicles are still emerging or immature (2). Wiping weedy rice plants with a dilute solution of glyphosate or cycloxydim using a rope-wick applicator substantially reduced weedy rice seed viability (33). Herbicides such as sethoxydim at panicle initiation or amidochlor (applied at 90% heading) also improved weedy rice control (61). These treatments are aimed at reducing seed set in weedy rice; however, they increase production costs and do not prevent rice yield losses already determined by the season-long interference by weedy rice.

Herbicide-resistant rice varieties have been introduced or are being tested in several parts of the world, mostly aiming to weedy rice control. Non-transgenic imidazolinone-herbicide resistant rice (IMI rice) was recently introduced in the U.S. and is being rapidly adopted as a tool to control weedy rice. It was planted in about 80,000 ha and is expected to at least double in area in 2004 (93). IMI rice is now also grown in Costa Rica, Colombia, and Brazil. Transgenic glufosinate-resistant rice has also been developed and field-tested and is expected to be released by the end of 2004 (24,94,98–100).

Glufosinate-resistant hybrid rice is also under development in China and Korea for the purpose of simplifying hybrid-seed production procedures with selective herbicides (i.e., culling of herbicide-sensitive pollen parents after fertilization) (18). Transgenic glyphosate-herbicide-resistant rice is also in development (8,94).

There are major concerns with herbicide-resistant varieties, especially regarding the likely flow of herbicide resistance genes to weedy rice: the problem of managing herbicide-resistant crop volunteers and new selection pressures imposed on rice weeds that could result in the evolution of new cases of herbicide resistance in weedy rice (85). Herbicide-resistant gene flow from cultivated rice to weedy rice is discussed in detail in Chapter 20.

17.5 THE SPREAD OF WEEDY RICE

17.5.1 Domestication and Weediness

What makes being congeneric and even conspecific weedy rice a troublesome weed in the rice crop? A handful of attributes differentiates cultivated rice from weedy forms. The weedy forms resemble wild types in many respects or crops gone feral. An essential difference between wild and domesticated plants is whether or not they can propagate themselves without the farmer's assistance (13). Weedy rice is basically domesticated. Wild plants exist without human intervention; few weeds can.

Wild and weedy rices typically shed and disseminate their mature seed, which have varying degrees of dormancy. This guarantees dispersion over time. Domestication of rice involved selection against seed shattering (disarticulation of the mature inflorescence) and abolishment of secondary dormancy. A small degree of seed shattering is considered desirable for easy threshing (especially where not done mechanically), and primary dormancy is needed to prevent preharvest sprouting of seeds on the plant after maturation, but before crop harvest (13,48).

Increased seed production and more determinate growth are also characteristics associated with domestication (14,49). Domestication of rice also involved selection against photoperiodic controls of flowering (domesticated rice is day-length independent), as well as reductions in height with the green revolution, and modifications of other aspects of plant architecture (90). Many domestication related traits are controlled by single Mendelian loci that are located on a few chromosomes (11,105,115). Therefore, it would not be surprising that weedy rices that emerge in association with rice cultivation could be the result of the rapid loss of a few key domestication alleles when they are no longer preserved by selection (90).

Similar to *O. rufipogon*, weedy rice disperses its seed by shattering as the seeds mature on the panicle. The seeds then germinate over long periods due to varying levels of secondary dormancy. There is ample variability in the degree of shattering among weedy rice biotypes (19,36). Shattering in *Oryza* species is quantitatively controlled by several genes located on different chromosomes (31,42,77,79,83,96). The major shattering gene *Sh3*, originally found in *O. glumaepatula*, is also present in *O. rufipogon*, *O. meridionalis,* and *O. glaberrima* (77,78). Loss of its function in *O. sativa* is a primary aspect of domestication and if domestic rice can back-mutate to shattering, the result would be a weedy rice that is a crop mimic, which could not be distinguished from rice during vegetative growth.

Seed dormancy and awnedness are determined by a large number of genes distributed on almost all rice chromosomes (13,63,101,112). Seed dormancy and red pericarp also appear to be associated (47,48). Seed dormancy in some weedy *O. sativa* strains is imposed by the hull and by both the hull and pericarp/testa in others. Shattered grains first have primary dormancy but later germination gradually improves (35,36). Removal of the seed hull breaks dormancy (80).

Weedy rices are also commonly distinguished in the field from modern cultivated varieties by a series of morphological traits. There is a generalized notion among agronomists and farmers along with experimental evidence that weedy rice is taller, produces more tillers, has lighter green leaves, and matures seed earlier than cultivated rice. It also frequently has awns and pubescent seeds (15,27,82,87–89), but there is tremendous variation in the field. Some weedy populations overlap in flowering time with cultivars. Weedy rice seems to rapidly evolve and adapt to particular locations or conditions. Crop mimicry in weedy rice renders its detection and control difficult in early growth stages. Farmers often need to wait until panicle emergence to differentiate weedy from cultivated rice, leaving little time and only a few tools to prevent seed set and shedding.

In my own experience, it is not uncommon to find 6 to 10 morphologically distinct types of weedy rice on a single farm or rice field in Latin American countries. Often, some weedy types are almost identical to cultivars to the naked eye, except for a shattering habit in the weedy form (72). There are also weedy crop mimics that resist shattering but possess red grains (88), similar to ecotypes described by Vaughan et al. (111) in the U.S. Latin American agronomists and farmers refer to these crop mimics as "varietal" weedy rices. The genetics of these crop mimics have not been studied, but they could represent partially dedomesticated forms arising from back mutation. Some Korean crop mimics with *indica* characteristics, including long grains, which have a restricted distribution, probably derived from old cultivars (20,21,103,104). Several historical records of old Korean literature on rice culture indicate that red rice had been widely cultivated and has survived as a volunteer weed in rice fields up to the present (51).

17.5.2 Contaminated Seed

It is not surprising that the spread of weedy rice is increasing given the similarity between weedy rice and the rice-crop. Contaminated crop seed is a primary factor in the dissemination of weedy rice. Unfortunately, even certified seed allows some level of contamination in many countries. The problem is exacerbated by the practice of saving seed. In Latin America, where there are high infestation levels and great variability in weedy rice morphology, farmers often produce their own seed or use commercial grain for sowing (81). The lack of geographical separation of accessions in our studies of weedy rice from five Latin American countries supports the view that weedy rice in this region frequently moves among rice growing areas (Skov, Valverde, Wellendorf, and Andersen, unpublished).

The allowance of weedy rice seed in certified rice seed is one seed per kg in Costa Rica. A lower category (also common in other Latin American countries) designated as authorized seed (first generation after certified seed) allows two weedy rice seeds per kg. In 1998–1999, about 26% of the areas dedicated to certified seed production were rejected, about one-fourth of that because of the presence of weedy rice. In 2003, about 38% of the area dedicated to certified seed production was rejected, about 43% of that because of presence of weedy rice (Alizaga 2004, Oficina Nacional de Semillas, Costa Rica, personal communication). Similarly, sampled seed lots being used by farmers in the Santa Maria region of the Rio Grande do Sul, Brazil were contaminated with two types of weedy rice (66). Only 17% of the samples were free of weedy rice; 53% of the samples contained more than the legally accepted maximum in Brazil (4 weedy rice seeds per kg) and of these, 13% had more than 40 weedy rice seeds per kg. More than half of the farmers used saved seed and only 10% planted certified seed; the rest acquired seed from neighbors.

An aggressive program of Brazilian certified seed production has improved quality to a point where about 75% of the certified seed lots contained no weedy rice and about 95% had 4 or less weedy rice seeds per kg (25). The weedy rice allowance in certified seed has now been reduced to none in Rio Grande do Sul (92) whereas in the mid-1980s it was 24 seeds per kg (26).

Even if strict standards are enforced, crop mimics differing from the crop only by their shattering habit would not be detected by seed analysis. Weedy rice also proliferates in Portugal because farmers save commercial grain to be used as seed for the following cropping season. The European Union (E.U.) certifies rice seed only if it contains no more than 6 weedy rice seeds per kg (22).

The results from analysis of 351 samples throughout Vietnam showed that contamination of crop seed with weedy rice seed was high (314 seeds per kg). In the Vinh Long Province, it could be as high as 500 seeds per kg; the lowest contamination found was 80 seeds per kg (64). Contamination by weedy rice has forced Philippine farmers who sow farm-saved seed to limit its use to three cropping seasons after which it is replaced with a commercial cultivar that again becomes progressively contaminated as seed is saved and used over successive harvests (75). This practice evidently is not sustainable, because it would not have a major impact on the soil seed bank.

The importance of informal seed exchange mechanisms is illustrated by the wide diffusion of varieties that were never officially released or the almost immediate introduction of newly released varieties from one country into another. In Panama, a variety known popularly as VIOAL was apparently taken from an adaptation trial from the breeding program that resulted in the liberation of cultivar L-7. VIOAL is the acronym in Spanish for International Observation Rice Nursery for Latin America, a program that evaluates advanced lines generated in the region by the National Agricultural Research Development Programs and by CIAT (Centro Internacional de Agricultura Tropical) (45). Within a year after the variety Fedearroz-50 was released in Colombia, some Costa Rican farmers were already planting small plots with this variety for seed increase. Informal importation of rice seed can aggravate weedy rice problems, especially if contaminated with new biotypes.

17.5.3 Farmer Attitudes about Weedy Rice and Its Dissemination

There is little published information on how farmers deal with new genetic combinations resulting from introgression and on their propensity to remove off-type plants, or even if they can distinguish weeds from cultivars (56). The majority of almost 5000 farmers surveyed in Vietnam, who primarily practice rice monoculture, can morphologically distinguish weedy rice, mostly because it is taller than their planted varieties, and they see the presence of awns and of easy shattering. They also recognize that weedy rice causes yield losses and reduces the quality of milled rice. Most weedy rice in Vietnam has a red pericarp. They take action to remove it from their fields, as 80% rogue weedy rice plants at booting to flowering. Still few attribute infestations to contaminated seed and rather believe that weedy rice evolves from cultivated rice (and they may be partially right) or emerges from the soil (19).

Farmers from areas of Malaysia where weedy rice had been present for a long period recognized it better than those from areas where the infestation was incipient. In other regions in the world, farmers also identified weedy rice by some of its most noticeable morphological characteristics, such as taller plants, pigmentation, and shorter grain compared to cultivated rice. Farmers less exposed to weedy rice problems were also less knowledgeable concerning possible control methods (113).

Farmers' perceptions differ among areas and cropping systems, and determine the control practices most likely to be adopted for weedy rice management. Thus the few Vietnamese farmers who positively value the impact of practices such as seeding method, changing rice cultivars, using certified seed, and legume cropping in rotation with rice are more prone to modifying their cropping practices. Farmers in Malaysia are more interested in finding chemical control options for weedy rice (114). Heavy weedy rice infestations in Brazil have been attributed to a high percentage of rice growing on rented land (26). The same occurs in Costa Rica, where farmers who sow rented land have no interest in the additional investments required to adopt mid- and long-term weedy rice control strategies for which there would be no guaranteed benefits.

17.6 GOING UNDETECTED

The evolution of weedy rice is closely associated with farming practices and its control and prevention of spreading will depend on local agroecological conditions and farmers' attitudes (114). The patterns of introduction and dissemination of weedy rice in areas of the world where no *O. rufipogon* is present or where it is confined to locations away from rice growing areas illustrate how feral and other weedy forms can go undetected. The appearance and dissemination of weedy rice in Malaysia, briefly discussed by Vaughan et al. (Chapter 16), is perhaps one of the clearest cases of endoferality in rice. This endoferal weedy rice arose from cultivated rice, probably as a result of volunteer seeding (use of seed dropped by cultivated rice for establishing next season crop) over large areas (1). Malaysian cultivars tend to shatter, and volunteer seeding probably led to the selection of shattering rice ecotypes, later recognized as weedy rice.

Even if weedy rice has a colored pericarp, it may go undetected in seed lots if the external seed morphology is similar to that of the variety; the red caryopsis may only be recognized after hulling as in Latin America (Figure 17.2). Hybridization and introgression may go undetected until new ecotypes of weedy rice become widespread. This is illustrated with a weedy rice type commonly found in Costa Rica that farmers know as *arrozon* ("very tall rice"). For years it has been easy to distinguish it from relatively early stages of growth because of its fast growth and pale green leaves. Later it develops a long panicle with rounded grains quite distinct from commercial varieties. Rice

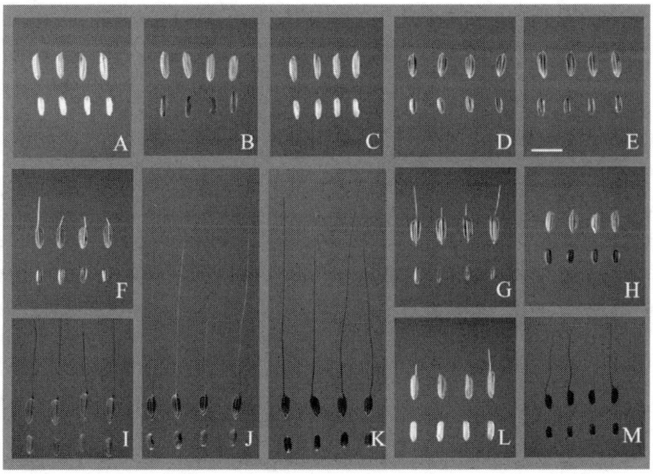

FIGURE 17.2 Variability in seed characteristics among selected weedy rice accessions from Costa Rica. Accessions B through K — (B) shattering red rice, (C) shattering "varietal" type with white grain, (D) typical "arrozon" weedy rice, (E) dwarf arrozon, (F) straw-awn arrozon, (G) reddish-awn arrozon, (H) red-grain arrozon, (I) medium, reddish-awn arrozon, (J) long, straw-awn arrozon, (K) black-hull weedy rice — were collected at a single field planted with rice cv CR-5272 (accession A) in Parrita, Puntarenas Province. Two additional accessions are included for comparison: (L) rice cv CR-1113, (M) *Oryza latifolia*. Bar = 10 mm, except for J and K, which were reduced to 0.65 of the rest.

farmers are beginning to complain about arrozon mimics that are the same height of the cultivar and mature almost simultaneously. In Surinam, it is common to find weedy rice with typical characteristics of their cultivated rice varieties, such as extra-long grains and smooth leaves (62).

Abandonment of old cultivars and landraces could continue generating locally adapted weedy rice populations. An estimated 29% of the rice area in Asia is still planted to landraces. In some countries, especially those where rice is cultivated under rainfed conditions, farmers still grow traditional varieties on a high proportion of the planted land (53). Farmers may abandon planting their landraces to take advantage of new opportunities in the changing rice cropping environment and to realize higher yields (10), providing an opportunity for these landraces to go feral and become weedy. Those landraces that were grown before the introduction of new varieties could have left shattered seed in the seed bank, as many are heavy shattering strains.

Volunteer landrace plants may not be initially seen as weeds by the farmer, who is more familiar with their characteristics than the modern varieties. They may also go undetected via the soil seed bank if rice production is abandoned. Two biotypes were found in Costa Rica, differentiated by the presence of anthocyanins in the apiculus in one of them, which were similar to and abandoned variety (97). The biotypes were found exclusively in a field that was planted to rice for the first time after 16 years, so the weed is probably a feral weedy rice that remained in the soil seed bank.

The worse-case scenario I have seen is that of farmers in the Chiriqui Province of Panama who believe or have been misadvised that a seeding crop seed that is heavily contaminated with weedy rice is good way to achieve acceptable yields even in bad years. Ironically, in a field destined for seed production, I easily assessed 21 apparently distinct weedy rice types based on morphological traits. Studies of the putative evolution of feral rice in recently abandoned fields in Sri-Lanka are described in Chapter 18.

An additional problem is arising in northern Latin America with the introduction of herbicide-resistant rice. Current IMI rice varieties are short-cycle *japonicas* that mature in about 85 days, whereas most conventional varieties have a longer cycle (115 to 125 days). In addition to possible overlapping flowering with early maturing weedy rices that would facilitate hybridization, managing IMI volunteers becomes critical when a conventional variety is subsequently planted as part of the herbicide resistance management practices.

The suggested stewardship programs are based on sequential application of imidazolinone herbicides aimed at eliminating all weedy rice individuals within the cropping season, rotation to conventional varieties every year or after a maximum of two cropping seasons of herbicide-resistant rice, roguing individuals that escape chemical control, and exclusion of weed saving. Given the characteristics of rice farming in areas such as Latin America, it is unlikely that these programs will be widely respected. Thus, the herbicide-resistant rice technology is prone to failure. Because elimination of rice volunteers is accomplished primarily by postemergence application of glyphosate prior to planting, this practice would be ineffective for volunteers of glyphosate-resistant rice once introduced. Also, hybridization between herbicide-resistant varieties and weedy rice will probably increase the diversity of weedy rice populations by the introgression of genes from *japonica* cultivars. Dedomestication of the herbicide-resistant varieties abandoned as a result of widespread resistance in weedy rice populations will add to the problem. Transgenes conferring traits that increase the fitness of hybrids between transgenic and weedy rice would surely aggravate weedy rice infestations, unless containment or mitigation technologies are instituted.

Weedy rice is a complex problem in rice production that will not be solved with short-term, individual tactics. The ability of weedy rice to adapt to varying cropping conditions and to mimic cultivars makes it tremendously difficult to control. Designing and implementing weedy rice management strategies, particularly integrated weed control tactics, is a real challenge given the bioecological characteristics of weedy rice and the need to profitably produce the crop. The cost of weed control is already elevated in high-input rice production systems. The added cost of weedy rice management and the possible yield penalties associated with modifications of the cropping system (e.g., delayed planting) could render rice production unfeasible at today's prices. The transgenic era has increased our awareness and interest in the genetic and agroecological relationships between cultivated and weedy rice. Hopefully, the basic and practical knowledge derived from this interest will be conveyed to the rice farmer, who has the final decision about implementing sound weedy rice control measures.

ACKNOWLEDGMENTS

The author acknowledges support from the Danish Council for Development Research to study weedy rice in Latin America.

LITERATURE CITED

1. Abdullah MZ, Vaughan DA, Watanabe H, Okuno K. 1996. Origin and diversity of weedy rice (*padi angin*). In *Ecology of major weeds and their control in direct seeding rice culture of Malaysia*, Watanabe H, Man A, Zuki M, Eds., pp. 167–181. Ibaraki, Japan: Japan International Research Center for Agricultural Sciences. Seberang Peray, Malaysia: Malaysian Agricultural Research and Development Institute. Kedah, Malaysia: Muda Agricultural Development Authority.
2. Agostinetto D, Fleck NG, Menezes VG, Costa ELN. 2002. Supressao da producao de sementes de arroz-vermelho pela aplicacao de herbicidas em arroz irrigado. *Pesq. Agropec. Bras.* 37:57–65.
3. Akimoto M. 1999. Bio-systematics of the *Oryza* AA genome wild taxa (*Oryza sativa* complex) by morpho-physiological traits, isozymes and RFLPs. PhD thesis. Hokkaido, Japan: Hokkaido University, 117 pp.
4. Andres A, Avila LA, Marchezan E, Menezes V. 2001. Rotação de culturas e pousio do solo na redução do banco de sementes de arroz vermelho em solo de várzea. *Rev. Bras. Agrociên.* 7:85–88.
5. Askew SD, Street JE, Shaw DR. 1998. Herbicide programs for red rice (*Oryza sativa*) control in soybean (*Glycine max*). *Weed Technol.* 12:103–107.
6. Avila LA, Marchezan E, Souto J, Siqueira CA. 1999. Interferência do arroz vermelho sobre o arroz irrigado. In: *Congresso brasileiro de arroz irrigado, 1; reunião da cultura do arroz irrigado*, Pelotas: Embrapa Clima Temperado, pp. 594–596.
7. Avila LA, Andres A, Marchezan E, Menezes VG. 2000. Banco de sementes de arroz vermelho em sistemas de semeadura de arroz irrigado. *Ciencia Rural* 30:773–777.
8. Baldwin FL. 1999. The value and exploitation of herbicide tolerant crops in the U.S. *Proc. Brighton Crop Prot. Conf. — Weeds,* Brighton, U.K., pp. 653–660.
9. Baltazar AM, Janiya JD. 2000. Weedy rice in the Philippines. In *Wild and weedy rices in rice ecosystems in Asia — a review*, Baki BB, Chin DV, Mortimer M, Eds., pp. 74–78. Manila, Philippines: International Rice Research Institute, Limited Proc. No. 2.
10. Bellon MR. 2004. Conceptualizing interventions to support on-farm genetic resource conservation. *World Dev.* 32:159–172.
11. Bres-Patry C, Lorieux M, Clément G, Bangratz M, Ghesquière A. 2001. Heredity and genetic mapping of domestication-related traits in a temperate *japonica* weedy rice. *Theor. Appl. Genet.* 102:118–126.
12. Buu BC. 2000. Collecting wild relatives of rice from the Mekong Delta, Vietnam: distinguishing wild rice from weedy rice. In *Wild and weedy rices in rice ecosystems in Asia — a review*, BB Baki, DV Chin, M Mortimer, Eds., pp. 5–6. Manila, Philippines: International Rice Research Institute, Limited Proc. No. 2.
13. Cai H-W,·Morishima H. 2000. Genomic regions affecting seed shattering and seed dormancy in rice. *Theor. Appl. Genet.* 100:840–846.
14. Cai H-W, Morishima H. 2002. QTL clusters reflect character associations in wild and cultivated rice. *Theor. Appl. Genet.* 104:1217–1228.
15. Canal RE. 2000. Caracterización de biotipos y evaluación de niveles de infestación de la maleza arroz rojo (Oryza sativa) en el distrito de riego del río Zulia, Norte de Santander, Colombia. Tesis MS. Colombia: Postgrado en Agronomía, Universidad Nacional Experimental del Táchira. 81 pp.

16. Catala MM. 1995. Chemical and cultural practices for red rice control in rice fields in Ebro Delta (Spain). *Crop Prot.* 14:405–408.
17. Catala MM, Querol S. 1999. Influencia del arroz salvaje (*Oryza sativa* L.) en el rendimiento en enteros. Logrono: *Actas VII Congr. Soc. Espanola Malherbol.*, pp. 177–182.
18. Chen LJ, Lee DS, Song ZP, Suh HS, Lu B-R. 2004. Gene flow from cultivated rice (*Oryza sativa*) to its weedy and wild relatives. *Ann. Bot.* 93:67–73.
19. Chin DV, Hach CV, Thanh NC, Tai NT. 2000. Weedy rice situation in Vietnam. *FAO global workshop on red rice control*, Cuba, Rome: Food and Agriculture Organization, pp. 67–74.
20. Cho Y-C, Chung T-Y, Suh H-S. 1995. Genetic characteristics of Korean weedy rice (*Oryza sativa* L.) by RFLP analysis. *Euphytica* 86:103–110.
21. Choi HC. 2000. Ecological and agronomical variation of Korean weedy rice accessions. *Proc. Int. Workshop Biol. Management of Noxious Weeds Sust. Labor Saving Rice Prod.*, Ibaraki, Japan: National Agriculture Research Center, Ministry of Agriculture, Forestry and Fisheries, pp. 22–39.
22. Da Silva LM, Valério-Menino JP, Rodrigues CF. 2000. Eficacia de herbicidas de post-emergencia contra el arroz rojo en Portugal. *FAO global workshop on red rice control*, Cuba, Rome: Food and Agriculture Organization, pp. 109–113.
23. Davies ELP. 1984. Mechanical and chemical methods for the control of annual wild rice (*Oryza barthii*). *Trop. Pest Manage.* 30:319–320.
24. Department for Environment Food and Rural Affairs. 2004. Notification C/GB/03/M5/3 from Bayer Cropscience Ltd. for consent to market LLRICE62. London: Department for Environment Food and Rural Affairs, 13 pp. Available http://www.defra.gov.uk/environment/gm/regulation/euconsent.htm.
25. De Souza JAB, Cordero EJ, Barros JA, Marques JB, Valério MG, et al. 2001. Arroz vermelho nas amostras de sementes analisadas nos laboratórios do IRGA de 1996 a 2000. *Anais Congr. Bras Arroz Irrigado*, pp. 744-746.
26. De Souza PR, Fisher MN. 1986. Arroz vermelho: danos causados à lavoura gaúcha. *Lav. Arroz.* 39:19–20.
27. Diarra A, Smith RJ Jr, Talbert RE. 1985. Growth and morphological characteristics of red rice (*Oryza sativa*) biotypes. *Weed Sci.* 33:310–314.
28. Diarra A, Smith RJ Jr, Talbert RE. 1985. Interference of red rice (*Oryza sativa*) with rice (*O. sativa*). *Weed Sci.* 33:644–649.
29. Eleftherohorinos IG, Dhima KV, Vasilakoglou IB. 2002. Interference of red rice in rice grown in Greece. *Weed Sci.* 50:167–172.
30. Estorninos LE Jr, Talbert RE, Gealy DR. 1998. Response of Stuttgart red rice to different population densities of three rice cultivars. *Proc. Southern Weed Sci. Soc.* 51:199.
31. Eiguchi M, Sano Y. 1990. A gene complex responsible for seed shattering and panicle spreading found in common wild rice. *Rice Genet. Newsl.* 7:105–107.
32. Federici MT, Vaughan D, Tomooka N, Kaga A, Wang XW, et al. 2001. Analysis of Uruguayan weedy rice genetic diversity using AFLP molecular markers. *Electr. J. Biotechnol.* 4:130–145.
33. Ferrero A. 2001. Red rice control in rice. In *The new development in rice agronomy and its effects on yield and quality in Mediterranean areas*, Chataigner J, Ed., Paris: Centre International de Hautes Etudes Agronomiques Mediterraneennes. Rome: FAO. Edirne (Turkey): Thrace Agricultural Research Institute. Available online www.ciheam.org/util/search/detail_numero.php?mot=771&langue=fr.
34. Ferrero A. 2003. Weedy rice: biological features and control. In *Weed management for developing countries. Addendum 1*, Labrada R, Ed., pp. 89–107. Rome: FAO, Plant Production and Protection Paper 120 Add1.
35. Ferrero A, Vidotto F. 1998. Germinability after flowering, shattering ability and longevity of red rice seeds. In *6th EWRS Mediterranean Symposium 1998*, Montpellier, France, pp. 205–211.
36. Ferrero A, Vidotto F. 1998. Shattering ability of red rice seeds in cultural conditions. *Med. Fac. Landb. Toegepaste Biol.* 63:839–843.
37. Ferrero A, Vidotto F, Balsari P, Airoldi G. 1999. Mechanical and chemical control of red rice (*Oryza sativa* L. var. *sylvatica*) in rice (*Oryza sativa* L.) pre-planting. *Crop Prot.* 18:245–251.
38. Fischer AJ. 1996. Integrated red rice management in Latin American rice fields. In *Proceedings Second International Weed Control Congress, Copenhagen, Denmark*, pp. 653–664.

39. Fischer AJ, Ramrez A. 1993. Red rice (*Oryza sativa*): competition studies for management decisions. *Int. J. Pest Manage.* 39:133–138.
40. Foloni LL. 1995. Adjuvant effects on sulfosate and glyphosate for control of red-rice in rice. *Proc. Brighton Crop Prot. Conf. — Weeds,* Brighton, U.K., pp. 743–746.
41. Franco DF, Magalhães AM, Petrini JA, Andres A. 2001. Viabilidade de sementes de arroz vermelho. *Anais Congr. Bras. Arroz Irrigado*, pp. 680–682.
42. Fukuta Y, Tadashi Y. 1998. Mapping of a shattering resistance gene in a mutant line SR-5 induced from an *indica* rice variety, Nan-jing11. *Breed. Sci.* 48:345–348.
43. García-Préchac F, Ernst O, Siri-Prieto G, Terra JA. 2004. Integrating no-till into crop-pasture rotations in Uruguay. *Soil Till. Res.* 77:1–13.
44. Gealy DR, Tai TH, Sneller CH. 2002. Identification of red rice, rice, and hybrid populations using microsatellite markers. *Weed Sci.* 50:333–339.
45. Guimarães EP, Borrero J, Ospina-Rey Y. 1996. Genetic diversity of upland rice germplasm distributed in Latin America. *Pesq. Agrop. Bras.* 31:187–194.
46. Gómez de Barreda D, Sendra J, Carreres R, Del Busto A, Biendicho E. 2000. Red rice in Spain. *FAO global workshop on red rice control*, Cuba, Rome: Food and Agriculture Organization, pp. 51–54.
47. Gu X-Y, Chen Z-X, Foley ME. 2003. Inheritance of seed dormancy in weedy rice. *Crop Sci.* 43:835–843.
48. Gu X-Y, Kianian SF, Michael E, Foley ME. 2004. Multiple loci and epistases control genetic variation for seed dormancy in weedy rice (*Oryza sativa*). *Genetics* 166:1503–1516.
49. Harlan JR. 1975. *Crops and man.* Madison, WI: American Society of Agronomy. 295 pp.
50. Harushima Y, Nakagahra M, Yano M, Sasaki T, Kurata N. 2002. Diverse variation of reproductive barriers in three intraspecific rice crosses. *Genetics* 160:313–322.
51. Heu MH, Cho YC, Suh HS. 1990. Cross affinity of Korean weedy rice to the cultivars. *Korean J. Crop Sci.* 35:233–238.
52. Holm L, Doll J, Holm E, Pancho J, Herberger J. 1997. *World weeds. Natural histories and distribution.* New York: John Wiley and Sons, Inc., pp. 531–547.
53. Hossain M, Gollin D, Cabanilla V, Cabrera E, Johnson N, et al. 2003. International research and genetic improvement in rice: evidence from Asia and Latin America. In *Crop variety improvement and its effect on productivity: the impact of international agricultural research*, Evenson RE, Gollin D, Eds., pp. 5–108. Wallingford, U.K.: CABI Publishing.
54. Ishikawa R, Yamanaka S, Kanyavong K, Fukuta Y, Sato Y-I, et al. 2002. Genetic resources of primitive upland rice in Laos. *Econ. Bot.* 56:192–197.
55. Itani T, Tamaki M, Arai E, Horino T. 2002. Distribution of amylose, nitrogen, and minerals in rice kernels with various characters. *J. Agric. Food Chem.* 50:5326–5332.
56. Jarvis DI, Hodgkin T. 1999. Wild relatives and crop cultivars: detecting natural introgression and farmer selection of new genetic combinations in agroecosystems. *Mol. Ecol.* 8:S159–S173.
57. Johnson DE. 1997. *Weeds of Rice in West Africa.* Bouake, Côte d'Ivoire: West Africa Rice Development Association, 312 pp.
58. Johnson DE, Riches CR, Kayake J, Sarra S, Tuor FA. 2000. Wild rice in sub-Saharan Africa: its incidence and scope for improved management. *FAO global workshop on red rice control*, Cuba, Rome: Food and Agriculture Organization, pp. 87–94.
59. Johnson DE, Wopereis MCS, Mbodj D, Diallo S, Powers S, et al. 2004. Timing of weed management and yield losses due to weeds in irrigated rice in the Sahel. *Field Crops Res.* 85:31–42.
60. Kim SY, Song Y, Chung NJ, Park ST, Kim HY. 2000. Cultural practices for weedy rice control in paddy field of Korea. *Proc. Int. Workshop Biol. Management of Noxious Weeds Sust. Labor Saving Rice Prod.*, Ibaraki, Japan: National Agriculture Research Center, Ministry of Agriculture, Forestry and Fisheries, pp. 198–208.
61. Kwon SL, Smith RJ Jr, Talbert RE. 1991. Interference durations of red rice (*Oryza sativa*) in rice (*O. sativa*). *Weed Sci.* 39:363–368.
62. Khodabaks RM. 2000. Red rice control in Surinam. *FAO global workshop on red rice control*, Cuba, Rome: Food and Agriculture Organization, pp. 51–54.
63. Lin SY, Sasaki T, Yano M. 1998. Mapping quantitative trait loci controlling seed dormancy and heading date in rice, *Oryza sativa* L., using backcross inbred lines. *Theor. Appl. Genet.* 96:997–1003.

64. Mai V, Chien HV, Van V, Thi V, Suong T, Thiet LV. 2000. Rice seed contamination in Vietnam. In *Wild and weedy rices in rice ecosystems in Asia — a review*, Baki BB, Chin DV, Mortimer M, Eds., pp. 17–19. Manila, Philippines: International Rice Research Institute, Limited Proc. No. 2.
65. Marchezan E, Coradini JZ, Avila LA, Segabinazzi T. 2001. Eficiência da avaliação do banco de sementes na predição da infestação por arroz vermelho e rendimento de grãos do arroz irrigado após dois anos de rotação de cultura e pousio do solo. *Rev. Bras. Agroc.* 7:15–17.
66. Marchezan E, Menezes WL, Siqueira CA. 2001. Controle da qualidade das sementes de arroz irrigado utilizadas em Santa Maria/RS. *Ciên. Rural* 31:375–379.
67. Marchezan E, Oliveira APBB, Avila LA, Bundt ALP. 2003. Dinamica do banco de sementes de arroz-vermelho afetado pelo pisoteio bovino e tempo de pousio da area. *Planta Daninha* 21:55–62.
68. MARDI (Malaysian Agricultural Research and Development Institute). 2003. Management of weedy rice (padi angin) in direct-seeded rice. Available online: http://www.mardi.my.
69. Menezes VG, Da Silva PRF. 1998. Manejo de arroz vermelho atraves do tipo e arranjo de plantas em arroz irrigado. *Planta Daninha* 16:45–57.
70. Menezes VG, Da Silva PRF, Carmona R, Rezera F, Mariot CH. 1997. Interferencia do arroz vermelho no rendimento de engenho de cultivares de arroz irrigado. *Cien. Rural* 27:27–30.
71. Montealegre FA, Clavijo J. 1991. Análisis de la competencia entre tres tipos de arroz rojo y la variedad Oryzica-1. *Agron. Colombiana* 8:342–349.
72. Montealegre FA, Clavijo J. 1991. Tipos de arroz rojo en Colombia. *Fedearroz* 40:16–23.
73. Montealegre FA, Vargas JP. 1989. Efecto de algunas prácticas culturales sobre la población de arroz rojo y los rendiientos del arroz comercial. *Arroz* (Colombia) 38:19–24.
74. Morishima H. 2001. Evolution and domestication of rice. In *Rice genetics IV. Proc. 4th Intl. Rice Genet. Symp.*, Khush GS, Brar DS, Hardy B, Eds., pp. 63–77. New Dehli: India: Science Publishers Inc. Los Baños (Philippines): International Rice Research Institute.
75. Mortimer M, Pandey S, Piggin C. 2000. Weedy rice: approaches to ecological appraisal and implications for research priorities. In: *Wild and weedy rices in rice ecosystems in Asia — a review*, Baki BB, Chin DV, Mortimer M, Eds., pp. 97–105. Manila, Philippines: Interntional Rice Research Institute, Limited Proc. No. 2.
76. Mouret JC. 2000. Strategies and effects of cropping practices on the levels of red rice infestation in the Camargue (France). In *Wild and weedy rices in rice ecosystems in Asia — a review*, Baki BB, Chin DV, Mortimer M, Eds., pp. 115–122. Manila, Philippines: International Rice Research Institute, Limited Proc. No. 2.
77. Nagai YS, Doi K, Yoshimura A. 2002. A seed shattering gene, *Sh3*, and its inhibitor gene found in an African cultivated rice, *Oryza glaberrima* Steud. *Rice Genet. Newsl.* 19:76–77.
78. Nagai YS, Sobrizal T, Sanchez PL, Kurazaku T, Doi K, et al. 2002. *Sh3*, a gene for seed shattering, commonly found in wild rices. *Rice Genet. Newsl.* 19:74–75.
79. Nagao S, Takahashi M. 1963. Trial construction of 12 linkage groups in Japanese rice. *J. Fac. Agr. Hokkaido Univ.* 53:72–130.
80. Naredo MEB, Juliano A, Lu BR, Guzman F, Jackson MT. 1998. Responses to seed-dormancy breaking treatments in rice species. *Seed Sci. Technol.* 26:675–689.
81. Noldin JA. 2000. Red rice status and management in the Americas. In *Wild and weedy rices in rice ecosystems in Asia — a review*, Baki BB, Chin DV, Mortimer M, Eds., pp. 21–24. Manila, Philippines: International Rice Research Institute, Limited Proc. No. 2.
82. Noldin JA, Chandler JM, McCauley GN. 1999. Red rice (*Oryza sativa*) biology. I. Characterization of red rice ecotypes. *Weed Technol.* 13:12–18.
83. Oba S, Kikuchi F, Maruyama K. 1990. Genetic analysis of semidwarfness and grain shattering of Chinese rice variety "Ai-Jio-Nana-Te." *Jpn. J. Breed.* 40:13–20.
84. Oka HI. 1991. Genetic diversity of wild and cultivated rice. In *Rice biotechnology,* Khush GS, Toenniessen GH, Eds., pp. 55–81. Wallingford: CAB Int.
85. Olofsdotter M, Valverde BE, Madsen KH. 2000. Herbicide resistant rice (*Oryza sativa* L.): global implications for weedy rice and weed management. *Ann. Appl. Biol.* 137:279–295.
86. Ortiz A. 2000. Efecto de algunos ecotipos de arroz rojo sobre el rendimiento en molino y transparencia de los cultivares de arroz Fonaiap 1, Cimarron y Zeta-15. *Agron. Trop.* 50:633–643.
87. Ortiz A, Lopez L, Lizaso J. 1999. Desarrollo y caracterización morfológica de ecotipos de arroz rojo y cultivares de arroz en Venezuela. *Agron. Trop.* 49:51–67.

88. Ortiz A, Lopez L, Lizaso J. 2000. Comparación de algunos componentes del rendimiento, latencia de semillas y dimensiones de los granos entre poblaciones de arroz rojo y variedades de arroz (*Oryza sativa* L.) en Venezuela. *Rev. Fac. Agron.* (*Maracay*) 26:39–51.
89. Ortiz A, Lopez L, Lizaso J, Lazo JV. 2002. Caracterización de poblaciones de arroz rojo y variedades de arroz en Venezuela. *Agron. Trop.* 52:23–44.
90. Paterson AH. 2002. What has QTL mapping taught us about plant domestication? *New Phytol.* 154:591–608.
91. Peña-Deyán JL, Ortiz A. 2001. Determinación de las especies involucradas en el complejo de maleza denominado arroz rojo en las zonas arroceras de Venezuela. *Agron. Trop.* 51:439–451.
92. Rio Grande do Sul. 2001. *Normas e padrões de produção de sementes para o Estado do Rio Grande do Sul 2000.* Porto Alegre (Brazil): Secretaria de Agricultura e Abastecimento, Departamento de Produção Vegetal, Comissão Estadual de Sementes e Mudas do Estado do Rio Grande do Sul. 160 pp.
93. Rood MA. 2004. Red rice update. *Rice J.* 107:17–20.
94. Rood MA. 2001. Almost here: herbicide-resistant rice varieties expected. *Rice J.* 104:16–18.
95. Ryang HS, Kim JK. 1998. Establishment of control system of weedy rice (*Oryza sativa*) and barnyardgrass (*Echinochloa crus-galli*) in direct-seeded rice. I. Effect of oxadiazon, molinate, thiobencarb on control of weedy rice and barnyardgrass in water-seeded rice. *Korean J. Weed Sci.* 18:106–115.
96. Sanchez PL, Kurazaku T, Hirata C, Sobrizal T, Yoshimura A. 2002. Identification and mapping of seed shattering genes using introgression lines from wild rice species. *Rice Genet. Newsl.* 19:78–79.
97. Sánchez-Olguín E. 2001. *Caracterización morfológica de las poblaciones de arroces maleza en fincas arroceras de Guanacaste y Parrita.* Tesis. Bachiller. Costa Rica: Escuela de Agronomía, Sede Regional de San Carlos, Instituto Tecnológico de Costa Rica. 54 pp.
98. Sankula S, Braverman MP, Jodari F, Linscombe SD, Oard JH. 1997. Evaluation of glufosinate on rice (*Oryza sativa*) transformed with the BAR gene and red rice (*Oryza sativa*). *Weed Technol.* 11:70–75.
99. Sankula S, Braverman MP, Linscombe SD. 1997. Response of BAR-transformed rice (*Oryza sativa*) and red rice (*Oryza sativa*) to glufosinate application timing. *Weed Technol.* 11:303–307.
100. Sankula S, Braverman MP, Linscombe SD. 1997. Glufosinate-resistant, BAR-transformed rice (*Oryza sativa*) and red rice (*Oryza sativa*) response to glufosinate alone and in mixtures. *Weed Technol.* 11:662–666.
101. Sato S, Ishikawa S, Shimono M, Shinjyo C. 1996. Genetic studies on an awnness gene *An-4* on chromosome 8 in rice, *Oryza sativa* L. *Breed. Sci.* 46:321–327.
102. Song ZP, Lu B-R, Zhu YG, Chen JK. 2003. Gene flow from cultivated rice to the wild species *Oryza rufipogon* under experimental field conditions. *New Phytol.* 157:657–665.
103. Suh HS, Sato Y I, Morishima H. 1997. Genetic characterization of weedy rice (*Oryza sativa* L.) based on morpho-physiology, isozymes and RAPD markers. *Theor. Appl. Genet.* 94:316–321.
104. Tang LH, Morishima H. 1996. Genetic characteristics and origin of weedy rice. In *Origin and differentiation of Chinese cultivated rice*, Wang X, Sun C, Eds., pp. 211–218. China: China Agricultural University Press. Available online: http://www.carleton.ca/~bgordon/Rice/papers/tang96.htm.
105. Thomson MJ, Tai TH, McClung AM, Lai X-H, Hinga ME, Lobos KB, et al. 2003. Mapping quantitative trait loci for yield, yield components and morphological traits in an advanced backcross population between *Oryza rufipogon* and the *Oryza sativa* cultivar Jefferson. *Theor. Appl. Gen.* 107:479–493.
106. Valverde BE, Itoh K. 2001. World rice and herbicide resistance. In *Herbicide resistance in world grains*, Powles SR, Shaner D, Eds., pp. 195–249. Boca Raton, FL: CRC Press.
107. Valverde BE, Riches CR, Caseley J. 2000. Prevention and management of herbicide resistant weeds in rice: experiences from Central America with *Echinochloa colona.* San Jose, Costa Rica: Camara de Insumos Agropecuarios. 123 pp.
108. Vaughan A, Morishima H. 2003. Biosystematics of the genus *Oryza*. In *Rice: origin, history, technology, and production*, Smith CW, Dilday RH, Eds., pp. 27–65. New York: John Wiley & Sons, Inc.
109. Vaughan DA, Morishima H, Kadowaki K. 2003. Diversity in the *Oryza* genus. *Curr. Opin Plant Biol.* 6:139–146.
110. Vaughan D, Tomooka N. 1999 Wild rice in Venezuela. *Rice Res. Genet.* 16:15–16.
111. Vaughan LK, Ottis BV, Prazak-Havey AM, Sneller C, Chandler JM, et al. 2001. Is all red rice found in commercial rice really *Oryza sativa*? *Weed Sci.* 49:468–476.
112. Wan J, Nakazaki T, Kawaura K, Ikehashi H. 1997. Identification of marker loci for seed dormancy in rice (*Oryza sativa* L.). *Crop Sci.* 37:1759–1763.

113. Watanabe H, Azmi M, Zuki M. 1996. Ecology of weedy rice (*Oryza sativa* L., locally called *padi angin*) and its control strategy. In *Ecology of major weeds and their control in direct seeding rice culture of Malaysia*, Watanabe H, Man A, Zuki M, Eds., pp. 112–166. Ibaraki, Japan: Japan International Research Center for Agricultural Sciences. Seberang Peray, Malaysia: Malaysian Agricultural Research and Development Institute. Kedah, Malaysia:Muda Agricultural Development Authority.
114. Watanabe H, Vaughan DA, Tomooka N. 2000. Weedy rice complexes: case studies from Malaysia, Vietnam, and Surinam. In *Wild and weedy rices in rice ecosystems in Asia — a review*, Baki BB, Chin DV, Mortimer M, Eds., pp. 25–34. Manila, Philippines: International Rice Research Institute, Limited Proc. No. 2.
115. Xiong LZ, Liu KD, Dai XK, Xu CG, Zhang Q. 1999. Identification of genetic factors controlling domestication-related traits of rice using an F_2 population of a cross between *Oryza sativa* and *O. rufipogon*. *Theor. Appl. Genet.* 98:243–251.
116. Zuki M, Watanabe H, Ho N-K. 1996. Weed infestation and problems in direct seeded rice fields of Malaysia. In *Ecology of major weeds and their control in direct seeding rice culture of Malaysia*, Watanabe H, Man A, Zuki M, Eds., pp. 13–34. Ibaraki, Japan: Japan International Research Center for Agricultural Sciences. Seberang Peray, Malaysia: Malaysian Agricultural Research and Development Institute. Kedah, Malaysia:Muda Agricultural Development Authority.

18 Properties of Rice Growing in Abandoned Paddies in Sri Lanka

Buddhi Marambe

18.1 INTRODUCTION

Cultivated rice is still the prime mover of the economy of Sri Lanka. The importance of the crop in the national economy dwindles, as the share of the agricultural sector in national income declines concurrently with faster growth of non-farm income. Rice, as the single most important crop, occupies 34% of the total cultivated area in Sri Lanka and contributes to approximately 5% of the gross domestic product (GDP). On average, 560,000 ha are cultivated during *maha* (main cultivating season — September to February, northeast monsoon) and 310,000 ha during *yala* (minor cultivating season — April to August, southwest monsoon). About 1.8 million farm families are engaged in paddy cultivation islandwide. Sri Lanka currently produces 2.7 million tons of rough rice annually and satisfies around 95% of the domestic requirement. Rice provides 45% of the total calories and 40% of the total protein requirement of an average Sri Lankan. The per capita consumption of rice fluctuates around 100 kg per year depending on the price of rice, bread, and wheat flour. It is projected that the demand for rice will increase at 1.1% per year. To meet this requirement, rice production should grow at the rate of 2.9% per year.

In Sri Lanka, rice is grown in diverse environmental and soil conditions: from drought prone areas in the dry zone to water logged areas in the wet zone, at elevations from sea level to mountains of about 3800 m above sea level, at maximum temperatures (during May to August) ranging from about 17°C in up country to about 40°C in the semiarid dry zone, and rain fed to irrigated conditions. The properties of rice-growing soils vary in texture, drainage, nutritional status, and edaphic problems. These soils are in various topographical, pedological, and hydrological conditions in various land forms. Most of the rice lands are located in inland valleys and valley bottoms where soils are heavy clay. The water table of these soils rises to the ground level during the rainy season. A limited portion of rice lands is located on the coastal plains, flood plains, terraced slopes, and in uplands. Because of this diversity, and different management levels, the average rice yields in these lands are ranging from about 2 to 8 tons per ha.

18.1.1 Land Use in Rice Cultivation in Sri Lanka

Continuous cropping with two rice crops per year on the same land is the predominant land use in Sri Lanka. Rice is cultivated either as a rainfed or as a supplementary or fully irrigated crop. The system of rice cultivation mainly depends on the available rainfall and its distribution. Except in semiarid areas where rice cultivation is marginal, average rainfall in rice growing areas of Sri Lanka can meet at least part of the water requirement for a rice crop during the cropping season. Rice is a semi-aquatic plant and does not necessarily need standing water for a successful crop. However, due to uncertainty of water supply through either irrigation or rain and the need to reduce weed infestation, rice is always cultivated as a crop with standing water. Not all the irrigable area in a paddy tract of a reservoir is cultivated in some seasons when there is a high probability for insufficient rainfall. Low rainfall experienced in different years has also resulted in farmers abandoning the rice fields or inclusion of non-rice crops in these irrigated as well as totally rainfed systems.

Abandoning of paddy fields has been common in irrigable areas under minor irrigation schemes (reservoirs with a catchment area less than 80 ha) and especially under village tank systems. Under the village tank systems of the dry zone, water storage or rainfall is not adequate even for cultivating low-water demanding field crops during the *yala* season. This has led to frequent fallowing of these lands between the *maha* rice crops. The irrigable area under village tank systems is not totally cultivated even in some *maha* seasons due to insufficient rainfall (10). Rainfall probability analysis shows that a village tank would attain 80% of the water storage only once in 5 years, and 33% of the storage is expected once in 3 years. If the reservoirs contain insufficient water for rice cultivation by the proper time expected by the farmers, cultivation is delayed or the fields are fallowed. There is a recent tendency of abandoning rice fields by farmers due to continuous shortage of water during successive cultivating seasons.

Historically, the dry zone of Sri Lanka was replete with functional irrigation systems. Most of these reservoirs are seen even today and some are abandoned. There are approximately 18,000 village tanks throughout the country (10), with 3119 operational village tanks in the dry zone and about 7000 village tanks abandoned. Apart from a scarcity of water for cultivation, a shift of labor from agriculture to industry, and youth migrating to the cities for better employment have also resulted in abandoning of rice fields in the rural dry zone of Sri Lanka. The increased input costs have resulted in rice farmers moving to various other ventures resulting in abandoning of rice fields. The extent of abandoned rice fields has not been estimated; however, it is probably many thousands of ha per season. These abandoned paddies are an ideal place to study the succession (if any) of cultivated rice to volunteer rice and to weedy rice.

18.1.2 Wild and Weedy Rices in Sri Lanka

The Plant Genetic Resources Centre of the Department of Agriculture, Sri Lanka has made an islandwide collection (except in north and northeast regions of the country) of rice strains. These include 74 accessions of *Oryza nivara*, 22 accessions of *Oryza rufipogon*, 13 accessions of *Oryza eichingeri*, 10 accessions of *Oryza granulate*, and 21 accessions of *Oryza rhozomatis* (16). These come from abandoned reservoirs and rice fields. However, none of these have been reported as weedy forms being troublesome to farmers in cultivated rice fields.

Weedy forms of rice (*Oryza sativa*) were first reported in Sri Lanka in 1997 in the Ampara district, in the southeastern part of the country. Ampara is a major rice growing region of the country covering about 67,000 ha of rice land (12) and the district accounts for about 9 to 10% of the total rice cultivated in Sri Lanka. Dry sowing is common in this district as the soil is non-calcic brown (Haplustalf) that can be easily plowed under dry conditions. In addition, the scarcity and high cost of labor has forced farmers from transplanting to dry sowing. Direct seeding began supplanting transplanting about 20 years ago in the Ampara district and has been the predominant practice for the last decade. In Sri Lanka, rice is typically cultivated as a monocrop.

Four different types of weedy rice were reported in Sri Lanka on the basis of awn length (12). All weedy rices shatter within 2 to 3 weeks after panicle emergence, further enriching the soil seed bank. The majority of the seeds are unfilled (70 to 85%), and the filled grains had a low ability to germinate 4 weeks after shattering. The presence of weedy rice, which is morphologically similar to cultivated rice, aggravates weed management problems in direct-seeded rice fields (12). The resistance of weedy rice to postplanting herbicides used in rice fields poses a serious threat to the country's rice production systems. Weedy rice infestation was more serious in dry seeded rice fields and less so in wet seeded fields, similar to reports elsewhere (4, Chapter 16, Chapter 17, Chapter 19, Chapter 20). Weedy rice species have become greater problems since farmers became more reliant on chemical means to control other weeds in rice (7).

The purported pathways of evolution of weedy rice are discussed at length in Chapter 16, Chapter 19, and Chapter 20, especially the possibilities of old rice varieties going feral, gene flow between wild and cultivated japonica types, as well as crosses between cultivated rice and wild rice such as *Oryza nivara* and *O. rufipogon* (4,6,7,8,17,18). The possibility of evolution of feral forms of rice by dedomestication of current species on its own (*endoferality*) cannot be excluded.

The first step toward ferality may be to become a volunteer weed. Persistence of these volunteers in the same fields for extended periods would provide opportunities for selection of back mutations of various feral traits and then for them to recombine (2). Haygood et al. (9) suggested that wild relatives and landraces of crops form a small fraction of plant species and varieties and are the most likely sources of genes for improving crops. Thus, gene flow from wild types to cultivated species would enhance the possibility of evolution of feral forms (*exoferality*). This would give devastating results in the case of rice, as the crop is mainly cultivated as a monocrop and frequently the wild relatives are found in close proximity, thus facilitating this process. The present study was carried out to identify the properties of rice plants naturally growing in rice fields that were abandoned in different agroecosystems of Sri Lanka. The results of these studies are expected to assist in analyses to identify the evolution of ferality within the agroecosystem and its implications to agriculture.

18.2 METHODOLOGY

A preliminary survey was conducted to identify rice fields abandoned for at least three consecutive cultivating seasons in different agroecological zones of Sri Lanka, excluding the north and eastern provinces. The sampling sites and other relevant information are described in Table 18.1. The landowners were consulted to confirm the fallow period. The landowners also confirmed that there have been no human interventions since fallowing started. The occasional entry of cattle for grazing was reported. Single rice plants were found scattered over all (except one) of the abandoned fields. The minimum distance between plants within a sampling site was approximately 4 m. In all cases, the abandoned paddy fields sampled were at least 90 m from the nearest cultivated sites. Broadleaf weeds such as *Portulaca oleraceae* and *Ageratum conyzoides*, upland grass weeds such as *Eleusine indica*, and few patches of the sedge *Cyperus rotundus* colonized all the abandoned sites.

The growth of rice plants in abandoned paddies was monitored by measuring the plant height, tiller and panicle number, number of spikelets, and filled seeds per plant. The plant height was measured from the base to the tip of the flag leaf. The seeds were collected after covering the panicle with polyethylene bags with ventilation openings. The results were compared with available long-term data (6 seasons) of a popular rice variety cultivated in Sri Lanka (BG 300, which matures in 3 months) with all the management practices recommended by Department of Agriculture (5).

Rice plants growing naturally in the selected abandoned paddy fields were selected. The sampling sites were selected >5 m away from the field boundaries. Soil samples were collected to

TABLE 18.1
Characteristics of the Accession Sites

Name	Agroecological Region	Continuous Fallow Period	Number of Plants Selected
Peradeniya	Mid-country wet zone (WM 1)	4 seasons (approximately 24 months)	8
Kegalle	Mid-country wet zone (WM 1)	4 seasons (approximately 24 months)	6
Ratnapura	Low country wet zone (WL 1)	3 seasons (approximately 18 months)	5
Mahailuppallama	Low country dry zone (DL 1)	4 seasons (approximately 24 months)	7

estimate the soil seed bank, using a 30 x 30 cm quadrant, 30 cm away from the base of the selected rice plants or populations. The depth of sampling was 0 to 5 cm.

Seed characteristics of the rice plants from abandoned paddy fields were recorded and germination was tested at the laboratory. Seeds of these rice plants, together with those of a weedy rice strain (*Oryza sativa*) seeds collected from Ampara district in the eastern province (low country dry zone) and a cultivated variety, BG 300, were used for the experiments. The weedy rice and cultivated rice were used as the controls. The experiment to assay secondary dormancy was performed on moistened filter paper laid down in Petri dishes in 6 replicates with 25 seeds per replicate. The Petri dishes with seeds were kept in a germination chamber at a light intensity of 50,000 lumens for 12 hours and at 28°C/24°C day/night temperatures. Tetrazolium tests were conducted to estimate the viability of the seeds. The tests were conducted with 3 replicates using 10 seeds per replicate.

Seeds of rice plants from the abandoned paddy fields, together with those of the control species, were soaked in water for 24 hours and then laid on a moist filter paper for 48 hours before planting. This method is recommended by the Department of Agriculture of Sri Lanka and also practiced by farmers to pre-germinate rice seeds for direct seeding. Hot water treatment (5) was imposed in an attempt to remove any suspected secondary dormancy of seeds. The germinated seeds were grown for one season in plastic pots (30 cm diameter x 30 cm height) filled with topsoil under plant house conditions at the Faculty of Agriculture, University of Peradeniya. Two plants were grown in each pot with 5 pots per treatment. Plant growth and morphological characteristics were recorded. Pots carrying weedy rice and cultivated variety BG 300 were used as control treatments. All the pots were provided irrigation up to field capacity and basal and top dressings of fertilizer as per recommendation of the Department of Agriculture (DOA) (5).

The results were subjected to statistical analysis using analysis of variance. The treatment means were compared with Duncan's multiple range test (DMRT) at $p = 0.05$.

18.3 RESULTS AND DISCUSSION

18.3.1 Field Observations of Morphological Characteristics

The plant height at heading of the rice plants observed in abandoned paddy fields varied among the sampling sites (Table 18.2) and ranged from 107 to 138 cm. The number of tillers varied from 14 to 18 per plant, with 6 to 9 panicles per plant. There were more unproductive tillers on weedy rice plants than on the cultivated variety. The number of seeds per panicle varied among sites and ranged from 39 to 67, and the proportion with filled grain was remarkably low. The presence or absence of awns was recorded, as weedy rice (with awns) has been reported in many parts of the

TABLE 18.2
Morphological Dissimilarity between Rice Plants Found in Abandoned Rice Fields and the Cultivar that Had Been Cultivated

Site	Plant Height (cm)	No. of Tillers per Plant*	No. of Panicles per Plant*	No. of Spikelets per Panicle*	Filled Grain %
Cultivar BG 300	90.5 a	6 a	6 a	88 a	85 a
Peradeniya	122.0 b	17 b	8 a	52 bc	32 b
Kegalle	110.0 b	17 b	7 a	37 c	26 b
Ratnapura	129.5 b	14 b	7 a	64 b	30 b
Mahailuppallama	136.0 b	16 b	6 a	42 c	36 b

* After square root transformation of data.

Note: Within each column, data followed by same letter are not statistically significant by the DMRT at $p = 0.05$.

TABLE 18.3
Dissimilarity of Seed Characters between Rice Plants Found in Abandoned Rice Fields and the Cultivar that Had Been Cultivated, and Similarity to Weedy Rice

Site	Germination %	Seed Viability %	Weight of 10 Filled Grains (g)
Cultivar BG 300	88 a	90 a	0.268 a
Weedy rice	22 c	24 c	0.269 a
Peradeniya	24 c	36 b	0.262 a
Kegalle	12 d	16 c	0.270 a
Ratnapura	36 b	44 b	0.261 a
Mahailuppallama	32 bc	42 b	0.264 a

Note: Within each column, data followed by same letter are not statistically significant by the DMRT at $p = 0.05$.

country (12). Awns were not observed in the rice plants from abandoned paddy fields. The seed bank studies indicated the presence 8 to 36 rice seeds per 900 cm^2 in the locations studied. However, 96% of these seeds were unfilled and did not germinate.

The morphological features of these plants from abandoned paddies clearly indicated that they profusely tiller ($p < 0.05$), and produce similar number of panicles ($p > 0.05$) when compared to the cultivated variety of rice (based on long-term experimental data). Shattering was observed, but could not be accurately estimated at the field level. Thus, pot experiments were carried out at the plant house of the Faculty of Agriculture, University of Peradeniya, Sri Lanka to study the biology of the plants.

18.3.2 Germination and Seed Viability

Germination of the seeds collected from all sites was low when compared to the cultivated variety of rice (BG 300) and was similar to the weedy rice samples (Table 18.3). The lowest germination and viability was reported in samples collected from Kegalle and the highest was reported from Ratnapura. The rice plants from the Kegalle site had the least filled grain as well as the lowest germination and viability. Further studies indicated that hot water treatment did not enhance the germination of rice seeds collected from abandoned paddies or those of weedy rice type. The seeds of weedy rice type had awns with lengths varying from 2 to 4 cm. Seeds from all other accessions were not awned.

18.3.3 Plant Growth Characteristics under Greenhouse Conditions

The number of tillers on the plants from abandoned paddies was significantly higher ($p < 0.05$) when compared to cultivar BG 300, which had the lowest tiller number per plant (Table 18.4). The plants from abandoned paddy at the Kegalle site produced a similar number of tillers as weedy rice, suggesting a possible dominant back mutation to a feral form. Li et al. (11) identified a rice mutant having a master molecular control of rice tillering. Genetic analysis of this *monoculm1* or *moc1* mutant revealed that the low tillering phenotype of domesticated rice was caused by a recessive mutation in a single nuclear gene. MOC1 acts as a master switch in a fine-tuned signaling network that controls tiller formation in rice (11).

The rice plants from abandoned paddies were taller than those of the cultivated variety, but similar to that of the weedy rice plants (Table 18.4). The shortest plants were at the Kegalle abandoned paddy, and the tallest were from Ratnapura. The relative growth rate of rice from

TABLE 18.4
Dissimilarity of Growth Characteristics of Plants from Seeds Collected from Abandoned Paddies and the Cultivar that Had Been Cultivated, and Similarity to Weedy Rice, When Cultivated in Uniform Plant-House Conditions

Site	Number of Tillers per Plant*	Plant Height (cm/plant)	Growth Rate (mg/g/day)	Net Assimilatory Rate ($g/m^2/day$)
Cultivar BG 300	8 a	86 a	181 a	15.9 a
Weedy rice	12 b	109 b	217 c	17.2 a
Peradeniya	16 c	113 b	220 cd	17.6 ab
Kegalle	12 b	102 b	142 b	14.3 a
Ratnapura	15 c	129 c	268 d	19.7 b
Mahailuppallama	17 c	121 c	241 d	17.6 ab

* Analysis done after square root transformation of data.

TABLE 18.5
Dissimilarity of Yield Characteristics of Plants from Seeds Collected from Abandoned Paddies and the Cultivar that Had Been Cultivated, and Similarity to Weedy Rice, when Cultivated in Uniform Conditions

Site	No of Panicles per Plant*	No of Seeds per Panicle*	Shattering %	Filled Grain %
Cultivar BG 300	8 a	82 a	0 a	84 a
Weedy rice	2 c	80 a	82 c	22 c
Peradeniya	4 b	48 b	82 c	32 bc
Kegalle	2 c	31 c	71 b	26 c
Ratnapura	4 b	44 bc	86 c	30 bc
Mahailuppallama	3 b	38 c	80 c	36 b

* Data after square root transformation.
Note: The data were collected when 50% of the flag leaf turned yellow.

abandoned paddies and weedy rice was significantly higher than the cultivated variety ($p < 0.05$). However, the net assimilatory rate of the majority of test plants was similar to that of the cultivated variety, except that from Ratnapura, which was significantly higher than that of BG 300 ($p < 0.05$). The lowest growth and assimilation rates were in the abandoned rice at Kegalle. The rice plants from abandoned paddies from Ratnapura are more efficient in dry matter production than the cultivated rice variety BG 300, despite the experiments being performed under plant house conditions with fertilizer and water management as per DOA recommendations (5), where a cultivar is expected to produce optimally.

The number of panicles per plant and the number of seeds per panicle in the pot experiment from abandoned paddies were less than those observed in the field study (Table 18.2 and Table 18.5). The lowest number of panicles per plant and seeds per panicle were from rice plants of abandoned paddies at Kegalle (Table 18.5). The cultivated rice variety, BG 300, had the highest number of panicles and seeds per plant. Shattering was observed at significantly higher rates ($p < 0.05$) from plants of abandoned paddies and from the weedy rice type than the cultivar (Table 18.5). The highest shattering was observed in the weedy rice plants, followed by those from Kegalle. The rice variety had the highest proportion of filled grains when compared to plants from abandoned paddies and weedy rice type ($p < 0.05$), but the lowest was observed from abandoned paddies at Kegalle.

Interestingly, weedy rice plants, with about 80% shattering as in the case of abandoned paddies, had a similar number of seeds per panicle as the cultivar.

The cultivated variety is a dwarf, high yielding variety, which matures in 95 days. There was no shattering observed even 1 week after maturity (the expected time of harvest). The weedy rice matured early, at 91 days from germination with 100% shattering within 2 days after maturity. All the rice plants from abandoned paddies took 101 to 106 days to mature with 100% shattering within 2 days of reaching maturity. Rice farmers in Sri Lanka also cultivate varieties that mature in 3.5 months (approximately 105 days) and these varieties are as popular as the 3-month old varieties. The owners of the sampled sites confirmed that they have been growing BG 300 during the last 2 seasons prior to abandoning the land. Further studies to confirm the growth duration of the plants from abandoned paddies under field conditions are needed prior to drawing any concrete conclusions.

Some of the conspecific red rice and other *O. sativa* types can be considered to be progenitors to or recently evolved feral forms of domestic rice (7, Chapter 16). Weedy rices shatter most of their seeds before cultivated rice is harvested, so the farmer loses rice yield while filling the soil seed bank with weeds (3). Here the weedy type matured before the cultivated variety and the plants from abandoned paddies matured about 10 days later than the cultivar. The actual dates to reach maturity have not been established yet for the plants from abandoned paddies under field conditions. Although the proportion of filled grain of weedy type and plants from abandoned paddies was significantly lower than the cultivar, the filled seeds of these species shattered within 2 days of maturity. This would enhance weediness of the species, as mature seeds, even at low numbers of viable seeds, would enrich the soil seed bank thus posing a threat to following rice crops.

Seed shattering is a primitive trait common in wild rice species, which has been eliminated in the process of rice domestication, as described in Chapter 16 and Nagai et al. (14,15). Weedy rice is clearly preadapted to the habit of direct-seeded rice, because it possesses many of the same life-history characteristics as the crop cultivar (13). The DNA fingerprinting has shown clear taxonomic differentiation of wild rices from weedy rice, which in turn was closely related to the prevailing cultivar (1). Mortimer et al. (13) suggested three mechanisms to explain the occurrence of weedy rice in the absence of introgression of genes conferring weedy traits from wild relatives to weedy rice:

1. Immigration of weedy rice by seed dispersal
2. Cultivar breakdown due to high mutation rates at loci conferring weediness traits and subsequent selection of weedy forms
3. Heterozygosity within cultivars used by farmers or conserved as on-farm seed

It is a known fact that there is less of a chance for ferality when the certified seeds are cultivated. In Sri Lanka, the supply of certified rice seeds by the state and private sector accounts for only 15% of the total requirement, thus making provision for other unregistered suppliers to produce seeds without proper management practices and certification. Although self-production of seed by the farmers is encouraged, in the majority of the cases, successful results have not been achieved due to poor supervision or monitoring of the production process. Further investigations are necessary to discern the underlying causes of the evolution of feral forms of rice under Sri Lankan conditions.

18.4 CONCLUDING REMARKS

In many places where weedy rices are problems, it is not known whether the weeds are the conspecific red and feral forms or the other species (3). The present study indicates that, based on the information provided by the landowners of the sampling sites, the rice plants in abandoned paddies have grown without human intervention. Rice plants grown without human intervention have led to the presence of feral forms of rice. These feral forms found in the abandoned paddies have increased tillering and especially increased shattering typical of weedy rice, as well as low seed viability and germinability. These feral forms in abandoned paddies differ from the prevalent form of weedy rice in Sri Lanka insofar as they are not awned and typically have a much later harvest date. Weedy forms resembling the feral forms have not been reported in active rice paddies. Thus, it is important to further study these mechanisms involved to ascertain whether the weedy forms have introgressed genes from cultivated rice (7). However, to date, there is a scarcity of field information with respect to natural introgression in rice cultivars with other weedy and wild *Oryza* spp.

The potential for the occurrence of "volunteer weeds" in the case of rice and for them to become feral rendering their control more difficult, is a possibility. It is not clear whether these weedy forms were there all along and could fill an ecological niche formed when rice was no longer transplanted, or whether they are a function of a slow endoferal dedomestication of volunteer rice that remained from previous seasons. The latter seems to be more likely due to the dissimilarity to the prevalent weedy rice in Sri Lanka. To study the evolution of ferality and its implications to agriculture, further studies on the levels of ferality in currently used and abandoned paddies in Sri Lanka, ascertainment of rapidity of transgene introgression from rice to weedy rice, and effectiveness of mitigation strategies would be important to establish what feral characteristics have evolved during dedomestication in paddies in Sri Lanka that have been abandoned for various reasons.

LITERATURE CITED

1. Abdullah MZ, Vaughan DA, Mohamad O. 1991. Wild relatives of rice in Malaysia: their characteristics, distribution, ecology and potential for rice breeding. MARDI Report No 145, Seberang Perai, Malaysia: Malaysian Agricultural Research and Development Institute, 28 pp.
2. Al-Ahmad H, Galili S, Gressel J. 2004. Tandem constructs to mitigate transgene persistence: tobacco as a model. *Mol. Ecol.* 13:697–710.
3. Balthazar AM, Janiya JD. 2000. Weedy rice in the Philippines. In *Wild and weedy rice in rice ecosystems in Asia — a review,* Baki BB, Chin DV, Mortimer AM, Eds., Manila, Philippines: International Rice Research Institute, pp. 75–78.
4. Chin D. 2001. Biology and management of barnyardgrass, red sprangletop and weedy rice. *Weed Biol. Manage.* 1:37–41.
5. DOA. 1997. *Technoguide*, Peradeniya, Sri Lanka: Department of Agriculture, 82 pp.
6. Ellstrand NC, Prentice HC, Hancock JF. 1999. Gene flow and introgression from domesticated plants to their wild relatives. *Ann. Rev. Ecol. Syst.* 30:539–563.
7. Gressel J. 2002. *Molecular biology of weed control.* London: Taylor and Francis.
8. Hancock JF, Grumet R, Hokanson SC. 1996. The opportunity for escape of engineered genes from transgenic crops. *HortSci.* 31:1080–1085.
9. Haygood R, Ives AR, Andow DA. 2003. Consequences of recurrent gene flow from crops to wild relatives. *Proc. R. Soc. London,* 270:1879–1886.
10. Karunasena KGJ, Marambe B, Sangakkara UR, Dharmasena PB. 1997. Productivity of rice and chilli under village tanks of Sri Lanka in maha season with respect to resource utilization. *Trop. Agric. Res.* 9:168–181.
11. Li X, Qian Q, Fu Z, Wang Y, Xiong G, Zeng D, Wang X, Liu X. Teng S, Hiroshi F, Yuan M, Luo D, Han B, Li J. 2003. Controlling of tillering in rice. *Nature* 422**:**618–621.
12. Marambe B, Amarasinghe L. 2000. Weedy rice in Sri Lanka. In *Wild and weedy rice in rice ecosystems in Asia — a review,* Baki BB, Chin DV, Mortimer AM, Eds., Manila, Philippines: International Rice Research Institute, pp. 79–82.
13. Mortimer M, Pandey S, Piggin C. 2000. Weedy rice: approaches to ecological appraisal and implications for research priorities. In *Wild and weedy rice in rice ecosystems in Asia — a review,* pp. 97–105, Baki BB, Chin DV, Mortimer AM, Eds., Manila, Philippines: International Rice Research Institute.
14. Nagai YS, Doi K. Yoshimura A. 2002. A seed shattering gene, *Sh3*, and its inhibitor gene found in an African cultivated rice, *Oryza glaberrima* Steud. *Rice Genet. Newsl.* 19:74–75.
15. Nagai YS, Sobrizal PL, Sanchez T, Kurakazu K, Doi Yoshimura A. 2002. *Sh3*, a gene for seed shattering, commonly found in wild rices. *Rice Genet. Newsl.*19:76–77.

16. PGRC. 2000. *Collection sites of crop germplasm in Sri Lanka 1987–1999: an atlas.* Peradeniya, Sri Lanka: Plant Genetic Resources Centre, Department of Agriculture, 37 pp.
17. Suh, HS, Sato YI, Morishima, H. 1997. Genetic characterization of weedy rice (*Oryza sativa*) based on morpho-physiology, isozymes, and RAPD markers. *Theor. Appl. Genet.* 94:316–321.
18. Vaughan DA. 1994. *The wild relatives of rice.* Manila, Philippines: International Rice Research Institute, 137 pp.

19 Coexistence of Weedy Rice and Rice in Tropical America — Gene Flow and Genetic Diversity

Zaida Lentini and Ana Mercedes Espinoza

19.1 INTRODUCTION AND DISSEMINATION OF RICE IN THE AMERICAS

Rice (*Oryza sativa*) was not part of the diet of the indigenous Americans during pre-Colombian times, with the exception of some Amazonian tribes that collected rice for consumption from native populations of *Oryza grandiglumis* and *Oryza glumaepatula* (called by local people *abati-uaupé*, which means water-maize). The utilization of these species did not lead to domestication as in Asia and Africa (52). With the rediscovery of America by Columbus in 1492 and the conquest of different territories, food supply was a critical issue. The massive utilization of indigenous people for gold and silver extraction during colonial times produced food shortages, creating the need for planting European staple crops like wheat and rice (Asian origin) in the Caribbean island *La Española* (now Dominican Republic and Haiti) that had begun in 1495. Several attempts were made at rice cropping but without success. The first reference to a successful harvest of rice is from Puerto Rico in 1535 and since then, rice cultivation has been disseminated to the Caribbean, in particular to Cuba and Jamaica. The Spanish colonizers introduced the crop into Mexico and Peru after the conquest of the Aztec and Inca empires around 1549, about 100 years prior to its introduction to southern U.S. in 1646 and 1685 (10). Since then, rice was distributed to other colonies and became part of the diet in many Latin American countries (9).

Recent reports suggest that cultivated rice was introduced not only through Europe, but also from western Africa in 1500s (10). The lack of workers during the conquest of the Americas led to a continuous introduction of African people from western Africa over more than 350 years. Africans, who were brought to America to work on sugarcane plantations and other duties in the Caribbean, favored rice cultivation due to their previous knowledge of this crop. This process was fundamental for the successful adaptation of rice to the New World. It appears that rice was introduced from western Africa through Brazil and the Caribbean from 1513 to 1515, and then through the lowlands of Colombia, Ecuador, eastern Nicaragua, Jamaica, and Cuba (10). Although *O. glaberrima* may have been introduced into tropical America from western Africa during colonial times, at present *O. sativa* is the species widely cultivated. Rice production from the cultivation of landraces is not significant in this region.

Rice gradually became one of the most important food grains in tropical America in the 20th century and currently supplies consumers with more calories than staples such as wheat, maize, cassava, and potatoes. It is surpassed only by sugar as a source of energy in the diet (12). Latin American and Caribbean per capita consumption of rice increased 3- to 6-fold over the last 70 years. Rice was mainly accessible to wealthy people at the beginning of the 1900s, and the per capita consumption was in the range of less than 10 kg per year. With the dissemination and cultivation of modern varieties adapted to tropical American ecosystems, the price of rice by the 1980s was reduced by about 60% allowing its accessibility to the poor population. Rice has become particularly important in the diets of the poor, who constitute about 40% of Latin America's total population. The present mean per capita consumption of rice in Latin America and the Caribbean is in the range of 30 kg and some countries consume of up to 60 kg per capita per year. Rice displaced traditional staples such as cassava and plantains, which are bulkier and more perishable, and rice is thus more amenable to the rapid urbanization throughout the region.

More efficient rice production is a matter of considerable urgency given the continued rapid growth of Latin America's metropolitan areas and of its population in general. One of the major obstacles to rice improvement in Latin America is the lack of stable yields due to several major diseases (rice blast, sheath blight, spotted grain, scald leaf, and RHBV (rice hoja blanca virus)) and pests (*sogata* plant hopper, stem borer, water weevil) (21). Weed control accounts for about 30% of the production cost. Weedy rice is the main constraint in the region compared to other weeds affecting rice production (*Brachiaria mutica*, *Diplachne fascicularis*, *Echinocloa spp.*, *Vigna vexillata*) significantly reducing yield as well as crop value at harvest (21, Chapter 17). The presence of weedy rice in paddy fields is frequently the result of the predominant use of farmer-saved seed instead of certified seed, as well as the direct seeding of rice cultivation by farmers. The problem is exacerbated by the lack of crop rotation and the common practice of several crop cycles per year (Chapter 17). In contrast to temperate irrigated rice production where *O. sativa f. spontanea* (red rice) is the main constraint (25, Chapter 16 and Chapter 20), wild/weedy/cultivated rices from numerous *Oryza* species (mostly annual and diploid), usually with feral traits (seed shattering) and varying degree of sexual compatibility, are present in rice as weeds throughout the world (29). The current work describes the composition of wild *Oryza* species in tropical America, with a major focus on Costa Rica and Colombia, and describes recent studies on diversity and gene flow analyses of the weedy rice complex in this region.

19.2 *ORYZA* SPECIES IN TROPICAL AMERICA

19.2.1 Overview of Species Composition and Distribution

The rice genus *Oryza* has a pan-tropical distribution and comprises approximately 23 species distributed in Asia, Africa, Australia, Central and South America, and the Caribbean (52). Four species have been recorded in tropical America — *O. glumaepatula*, *O. grandiglumis*, *O. alta*, and *O. latifolia* (Table 19.1). *Oryza glumaepatula* (diploid, AA genome) classifies within the primary gene pool (3,12,52), whereas *Oryza latifolia* and *O. alta* are allotetraploid (CCDD) and included in the secondary gene pool.

Oryza glumaepatula (2n = 24) is a perennial, tufted and spreading grass with brittle culms near the base of the plants. It occurs in swamps and marshes in open ditches and pools, beside rivers, and near cultivated rice fields, usually in deep water, and open habitats (31). It is distributed from Mexico to southern Brazil and Bolivia (Table 19.1). Due to its morphological similarity with *O. rufipogon*, *O. glumaepatula* was originally classified as the American strain of *O. rufipogon* (52,53). Nevertheless, interspecific crosses between these two species have incompatibility barriers, suggesting reproductive isolation (14,28). *Oryza glumaepatula* is a distinct AA species based on morphological traits (33) as well as molecular markers (48). More recently, molecular markers have demonstrated that *O. glumaepatula* has a higher genetic relatedness to the African species *O. glaberrima*, *O. barthii*, and *O. longistaminata* than to the Asian *O. rufipogon* (3,22).

Oryza latifolia (2n = 48) is a perennial and short to tall (usually 1 to >2 m) grass with broad leaves (up to 5 cm) distributed from tropical Mexico and the Caribbean to southern Brazil and northern Argentina (52). This species has great morphological and phenological diversity and it is found in open or semiopen habitats in the different life zones and ecosystems (Table 19.1). *Oryza*

TABLE 19.1
Distribution of *Oryza* Species in the Americas

Species	2n	Genome	Distribution	Habitat
O. glumaepatula	24	AA	Bolivia, Brazil, Colombia, Costa Rica, Cuba, Dominican Republic, French Guiana, Guyana, Honduras, Mexico, Panama, Surinam, and Venezuela	Swamps and marshes, in open ditches and pools, beside rivers, and near to cultivated rice fields, usually with deep water. Grows in open habitats. Perennial.
O. alta	48	CCDD	Belize, Brazil, Colombia, Guyana, and Paraguay	Savanna and woodland, wet places along streams, rivers, lake edges, or canals in deep water. Open, sunny locations. Perennial.
O. grandiglumis	48	CCDD	Argentina, Bolivia, Brazil, Colombia, Costa Rica, Ecuador, French Guiana, Paraguay, and Peru	Savanna and woodland. Grows in water at river's edges or wet places having clay and alluvial soils. Found in open and shaded habitats. Perennial.
O. latifolia	48	CCDD	Argentina, Belize, Bolivia, Brazil, Colombia, Costa Rica, Cuba, Dominican Republic, Ecuador, El Salvador, French Guiana, Guyana, Guatemala, Haiti, Honduras, Mexico, Nicaragua, Panama, Paraguay, Peru, Puerto Rico, Surinam, Trinidad, and Venezuela	Low forest, rainforest, secondary growth forest, open woodland, undulating savanna, pasture, cultivated fields, open swamp, hill slopes, high ridges, coastal belts. Grows in or near water such streams, riverbanks or pool edges, or springs. Grow in open or semiopen habitats. Perennial.
Osp1 (*O. glumaepatula* *O. grandiglumis*) (57)	ND	ND	Costa Rica	Hybrid nature confirmed by microsatellites. Shows intermediate traits between *O. grandiglumis* *(CCDD)* and *O. glumaepatula* (AA) (57).
Osp2 (57)	ND	ND	Costa Rica	Intermediate traits between Oryza *latifolia* and *Oryza grandiglumis* (57).

alta (2n = 48) is a perennial, tall (up to 4 m), and erect grass with broad leaves distributed from Belize to Paraguay. It is found in open, sunny places, savanna and sometimes in woodland, in wet places such as along streams, rivers, lake edges, or canals in deep water (Table 19.1). *Oryza grandiglumis* (2n = 48) is a perennial and tall grass, with broad leaves and pubescent ligules in Costa Rica and is distributed mainly in the Amazon basin from Colombia and French Guiana to Argentina and Paraguay (Table 19.1). *Oryza grandiglumis* and *O. alta* are distinct monophyletic species based on RFLPs (restriction fragment length polymorphisms). DNA hybridization and phylogenetic analyses of DNA sequences indicate that *O. latifolia* is the most divergent of the group (2,22). Knowledge about *Oryza* species in tropical America has greatly increased since a series of germplasm collecting missions in Brazil, Paraguay, and Argentina in the 1990s, and subsequent research on collected germplasm from Colombia, Costa Rica, and Venezuela (53). Regional demographic growth in Asia has caused an erosion of biodiversity of *Oryza* species (i.e., *O. rufipogon* and *O. nivara,* ancestors of *O. sativa*), which highlights the relevance for documenting the diversity of tropical American *Oryza,* as they are new sources of diversity for breeding. Initial attempts to use this diversity are in progress (8). A thorough distribution mapping of *Oryza* species has been conducted nationwide in Brazil and Costa Rica (6,57). Below we describe the status of knowledge in Costa Rica and discuss evidence for gene flow among *Oryza* species.

19.2.2 Distributions and Genetic Diversity of Wild *Oryza* Species in Costa Rica

A total of 312 populations of *Oryza* were identified and collected in the lowlands of Costa Rica in different ecosystems up to 650 m above sea level ranging from the dry to the humid tropical forest (57). Forty percent of the samples were found in nature reserves (Figure 19.1). Principal component analysis revealed the presence of 3 different discrete groups corresponding to *O. glumaepatula, O. grandiglumis,* and *O. latifolia* (Figure 19.2). About 75% of the populations corresponded to *O. latifolia,* 11% to *O. grandiglumis,* and 3% to *O. glumaepatula. O. glumaepatula* and *O. grandiglumis* are restricted to habitats with marked changes in water level. These ecosystems are fragile and they are vulnerable to extinction. *Oryza latifolia* is the only native wild rice species that grows in both flooded and not flooded areas, including slopes.

Two of the populations collected in Costa Rica did not fit within the taxonomic descriptions of any *Oryza* species described in tropical America so far, and they were recorded as *Osp1* and *Osp2* (57). These plants are sterile, propagate asexually by fission of the culms, and are dispersed by water currents. The *Osp1* shows intermediate traits between *O. grandiglumis* (CCDD) and *O. glumaepatula* (AA) (56, Figure 19.2); it is sympatric and has synchronized flowering with these two species. Molecular analyses using microsatellite markers confirmed that *Osp1* is indeed the product of hybridization between *O. glumaepatula* and *O. grandiglumis. Osp2* is found in the same habitats as *O. grandiglumis* (and it has intermediate traits between *Oryza latifolia* and *Oryza grandiglumis*) (Figure 19.2) but the molecular identity of the progenitors for *Osp2* is not yet clear (57). Hybridization events seem to have played an important role in the diversification of the genus *Oryza* (22), so these hybrids could be part of this process.

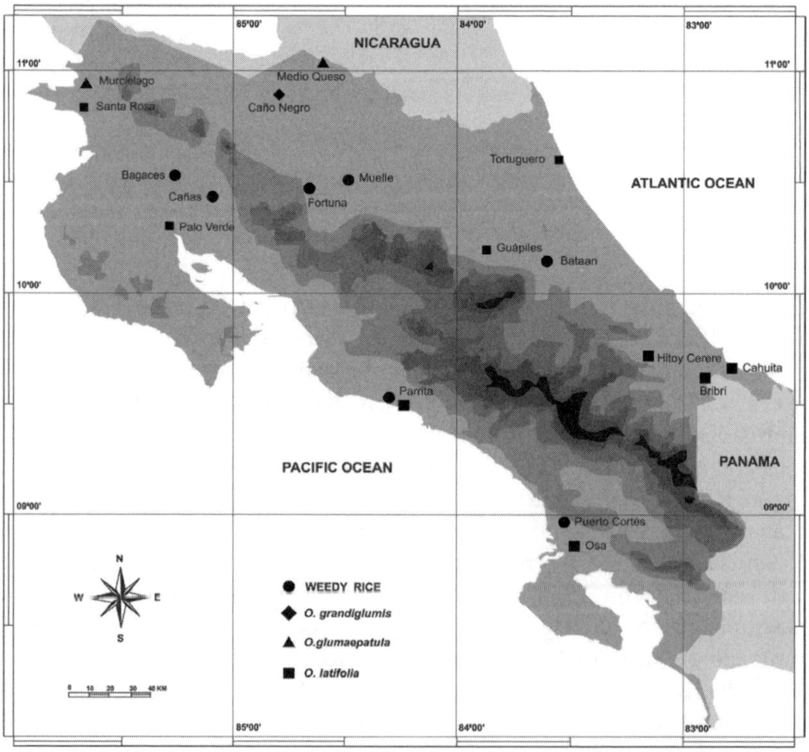

FIGURE 19.1 Distribution of wild *Oryza* species (*O. glumaepatula, O. grandiglumis, O. latifolia)* and weedy rice accessions in Costa Rica.

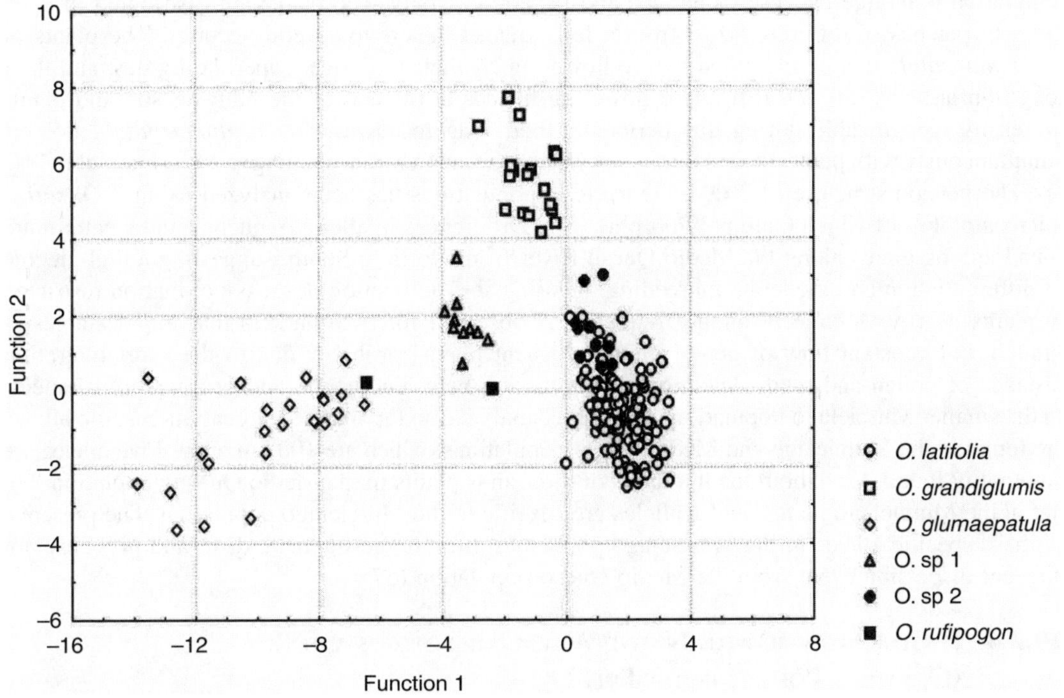

FIGURE 19.2 Scatterplot of all individuals of Costa Rican wild *Oryza* species (according to Table 19.1), graphed for the first two values of a principal component analysis.

Oryza latifolia is the most abundant and widespread wild *Oryza* species in the lowlands of Costa Rica, with populations at altitudes from sea level up to 650 m (Figure 19.1). This species has the most diverse morphology among plants from different geographic regions, climatic zones, and microhabitats (57). Populations from the Pacific slope are significantly smaller than those from the Atlantic. Plants from open flooded habitats are significantly taller than those from drier semiopen areas (57). Genetic isozyme analyses of 9 populations from the different zones reveal that 8 of the populations from the Pacific are the most diverse (45). Most populations from the Atlantic grouped together in 1 cluster and likewise those from the Pacific in another cluster when consensus dendrograms were generated. There is a high level of interpopulation diversity but most populations are monomorphic for at least 1 genotype, suggesting reduced gene flow between populations. No variation is observed within progeny when progeny groups are analyzed to study the mating system in this species, which is characteristic of autogamous species or with clonal reproduction. However, the high frequency of heterozygouslike patterns may suggest that the reproductive system of *O. latifolia* (CCDD) might be more complex (45). This knowledge could aid in designing conservation strategies for this species and designing gene flow studies between rice and *O. latifolia*. This is particularly relevant because *O. latifolia* is frequently found near or associated with rice fields in Costa Rica and this species flowers as long as water is available; thus its contribution to the weedy rice complex needs to be examined. *Oryza grandiglumis* is found in 3 localities in northern Costa Rica, near the border with Nicaragua (Figure 19.1) (57), including 1 wetland site where the largest population of *O. glumaepatula* is also found.

19.2.3 *Oryza glumaepatula*, the AA Genome Wild Relative of Rice in Tropical America

Oryza glumaepatula was found in two sites in Costa Rica (Figure 19.1). The larger population is near the Nicaraguan border and is composed of hundreds of thousands of plants. This perennial population is unique because of its size and genetic diversity. The plants are distributed in open areas in patches of different sizes, from a few squares meters to several hectares. The plants of *O. glumaepatula* can be identified before flowering by their long rice, paperlike ligules, and their long thin auricles (56,57). It has two flowering flushes at the end of the rainy season and plants are easily recognizable during this period by their long red awns. *Oryza glumaepatula* flowers simultaneously with plants from natural ecosystems, even after transplanting to experimental fields.

The genetic structure of 2 *O. glumaepatula* populations has been analyzed using 7 *O. sativa* microsatellites, of which 6 are polymorphic (57). Differences in allelic frequency and s pattern are found among plants along the Medio Queso River from North to South, suggesting a high degree of differentiation and moderate inbreeding, favoring self-pollination or cross-pollination restricted to plants nearby. A cline of allelic frequency is observed for two markers that might suggest a limited, but constant flow of genes among adjacent plants, probably due to the short migration distance of pollen and seeds, leading to a correlation between genetic and geographic distances. In the smaller Murciélago population, all the loci analyzed so far are fixed. Location-specific alleles are found in the Murciélago and Medio Queso populations, which are 100 km apart. Five alleles are commonly found throughout the hundreds of thousands plants in the Medio Queso population, but not in the Murciélago plants, and 2 alleles are specific to the Murciélago population. The presence of these specific alleles in the Murciélago population may suggest genetic drift, and probably not a recent dispersion event from the Medio Queso population (57).

19.2.4 Is Gene Flow between Native Wild *Oryza* Species and Rice (*O. sativa*) Possible in the Field?

Of the four native wild *Oryza* species of tropical Americas, *O. glumaepatula* is the species more prone to hybridize with domestic rice, as both species are diploid and have the same genome type. The reproductive biology of *O. glumaepatula* suggests that although this species is predominantly autogamous, it could act as facultative allogamous, because a few outcrossing events are found in the populations collected from the natural environment in Costa Rica. Comparing the genetic profile of mother plants and their corresponding progeny using fluorescent microsatellite analysis elucidated the mating-type system of accessions of *O. glumaepatula* from Costa Rica. The results indicate a high proportion of homozygous individuals with low allelic variation within progeny derived from the same mother plant. Progeny derived from heterozygous mothers have the typical segregation of self-pollinated plants. However, few heterozygous individuals are detected from homozygous mother plants (T. Quesada, R. Trejos, J. Lobo, and A. Espinoza, unpublished results, CIBCM, Costa Rica). These findings suggest that *O. glumaepatula* could be pollinated by *O. sativa*, and if hybrids are fertile, they could become a potential bridge for gene flow between these two species.

Oryza sativa and *O. glumaepatula* are sympatric in some areas at the Medio Queso wetland, because some farmers occasionally crop rice within the *O. glumaepatula* population and thus there is the possibility of hybridization under natural conditions. Nevertheless, *O. glumaepatula* flowers once a year, narrowing the possibilities of cross-pollination with *O. sativa* to the end of October. Rice farmers in nearby areas where natural populations of the wild species occur usually plant rice at the beginning of the rainy season in early June and commercial rice fields flower mid-August. Thus flowering overlap of both species is not expected to occur. However, climatic variation (e.g., el Niño) may delay the planting season of rice and affect the reproductive biology of *O. glumaepatula,* so overlapping flowering synchronization could still occur occasionally. Commercial rice plantations are often 20 to 100 m from the *O. glumaepatula* populations in Costa Rica. Although rice breeders recommend 5 to 20 m between different rice lines to produce certified seeds, some recent experiments of gene flow in Asia have shown that hybridization events could occur at distances as far as up to 42 m between *Oryza* species (59). These results can have implications in gene flow if wind direction favors pollen dispersal from one species to the other. If *O. glumaepatula* *O. sativa* hybrids are fertile, they could become a potential bridge for gene flow between these

2 species. Preliminary controlled hybridization experiments in the greenhouse between some individuals of the *O. glumaepatula* ecotype from Costa Rica and rice varieties demonstrate that the F_1 hybrids between these two species are completely sterile. However, fertility of hybrids derived from crosses using a broader range of genetically distinct *O. glumaepatula* individuals from Costa Rica needs additional examination.

19.3 COEXISTENCE OF WEEDY RICE WITH DOMESTIC RICE IN FIELDS

As in other parts of the world, weedy rice is readily found in tropical America. Weedy rice appears to be mainly composed of annual *Oryza* spp. with feral traits including seed shattering. In contrast to Asia where manual transplanting is still predominant, direct seeding of weedy rice-contaminated seed is common for a high proportion of rice farmers in tropical America, ensuring field reinfestations and making it one of the most serious weed problems in this region (21, Chapter 17). Weedy rice populations are commonly classified into two major groups based on hull color of mature seeds, such as strawhull and blackhull types (41, Chapter 20). Weedy rice is often referred to as red rice because of the red color of its pericarp, and it has been botanically classified as *O. sativa f. spontanea,* the same species as cultivated rice (14,17,18,35,42). These classifications of weedy rice still need to be clarified, as weedy rice may have different origins in the same or different locations. Reports suggest that weedy rice may include other *Oryza* species including *O. barthii*, *O. glaberrima*, *O. longistaminata*, *O. nivara*, *O. punctata*, *O. sativa*, and *O. latifolia* (an American tetraploid) (29). Hybrid swarms between the American form of *O. perennis* and *O. sativa* have been found in Cuba (13). Weedy rice can have intermediate characteristics between wild *O. rufipogon* and cultivated *indica* or *japonica* varieties of *Oryza sativa* (7,36). Another hypothesis is that weedy rice may have endoferally evolved through the dedomestication of domestic rice to weedy types, where wild rice is not present (54). In addition to seed shattering, weedy rice seeds may possess secondary dormancy, the plants typically are tall, late maturing, and have pubescent leaves and hulls (35). However, some types are morphologically indistinguishable from rice varieties yet still shatter seed (36), as described below. Seed shattering and secondary dormancy favor the persistence of the weedy rice in rice fields. These characteristics in addition to the vigorous growth and other plant traits make this weed highly competitive with domestic rice, and a potential candidate to be a gene receptor from the cultivated species. Natural gene flow estimates in the field from herbicide-resistant rice into weedy rice under temperate conditions indicate hybridization rates of <1% (15,19,39,58), as confirmed by PCR (polymerase chain reaction) analysis. However, a cumulative hybridization rate (over a 3-year period) under temperate conditions may be from 1 to 52% (23), indicating that genes from rice varieties may transfer and be quickly fixed into weedy rice if they have a selective value. The cumulative rate of introgression may be even higher under tropical conditions because of the lack of crop rotation and several crop cycles per year. Several biological, genetic, and environmental factors affect the level of outcross compatibility, including temperature, humidity, genotype, flower morphology, stigma receptivity, pollen viability, pollen germination, and tube development (32). Several studies are in progress in temperate rice regions to understand the rice/weedy complex gene flow/introgression dynamics (Chapter 20), and the work presented herein is the only one so far establishing baseline gene flow information for tropical America (25). Gene flow rates and introgression are significantly affected by genotype environmental interactions (22); thus the knowledge under tropical conditions is key to understanding gene flow and introgression dynamics in the rice/weedy complex at large.

Some effort has recently been directed to understanding the genetic structure of weedy rice from temperate regions by using molecular markers. An AFLP (amplified fragment length polymorphism) analysis of 26 weedy rice accessions collected in Uruguay and 6 Uruguayan varieties showed a clear relationship between molecular markers and morphological traits such as seed traits including awn and hull colors (20). An analysis of 36 weedy rice accessions collected in Brazil with RAPD (random amplified polymorphic DNA) markers delineated 6 groups (5). The 2 most distinct groups were composed of awned weedy rice accessions, and another group including commercial varieties and some weedy types with seed traits similar to the varieties. A microsatellite marker analysis of the genetic diversity of 79 accessions of weedy rice from the U.S. delineated 3 distinct genotypic groups consisting of awnless, strawhull weedy rice types; weedy rice with awns and black hulls; and a third group including all the weedy rice/rice hybrids and rice cultivars. However, there are too few studies in this area and a better understanding of the genetic structure of weedy rice and its relationship with rice varieties and other AA genome *Oryza* species is needed.

The present work is part of a project to analyze the gene flow from non-transgenic or transgenic rice into wild/weedy relatives in the neotropics, and its effects on the population genetic structure of the recipient species (36). Here we present morphological, phenological, and molecular characterization of weedy rice populations collected from farmers' fields in Colombia and Costa Rica. In addition to the commercial *indica* type of rice varieties commonly grown in these regions, the American AA and CCDD genome *Oryza* wild species and the AA genome wild species *O. rufipogon* (Asian origin), *O barthii,* and *O glaberrima* (African origin) were included as controls to have a better understanding of the potential origin of the weedy rice complex in tropical America. A selection of types representing the biodiversity in these weedy rice populations was used to conduct gene flow analyses and identify indicators for easy tracing and monitoring of genetic introgression from rice into weedy rice. These indicators will be used to study the persistence of domestic genes in weedy populations under field conditions through subsequent generations.

A total of 900 weedy rice accessions from Colombia and Costa Rica were analyzed based on a 3-digit code for seed color traits (awn, apiculus, and hull) (31) substituting local names used in previous studies performed in Colombia (40), Venezuela (43), Brazil (1), and Mexico (27). The seed code offers a nomenclature standard that will allow the morphological comparison of weedy rice accessions among researchers and will facilitate future analysis (4). Little is known about the phenological diversity of weedy rice types under field conditions. Knowledge about weedy rice phenology is key for gene flow analysis. In addition to the characterization by seed traits commonly used in most studies intended to analyze the diversity of weedy rice, the accessions were subjected to comparative phenology field trial analysis in both countries, and genetic characterization by molecular markers.

19.3.1 Costa Rican Weedy Rice

The morphometric relationships of weedy rice accessions from Costa Rica were established by comparing 27 morphological traits with commercial rice cultivars, landraces, and wild *Oryza* species found in Costa Rica, Asia, and Africa, using multivariate analysis (Figure 19.3). Three principal components explained 66% of the variation observed. The first principal component accounted for 36% of the variation and separated CCDD genome type *O. latifolia* and *O. grandiglumis* from AA genome species *O. sativa*, *O. glumaepatula*, *O. rufipogon,* and *O. glaberrima.* The second and third principal components, explaining 18 and 11%, respectively, of the variation, separated the weedy morphotype groups from the AA genome species *Oryza sativa*, *O. glaberrima,* and *O. rufipogon.* The weedy morphotypes were scattered between the *indica* commercial rice varieties and the cluster landraces, *O. glaberrima* and *O. rufipogon.* None of the morphotypes collected in Costa Rica clustered either with the allotetraploids CCDD species or with *O. glumaepatula* (4). The cluster formed by cultivated landraces, *O. glaberrima,* and some weedy rice types from Costa Rica (Figure 19.3) suggests an interesting relationship among them. Although this could be due to morphometric convergence, a past influence of *O. glaberrima* on the development of both landraces and weedy rice could also be possible.

A comparative phenology field trial was conducted in Costa Rica, using the 21 most representative weedy types and the 5 most popular commercial rice varieties and accessions of *O. glaberrima*

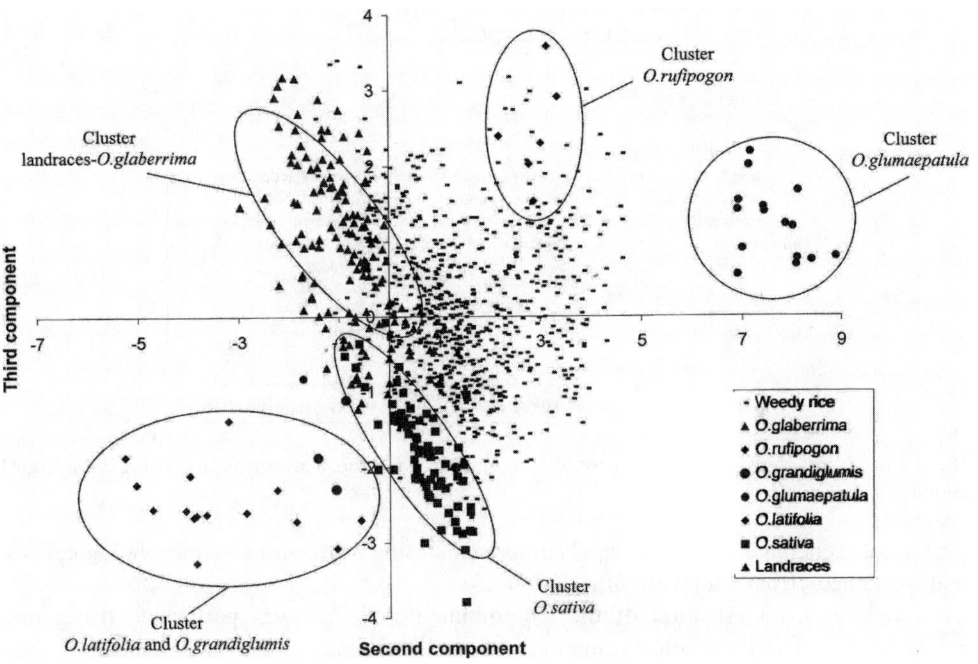

FIGURE 19.3 Distribution of individuals of Costa Rican weedy rice accessions, *O. sativa* (rice commercial varieties and landraces), *O. glaberrima*, *O. rufipogon*, *O. glumaepatula*, *O. grandiglumis*, and *O. latifolia*, according to the second and third principal components.

and *O. rufipogon*. The single annual flush of *Oryza glumaepatula* flowers at the end of the rainy season prevents comparison with other materials under controlled conditions. Three groups of weedy rice types were identified according to the number of days after seeding required to reach 50% anthesis (early 80 to 85 days, intermediate 98 to 107 days, and late flowering 111 to 118 days). Most of the weedy types (71%) had overlapping flowering with the variety Setesa-9 (an early variety), whereas 28% of the weedy rice accessions had overlapping flowering with the varieties CR-5272 and CR-4338 (intermediate varieties), and just 1% overlapped at anthesis with CR-1821 (a late variety) (E. Sánchez, G. Arrieta, and A. Espinoza, unpublished results, CIBCM, Costa Rica). These previous results suggest that early and intermediate varieties are more prone to hybridize with weedy rice morphotypes under Costa Rican standard planting date conditions than the late variety.

19.3.2 Colombian Weedy Rice

The accessions of weedy rice from Colombia were analyzed by comparing 20 morphological traits commonly used to characterize *Oryza* species, with the most popular Colombian commercial rice varieties and the AA genome species *O. glumaepatula*, *O. rufipogon*, *O. glaberrima*, and *O. barthii*. Principal coordinate analysis using qualitative morphological parameters grouped the weedy types into 3 major clusters explaining 67% of the variation observed in the population (Figure 19.4). The cluster including awnless-strawhull weedy types and varieties was the largest, with half of the weedy accessions. This was followed by a cluster of 37% weedy accessions with straw-colored awns and hulls and types that are intermediate between varieties and the wild species (36,46,47,51) (Figure 19.4). A third distinct cluster with 13% weedy accessions included the wild species *O. rufipogon*, *O. glumaepatula*, and *O. barthii* along with the majority of brown-black awned and hulled types (Figure 19.4). A tight association was noted in this cluster between various weedy rice accessions and the wild *Oryza* species, especially with *O. rufipogon* (36). A similar cluster

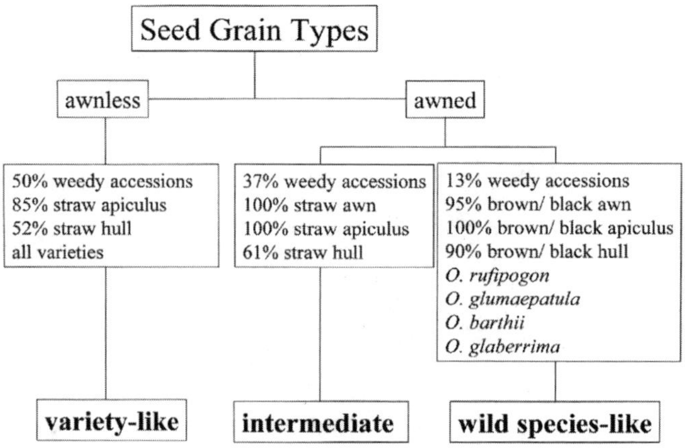

FIGURE 19.4 Major seeds type clusters of Colombian weedy rice accessions according to a coordinate component analysis.

association was generated with 5 principal components using 16 quantitative morphological descriptors that explained 70% of the variability (46,47,51).

Phenological characterization of the Colombian population was performed using progeny derived from 2 consecutive self-pollinations of each original accession collected in the field (46,47,51). The 9 most popular *indica* commercial varieties grown in the region, the transgenic rice Cica 8 line (containing a transgene conferring resistance to the RHBV virus) (37), and the wild species *O. rufipogon*, *O. glaberrima*, and *O. barthii* were used as controls. Three major weedy rice groups were discriminated ($p = 0.005$): early flowering (from 85 to 88 days) including *O. glaberrima* and *O. barthii*; intermediate flowering (from 91 to 112 days) including most of the weedy rice accessions, rice varieties, and an *O. rufipogon* accession from India; and late flowering (from 116 to 120 days) with few weedy rice accessions, the varieties Cica 8 and Fedearroz 50, the Cica 8 transgenic rice line, and the *O. rufipogon* accession from China. Flowering of the weedy types overlaps with more than 1 variety when there is intermediate to late flowering. Most progeny derived from the same weedy accession had a 10-day variation in 50% anthesis, similar to the variation between individual plants within cultivars suggesting no variation in flowering QTLs (quantitative trait loci). In contrast, a few (<10%) accessions had 15 to 25-day variations in flowering time between progeny-derived sister plants, suggesting that they were products of outcrossing (46,47,51). Despite the broad overlap of flowering between weedy types and rice varieties, most individual weedy accessions bred true to type suggesting that weedy rice is predominantly autogamous.

Our research group has completed a microsatellite marker molecular characterization of 148 accessions of weedy rice collected from farmers' fields in the main rice producing areas of Colombia (26,36,47,51). Their genetic profile was generated using a set of 50 rice-microsatellite markers (4 per chromosome) (38) and compared with 9 Latin American commercial *indica* rice varieties, manual hybrids between weedy rice accessions and rice varieties, and the AA genome species *O. barthii, O. glaberrima, O. glumaepatula,* and *O. rufipogon*. Highly polymorphic and genotype-specific microsatellites distinguishing rice varieties, weedy rice types, and *Oryza* species were identified, allowing the differentiation of manual hybrids from individual genotypes, the determination of diversity of weedy rice types, and the association with plant morphological traits such as seed characteristics. A total of 146 alleles were scored with the 19 most polymorphic markers. Most weedy rice accessions (87%) were homozygous, and the remaining 13% were heterozygotes for each microsatellite. The low number of heterozygotes indicated outcrossing.

Multiple correspondence analyses using the microsatellites generated 5 groups, of which 3 groups include all the weedy rice accessions (26) (Group V, Group R, and Group OR, Figure 19.5). *Oryza glumaepatula* is the most distant group separated in 1 cluster, followed by *O. glaberrima*

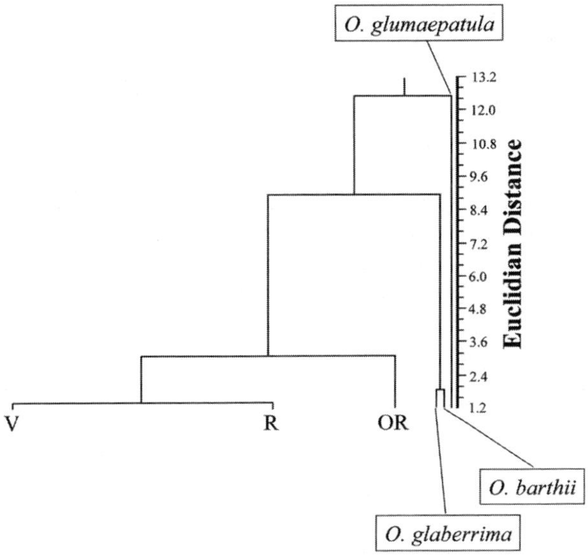

FIGURE 19.5 Multiple correspondence analysis of Colombian weedy rice accessions using microsatellite marker data and seed morphological traits. V, varietylike; R, intermediate type; OR, *O. rufipogon*-like type.

and *O. barthii*, which are grouped in another cluster closer to the weedy rice and *O. rufipogon* (Figure 19.5). *Oryza rufipogon* is clearly distinct from the other *Oryza* species and rice varieties, but it is clustered within 1 of the weedy rice groups (Group OR, Figure 19.5). Multiple correspondence analyses combining the seed morphological traits with the molecular profile allowed a better discrimination between the 3 main groups within the weedy rice population: 1 varietylike Group V, 1 *O. rufipogon*-like Group OR, and 1 clearly distinct group with intermediate traits R (Figure 19.5) (26). The largest group includes 58% of all the weedy accessions and is composed of awnless or awned strawhulls and all the rice varieties (Group V, Figure 19.5). Some of the weedy types were genetically more closely related to the varieties than to the other weedy rice accessions within this group. This cluster includes all the awnless straw hull weedy types collected, commonly known in Colombia as weedy rice-variety types resembling commercially grown varieties. The second cluster contains 7% of the weedy accessions and more than 80% of the awned blackhull seeds and includes *O. rufipogon* (Group OR, Figure 19.5). Some weedy types were genetically and morphologically more similar to *O. rufipogon* than to other awned black hull weedy types. The third cluster (35% weedy accessions) is clearly genetically distinct from the other 2 groups, is composed of the remaining weedy rice accessions that did not fall in either of the other 2 groups, and is characterized by awnless or awned seeds with either straw colored, golden, brown, or with golden furrows (Group R, Figure 19.5). The highly polymorphic- and genotype-specific markers also allowed the differentiation of manual hybrids from individual genotypes, as well as the estimation of the diversity of weedy rice types, and their association with plant morphological traits such as seed characters (26). The weedy types that morphologically and genetically resemble the varieties appear to be indicators of hybridization between weedy rice and the crop and are thus candidates for being a reservoir of gene flow in the rice/weedy rice system. The use of molecular markers in addition to morphological phenotyping would allow a better understanding of the gene flow dynamics between the weedy complex and rice and the potential consequences.

19.3.3 Weedy Rice Resemblance across Two Countries

The weedy rice complexes in Colombia and Costa Rica are broadly diverse. Three clearly distinct major weedy types are distinguished at the morphological and genetic levels. These major types are found in farmers' fields in both countries suggesting that they may be representative of the diversity found in tropical America. The most abundant types resemble rice varieties (varietylike type), followed by types resembling wild *Oryza* species (*O. glaberrima*-like type in the case of Costa Rica and *O. rufipogon*-like type in both cases of Colombia and Costa Rica) and intermediate types with characteristics between *O. sativa* and *O. rufipogon*. Most weedy rice accessions in Colombia and Costa Rica grouped morphologically close to the commercial AA genome rice varieties, consistent with other weedy rice morphometric studies that classified the weedy rice plants as *Oryza sativa* (1,41,43). However, some accessions were also located morphologically close to *O. rufipogon,* sharing similar seed size and color traits, culm number, and plant height traits with this wild species (4,46,47,51). Additionally, a group of morphotypes bore intermediate characteristics between *O. sativa* and *O. rufipogon,* suggesting that hybridization could have taken place in the past between these species.

In both countries, weedy rice types were also broadly diverse in grain size. Short (6.0 to 7.5 mm) and medium (7.6 to 9.0 mm), as well as long (9.1 to 10.0 mm, the same as commercially grown varieties) and extra long (>10 mm) grain types were found in the weedy populations. Most weedy accessions have either short, slender grains or short, thick grains similar to *O. rufipogon* and *O. glaberrima*. Some types have long, slender grains similar to the cultivars; and a few have extra long and thick grains similar to *Oryza barthii* (46,47,51). More than 90% of the weedy accessions collected in Costa Rica and Colombia have red pericarp with varying degrees of anthocyanin content and the rest have white pericarp similar to the varieties (4,46,47,51). The morphological, phenological, and genetic characterization of the Costa Rican and Colombian weedy rice complex is an initial step to understanding its diversity and complexity. This characterization provides additional information about potential origin, genetic structure, and reproductive biology of the weedy rice types, as well as background information to understand the gene flow dynamics between weedy rice and the crop as described below.

19.4 RICE–WEEDY RICE GENE FLOW IN TROPICAL AMERICA

Gene introgression pre-supposes physical proximity of the crop to its wild/weedy relatives and overlapping flowering so that pollination can effect gene transfer. Introgression also requires genetic compatibility between the crop and its immediate wild/weedy relatives, as well as a fitness of the hybrid that will allow its survival. In tropical America, weedy rice is broadly diverse and overlapping flowering with the crop occurs in different environments, as seen above.

As described in Section 19.3.3, we have identified genotype-specific microsatellite markers that allow discrimination between of individual genotypes from manual hybrids derived from crosses between rice and diverse weedy types (26). These are now being used to characterize the genetic structure of the experimental populations prior to gene flow and to detect outcrossing rates in the field. The spatial distribution of alleles is used to study local gene flow, including pollen dispersal distances. Weedy rice accessions representing the diversity of types found in the Costa Rican and Colombian populations were selected to conduct gene flow analysis in the field and identify indicators for easy tracking and monitoring of genetic introgression in the crop/weedy rice complex.

Gene flow studies between rice and weedy rice were conducted in Costa Rica by tracing the dominant presence of anthocyanins from the local rice variety Setesa-9 as a morphological marker, as well as glufosinate resistance in transgenic rice. The experimental field design uses mixed seeds of pollen donor plants containing the marker trait with each one of the weedy rice types selected to simulate infestation levels of 10, 30, and 60%, with each type. In the case of the near neutral anthocyanin marker gene, the hybridization rate varied from 0.2 to 3.8% in 9000 progeny plants evaluated per each weedy type. Similarly, in the case of the transgenic glufosinate resistance trait, rates of 0.1 to 0.4% were recovered in 1000 progeny plants analyzed per weedy type. In contrast to field assays, manual crosses in greenhouse conditions between the transgenic or non-transgenic rice (pollen donor) and the weedy types achieved significantly higher hybridization rates (15 to 30%), as expected.

Similarly and expectedly, higher crossing rates were observed in manual crosses under greenhouse conditions than the natural flow in the field (from 10^{-2} to 10^{-3}) in Colombia. Higher hybridization rates in manual crosses are observed (at least 2-fold) when rice is used as the male parent (pollen donor) and weedy rice as the female parent (pollen recipient) (44). A lower hybridization rate was noted in the reciprocal crosses (using weedy rice as the pollen donor), probably suggesting a preferential gene flow rate from rice into the weedy rice. Hybridization rates of 7 to 42% in manual crosses were confirmed by microsatellite analysis (44).

Gene flow analyses with weedy rice under Colombian experimental field conditions were conducted using virus (RHBV)-resistant transgenic rice (37) and *gus*-marker transgenes, and a non-transgenic rice variety locally known as Purple, characterized by dominant purple (anthocyanin containing) leaves, tillers, and grain apiculus. Randomized plot designs planting rice intermingled with 20% weedy rice, simulating farmers' field conditions and reflecting the economic threshold level for weedy rice infestation in Colombia. Hybridization rates ranged from 0.03 to 0.3% when either transgenes or anthocyanin marker genes were used to trace gene flow in about 23,000 derived progeny plants, which were confirmed by microsatellite markers (44).

The scoring of phenotypic traits alone (i.e., transgenic herbicide resistance, non-transgenic anthocyanin color in flowers, stems, and leaves) appears to overestimate the level of hybridization. Similar results have been found by others using herbicide resistance (25), probably indicating the use of a sublethal herbicide dose or non-uniform herbicide application. Because of these potential errors, it is advisable that phenotypic data from putative outcrossing events in a particular crop or year be confirmed using molecular techniques that can specifically identify the original parents or trait or gene itself. Microsatellites are valuable genetic markers because they are simple and co-dominant, allow detection of high levels of allelic diversity, and are easily and economically assayed by PCR (10,30).

In addition to microsatellite markers, polymorphic non-coding regions of mtDNA (mitochondrial DNA) and cpDNA (chloroplast DNA) in plants can also be used as an approach to discern polymorphism among rice and wild or weedy relatives. The cpDNA and mtDNA polymorphism analyses of maternal inheritance complements the information generated with microsatellite markers of nuclear co-dominant inheritance. Microsatellites allow the identification of potential hybrids, but do not give any information on how those hybrids were generated (i.e., the direction of gene flow). However, by adding markers of maternal inheritance (such as cpDNA or mtDNA), it is possible to discern the female and male parents, deduce the preferential direction of pollen flow, and reconstruct the hybridization and introgression dynamics taking place in farmer's fields over generations. Preliminary analysis of the weedy populations from Colombia, using a set of universal primers for amplification of polymorphic non-coding regions of mtDNA and cpDNA in plants (16) indicates that 9 of the 12 tested cpDNA primers amplified the corresponding non-coding sequence regions of 1 accession of weedy rice, 2 rice varieties, manual crosses between transgenic rice, and 2 varieties, *Oryza glumaepatula* and *Oryza rufipogon* (L.F. Fory and Z. Lentini, unpublished results, CIAT, Colombia). These primers had been designed to amplify the complete sequences of chloroplast genome of *Oryza sativa* and *Nicotiana tabacum* (16). The sizes of amplified fragments range between 1100 and 2800 bp. The digested PCR products of 8 fragment sizes with 4 restriction enzymes (PstI, RsaI, DraI, HaeIII), revealed cpDNA polymorphisms in the region CP8 corresponding to the non-coding regions between the amino acid trnS [tRNA-Ser- (GGA)] and trnT [tRNA-Thr (UGU)]. A polymorphic fragment was observed for *O. glumaepatula* after the digestion with Dra I. This preliminary work identified polymorphic non-coding regions of chloroplast DNA (cpDNA) distinguishing *O. glumaepatula* from *O. sativa*, suggesting the possibility of using organelle (maternal inheritance) polymorphism jointly with genotype-specific microsatellites as tools to trace gene flow and deduct the potential direction of pollen flow in rice. Work is in progress to identify specific polymorphic cpDNA and mtDNA patterns between varieties, weedy rice and wild *Oryza* species.

19.5 CONCLUSIONS

None of the weedy types from Colombia and Costa Rica analyzed so far clustered with any of the American CCDD species or *O. glumaepatula*, discarding at present the possibility of exoferal origin of weedy rice involving these species in this region. However, our results suggest that few outcrossing events are found in natural populations of *O. glumaepatula* collected in Costa Rica, suggesting this species could act as facultative allogamous. Sympatric populations of *O. glumaepatula* and *O. sativa* are found at the Medio Queso wetland in Costa Rica; thus gene flow from rice into natural populations of *O. glumaepatula* is feasible. So far, manual interspecific crosses using the Costa Rican ecotype have yielded completely sterile hybrids. However, it has been demonstrated that vigorous growing plants can be recovered from interspecific artificial crosses between the Brazilian *O. glumaepatula* ecotypes and *O. sativa* when hybrids are backcrossed to rice and weak plants when hybrids are backcrossed to *O. glumaepatula* (30,34). Paternity analysis of these F_1 interspecific progeny demonstrated introgression of *O. glumaepatula* (11). Transgressive segregation is recovered in these populations outperforming the *O. sativa* variety including some relevant traits that can effectively change plant architecture and increase grain yield, such as the number of tillers and panicles per plant and feral traits such as shattering that segregate in the derived progeny plants (8). Thus, interspecific hybrids between *O. sativa* and *O. glumaepatula* spontaneously backcrossing to rice in the field may have a comparative weediness advantage in comparison with cultivars, thus being potential candidates for exoferal complexes in rice fields. The potential contribution of *O. glumaepatula* in the weedy rice complex throughout tropical America should be determined.

Our current data do not rule out the most likely possibility of exoferal origin of weedy rice in tropical America involving interspecific crosses between rice varieties and either *O. rufipogon* or *O. glaberrima*. The presence of *O. rufipogon* (Asian origin) has been documented in herbarium records in tropical America (53). Accessions that are morphologically, phenologically, and genetically indistinguishable from *O. rufipogon* have also been identified in Colombia, as described above. *Oryza rufipogon* could have been introduced as a seed contaminant of Asian rice varieties. Reports suggest that *O. glaberrima* was introduced and cultivated in tropical America during colonial times. African varieties could have been cultivated as landraces and become part of the weedy complex. Despite a possible exoferal origin of weedy rice in tropical America, the significant predominance of weedy types resembling rice varieties (in some cases indistinguishable from currently grown commercial varieties except for seed shattering) supports the hypothesis of a possible endoferal origin as a result of a dedomestication process as well. A few mutations could revert a variety to a feral phenotype (shattering, awning, dark hull). However, varietylike weedy types could also be indicators of recurrent hybridization between weedy rice and the crop. Weedy types having high morphological resemblance to the crop are found in farmers' fields across temperate and tropical regions. Thus, major weedy types appear conserved across different agroecosystems. The morphologically distinct Colombian weedy types are also genetically distinct from each other when characterized by microsatellite molecular markers. The significance of the phenotypic similarities found between weedy rice accessions from different regions (Colombia and Costa Rica with respect to those from temperate zones (Chapter 17 and Chapter 20) deserves a more in-depth examination of the origin or nature of the weedy rice complex.

Most weedy rice accessions bred true (from self-pollination) for the quantitative traits evaluated including flowering; the progeny of a few accessions had significant segregation, suggesting that they are products of outcrossing. Low but significant natural hybridization rates (from 0.01 to 0.1%)

from rice into weedy rice were noted in the field when weedy rice was intermingled either with the crop or transgenic lines. Cumulative hybridization rates (in consecutive years or periods) and rates of introgression may be even higher under tropical respect to temperate conditions because of lack of crop rotation and several crop cycles per year. Thus, any trait deployed into varieties may readily introgress into weedy rice. Some of these traits, in particular resistance to biotic and abiotic stresses, may increase fitness under cropping conditions and when combined with feral traits, enhance competitiveness of the weedy respect to the crop. In this context, herbicide resistance and drought tolerance traits deserve a careful assessment. No different consequences are anticipated with respect to increased weediness as result of introgression of novel traits from a transgenic vs. non-transgenic crop. Increased weediness, if at all, is the result of the trait introgressed, not the source or nature of the gene itself. Weedy rice is clearly a consequence of crop mismanagement. The use of certified seeds jointly with integrated weed management approaches may allow the cultivation of transgenic or non-transgenic crops with broad abiotic or biotic adaptation under tropical conditions without exacerbating ferality in rice fields. A better understanding of gene introgression dynamics over time and hybrid fitness analyses between rice and weedy rice under farmers' fields in tropical America would allow designing mitigation practices adapted to this region. The analysis of gene flow or introgression dynamics in the crop/wild/weedy complex using microsatellites complemented with organelle (maternal inheritance) polymorphism is highly relevant as a tool to trace and discern the predominant direction of gene flow in rice and the puzzling origin of weedy rice.

ACKNOWLEDGMENTS

The authors are grateful to BMZ and GTZ for their financial support through project No. 99.7860.2-001.00 "Gene Flow Analysis for Assessing the Safety of Bio-Engineered Crops in the Tropics" convened by CIAT, Cali, Colombia. We would also like to thank Griselda Arrieta, Elena Sánchez, Tania Quesada, Raúl Trejos, Jorge Lobo, and Alejandro Zamora at CIBCM (Costa Rica), and Luisa F. Fory, Eliana Gonzalez, Paola Ruiz, Juan J. Vásquez, Rosana Pineda, Edgar Corredor, and Myriam C. Duque at CIAT (Colombia) for their technical support and dedication.

LITERATURE CITED

1. Agostinetto D, Fleck NG, Rizzardi MA, Junior AM, Vidal RA. 2001. Arroz Vermelho: Ecofisiologia e Estratégias de controle. *Ciência Rural* 31:341–349.
2. Aggarwal RK, Brar DS, Huang N, Khush GS. 1996. Differentiation within CCDD genome species in the genus *Oryza* as revealed by total genomic hybridization and RFLP analysis. *Rice Genet. Newsl.* 13:54–57.
3. Akimoto M. 1998. Population genetic structure of wild *Oryza glumaepatula* distributed in the Amazon flood area influenced by life history traits. *Mol. Ecol.* 7:1371–1381.
4. Arrieta-Espinoza G, Sánchez E, Vargas S, Lobo J, Quesada T, Espinoza AM. 2004. The weedy rice complex in Costa Rica. I. Morphologic study and relationship with commercial rice varieties and wild *Oryza* relatives. *Genet. Resour. Crop Evol.* (in press).
5. Avozani A, Barbosa J, Maciel J, Gaedka L, Haa L. Concerçao XIX. 2001. Caracterizaçäo genética de biotipos de arroz vermelho com marcadores moleculares-RAPD. II Congreso Brasilero de Arroz Irrigado. XXIV Reuniao la cultura do arroz irrigado. Porto Alegre-Rio Grande do Sul:Instituto Rio Grandense do Arroz-IRGA.
6. Buso GSC, Rangel PHN, Ferreira ME. 1998. Analysis of genetic variability of South-American wild rice populations (*Oryza glumaepatula*) with isozymes and RAPD markers. *Mol. Ecol.* 7:107–117.
7. Bres-Patry C, Lorieux M, Clément G, Bangratz M, Ghesquière A. 2001. Heredity and genetic mapping of domestication-related traits in a temperate *japonica* weedy rice. *Theor. Appl. Genet.* 102:118–126.

8. Brondani C, Brondani RPV, Rangel PHN, Ferreira E. 2001 Development and mapping of *Oryza glumaepatula*-derived microsatellite markers in the interspecific cross *O. glumaepatula* *O. sativa*. *Hereditas* 134:59–71.
9. Cabezas E., Espinoza AM. 2000. El arroz en América: su introducción y primeras siembras. *Rev. Historia América* 126:7–18.
10. Carney JA. 2001. *Black rice. The African origins of rice in the Americas.* Cambridge, MA: Harvard University Press, 240 pp.
11. Cavalheiro ST, Brondani C, Rangel PHN, Ferreira ME. 1996. Paternity analysis of F1 interspecific progenies of crosses between *O. sativa* varieties and its wild relative *O. glumaepatula* using SSR and RAPD markers. *Bras. J. Genet.* 19:225–232.
12. CIAT. Rice: Latin America's Food Grain of Choice. CIAT in Focus. Crop Commitments. p. 8. http://www.ciat.cgiar.org/ciatinfocus/pdf/ricefocus.pdf.
13. Chu YE, Oka HI. 1969. Introgression across isolating barriers in wild and cultivated Oryza species. *Evolution* 24:344–355.
14. Chu YE, Morishima H, Oka HI. 1969. Reproductive barriers distributed in cultivated rice species and their wild relatives. *Jpn. J. Genet.* 44:207–223.
15. Chen LJ, Lee DS, Song ZP, Suh HS, Lu BR. 2004. Gene flow from cultivated rice (*Oryza sativa*) to its weedy and wild relatives. *Ann. Bot.* 93:67–73.
16. Demuesure B, Sodzi N, Petit R. 1995. A set of universal primers for amplification of polymorphic non-coding regions of mitochondrial and chloroplast DNA in plants. *Mol. Ecol.* 4:129–131.
17. Diarra A, Smith R, Talbert R. 1985. Growth and Morphological Characterization of Red Rice (*Oryza sativa* L.) Biotypes. *Weed Sci.* 33:310–314.
18. Ellstrand NC, Prentice HC, Hancock JF. 1999. Gene flow and introgression from domesticated plants into their wild relatives. *Ann. Rev. Ecol. Syst.* 30:539–563.
19. Estorninos LE Jr, Gealy DR, Dillon TL, Baldwin EL, Burgos RR, Tai TH. 2002. Determination of hybridization between rice and red rice using microsatellite markers. *Proc. South Weed Sci. Soc.* 55:197–198.
20. Federici MT, Vaughan D, Tomooka N, Kaga A, Wang XW, Doi K, Francis M, Zorrilla G, Saldain N. 2001. Analysis of Uruguayan weedy rice genetic diversity using AFLP molecular markers. *Electron. J. Biotechnol.* 4:130–145.
21. Fisher AJ, Ramírez A. 1993. Red rice (*Oryza sativa)*: competition studies for management decisions. *Int. J. Pest Manage.* 39:133–138.
22. Ge S, Sang TLB, Hong DY. 1999. Phylogeny of rice genomes with emphasis on origin of allotetraploid species. *Proc. Natl. Acad. Sci. USA* 96:14400–14405.
23. Guadaggnuolo R, Savova-Bianchi D, Felber F. 2001. Gene flow from wheat to jointed goat grass, as revealed by RAPD and microsatellite markers. *Theor. Appl. Genet.* 103:1–8.
24. Gealy D, Tai T, Sneller C. 2002. Identification of red rice, rice, and hybrid populations using microsatellite markers. *Weed Sci.* 50:333–339.
25. Gealy D, Mitten DH, Rutger JN. 2003. Gene flow between red rice (*Oryza sativa*) and herbicide-resistant rice (*O. sativa*): Implications for weed management. *Weed Technol.* 17:627–645.
26. González E, Vasquez JJ, Corredor E, Fory LF, Mora A, Silva J, Duque MC, Lentini Z. 2002. Molecular characterization of rice and rice wild/weedy relatives using microsatellites and their use to assess gene flow in the Neo-Tropics. Beijing, China. Proceedings 7th International Symposium on the Biosafety of Genetically Modified Organisms.
27. Hernández L, Armenta JL, Nieves T. 1979. Arroz rojo, amenaza de las variedades comerciales en México. Programa de divulgación, Culiacán: CIAS, 37 pp.
28. Hinata K, Oka HH. 1962. A survey of hybrid sterility relationships in the Asian forms of *Oryza perennis* and *Oryza sativa*. *Jpn. J. Genet.* 37:314–328.
29. Holm L, Doll J, Holm E, Pancho J, Herbeger J. 1997. The wild rices. In *World weeds: natural history and distribution*. New York: John Wiley & Sons. pp. 531–547.
30. Ikeda K, Sobrizal PL, Sanchez H, Yoshimura A. 1999. Hybrid weakness restoration gene (*Rhw*) for *Oryza glumaepatula* cytoplasm. *Rice Genet. Newsl.* 16:62–64.
31. IRRI (International Rice Research Institute), 1996. *Standard evaluation system for rice*, 4th ed., Los Baños, Philippines: IRRI, 51 pp.
32. Jensen W, Salisbury F .1988. *Botánica*. 2a edición. México City: McGraw-Hill. 762 pp.

33. Juliano AB, Naredo NE, Jackson, MT. 1998. Taxonomic status of *Oryza glumaepatula* Steud. I. Comparative morphological studies of the New World diploid and Asian AA genome species. *Genet. Res. Crop Evol.* 45:197–203.
34. Kwon S, Smith RJ, Talbert RE. 1991. Interference durations of red rice (*Oryza sativa* L.) in rice (*O. sativa*). *Weed Sci.* 39:363–368.
35. Langevin S, Clay K, Grace J. 1990. The incidence and effects of hybridization between cultivated rice and its related weed red rice (*Oryza sativa* L.) *Evolution* 44:100–108.
36. Lentini Z. 2001. Gene flow assessment of LMOs in the neo-tropics. In *LMOS and the Environment*. Roseland CR, Ed., Proceedings of an International Conference. November 27–30, 2001. Raleigh, NC, Paris: OECD, pp. 81–88.
37. Lentini Z, Lozano I, Tabares E, Fory L, Domínguez J, Cuervo M, Calvert L. 2003. Expression and inheritance of hypersensitive resistance to rice hoja blanca virus mediated by the viral nucleocapsid protein gene in transgenic rice. *Theor. Appl. Genet.* 106:1018–1026.
38. McCouch SR, Chen X, Panaud O, Temnykh S, Xu Y, Cho YG, Huang N, Ishii T, Blair M. 1997. Microsatellite marker development, mapping and applications in rice genetics and breeding. *Plant Mol. Biol.* 35:89–99.
39. Messeguer J, Marfa V, Catala MM, Guiderdoni E, Mele E. 2004. A field study of pollen-mediated gene flow from Mediterranean GM rice to conventional rice and the red rice weed. *Mol. Breed.* 13:103–112.
40. Montealegre F, Clavijo J. 1991. Caracterización morfo-fisiológica de algunos tipos de arroz rojo (*Oryza sativa* L.) en Colombia. Federación Nacional de Arroceros, FEDEARROZ, Santafé de Bogotá, Colombia. *Rev. Arroz* 41:18–25.
41. Noldin JA, Chandler J, McCauley G. 1999. Red rice (*Oryza sativa*) Biology. I. Characterization of red rice ecotypes. *Weed Technol.* 13:12–18.
42. Oka HI, Chang W. 1961. Hybrid swarms between wild and cultivated rice species, *Oryza perennis* and *O. sativa*. *Evolution* 15: 418–430.
43. Ortiz AD, López L, Lizaso J. 1999. Desarrollo y Caracterización morfológica de ecotipos de arroz rojo y cultivares de arroz en Venezuela. *Agron. Trop.* 49:51–67.
44. Pineda RP. 2004. Gene flow evaluation from transgenic rice virus resistant in weedy rice. M.S. Thesis. National University of Colombia. Medellín.
45. Quesada T, Lobo J, Espinoza AM .2002. Isozyme diversity and analysis of the mating system of the wild rice *Oryza latifolia* Desv. in Costa Rica. *Genet. Res. Crop Evol.* 49:633–643.
46. Ruiz P. 2003. Phenotypic and genetic characterization of weedy rice from Tolima and Huila Departments. PhD Thesis. Javeriana University. 186 pp.
47. Ruiz P, Vasquez JJ, Corredor E, González E, Fory LF, Mora A, Silva J, Duque MC, Lentini Z. 2002. Gene flow analysis from rice into wild/weedy relatives in the neo-tropics: morphological and phenological characterization of red rice. Proceedings 7th International Symposium on the Biosafety of Genetically Modified Organisms. Beijing, China. October 10–16, 2002.
48. Sano Y, Sano R. 1990. Variation in the intergenic spacer region of ribosomal DNA in cultivated and wild rice species. *Genome* 33:209–218.
49. Schmit V, Jardin P, Baudoin J, Debouck D. 1993. Use of chloroplast DNA polymorphisms for the phylogenetic study of seven *Phaseolus* taxa including *P. vulgaris* and *P. coccineus*. *Theor. Appl. Genet.* 87:506–516.
50. Sobrizal K, Ikeda PL, Sanchez EK, Doi K, Angeles R, Khush GS, Yoshimura A. 1999. Development of *Oryza glumaepatula* introgression lines in rice, *Oryza sativa* L. *Rice Genet. Newsl.* 16:107–108.
51. Vasquez JJ. 2002. Morphological, phenological and genetic characterization of red rice collected in the Saldaña County (Tolima Department). The Andes University. Bogotá. 210 pp.
52. Vaughan DA. 1994. *The wild relatives of rice. A genetic resources handbook.* Los Baños, Philippines: International Rice Research Institute, 137 pp.
53. Vaughan DA, Tomooka N. 1999. Wild rice in Venezuela. *Rice Genet. Newsl.* 16:15–17.
54. Vaughan D, Morishima H, Kadowaki K. 2003. Diversity in the *Oryza* genus. *Curr. Opin. Plant Biol.* 6:139–146.
55. Xiao J, Li J, Grandillo S, Ahn SN, Yuan L, Tanskley SD, McCouch, SR. 1998. Identification of trait improving quantitative trait loci alleles from a wild rice relative, *Oryza rufipogon*. *Genetics* 150:899–909.

56. Zamora A. 2001. Diversidad morfológica y genética de las especies de *Oryza* (Poaceae) nativas de Costa Rica. Tesis Magister Scientiae in Biology. Universidad de Costa Rica. 178 pp.
57. Zamora A, Barboza C, Lobo J, Espinoza AM. 2003. Diversity of native rice (Poaceae: *Oryza*) species of Costa Rica. *Genet. Res. Crop Evol.* 50:855–870.
58. Zhang N, Linscombe S, Oard J. 2003. Out-crossing frequency and genetic analysis of hybrids between transgenic glufosinate herbicide-resistant rice and the weed, red rice. *Euphytica* 130:35–45.
59. Zhi Ping Song, Bao-Rong Lu, Ying Guo Zhu, Jia Kuan Chen. 2003. Gene flow from cultivated rice to the wild species *Oryza rufipogon* under experimental field conditions. *New Phytol.* 157:657–665.

20 Gene Movement between Rice (*Oryza sativa*) and Weedy Rice (*Oryza sativa*) — a U.S. Temperate Rice Perspective

David R. Gealy

20.1 INTRODUCTION TO U.S. TEMPERATE RICE PRODUCTION

20.1.1 Localization and Production Practices

The paradigm in the U.S. rice industry is one that demands economical production of high-yielding, high-quality rice. From 1999 to 2003, annual U.S. rice production has averaged 9.3 million metric tons from about 1.3 million ha in Arkansas (approximately 49% of total ha), Louisiana (approximately 17%), California (approximately 14%), Mississippi (approximately 8%), Texas (approximately 6%), and Missouri (approximately 6%), (53) (Figure 20.1). More than 40% of the U.S. production is exported, ranking fourth in the world (8).

Rice production in the U.S. is conducted under temperate climatic conditions and consists entirely of irrigated lowland systems. California, with its relatively cool climate, produces primarily *japonica* (42) medium-grain rice in water-seeded, aerial planting systems (80). These planting practices are particularly well suited to California's monoculture rice production on fields that have been precision leveled and bordered with permanent levees. The southern U.S. with its relatively warmer climate, produces mostly tropical *japonica* (42), long-grain rice (80). In Missouri, Arkansas, Mississippi, north Louisiana, and Texas, rice is typically drill-seeded, whereas much of the Gulf Coast area in southwest Louisiana is water-seeded. Rice in Texas and southwest Louisiana is typically ratoon-cropped (80). Most rice produced in the southern U.S. is rotated with other crops such as soybean, wheat, corn, or cotton, and occasionally with aquaculture of crawfish.

20.1.2 History of Cultivar Development

By world standards, the rice industry in the U.S. is young. The history, localities, and processes of rice domestication worldwide are presented in detail elsewhere in this volume (Chapter 16). Moldenhauer et al. (51) have recently reviewed the history of rice varieties grown in the U.S. Among the earliest rice varieties was the long-grain, Carolina Gold, grown in South Carolina in the 17th and 18th centuries. At the turn of the 20th century, rice production moved from North Carolina, South Carolina, and Georgia to Louisiana and Texas where higher yielding long-grain varieties were introduced or bred. Approximately 140 public rice varieties have been released in the U.S. since 1911. These have incorporated traits for higher productivity such as shortened plants, better resistance to lodging, increased yield potential, hastened maturity, and more tolerance to disease and low temperature, as well as improved cooking quality (51).

Southern medium-grain rice varieties that combined the desirable high yields of short-grain japonicas and the cooking characteristics of long-grains were first grown in 1911 in Louisiana. In

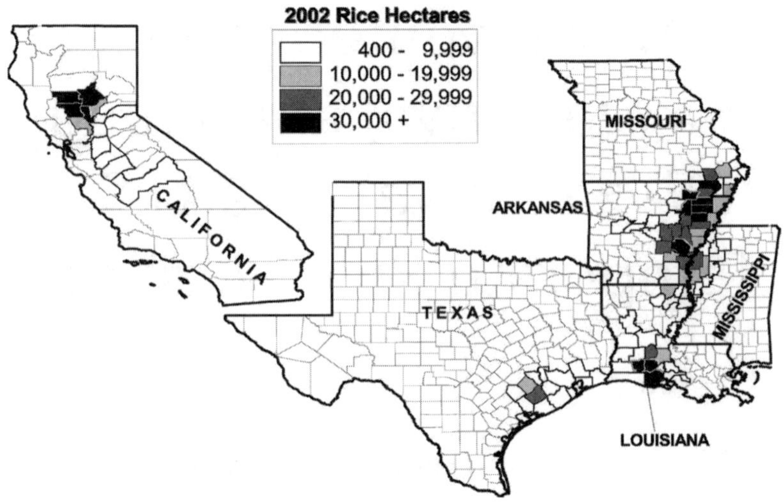

FIGURE 20.1 Rice production by county estimated for each rice producing state in the U.S. (Data adapted from National Agricultural Statistics Service — United States Department of Agriculture (53) with permission.)

California, short-grain varieties accounted for most of the rice grown in the 20th century until the late 1950s when production shifted to medium-grains, such as Calrose, and their semidwarf progeny. Medium-grain rice varieties now occupy more than 80% of the California rice area, followed by short-grain rice varieties at 10%, and long-grain rice varieties at <5% (51). During the past 5-year period, long-grain cultivars have occupied 74% of the rice area in the U.S., followed by medium-grains at 25%, and short-grains at 1% (53). Although numerous rice cultivars have been developed for the U.S., their genetic base is rather narrow. Nearly all of the cultivars grown in the U.S. before 1990 were traceable to fewer than 50 introductions into the southern states and California (12). *Indica* rice, which was domesticated in tropical Asia, has not been commercially produced in the U.S. in pure form (51). However, *indica* rice has been a source of useful genes for traits such as semidwarfism in U.S. cultivars and has great yield potential (43).

20.2 WEED PROBLEMS — THE RED RICE DILEMMA

20.2.1 Introduction and Distribution in the U.S.

Weeds are among the most economically important pests of rice worldwide and are also a major constraint to its production in the U.S. Herbicides are integral to weed control and rice production in U.S. rice, with about 20 different herbicides presently available or under development for use (38). More than 70 species of weeds infest rice fields in the U.S. (76), and *Echinochloa* species are major problems in all areas (26). Red rice (*Oryza sativa* L.) is thought to be the only weedy rice species in the U.S. Apparently introduced from India, it was first acknowledged as a weed in the U.S. in 1846 (9). It is exclusively and particularly troublesome to rice in the southern U.S. (5,13,88). It is an aggressive, annual, crop mimic, and feral form of cultivated rice (*Oryza sativa* L.) capable of experiencing bidirectional gene flow with this crop (60) (Figure 20.2). The lone instance of a wild rice infestation in the U.S. was *O. rufipogon*, which was documented to have been introduced into a localized area of the Florida Everglades as late as the 1950s (85). Red rice infestations were widespread in California from the 1920s through the 1940s, persisting to some extent even after water-seeding methods were adopted (91). In the late 1940s, accelerated production and use of red rice-free certified seed was implemented (91). Used in conjunction with continuous, intensive water-seeding, these practices caused the red rice problem to essentially disappear from California (91). Although there have been periodic findings of red rice since that time (e.g., in 1981), no populations

FIGURE 20.2 Red rice (tall plants) infesting rice (shorter plants) at Stuttgart, AR.

have become established (Hill, 2004, University of California — Davis, Davis, CA, personal communication). Recently, red rice from an unknown source has again been discovered in a localized area of California (66). Infested fields in Glenn County are being taken out of rice production (for several years), flushed periodically, and sprayed with glyphosate to eliminate the red rice (Hill, 2004, ibid, personal communication). If farmers begin to adopt dry-seeding practices as a means to manage herbicide-resistant weed populations in California (67), red rice could easily become a more widespread problem. Red rice has been estimated to infest 30 to 40% of rice production areas in Arkansas, about 50% in Mississippi, 40 to 50% in Texas, and nearly 100% in Louisiana (10).

Weedy and wild rice in other areas of the world are discussed in depth elsewhere in this volume (Chapter 16, Chapter 17, Chapter 19, Chapter 21). In contrast to the relatively simple situation in the U.S., diverse populations of weedy rice and a large number of wild rice species typically infest tropical America and most rice producing areas of the world (Chapter 16, Chapter 17, Chapter 19).

20.2.2 Economic and Agronomic Impacts of Red Rice

Annual losses due to weeds in rice in the southern U.S. have been estimated at $45 million under best management practices and hypothetically would increase to $640 million if herbicides were entirely omitted from these systems (5). However, heavy reliance on herbicides has selected for numerous herbicide-resistant weed populations (34,35,83,84).

Red rice is highly competitive, and the competitive ability of one red rice plant can be equivalent to as many as four rice plants (63). Thus, high populations of red rice can lead to near complete crop failure (11). The interaction between rice and red rice can vary depending on their biological characteristics. In recent greenhouse studies, a tall red rice line reduced growth of rice cultivars more than did a shorter red rice hybrid derivative, while in the same studies, a high-tillering weed-suppressive rice cultivar (PI 312777) reduced growth of the red rice lines more than did a long-grain commercial rice cultivar (Kaybonnet) (18). Results from related field studies were similar (15).

Red rice can be physically separated from long-grain rice at rice mills because of its medium-grain size. However, this process involves added costs and does not effectively remove medium-grain red rice from medium-grain rice or the occasional long-grain red rice from long-grain rice. Rice mills also can remove red seeds from dehulled (brown) rice using optical sorters, which further increases costs and reduces throughput rates. Thus, the value of rice decreases dramatically with increasing levels of red rice contamination. U.S. long-grain rough rice prices can be discounted

4.5 and 82%, respectively, for red rice contamination levels of 2.5 and 15.5% (81,82; Watkins, University of Arkansas, Stuttgart, AR, 2004, personal communication).

Although relatively rare, long-grain red rice seeds are sometimes observed, apparently derived from hybrids of long-grain cultivars and medium-grain red rice. Thus, from the standpoint of visual identification, long-grain red rice probably has a selective advantage over medium-grain red rice in southern U.S. rice, because it mimics the long-grain seeds (i.e., in the rough rice form when the red pericarp is not visible) produced by rice cultivars predominantly grown in the region. Likewise, there is selection pressure in favor of red rice plant types that are the same height or shorter than the rice cultivars they infest because people cannot easily distinguish these red rice types from rice plants during rouging operations. Short blackhull red rice types can be as difficult to spot as the strawhull types because their dark panicles blend in well with the shaded areas below the canopy. Such mimicry is especially problematic in the production of foundation seed rice where the tolerance for red rice plants and seeds is zero.

In much of southwest Louisiana, tillage is traditionally performed in muddy water (water tillage) in an attempt to control red rice before rice planting (80). However, all attempts to control red rice in the southern U.S. using traditional methods have resulted in unsatisfactory weed control.

Although herbicides are crucial to U.S. rice production, maintaining an irrigation flood throughout the growing season may account for about half of the overall weed control in these systems (38). Continuous flooding (e.g., water-seeding) generally provides greater control of non-aquatic weeds including red rice than do delayed flooding systems (e.g., drill-seeding). This difference primarily results from the low oxygen levels available to weed seeds at the soil surface under water and the lower oxygen levels that are available under soil that has been covered by water. Reduced oxygen levels inhibit germination even in flooding tolerant species, such as *O. sativa*, which germinate slowly if at all when seeds are covered by both soil and water. Red rice covered continuously by both soil and water normally will not germinate. However, practices that increase oxygen in the weed seed zone, even for brief periods during the planting process or shortly thereafter (e.g., removing the flood for 1 or 2 days to allow rice seedlings on the soil surface to anchor into the soil), can stimulate germination of red rice.

Reduced tillage practices, which comprise a small fraction of total rice production in southern U.S., can help to reduce red rice infestations because they keep buried red rice seeds in the soil (and non-germinable), while new seed remains on the soil surface where it is subject to predation or to effective control measures in rotational crops. Reduced tillage systems can become susceptible to red rice infestation if tires on farm implements are allowed to slice deeply through the soil surface, which exposes buried red rice seeds to oxygen and can stimulate their germination. Such interactions between tillage and production systems and red rice seed dormancy could complicate the management of red rice hybrid derivatives and the extent to which they could become feral over time.

20.2.3 Phenotypic and Genetic Characterization of Red Rice Populations

Numerous ecotypes of red rice infest U.S. rice fields. Most of these emerge with greater vigor and are much taller than the commercial cultivars they infest; have medium-grain size, awnless-straw-colored or awned-black-colored hulls, and high tillering capacity; and can flower earlier than, later than, or synchronously with rice cultivars (18,25,27,55,72).

Ferrero (21) showed that in the Mediterranean region, where *japonica* rice has traditionally been grown, red rice was most closely related to *japonica* rice, and red rice in Brazil, where *indica* has been primarily grown, was closely related to *indica* rice. Federici et al. (20) showed that strawhull awnless and blackhull awned red rice in Uruguay clustered into genetically separate groups, and another group clustered with cultivated rice. There was also a distinct genetic clustering among U.S. strawhull awnless and awned red rice types, U.S. rice cultivars, and red rice crosses (27). Many U.S. red rice types (particularly strawhull awnless) appear to be genetically closely

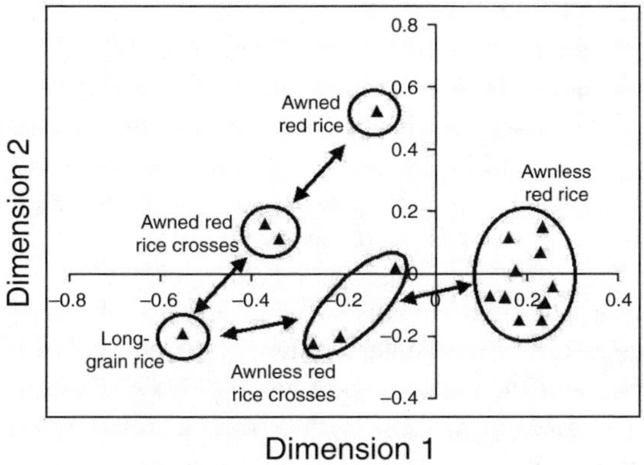

FIGURE 20.3 Genetic distances between rice and various red rice types. The multidimensional scaling (MDS) plot depicts relative genetic distances between long-grain rice, awnless strawhull red rice, and awned blackhull red rice. The data points located at intermediate distances between rice and awnless red rice represent plants with an awnless phenotype, whereas those located between rice and awned red rice represent plants with an awned phenotype. Thus, these are presumed to be hybrid derivatives from awnless and awned red rice parents, respectively. (From Gealy, unpublished data.)

related to *indica* rice based on similarities between simple sequence repeat (SSR) marker profiles (28,86) and seed amylose content (24), suggesting little gene introgression from the cultivated *japonica* rices. Some blackhull awned U.S. red rice types appear to be genetically similar to some *O. rufipogon* lines based on an evaluation of 19 microsatellite markers (86).

Presently, we are evaluating genetic diversity, outcrossing, and gene flow using SSR markers with standard polymerase chain reaction (PCR) and polyacrylamide gel electrophoresis (PAGE) or automated microcapillary array systems. These procedures are followed by statistical (cluster) analysis and development of multidimensional scaling graphs to visualize genetic distances and groupings between germplasm lines of rice or red rice (27). Unknown entries can be readily genotyped by comparing them to standard databases of *Oryza* samples in local or world rice collections. Genetic comparisons are graphically presented on a simplified multidimensional scaling plot derived from analysis of 130 SSR markers (Figure 20.3; Gealy, unpublished data). As shown on the multi-dimensional scaling plot, putative hybrids will typically be positioned equidistant from both parents, if present. These analyses provide a quick genetically based verification of putative hybrids. Applications of this technology include confirmation of the identity and potential sources of suspected red rice types or their hybrid derivatives in red rice infested U.S. rice fields, non-agricultural areas, or in contaminated seed lots. SSR and other useful molecular tools available for the study of weed biology have been recently reviewed (30).

20.2.4 Differential Herbicide Resistance in Red Rice Populations

In anticipation that herbicide-resistant rice systems would be used to control red rice in the southern U.S., numerous red rice accessions have been tested, and widely vary in their natural susceptibility to imazethapyr and glufosinate. Although these accessions typically have not survived recommended application methods and rates, some biotypes might be substantially favored at suboptimal herbicide rates or in areas of misapplication in herbicide-resistant rice systems. In a greenhouse test, 1 of the 10 accessions tested (TX4) was particularly resistant to glufosinate, surviving post-emergence applications of 1.12 kg/ha at the 4-leaf stage and then producing more than 2000 seeds per plant (54). Although no follow-up glufosinate response studies were reported for seedlings derived from the seeds from these surviving plants, similar field and greenhouse studies in Arkansas with up to 26 accessions have confirmed that several subsequent generations of TX4 plants have tolerated approximately double the rate of glufosinate compared to other red rice accessions (22; Gealy, unpublished data). In field tests with approximately 50 red rice accessions, several marginally tolerant blackhull types, including TX4 (as above) and 1995-8, survived 0.071 kg/ha postemergence applications of imazethapyr at the 3- to 4-leaf stage, but did not produce seeds (26). In a recent evaluation of herbicide tolerance in 112 red rice accessions obtained from 30 rice growing counties in Arkansas, 6 (5.4%), 9 (8%), and 10 (8.9%) accessions survived single applications of 1x rates of glufosinate, imazethapyr, and glyphosate, respectively (74). The ability to survive rates of these herbicides at or near commercially recommended levels would be a strong selective advantage for red rice accessions. A weed can be effectively resistant to an herbicide, even if inhibited, as long as its cohorts are controlled. Recent molecular evaluation of red rice accessions from Arkansas has revealed the presence of several nucleotide mutations present in regions of the ALS gene (site of imidazolinone resistance in rice) consistent with herbicide resistance in other species (72). In combination, these natural mutations in red rice and the earlier herbicide screening studies indicate that selection pressure by imazethapyr and related herbicides is likely to be substantial in imidazolinone (IMI) rice systems and should be monitored closely.

20.3 HERBICIDE-RESISTANT CULTIVARS

20.3.1 Background

Following the agronomic and commercial successes of glyphosate-resistant soybean, development of herbicide-resistant rice varieties has received increasing interest in recent years, especially as a potential means to control red rice. Thus, rice cultivars with resistance to one of three different chemical families of herbicidal action have been developed in the past decade. Imidazolinone-resistant (IMI-resistant) cultivars were developed by mutation techniques, whereas glufosinate- and glyphosate-resistant cultivars are transgenic (26). Because red rice is such a major problem in the rice industry, farmers with infestations of this weed were expected (1) to become the first to adopt herbicide-resistant rice systems. This has, in fact, occurred.

The efficacy of red rice control in these herbicide-resistant rice systems under field conditions has been impressive in the U.S. (Figure 20.4), but concerns about gene flow to weedy rice and a perceived incomplete public acceptance of such products for human food have sometimes delayed or prevented development and release of herbicide-resistant rice cultivars. Thus, of the three systems, only IMI-resistant rice is commercially grown at this time. Introduction of herbicide-resistant rice systems may also lead to longer term problems with herbicide-resistant volunteers in northern Latin America (Chapter 17) as well as the U.S.

20.3.2 IMI-Resistant Rice

IMI-resistant rice varieties have been commercially grown in the southern U.S. since 2002 (26). They are popular in the dry-seeded systems in Arkansas, facilitating a shift from water seeding to dry seeding. IMI-resistant varieties were grown on 33,000 ha in 2002 (85% as CL121) (Wells, 2004, Horizon Ag, Memphis, TN, personal communication). CL161, which has a higher level of herbicide resistance than did previous IMI-resistant varieties, became available the following year. In 2003, these varieties were grown on more than 80,000 ha (95% as CL161) (Wells, 2004, ibid, personal communication). For the 2004 crop year, up to 15% of all rice in the southern U.S. is expected to be planted to CL161 (Wells, 2004, ibid, personal communication). The IMI resistance trait appears to be inherited as a single dominant gene in CL161, which is expressed in a 3:1 (resistant:susceptible) ratio in the F_2 generation. This gene is expected to be in all related IMI-resistant varieties that are expected to be grown in the foreseeable future (Linscombe, 2004, Louisiana State University, Crowley, LA, personal communication).

FIGURE 20.4 Example of excellent red rice control from various postemergence glufosinate applications in transgenic glufosinate-resistant rice. Red rice plants are infesting areas in research plots in Stuttgart, AR between rows and plots of rice in the untreated areas. Glufosinate-resistant rice plants are growing in rows.

IMI-resistant hybrid rice (heterozygous) (e.g., CL XL8) was grown for the first time in the southern U.S. in 2003 on a limited acreage, which was expected to increase several fold for 2004 (Thompson, 2004, RiceTec, Alvin, TX, personal communication). Up-to-date areas of this and other rice cultivars are available from the Cooperative Extension Service in rice producing states such as Arkansas (<http://dd50.uaex.edu/dd50Logon.asp> and Louisiana <http://www.agctr.Isu.edu/subjects/rice/RiceAcreage.htm>). The IMI-resistance gene has been purposely used in only one of the two strip-planted parents cross-pollinated for the production of IMI-resistant hybrid seed. The male parent (pollen donor) bears the IMI-resistance gene and its strips are selectively managed following the established system for IMI-resistant rice production (i.e., they receive IMI-herbicide). This approach prevents red rice or non-IMI-resistant rice from growing within the male strips and inadvertently supplying pollen to the pool available to pollinate the female strips. Seed carried by the female strips (male sterile) result from cross pollination and represent the hybrid seed harvested for further planting. As the female parent does not carry the IMI-resistance gene, any accidental fertilization of their flowers by foreign red rice pollen would result in IMI-susceptible F_1 plants that can be easily killed by routine application of IMI herbicide. The hybrid seed would carry a single copy of the gene (as compared with two copies in traditional IMI-resistant varieties), thus reducing the probability of transferring the gene to undesirable rice by half. In addition to facilitating the control of red rice and producing high yields, these hybrid cultivars can be highly competitive against other weeds even with little or no herbicide applied (23,62,75), which may give them an added advantage in U.S. rice systems.

Red rice control can be optimized in drill-seeded (61,79) and water-seeded (65) IMI-resistant rice systems by applying imazethapyr in sequential treatments, which typically result in >94% control. This multiple application approach is intended to control early emerging red rice plants as well as those that may have escaped control by the first application or emerged afterward. The initial IMI-resistant cultivars (e.g., AS-3510, CL121, and CL141) had relatively low levels of resistance to imazethapyr and thus were prone to herbicide injury (46,61,87). Although the subsequent group of IMI-resistant rice cultivars (e.g., CL161) tolerates higher levels of imazethapyr, red rice control levels in these systems seldom reach 100%. Thus, small populations of red rice escapes will typically remain in fields as potential outcrossing partners with IMI rice. Models of weedy rice infestation dynamics suggest that even 98% control efficacy from preplant herbicides in stale seedbed systems in Europe can increase weedy rice seed populations in the soil seed bank over time (Chapter 21). Conversely, these models suggest that effective postemergence control of weedy rice, made feasible in herbicide-resistant rice, can deplete weedy rice levels in the seed bank. Numerous research efforts in the U.S. are currently underway to improve late-season control of red rice survivors.

20.3.3 Glufosinate-Resistant Rice

Transgenic glufosinate-resistant rice cultivars have been under development at Louisiana State University beginning in the mid-1990s (37,58). The glufosinate resistance trait in red rice hybrids is inherited as a single dominant gene, which is expressed in a 3:1 ratio in the F_2 generation (70,94). These studies also indicated that the presence of the *bar* gene (responsible for glufosinate resistance) did not change reproductive biology of rice or red rice, and that this gene functioned in F_1 and F_2 progeny from reciprocal crosses (70). Growth and development characteristics of glufosinate-resistant rice and the hybrid derivatives it forms with red rice were substantially equivalent to their non-transgenic counterparts (57).

As with IMI-resistant rice, excellent red rice control is achievable in glufosinate-resistant rice systems. Red rice control has typically been better when glufosinate is applied sequentially at low rates compared to single applications at higher rates and can reach nearly 100% (89). Tank mixes of glufosinate with propanil, quinclorac, or other rice herbicides also achieved similarly high levels of red rice control (68,89). The transformation events present in these earlier glufosinate-resistant cultivars have been discontinued, and a newer transformation event, LLRICE62, is being exclusively incorporated into rice cultivars for potential commercial release (50). Numerous large-scale, multi-year field evaluations have helped establish optimum glufosinate treatments and procedures for control of red rice (14).

20.3.4 Special Management Considerations

To minimize risks from herbicide carryover to non-resistant rice cultivars, and to mitigate against the gene flow-induced and pre-existent biotype-induced selection pressure problems, the manufacturer suggests that IMI rice should be planted on the same field no more than once every other year (71). Additional efforts that prevent uncontrolled red rice plants from flowering synchronously with the IMI rice crop, such as selecting cultivars with appropriate flowering dates and rouging red rice before flowering, should help reduce the possibility that herbicide-resistant red rice hybrid derivatives will become established and persist in these fields (26,71). However, secondary seed dormancy will probably ensure that some of the red rice hybrid derivatives produced in a particular year will be available to infest rice crops that are planted 2 or more years in the future. Similar considerations will also apply to glufosinate-resistant rice systems (26). Implications of gene flow between red rice and herbicide-resistant rice have been recently reviewed, defining the conditions for gene flow and proposing agronomic practices that can prolong the lifespan of these herbicide-resistant systems (26).

In the typical crop rotations with soybean, rice volunteers are readily controlled in the soybean crop by applying imazethapyr, glyphosate, or other grass-control herbicides. But herbicide-resistant volunteers could persist to a much greater extent if they were already resistant to the same herbicides (e.g., IMI rice or glyphosate-resistant rice) or those with similar modes of action.

20.4 OUTCROSSING CAUSES, RATES, AND CONSEQUENCES

20.4.1 Biological Basis for Outcrossing in *Oryza sativa*

Oryza sativa is primarily self-pollinated, because anthers dehisce at or just before the opening of the lemma and palea. Outcrossing can occur because anthers then elongate and remain outside of the spikelets after they close (93). Pollen exchange (*outcrossing*) between different *Oryza* plants is the key first step involved in gene flow. In addition to pollen exchange and production of F_1 seeds, hybrid derivatives require adequate levels of fitness and self-fertility for their traits to introgress into the red rice population (e.g., herbicide-resistant rice cultivars to weedy red rice). Outcrossing rates between herbicide-resistant and non-resistant rice cultivars and weedy rice species have been estimated from groups of thousands to millions of seeds produced in field experiments using a variety of methods (Table 20.1).

20.4.2 Historic Estimates and Limits of Rice-Red Rice Outcrossing Rates

Outcrossing rates of 4 cultivars in large-scale field tests of commercial cultivars were reported nearly 70 years ago in U.S. rice-producing states (3). Outcrossing rates averaged nearly 0.5% and reached a maximum of nearly 3.5%, but varied widely from year to year. During a 6-year period in Texas, outcrossing rates over distances of 0.3, 0.6, and 1 m ranged from 0.05% in 1936 to 1.4% in 1932 (approximately a 28-fold difference) for the cultivar pair Asahi Mochi-Colusa, while the proportion of plants producing at least 1 hybrid seed during these 2 years was 4 and 100%, respectively. Outcrossing was greater (0.52%) in the southern U.S. than in California (0.16%), presumably because environments were warmer and more humid in the south and were cooler and drier in California. The authors concluded (3) that the outcrossing observed throughout these studies probably represented the maximum levels possible for these varieties under the particular conditions present because the paired varieties flowered at the same time.

By contrast, outcrossing rates up to 34% were measured in male-sterile rice systems (2,36,47). The 1938 study suggests a biologically realistic upper limit to outcrossing rates that would be expected between common U.S. rice plants under optimum outcrossing conditions (e.g., same cultivar and synchronized flowering). The male-sterile studies provide estimates of maximum feasible outcrossing rates for U.S. rice under conditions in which the pollen from donor plants experiences no competition from pollen produced by acceptor plants, a situation that would rarely occur in nature with red rice as the acceptor. Still, the same rice/red rice hybrid could occur with red rice as the pollen parent. Rice-rice outcrossing scenarios should provide helpful insights into the dynamics of outcrossing between weedy rice and rice (26).

The male and female (male-sterile) parents used for hybrid rice production are chosen for their profuse pollen production and large, receptive stigmas, respectively, as high seed set is essential for production of hybrid rice seed. Pollen production by F_1 hybrid rice plants used in hybrid rice systems would probably be similar to or greater than the levels produced by conventional non-hybrid cultivars (Yan, 2004, USDA-ARS, Stuttgart, AR, personal communication). Thus, the pollen load produced by herbicide-resistant hybrid rice cultivars is likely to be as great or greater than that produced by normal selfed rice cultivars.

20.4.3 Biotic and Abiotic Factors Affecting Rice-Rice and Rice-Weedy Rice Outcrossing Rates

In numerous studies involving synchronously flowering herbicide-resistant rice, susceptible rice, or red rice (*O. sativa*), maximum outcrossing rates have been estimated at 0.7% or less, with the vast majority being less than 0.3% (Table 20.1). In these tests, various factors sometimes influenced outcrossing rates as described below.

Floral synchronization is the most critical factor affecting *O. sativa* outcrossing. The flowering periods of 2 plants must overlap such that viable pollen from the pollen donor and a receptive stigma on the pollen acceptor are available at the same time. Due to the relatively short life span of *O. sativa* pollen (minutes) and stigma (days) (93), plants that flower more than several days apart have virtually no chance to outcross with one another. Reduced or undetectable outcrossing between rice and weedy *O. sativa* accessions has been attributed to lack of floral synchronization (7,16,90) (Table 20.1). It must be noted, however, that exact synchronization does not necessarily

TABLE 20.1
Estimated Outcrossing Rates between Various Herbicide-Resistant and Susceptible Rice (*Oryza sativa*), Red Rice (*Oryza sativa*), and Other *Oryza* spp. in Pollen Dispersal Studies[1]

Outcrossing Species Pairs[2] (Pollen Acceptor/ Pollen Donor)	Location of Study	Brief Description of Study	Outcrossing Rates	Comments	Reference
			Tests with Hybrid Rice		
O. sativa rice cultivar/male-sterile *O. sativa* rice cultivar.	Davis, CA	Efficacy of hybrid rice production using male-sterile rice as pollen acceptor.	30 to 34% outcrossing detected	Relatively low outcrossing rates were achieved, even though pollen from donor plants experienced no competition from male-sterile florets.	Azzini and Rutger 1982 (2); Hu and Rutger 1991 (36); Mese et al. 1984 (47).
			Historic Test with U.S. Rice Cultivars		
O. sativa rice cultivar/ *O. sativa* rice cultivar.	Stuttgart, AR; Crowley, LA; Beaumont, TX; Biggs, CA	Gene flow rates and distances from glutinous-endosperm rice to non-glutinous rice (outcrossing visually detectable in harvested seed).	>2,000,000 seed evaluated across 4 varieties, 4 locations, over 4 to 4 years. Outcrossing at 0.3 to 1 m: 0.45% overall average. 3.4% maximum 0.16% in CA Outcrossing at 9 m (TX only): 0.0 to 0.3%	Seasonal and environmental conditions influenced outcrossing. Lower outcrossing rates in CA may result from higher temperature and lower humidity vs. southern states.	Beachell et al., 1938 (3).
			Tests with Glufosinate-Resistant Rice		
O. sativa susceptible rice cultivar (Cypress)/ *O. sativa* glufosinate-resistant rice cultivar (LLRICE62; variety, LL401).	Crowley, LA, and Beaumont, TX	Gene flow rates and distances from glufosinate-resistant rice to susceptible rice. Approximately 117,000 seeds screened in germination tests. Outcrossing confirmed using glufosinate screening and molecular techniques.	Sampled at 0 to 21 m from edge of glufosinate-resistant rice. Outcrossing rate: undetectable. Sampled within area of mixed resistant and susceptible rice (adjacent plants). Outcrossing rate: 0.08% (6/7700 seed tested) Overall average: 0.0051% (6/117000 seeds tested.	Outcrossing detected when florets of two rice types in near-direct contact.	Shannon Pinson, USDA — ARS, Beaumont, TX and Steve Linscombe, Louisiana State University, Crowley, LA, 2004, personal communication.

O. sativa susceptible rice cultivar (M202)/ *O. sativa* glufosinate-resistant rice cultivar (M202).	Davis and Biggs, CA	Gene flow rates and distances from glufosinate-resistant rice to susceptible rice. Outcrossing confirmed using glufosinate screening and molecular techniques.	Maximum outcrossing rate = 0.010 to 0.42%. Outcrossing averaged over all distances to 15 m = 0.007 to 0.11% (approximately 3,300,000 seed). Maximum detection distance = 2 m.	Ongoing	Albert Fischer, University of California — Davis, 2004, personal communication.
O. sativa susceptible rice cultivar/*O. sativa* glufosinate-resistant rice cultivar.	Robbins, CA (Davis, CA)	Gene flow rates and distances from glufosinate-resistant rice to susceptible rice. Outcrossing confirmed using glufosinate screening and molecular techniques.	0.1% outcrossing in adjacent plants (287,000 seeds) 0.01% outcrossing at 1.5 m (760,000 seed) Undetectable at >1.5 m		K. Johnson, S. Roberts, and D. Mitten, D. Bayer CropScience 2001 (cited in (26)).
O. sativa glufosinate-resistant rice cultivar (CPB6) and *O. sativa* susceptible rice cultivar (Purple Haze; has recessive purple leaf color trait)/weedy red rice (strawhull); also their reciprocal crosses.	Baton Rouge, LA	Reciprocal gene flow rates between glufosinate-resistant or susceptible rice and red rice. Outcrossing confirmed using glufosinate screening and molecular techniques, leaf pubescence, or green vs. purple leaf phenotype as markers. Plots drill-planted with 50:50 mix of red rice and rice.	Outcrossing rate at 8 to 25 cm distances between red rice and rice plants. Glufosinate-resistant rice/red rice = 0.33% in 6600 seeds. Purple rice/red rice = 0.7% in 4700 seeds. Red rice/glufosinate-resistant rice = 0.0% in 8000 seeds. Red rice/purple rice = 0.0% in 8200 seeds.	Outcrossing from rice plants to red rice plants (which were 18 and 30 cm taller than CPB6 and Purple Haze, respectively) was undetectable. F_1 plants exhibited hybrid vigor (not due to resistance gene): but also produced fewer, less fertile spikelets than parents, and had flowering delayed approximately 50 days later than either parent. Likely to prevent selfed seed production or backcrossing. Cultivation and reduced fitness of F_1 should select against feral hybrid derivatives.	Zhang et al. 2003(94).
O. sativa weedy red rice ecotypes (8 different types)/*O. sativa* glufosinate-resistant rice cultivars (Bengal, Gulfmont, or Cypress).	Fayetteville, AR	Gene flow rate from glufosinate-resistant rice to red rice ecotypes with different flowering dates. Outcrossing confirmed using glufosinate screening and molecular techniques.	Overall average at <0.25 m: 0.0146%. Maximum outcrossing at <0.25 m: 0.37% (blackhull red rice 10A/glufosinate-resistant Bengal).	Outcrossing most common with red rice ecotypes that flowered synchronously with glufosinate-resistant rice cultivar.	Wheeler and TeBeest 2002 (90).

TABLE 20.1 (continued)
Estimated Outcrossing Rates between Various Herbicide-Resistant and Susceptible Rice (*Oryza sativa*), Red Rice (*Oryza sativa*), and Other *Oryza* spp. in Pollen Dispersal Studies[1]

Outcrossing Species Pairs[2] (Pollen Acceptor/ Pollen Donor)	Location of Study	Brief Description of Study	Outcrossing Rates	Comments	Reference
O. sativa susceptible rice cultivar (Ariete)/*O. sativa* glufosinate-resistant Ariete rice.	Mortura, Italy	Gene flow rates, distances, and directions from glufosinate-resistant rice to susceptible rice.	0.08% outcrossing at 0.2 m, undetectable at 2.4 m, in approximately 57,000 seeds.	Most hybrids produced in direction of prevailing wind.	Messeguer et al. 2001 (48).
O. sativa susceptible rice cultivar (Thaibonnet)/*O. sativa* glufosinate-resistant Thaibonnet rice.	Amposta, Spain	Outcrossing confirmed using glufosinate screening and molecular techniques. Italy design: side-by-side rows (perpendicular to prevailing winds) at several separation distances. Spain design: concentric circles with central population of resistant rice.	0.09% outcrossing at 1 m, 0.01% outcrossing at 5 m, in approximately 215,000 seeds. Maximum outcrossing rate = 0.53% at 1 m.		
O. sativa weedy red rice or *O. sativa* susceptible rice cultivar (Senia, japonica)/*O. sativa* rice cultivar (with *gusA* and *bar* gene markers).	Amposta, Spain	Gene flow rates and distances from glufosinate-resistant rice to red rice or susceptible rice. Outcrossing confirmed using glufosinate screening and molecular techniques. Design: concentric circles of rice and red rice at several distances.	Red rice/transgenic rice: 0.036% outcrossing at 0.5 m in 125,000 seeds. Max rate: 0.06% Susceptible rice/transgenic rice: 0.086% outcrossing at 0.5 m in 165,000 seeds. Max rate: 0.11%. Max detection distance =10 m.	Most hybrids produced in direction of prevailing wind. The rice to rice outcrossing rate was 2.4 times greater than the rice to red rice rate.	Messeguer et al. 2004 (49).

O. sativa weedy red rice (strawhull or blackhull)/*O. sativa* glufosinate-resistant rice cultivar. Also their reciprocal crosses.	Brazil	Reciprocal gene flow rates between glufosinate-resistant or susceptible rice and red rice. Outcrossing confirmed using glufosinate screening and molecular techniques.	Conducted in adjacent 1 by 1 m field plots; 6000 seeds tested from each outcrossing combination. Transgenic rice/strawhull red rice: 0.22% outcrossing. Transgenic rice/blackhull red rice: 0.02% outcrossing. Strawhull red rice/transgenic rice: 0.26% outcrossing. Blackhull red rice/transgenic rice: 0.14% outcrossing.	Outcrossing with strawhull red rice > blackhull red rice. Outcrossing in both directions was similar with strawhull red rice as a parent. With blackhull red rice, outcrossing was greatest when red rice was pollen acceptor.	Noldin et al. 2002 (56).
O. sativa susceptible rice cultivar (Bengal)/ *O. sativa* glufosinate-resistant rice cultivar (LLRICE62).	Rio Grande do Sul, Brazil	Gene flow rates and distances from glufosinate-resistant rice surrounding susceptible rice.	0.07% outcrossing at 0.25 to 1 m in 9000 seeds tested.	Density of transgenic pollen grains is likely high in central area containing susceptible rice.	Magalhães et al. 2001 (45).
		Gene flow rates and distances to susceptible rice surrounding glufosinate-resistant rice: Outcrossing confirmed using glufosinate screening and molecular techniques.	0% outcrossing at 0 to 3 m in 64,000 seeds tested.	Density of transgenic pollen grains is likely diluted as distance from central pollen production area increases.	
Weedy *O. sativa f. spontanea* (13 accessions)/ *O. sativa* cultivar (Nam29/TR18; F_5).	Kyongsan, South Korea.	Gene flow from glufosinate-resistant rice to weedy rice. Confirmed using molecular techniques. Design: interplanted rice and weedy rice mixture.	*O. sativa f. spontanea/O. sativa:* 0.011 to 0.046% outcross frequency at 0.15 m in 120,000 to 240,000 seeds tested.	8 of 13 accessions flowered synchronously; outcrossing detected in only 3. Outcrossing differences attributed to flowering period and height differences.	Chen et al. 2004 (7).

TABLE 20.1 (continued)
Estimated Outcrossing Rates between Various Herbicide-Resistant and Susceptible Rice (*Oryza sativa*), Red Rice (*Oryza sativa*), and Other *Oryza* spp. in Pollen Dispersal Studies[1]

Outcrossing Species Pairs[2] (Pollen Acceptor/ Pollen Donor)	Location of Study	Brief Description of Study	Outcrossing Rates	Comments	Reference
Wild (weedy) *O. rufipogon*/*O. sativa* (Minghui-63) rice cultivar (Nam29/TR18; F_5).	Chaling, Hunan Province, China	Gene flow from susceptible rice to wild rice. Confirmed using molecular techniques. Design: alternating rows of rice and wild rice.	*O. rufipogon*/*O. sativa:* 1.21 to 2.19% outcross frequency at 0.5 m in 2200 seeds tested.	*O. sativa* outcrossing to wild rice is relatively high (approximately 50 times greater than to *O. sativa f. spontanea* weedy rice.	Chen et al. 2004 (7).
Tests with Imidazolinone-Resistant Rice					
O. sativa weedy red rice cultivar/*O. sativa* imidazolinone-resistant rice cultivar.	Stuttgart, AR	Gene flow rates and distances from imidazolinone-resistant rice to weedy red rice. Outcrossing confirmed using imazethapyr screening and molecular techniques.	Overall average outcrossing rate from 0 to 10 m, approximately 0.008%.	Ongoing.	Shivrain et al. 2004B (73).
O. sativa weedy red rice/*O. sativa* imidazolinone-resistant rice cultivar.	Stuttgart, AR	Gene flow rates between imidazolinone-resistant (Clearfield) rice and red rice ecotypes. Outcrossing confirmed using imazethapyr screening and molecular techniques.	0.012% outcrossing from 25,000 seeds.	Sampled from machine-harvested seed or bagged panicles. Hybrids detected only in machine-harvested seed.	Estorninos et al. 2002A (17).

O. sativa weedy red rice/*O. sativa* imidazolinone-resistant rice cultivar.	Louisiana	Gene flow rates between imidazolinone-resistant (Clearfield) rice and red rice in commercial fields. Outcrossing confirmed using imazethapyr screening and molecular techniques.	When single imidazolinone application did not control red rice, up to 30 seedlings/12,000 seeds sampled were from outcrossing (O.C. rate = 0.25%); but the 30 seedlings originated from seed on only 100 plants.	Natural outcrossing can produce high numbers of hybrids if red rice control efforts in Clearfield rice fail.	Zhang et al. 2004 (95).
O. sativa weedy red rice (Stuttgart Strawhull)/ *O. sativa* imidazolinone-resistant rice cultivar (CL121, CL141, or CF0051).	Stuttgart, AR	Gene flow rates between imidazolinone-resistant (Clearfield) rice and red rice ecotypes. Outcrossing confirmed using imazethapyr screening and molecular techniques.	In 2000, some IMI-resistant rice and red rice flowered nearly synchronously. In 2001, area sprayed with imazethapyr. CL121 (coincident flowering), 49 probable F_1 hybrids = 0.0013% outcrossing. CF0051 (partially coincident flowering), 36 probable F_1 hybrids = 0.00097% outcrossing. CL3291 (flowering non-coincident), 6 probable F_1 hybrids = 0.00016% outcrossing. Red rice seed production in 2000, approximately 3,700,000 seeds/field.	Coincident flowering produced highest outcrossing rates. Seeds sampled and counted indirectly (may over- or under-estimate seed numbers leading to over- or under-estimations of outcrossing rate.	Estorninos et al. 2003A (16).

TABLE 20.1 (continued)
Estimated Outcrossing Rates between Various Herbicide-Resistant and Susceptible Rice (*Oryza sativa*), Red Rice (*Oryza sativa*), and Other *Oryza* spp. in Pollen Dispersal Studies[1]

Outcrossing Species Pairs[2] (Pollen Acceptor/Pollen Donor)	Location of Study	Brief Description of Study	Outcrossing Rates	Comments	Reference
			Other Tests with Susceptible Rice		
O. sativa weedy red rice (blackhulled and strawhulled, respectively)/*O. sativa* susceptible rice cultivars (Kaybonnet and Starbonnet, respectively). Also their reciprocal crosses.	Stuttgart, AR	Reciprocal gene flow rates between susceptible rice and red rice ecotypes. Confirmed using molecular techniques.	In 2000, rice and red rice pairs with coincident flowering grown in adjacent rows. >2300 seedlings screened from seed. Outcrossing rate: Kaybonnet/blackhulled red rice 0.10%, blackhulled red rice/Kaybonnet 0.0%. Starbonnet/strawhull red rice 0.23%, strawhull red rice/Starbonnet 0.14%.	Outcrossing in the strawhull red rice pair was slightly greater than in the blackhull red rice pair. Outcrossing from red rice (taller plant) to rice (shorter plant) was greater than from rice to red rice.	Estorninos et al. 2003B (19).
Wild (weedy) O. *rufipogon*/*O. sativa* cultivar Minghui-63 (common paternal line for hybrid rice; prolific pollen producer).	Hunnan Province, China	Competition and effect of arrival times between foreign rice pollen and conspecific wild rice pollen.	Hand-fertilizing *O. rufipogon* with 1:1 mix of pollen from two species produced only 2% hybrids (as compared to an expected 50% level in which pollen preference was neutral). Thus, foreign *O. sativa* pollen is at severe disadvantage compared to conspecific *O. rufipogon* pollen.	Pollen competition may slow down (but not stop) gene flow from crop rice to wild rice. Hybridization was greatest when pollen from the foreign rice cultivar arrived earlier than did the wild rice pollen.	Song et al. 2002 (78).

Wild (weedy) *O. rufipogon*/*O. sativa* cultivar Minghui-63 (common paternal line for hybrid rice; prolific pollen producer).	Hunnan Province, China	Gene flow rates and distances from rice to wild rice. Confirmed using molecular techniques. Several field designs: central population of rice pollen donor (CPC); encircling population of rice pollen donor (EPC); alternating rows of rice pollen donor and wild rice (APC); single plot of rice pollen donor with wild rice planted at various distances unidirectionally (downwind) (UPC).	Maximum outcrossing frequencies observed. CPC: 2.8% at 1.2 m distance. EPC: 2.9% at 3.6 m distance. UPC: 1.2 to 1.5% at 5 to 43 m downwind. APC: 2.2% at 0.5 m. Outcrossing average over all experiments = 1.2%.	*O. sativa* outcrossing to wild rice (*O. rufipogon*) is many times higher than reported values for *O. sativa* to weedy rice (*O. sativa*). Pollen from rice cultivar moved downwind more than 40 m to fertilize wild rice.	Song et al. 2003 (77).
Various *Oryza* spp.	Several	Gene flow between various cultivated and wild rice species.	Rice outcrossing reported in other areas of the world ranged from 0 to 6.8%.		OECD 1999 (59).

[1] All reports indicate significant overlapping of flowering periods in the species pairs tested.

[2] Using conventional notation, a hybrid produced from the cross Plant A/Plant B indicates that Plant A was the female (pollen acceptor) and Plant B was the male (pollen donor).

produce greatest outcrossing rates. Outcrossing from a paternal hybrid *O. sativa* rice line to *O. rufipogon* was greatest when the hybrid rice pollen arrived 10 to 50 minutes earlier than the conspecific *O. rufipogon* pollen (77, Table 20.1).

Maximum outcrossing tends to occur in the direction of the prevailing winds during the flowering period because pollen from the donor plant readily moves with air currents. Outcrossing from herbicide-resistant rice to susceptible rice in Spain was 0.53% 1 m downwind from the pollen source, but only 0.015% in the opposite direction (48). In similar studies, outcrossing to rice and red rice was also greatest in the direction of prevailing winds, and rice/rice hybrids were detected at distances up to 10 m (49). Likewise, outcrossing from a paternal hybrid rice line to *O. rufipogon* was greatest in the downwind direction and was detectable at a distance of 43 m (77).

Densities and placement of pollen donor plants (spikelets) relative to pollen acceptor plants may affect the measured outcrossing rate in pollen dispersal studies with *O. sativa*. Greater densities of donor pollen near the spikelets of acceptor plants during anthesis should increase the probability that outcrossing will take place on the acceptor plant. These conditions may occur when the densities of donor spikelets (*ca.* proportional to plant population) or pollen grains produced per spikelet are high relative to those produced by the acceptors in close proximity. However, these assumptions seem to be contradicted by studies in southern China, in which outcrossing to *O. rufipogon* was not affected by the cross-sectional area of a centrally located cultivated rice pollen source (77).

20.4.4 Directionality of Outcrossing

Outcrossing between rice and red rice can occur in both directions (Table 20.1). Often, outcrossing from red rice (typically taller) to rice (typically shorter) is greater than in the reverse direction. In Arkansas, outcrossing rates between pairs of red rice and herbicide susceptible rice were 0.1 and 0.23% with red rice as the pollen donor and were 0.0 and 0.14%, respectively, with rice as pollen donor (19). Similarly, in Louisiana, outcrossing rates of 0.33 and 0.7% were determined from red rice to glufosinate-resistant rice and purple marker rice, respectively, and outcrossing from both rice types to red rice was undetectable (94). In Brazil, outcrossing between glufosinate-resistant rice and red rice also occurred in both directions, but outcrossing from the rice to red rice was as great or greater than from red rice to rice (56). Overall, the probabilities of outcrossing from red rice to rice seem to be somewhat greater than in the reverse direction, but both plant types can readily serve as pollen donor or acceptor. If hybrid seed production fields were inadvertently infested with red rice (such fields are disallowed, but cannot be avoided with 100% certainty), red rice pollen could presumably fertilize the male-sterile parental rice plants at relatively high rates (2,36,47), thus creating another avenue that favors the development of red rice hybrids with red rice as pollen donors.

Subsequent to the formation of a hybrid seed, the directionality of the original pollen flow can have a large practical impact on the development of hybrid derivatives as weedy or feral populations in the rice field. The non-embryonic tissues in *O. sativa* seed and spikelets produced from cross-pollination retain the characteristics of the mother plant. Thus, hybrid seeds that develop on rice panicles as the result of cross-pollination with red rice will likely remain attached to the rice plant until they are removed from the field during rice harvest. With the high harvest efficiencies of modern combines, as much 95% of rice seeds can be removed from fields. However, the relatively small proportion of seeds remaining in the field can equate to 300 kg/ha (e.g., 5% of typical 6000 kg/ha yield) or 1250 seeds/m^2, which is three times the typical seeding rate of 100 kg/ha for drill-seeded rice. Thus, the quantities of seed remaining on the soil surface are greatly reduced from the maximum levels possible, but large numbers remain available for production of volunteer plants or feral hybrid derivatives. Hybrid seeds that develop on red rice panicles after cross-pollination with rice are likely to fall to the soil surface, thus reseeding the hybrid. Shattering rates of 25 to 80% are common among U.S. red rice types (11,55,57). Thus, selection pressure favors introgression from rice to red rice compared to the reverse direction, because hybrid seed formed on red rice plants are much more likely to remain in the field. Consistent with this contention, models developed for European production systems predict that seed shattering rates of 20% or more will increase levels of herbicide-resistant weedy rice in the soil seed bank over time (Chapter 21).

20.4.5 Average Outcrossing Estimates for Rice, Herbicide-Resistant Rice, and Weedy Rice Combinations

Overall averages of maximum outcrossing rates from recent studies involving herbicide-resistant rice, susceptible rice, or *Oryza sativa* red rice were calculated from the information in Table 20.1 (excluding studies on male-sterile rice, *O. rufipogon,* and other wild rice species, and the Beachell et al. 1938 study (3)). These rates can be considered to be rough estimates of worst-case scenarios. Maximum outcrossing rates from all rice-rice and rice-red rice studies shown in Table 20.1 averaged 0.17%, ranging from 0 (undetectable) to 0.7%. There were no obvious factors that accounted for the wide variation in outcrossing rates observed among the numerous plant types, cultivars, locations, and experimental designs involved (Table 20.1). However, it seems likely that variations in floral synchrony and differences in height between pollen donors and acceptors and environmental variables, such as wind speed and turbulence, relative humidity, temperature, and solar radiation, could significantly affect outcrossing. For instance, a short rice plant would probably not transfer pollen to a tall plant in calm air, but could more easily accomplish the same transfer under turbulent wind conditions. Warm, humid, sunny conditions are ideal for manual cross-pollination of rice flowers by breeders because these conditions maximize floret opening and pollen survival (26). The same conditions probably also favor pollen flow to and from weedy rice. Conditions that deviate substantially from these would be expected to reduce natural outcrossing rates. Additionally, the numbers of fertile pollen donor and pollen acceptor florets could affect the pollen load that is available to fertilize the acceptor florets. All other factors being equal, the greater the pollen load (per land area) produced by the donor plants (e.g., the plant population is high or individual plants produce large amounts of pollen) relative to the number of acceptor florets (e.g., the plant population is low or individual plants produce few florets), the greater the probability that the donor plant will successfully fertilize the acceptor floret. With relatively few florets (seeds) involved, the equation for outcrossing rate (% outcrossing rate = 100 (hybrids detected/seeds evaluated)) will produce an inflated value. In the reverse situation, fertilization rates would be reduced and the calculation would likewise produce a deflated value. Such results could be subject to erroneous interpretation of gene flow rates, especially if populations of pollen donor and acceptor plants differ greatly, or if pollen flow in only one direction is evaluated.

Maximum outcrossing rates for the rice/herbicide-resistant rice studies only (obtained from Table 20.1) averaged 0.17%, ranging from 0 to 0.53%, and those for only the rice-red rice studies also averaged 0.17%, ranging from 0 to 0.7%. These average maximum outcrossing rates were strikingly similar given the broad range of conditions, locations, and rice or red rice types involved in these studies. Maximum outcrossing rates in glufosinate-resistant rice-red rice studies and IMI-resistant rice-red rice studies (obtained from Table 20.1) averaged 0.24 and 0.07%, respectively. Although these values indicate a possible difference between outcrossing potential in IMI-resistant rice and glufosinate-resistant rice, they may not differ in fact. IMI estimates were obtained from only four values, IMI studies were all conducted in two southern states in the U.S., and IMI and glufosinate estimates were never compared directly in common studies.

To demonstrate the potential impacts of low outcrossing rates over time, we can make simplified estimates of red rice seed production based on earlier field experiments in which a rice field heavily infested with red rice (40 red rice plants/m^2) produced 5000 kg red rice seeds/ha, or (5,000,000 g red rice seeds/ha)/(0.023 g/seed) = 217,000,000 red rice seeds/ha, with 540 red rice seeds/plant (15). Thus, with an initial outcrossing rate of only 0.07%, there could be up to (217,000,000 seeds/ha) (0.0007) = 152,000 F_1 hybrid herbicide-resistant plants/ha the first year after hybridization, and up to (152,000) (540) (3/4) (assuming a 3:1 segregation ratio for herbicide resistance in F_2 plants) = 62,000,000 F_2 herbicide resistance plants/ha by the second year. Using similar assumptions, but with a light red rice infestation level of 1/10,000 of that depicted above (e.g., 40 red rice plants/ha), an outcrossing rate of 0.07% could still result in up to 15.2 and 6200 hybrid herbicide-resistant plants/ha in the first (F_1) and second (F_2) year after hybridization, respectively. Thus, even low yet practical outcrossing rates with a gene that provides high Darwinian survival value (herbicide resistance) gene can result in rapid establishment of the gene.

The biotype or accession of red rice or weedy rice can affect outcrossing rates. Several reports on *O. sativa* red rice have indicated that outcrossing from rice to strawhull red rice was moderately greater than to blackhull red rice (19,56), and another indicated that outcrossing to a blackhull red rice type was greater than to other red rice types (90). Similarly, outcrossing from a rice cultivar to several weedy (red) rice accessions in Korea was detected in both *indica* and *japonica* accessions and ranged from 0 to 0.05% (7). Work reported from Colombia (Chapter 19) indicates that hybridization rates between rice and weedy rice under farm field conditions ranged from 0.03 to 0.3%. Although other factors may also contribute to the differences observed between biotypes in these studies, these authors most often cited differences in floral synchrony and plant heights.

20.4.6 Rice-Wild Rice Outcrossing Rates Are Relatively High

In contrast to the low outcrossing rates (approximately 0.17% average maximum) in *O. sativa*, outcrossing rates between *O. sativa* and wild *O. rufipogon* averaged approximately 2.2% (Table 20.1). Similarly, the maximum detection distance for *O. sativa*/*O. sativa* outcrossing ranged about 1 to 10 m, whereas *O. rufipogon*/*O. sativa* outcrossing was detectable at 40 m (Table 20.1). These relatively high outcrossing rates and detection distances are consistent with *O. rufipogon*'s greater reliance on outcrossing as a reproductive strategy and suggest that initial risks of pollen flow to this wild, AA genome, perennial *Oryza* relative in non-agricultural areas will be greater than pollen flow to the *O. sativa* red rice found in the U.S. and other temperate rice producing areas of the world (26). However, *O. rufipogon* pollen can greatly outcompete *O. sativa* pollen on *O. rufipogon* stigmas (78) (Table 20.1), which should mitigate against pollen flow from rice to wild *O. rufipogon,* but effects on pollen flow in the opposite direction were not addressed and thus are uncertain. Maximum distances of outcrossing detection are primarily relevant when considering field-to-field gene flow, but are essentially irrelevant within rice fields where weedy rice or wild rice plants, regardless of population density, are exposed to abundant levels of crop rice pollen (e.g., if 95% of all weedy rice plants were killed, separation distances from crop rice plants would remain extremely small).

20.4.7 Introgression Rate Considerations

Multiyear studies have shown that genes conferring increased fitness can quickly spread through a population of red rice comprising as much as 50% of the population after 2 crop years (39). In ongoing studies at the University of Arkansas, the spread of herbicide resistance from imazethapyr-resistant rice cultivars into red rice also has been confirmed, but to date, the spread of herbicide resistance has occurred to a lesser degree than that reported by Langevin et al. (39) (Burgos, University of Arkansas, Fayetteville, AR 2003, personal communication).

A model adapted for rice in Central American rain-fed systems was used to predict buildup of glufosinate-resistant weedy rice (44). Combining tillage and glufosinate depleted weed seeds from soil until resistance occurred. At a 1% outcrossing rate, the model predicted that predominantly glufosinate-resistant weedy rice populations would evolve within 5 to 8 crop seasons. Increasing the level of outcrossing to 5% decreased the time to evolution of resistance by 1 to 3 seasons.

Increasing annual weedy rice predation delayed onset of resistance. Although the model requires verification and is not directly applicable to U.S. rice systems, it provides insights into how production practices and pollen exchange can influence evolution of herbicide-resistant weedy rice. The model's assumption that resistant F_1 hybrids and their weedy rice parents will produce similar numbers of seed will probably not be appropriate for U.S. red rice types. F_1 hybrids from glufosinate-resistant rice were less productive than their awnless strawhull red rice parents, producing 34% fewer tillers, 39% fewer seeds per panicle, and seeds with 71% lower fertility levels (94). The parental red rice type can affect the productivity of rice/red rice hybrids. F_1 hybrids derived from strawhull red rice often flowered much later than either parent and frequently do not flower under field conditions (29,39,57,94). However, F_1 hybrids from Kaybonnet rice/awned blackhull red rice typically flowered during the same time frame as their parents (29) and have produced four times the seed yields of F_1 hybrids as Kaybonnet rice/awnless strawhull red rice, but will not do so if they are not herbicide resistant in the presence of herbicides (Gealy, 2003, unpublished data). Even with low numbers remaining, the backcrosses to red rice will restore the productivity, especially if the herbicide is continuously used.

20.5 PHENOTYPIC TRAITS OF RICE/RED RICE HYBRIDS IN THE U.S.

20.5.1 Key Traits in F_1 Hybrids

The phenotypic characteristics of long-grain rice/medium-grain red rice hybrids that are typically found in the southern U.S. appear to be generally predictable, but can be affected by the ecotype of red rice present (29). Some key phenotypic traits of hand-crossed hybrids between smooth leaf long-grain rice cultivars (Cypress and Kaybonnet) and four pubescent leaf red rice ecotypes were recently evaluated under Arkansas conditions (29). In these tests, plant types of all F_1 hybrids were taller, bushier, and more open than the rice parents. Likewise, all hybrids had pubescent leaves and red seeds (Figure 20.5A and Figure 20.5B), traits that are dominant over their recessive glabrous-leaf

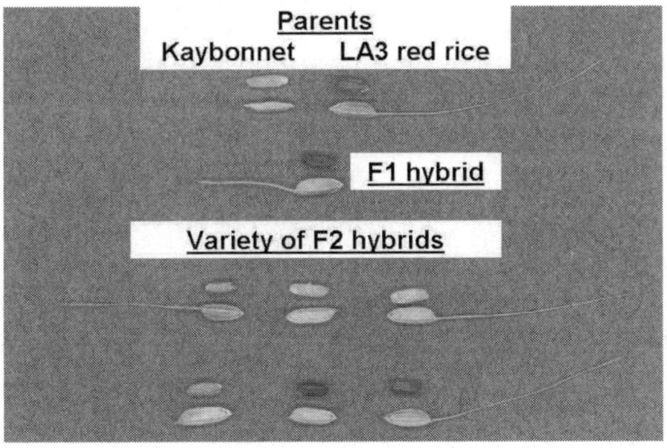

(A)

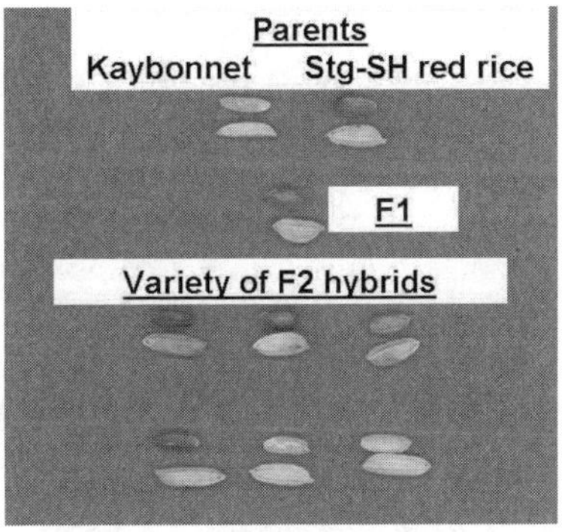

(B)

FIGURE 20.5 Inheritance and segregation of traits in F_1/F_2 of rice/red rice crosses. A: Parental, F_1, and F_2 seeds with and without hulls from a hand-crossed hybridization of Kaybonnet (female)/awned LA3 red rice. Kaybonnet rice parent (at top left) has long-grain seeds with light brown pericarp coloration. Red rice parent (at top right) has awned medium-grain seeds with red pericarp coloration. All F_1 seeds have medium-grain length with red pericarp coloration and are awned. Segregating F_2 hybrids have various combinations of seed and awn length and can have red or light brown pericarp coloration. B: Parental, F_1, and F_2 seeds with and without hulls from a hand-crossed hybridization of Kaybonnet (female)/awnless Stg-SH red rice. Kaybonnet rice parent (at top left) has long-grain seeds with light brown pericarp coloration. Red rice parent (at top right) has awnless medium-grain seeds with red pericarp coloration. All F_1 seeds have medium-grain length with red pericarp coloration and are awnless. Segregating F_2 hybrids have various combinations of seed length and can have red or light brown pericarp coloration.

and white-seed counterparts found in essentially all cultivated rice in the U.S. Thus, F_1 hybrids in the U.S. are expected to have neither glabrous leaves nor white seeds and are unlikely to be as short and erect as the common cultivated varieties. F_1 hybrids were awnless when the red rice parent was awnless (Figure 20.5B), but had awns as long or nearly as long as awned red rice parents (Figure 20.5A). The lower stem leaf sheath was light green (similar to red rice) when F_1 hybrids were derived from awnless red rice parents, and was reddish-purple when derived from awned red rice (color photos are printed in Rood (66)). Samples from Arkansas rice farms where putative hybrids were later shown to be genetically consistent with known hybrids of the appropriate ecotype had similar stem color characteristics (28). F_1 hybrids derived from awned red rice flowered during the same time range as both of the parents. However, those derived from awnless red rice flowered several weeks later than either parent (30). Similar delays in flowering of some rice/red rice hybrids have been reported (39,57,94). The differential flowering patterns among F_1 hybrids derived from different red rice types could influence rates of introgression in the U.S. It is conceivable that extreme delays in flowering experienced by the awnless F_1 hybrids may greatly reduce their ability to produce viable seed before rice harvest that can reinfest rice, especially in the northern part of the rice belt where growing seasons are shorter and chilling temperatures occur more frequently.

20.5.2 Differentiating between F_1 and F_2 Hybrids

F_2 phenotypic traits such as culm angles, coloration of lower stem, flowering dates, and awn lengths ranged broadly compared to those in the F_1, encompassing and sometimes exceeding the range of traits observed across the rice and red rice parents (29). In the F_2 phenotypes possessing a collection of recessive traits such as white seed, glabrous leaves, and awnless hulls were quite rare, but if they possessed the combination of traits that is desirable (together with the transgene), they could be the founder of a problematic population. Combinations that include red seeds, pubescent leaves, and awns (e.g., dominant traits) were most prevalent. Other traits such as purple stems (for awned types) and open culm angle were prominent in segregating F_2 populations. An understanding of the phenotypic traits expected in F_1 and F_2 populations can assist growers with red rice identification and management decisions. For instance, a large, bushy plant with pubescent leaves, red seeds, or much later maturity than commercial rice or red rice plants in the field may be consistent with an F_1 hybrid of rice and red rice. Although identification of the occasional F_1 hybrid plant in a field dominated by thousands of normal red rice plants can be tedious, growers should be on the lookout for these F_1 plants so that they can be promptly removed. A more common scenario is a group of plants consisting of various combinations of both dominant and recessive phenotypic traits (e.g., pubescent leaves and white seeds, erect growth habit and red seeds, pubescent leaves and short plant type, etc.), which may indicate that this is a segregating population in the F_2 or later generation and should impart a sense of urgency for initiation of remediation procedures such as roguing (26).

20.6 BACKCROSSING CONSIDERATIONS

Once formed, small numbers of fertile F_1 and F_2 hybrids could conceivably backcross with either herbicide-resistant rice or red rice parental plants. Crossing a heterozygous F_1 back to one of the homozygous parents creates a 1:1 segregation for polymorphic alleles from that parent (64). Backcrossing of heterozygous F_1 plants to resistant rice (with synchronized flowering) would probably result in a rapid evolution toward weedy derivatives that are homozygous (e.g., fixed) for the dominant herbicide-resistant trait. Segregants from these herbicide-resistant hybrids that also retain combinations of weedy traits such as tall plant height, high tillering, seed shattering, and dormancy could pose formidable weed problems in a rice agroecosystem in which use of the same herbicide-resistant rice system was continued in subsequent cropping cycles. Backcrossing F_1 plants to red rice would favor rapid evolution toward weedy derivatives that are homozygous for dominant traits such as red pericarp. Similar backcrossing scenarios with F_2 and parental plants would also produce phenotypes that converge toward one of the two parents, but the segregation among these phenotypes would be more complex due to the genetic variability in the F_2. Still, it is this F_2 variability that can provide material for founder populations of problem weeds. The importance of identifying weedy rice populations and the great extent to which they can remain undetected is detailed elsewhere in this volume (Chapter 17).

20.7 DORMANCY, SHATTERING, AND OTHER KEYS TO DOMESTICATION/DEDOMESTICATION

20.7.1 Dormancy and Shattering

Seed dormancy and shattering are both important traits that contribute to weediness of red rice and potential ferality of hybrid derivatives in rice agroecosystems. Shattering ensures that a large number of red rice seeds remain in the crop field and are not removed with harvested rice seed. Dormancy ensures that red rice seeds produced in 1 year are available for potential recruitment for weed infestations in multiple future years. All F_1 plants derived from reciprocal crosses between glufosinate-resistant rice and strawhull awnless red rice in Louisiana produced panicles with non-shattering seeds, suggesting that shattering is a recessive trait (69). By contrast, other genetic studies with strawhull awnless red rice and non-transgenic rice indicated that shattering behaved as a dominant trait and that segregation ratios were affected by the genetic background of the rice parent (41). In these studies, F_2 hybrids derived from Bellmont or Lemont rice parents segregated in 3:1 ratios for shattering (single dominant gene model), whereas F_2 hybrids derived from Labelle segregated in a 15:1 ratio (fits model of two epistatic genes). Crosses between IMI-resistant rice and strawhill red rice indicated that shattering was dominant over non-shattering, yet involved more than one gene (Shivrain, 2004, Univ. Arkansas, Fayettville, personal communication). Thus shattering inheritance appears to be somewhat complex, which is supported by other reports that it is a quantitative trait (4,6).

Shattering levels among red rice types in the U.S. vary greatly. Among 19 red rice types selected non-randomly from the southern U.S., more than half had shattering levels greater than 25%, but several did not shatter (55). Seed germination of these red rice types at harvest ranged from 0 to 28%. Occasionally, we have found red rice types that are so highly dormant that field experimentation with them has been difficult unless seeds were treated with dormancy breaking procedures such as high temperatures (45YC for 7 days) or KNO_3 (Gealy, unpublished data). Although the dormancy levels of hybrids derived from these red rice types and herbicide-resistant rice are not known, presumably they would be high also, and could affect weed seed-bank dynamics.

Germination levels as well as shattering of seeds produced by F_2 plants ranged broadly, were typically intermediate between those of the rice or glufosinate-resistant transgenic rice parent and the red rice parent, and were not more dormant or shatterable than the red rice parent (57). Thus, these segregating red rice hybrid populations apparently did not exhibit dormancy or shattering levels that could increase their ferality or weediness. Quantitative trait loci (QTLs) involved in non-shattering and reduced dormancy were evaluated for the wide cross, Taiwan *indica O. sativa rice* (non-shattering, non-dormant seed)/*O. rufipogon* wild rice (shattering, dormant seed) (6). The results suggested that shattering in wild rice is controlled by at least 4 genes, found on chromosome 1, chromosome 4, chromosome 8, and chromosome 11. Seed dormancy (hull-imposed or kernel dormancy) was controlled by numerous genes. Except for chromosome 4 and chromosome 10, all 12 chromosomes contained genes for seed dormancy, and chromosome 3, chromosome 5, chromosome 6, chromosome 9, and chromosome 11 contained more than 2 independent dormancy loci. Of the 4 shattering QTLs, 3 showed weak linkages to dormancy genes, but the shattering QTL with the greatest effect was not linked to dormancy. The linkage regions could recombine causing weak correlation between shattering and dormancy in segregating populations, which could somewhat increase the probability that these 2 traits would be inherited in combination, and thus enhance weediness or ferality of some hybrid derivatives. This trait association was thought to be due to natural selection for the 2 coadapted traits and not to a pleiotropic effect of a single gene or a close linkage to 2 genes. Similarly, 6 seed dormancy QTLs were identified from a mapping population derived from crosses between a non-dormant rice and a dormant weedy (red) rice strain (33). One QTL was tightly linked to red pericarp color. The results again suggested that genetically complex networks of loci regulate variation for seed dormancy in natural populations.

Although the probability is low that any single weedy rice hybrid will become a dominant weedy phenotype due to high levels of dormancy and shattering, QTLs quantitatively control both the extent of shattering and the average span of dormancy. Thus, these QTLs from an initial moderately shattering, moderately dormancy extended hybrid could be slowly enriched from additional crossing or mutation events to produce a phenotype with a high level of seed shattering and degree and duration of dormancy. Such optimized plant types could be highly formidable and adaptable in a variety of rice cropping systems.

20.7.2 Reproductive and Other Traits

In a mapping population of F_2 hybrids derived from a cross between *indica O. sativa* and *O. rufipogon*, the trait of extruding stigma was attributed to a single locus (92). This trait could be efficiently transferred, and plant types with the trait would be marginally favored as pollen acceptors in an outcrossing situation. This could impart greater outcrossing rates to male-sterile lines in hybrid breeding programs. Weedy rice types or hybrid derivatives with extruding stigmas could be favored as pollen acceptors in outcrossing situations and potentially become weedier. Conversely, lengthened anthers could increase the amount of pollen released from paternal lines that is available to fertilize male-sterile plants in hybrid rice systems, and weedy rice plants with this trait may be favored as pollen donors in outcrossing situations. Because the trait is complex, requiring 7 QTLs (all with small effects) to explain 47% of the variation, it will not be efficiently transferred between rice and weedy rice (92). Thus, it seems that anther length probably will not play a significant role in evolution of feral hybrid derivatives in U.S. rice, but rare hybrid derivatives with significantly lengthened anthers could potentially serve as the founder of a new problem weed population.

A genetic analysis of domestication related traits was conducted using a double haploid mapping population derived from a cross between a *japonica, O. sativa* rice cultivar, and an *O. sativa* weedy red rice (awned, shattering, red seeds) (4). Pericarp coloration was controlled by two complimentary genes (Rc on chromosome 1 and Rd on chromosome 7). Relatively few chromosome regions (on chromosome 1, chromosome 3, chromosome 4, and chromosome 7) accounted for most of the differences between rice and weedy red rice. Thus, genetic control of domestication traits was relatively simple. These were explainable by single QTLs with pleiotropic effects on more than one trait, by multiple-linked QTLs affecting individual traits, or by a combination of the two. Thus, co-localization of QTLs may lead to substantial correlation between traits. Chromosome 3 housed co-localized QTLs that included traits such as plant height, heading date, leaf width, panicle length, and total number of spikelets.

20.8 ANECDOTAL EVIDENCE OF GENE FLOW BETWEEN RED RICE AND RICE IN THE U.S.?

Although direct evidence is scarce, below I present several cases in which gene flow and introgression have apparently occurred between rice cultivars and red rice in U.S. rice fields. Presumably, the original hybrids resulted from natural crossing events in rice fields, but the possibility that hybrid seeds were imported into these fields from some other area cannot be ruled out completely.

20.8.1 Aroma Chemicals Detected in Red Rice Accessions at Low Frequencies

The aroma compound 2-acetyl-1-pyrroline is present in seeds of aromatic rice varieties (51). In a chemical evaluation of approximately 150 U.S. red rice accessions and non-aromatic rice varieties, 2-acetyl-1-pyrroline was detected only in 2 accessions, which were putative red rice hybrid derivatives that had been sampled from a Louisiana research field (24). These hybrid derivatives also produced long-grain, red seeds, robust plant types, and DNA marker profiling was consistent with known red rice/rice hybrids (27). Although the presence of 2-acetyl-1-pyrroline in these red rice plants cannot prove conclusively that the plants are hybrid derivatives, these observations are consistent with a scenario in which the plants are indeed hybrid derivatives from a cross between red rice and an aromatic rice cultivar. The first such aromatics were grown in Louisiana (e.g., Della) in the early 1970s and have been produced on a small scale since that time (51). Thus, the 2-acetyl-1-pyrroline trait appears to have introgressed into these red rice hybrid derivatives during the last 30 years. This trait has probably been neutral in southern U.S. rice systems, but traits with high selective advantages could be retained and detected at much higher frequencies in the population.

20.8.2 Blast Resistance Detected in Red Rice Accessions at Low Frequencies

Blast, caused by *Pyricularia grisea*, is a major rice disease in the U.S. that has been most effectively controlled by growing resistant cultivars (32). Susceptibilities of numerous red rice lines from the southern U.S. to the common races of *P. grisea* have been evaluated in greenhouse tests at Stuttgart, AR (40). The most blast-resistant red rice lines tested thus far appear to be hybrid derivatives, suggesting that the resistance may have introgressed into the red rice population from blast-resistant cultivars. To date, we have seen no evidence that these blast-resistant red rice lines are becoming more weedy and feral in U.S. rice fields, but periodic heavy blast infestations would probably favor their survival in those years. If the gene conferring blast resistance in U.S. cultivars (*Pi-ta*) were the basis for blast resistance in red rice, this would strongly suggest that gene flow from rice to red rice was the cause (Y. Jia, USDA—ARS, Stuttgart, AR, 2004, personal communication).

20.8.3 Short Stature, Short Awn Red Rice Plants Detected at Low Frequencies

Field surveys conducted in semi-dwarf rice fields have also been informative. Semi-dwarf rice (e.g., Lemont) was initially released in the southern U.S. in the mid-1980s (51). In 1992, Rutger and Kanter (cited in (26)) inspected isolated circular areas infested with red rice (apparently originating from a single seed of red rice) in rice production fields in Mississippi that had been planted to Lemont for 2 or more years. Because semi-dwarfism and pubescence are recessive traits and are associated with commercial cultivars, but not red rice, they could be used as a phenotypic indicator of introgression of cultivar traits into red rice. Various rice-red rice intermediate plant types (presumed red rice/rice hybrid derivatives) were observed, but only one glabrous, semi-dwarf hybrid derivative with red seed was observed among 2000 plants evaluated. Thus, introgression of these cultivar traits into red rice appears to be uncommon, but probably took place over a period of less than a decade. Kanter (cited in (26)) has also observed short red rice plants (apparent hybrid derivatives from semi-dwarf rice and red rice) in Mississippi foundation rice fields, but the frequency of this trait was not known. His observations are consistent with our own findings in Arkansas that a small number of red rice accessions (3 out of 438) randomly sampled from rice produced throughout the state in 2000 were no taller than the semi-dwarf cultivars, Lemont and Bengal, when grown in a common nursery at Stuttgart, AR in 2003 (unpublished data). Over all accessions, red rice heights ranged from 93 to 177 cm. The short red rice types were all less than 110 cm tall, with pubescent leaves and red, medium-grain seeds. These may have been hybrid derivatives that developed from a cross between red rice and a short rice cultivar. Heavy selection pressure against tall red rice ecotypes from repeated rope-wick application of non-selective herbicides could also select in favor of short red rice types, but the extent to which this practice has been used in Arkansas rice is not clear. DNA analysis could be used determine which if either of these explanations is correct. In the same experiment, less than 25% (99/438) of the red rice accessions were awned. A small subset of these plants (10/438 = 2.3%) had short awns and were no taller than the medium height, long-grain cultivar, Kaybonnet (125 cm). The combination of short stature and short awns suggests that these accessions may have been derived from crosses between rice cultivars and awned red rice types. Similar phenotypes have been observed in F_2 hybrids derived from awned red rice and long-grain rice (29). Although short stature has probably limited the ferality risk in these apparent red rice hybrid derivatives, presence of such traits in the red rice populations of Arkansas suggests that gene flow and introgression between commercial cultivars and red rice had been occurring well before herbicide-resistant varieties were grown by farmers. A more crucial consideration for risks of ferality is whether there has been a history of introgression between rice and red rice that has resulted in highly competitive and productive red rice. The presence of tall, highly tillering red rice plants with long-grain seeds and glabrous leaves would be highly suggestive that such introgression had occurred.

20.9 PROSPECTS FOR VOLUNTEERISM

In traditional non-transgenic or non-herbicide-resistant rice systems, harvest efficiencies can never be 100% (see Section 20.4.4). Thus, some volunteer commercial rice is essentially assured in the following spring crop. The density of potential volunteers arising from a typical commercial rice crop is usually reduced substantially through suicidal germination during the fall, seed predation by waterfowl during the winter, or preplanting tillage/weed control operations in the spring. Although volunteers can be present in significant numbers, current herbicides can effectively control them in rotational crops like soybean. The volunteer problem is likely to be more challenging in a no-till rotation of rice after rice, which favors preservation of viable rice seeds during fall and winter, and favors germination and development of rice in the spring. Volunteerism can also be a significant problem when rice cultivars produce moderately to highly dormant seeds. Moderately dormant rice cultivars such as Jasmine85 (a medium-tall aromatic cultivar released 1989) and various *Oryza* spp. germplasm lines in the USDA–ARS collection grown at the University of Arkansas Rice Research and Extension Center in Stuttgart, AR have sometimes become volunteer problems in rotational soybean or rice crops. Thus, *Oryza* plant types with dormancy (and other weedy characteristics) are now restricted to specific areas within the Center.

The rice volunteer problems in herbicide-susceptible rice noted above will probably become instantly more troublesome and less manageable once herbicide-resistant rice is added to the crop system. Because the volunteer rice plants are now resistant to an herbicide, this herbicide and others with similar modes of action cannot be used to control the volunteers. Similar concepts relative to maize, soybean, and other crops are expressed elsewhere in this volume (Chapter 10). Depending on the herbicide in question, management options for controlling a broad spectrum of weeds could become restricted. The threat of losing glyphosate as an effective tool to control volunteer IMI rice in non-resistant rice is as a key concern for possible future introduction of glyphosate-resistant rice to northern Latin America (Chapter 17).

20.10 CONCLUDING REMARKS

Classical thinking has held that because domestication is an evolutionary process operating under the influence of humans, it should progress gradually from the wild state toward domestication, but recent genetic studies indicate that domestication events may actually have occurred rapidly through the inheritance of tightly linked clusters of genes (64). If such events were to be run in reverse, naturalization processes could occur with similar speed and impact, which could have serious ramifications for gene flow interactions between rice and weedy rice.

Morishima (52) indicated that populations of perennial *Oryza* (*O. perennis;* reclassified as *O. rufipogon*) populations found in deep swamps were characterized by low reproduction, high outcrossing, and late flowering, and were more persistent in stable habitats compared to their annual counterparts growing in shallow or temporary swamps. This suggests that red rice hybrid segregants possessing an abundance of these perennial traits may be more likely to become successful feral populations after escaping from agricultural fields. Of these three traits, only late flowering seems to be prevalent in red rice populations of the U.S. A general absence of herbicide selection pressure will probably render the herbicide resistance gene irrelevant in these non-agricultural areas (with the possible exception of road ditches that are routinely sprayed with herbicide). Anecdotal evidence from Louisiana (D. Sanders, Louisiana State University, Clinton, LA, 2004, personal communication) indicates that red rice occasionally escapes from agricultural rice fields to waterways, bayous, swampy areas, and intermittently flooded areas. Some land managers have attempted to establish rice or red rice on or near bayous, swamps, or lakes as a source of food for ducks that would ostensibly improve hunting opportunities. However, this annual red rice species apparently has not become dominant in any of these habitats.

IMI-resistant rice varieties have been grown on a large scale in the U.S. since 2002. Evidence to date suggests that outcrossing has led to the formation of some early generation IMI-resistant red rice (*O. sativa*) hybrids. The degree to which these events may eventually affect the usefulness of IMI-resistant rice in rice is not yet known. It seems likely that observable gene flow and introgression into red rice as well as artificial selection for naturally occurring IMI-resistant red rice biotypes will occur in rice agroecosystems in the near term. However, significant herbicide-resistant red rice problems could develop in only a few cropping cycles in areas in which concentrated infestations of red rice experience herbicide failures during the initial period of IMI rice use. The time period over which these processes may prove to be economically or environmentally significant will probably be delayed further into the future and may somewhat depend on the success of grower scouting/roguing operations and whether herbicide-resistant rice systems based on alternate modes of action (e.g., glufosinate or glyphosate) are available to farmers. Exclusive dependence on a single herbicide-resistant rice system will almost certainly shorten the useful life-span of the technology compared to a more flexible situation in which two or more alternatives are available to growers. Because of the general habitat isolation and lack of herbicide selection pressure outside the rice agroecosystems of the U.S, gene flow and introgression of the herbicide resistance genes into non-flooded agricultural systems or non-agricultural areas in the form of feral populations seem remote, but not impossible. Conversely, in the less temperate to more tropical rice areas of the world, species of red, weedy, and wild rice often have traits such as high outcrossing that are more typical of perennial or wild plants. Thus, gene flow in these areas may be more likely to result in feral hybrid populations that could alter the makeup of non-agricultural ecosystems. Even so, the rate of gene flow may play less a part than high selection pressures that favor herbicide resistance genes already present in weedy rice populations. Akin to evolution of herbicide resistance where the mutation rate (analogous to rate of introgression) plays less of a role than the selective advantage of the gene, this situation was modeled (31, Chapter 1 this volume) and was eventually borne out in practical terms by the widespread appearance of triazine-resistant weeds despite initial resistance frequencies that were below limits of detection.

Based on the results from controlled experiments and other evidence presented, the rice-red rice systems in the U.S. have probably been experiencing low levels of gene flow for many decades and perhaps centuries. However, these interactions have apparently not had a major detrimental impact on ferality or weediness of red rice within or outside of these rice agroecosystems. Introduction of herbicide resistance genes into U.S. rice may establish a paradigm shift in these systems, but the potential new risks and hazards are probably primarily confined to rice production and related rotational crops occupying the same agroecosystems. In addition to development of herbicide-resistant rice, active test programs for transgenic fungal-, bacterial-, and insect-resistant strains of rice are underway. Their potential adverse effects on off-target organisms outside rice agroecosystems may warrant close scrutiny. Likewise, ongoing programs that seek to develop rice systems for large-scale production of pharmaceuticals or nutraceuticals may be of greater interest because of their potential impacts on human health and nutrition.

ACKNOWLEDGMENTS

The author thanks Junda Jiang and Jonathan Gressel for their reviews of an earlier version of this chapter, and Howard Black, Angela Rosencrantz, Kenton Bradley Watkins, Jason Hill, and Pam Smith for their helpful assistance with this project. Useful inputs into this chapter by colleagues at the workshop and their stimulating interactions are highly appreciated. Research was supported in part by the Arkansas Rice Research and Promotion Board.

LITERATURE CITED

1. Annou MM, Wailes EJ, Cramer GL. 2000. Economic analysis of adopting Liberty Link rice. In *Rice situation and outlook yearbook*, Evans MR, Ed., pp. 55–61. Washington, DC: United States Department of Agriculture Economic Research Service.
2. Azzini LE, Rutger JN. 1982. Amount of outcrossing on different male steriles of rice. *Crop Sci.* 22:905–907.
3. Beachell HM, Adair CR, Jodon NE, Davis LL, Jones J. 1938. Extent of natural crossing in rice. *Agron. J.* 30:743–753.
4. Bres-Patry C, Lorieux M, Clement G, Bangratz M, Ghesquiere A. 2001. Heredity and genetic mapping of domestication-related traits in a temperate japonica weedy rice. *Theor. Appl. Genet.* 102:118–126.
5. Bridges DC, Baumann PA. 1992. Weeds causing losses in the United States. In *Crop losses due to weeds in the United States*, Bridges DC, Ed., pp. 75–147. Champaign, IL: Weed Sci. Soc. of America.
6. Cai HW, Morishima H. 2000. Genomic regions affecting seed shattering and seed dormancy in rice. *Theor. Appl. Genet.* 100:840–846.
7. Chen LJ, Lee DS, Song ZP, Suh, HS, Lu B-R. 2004. Gene flow from cultivated rice (*Oryza sativa*) to its weedy and wild relatives. *Ann. Bot.* 93:67–73.
8. Childs N, Burdett A. 2000. The U.S. rice export market. In *Rice situation and outlook yearbook*, Evans MR, Ed., pp. 48–54. Washington, DC: United States Department of Agriculture Economic Research Service.
9. Craigmiles JP. 1978. Introduction. In *Red rice: research and control,* Eastin EF, Ed., pp. 5–6. College Station, TX: Texas Agricultural Experiment Station Bulletin 1270.
10. Deshaies M. 1996. Growing better rice (Special Report). *Rice Farming* 30:4–6.
11. Diarra A, Smith RJ Jr, Talbert RE. 1985. Interference of red rice (*Oryza sativa*) with rice (*O. sativa*). *Weed Sci.* 33:644–649.
12. Dilday RH. 1990. Contribution of ancestral lines in the development of new cultivars of rice. *Crop Sci.* 30:905–911.
13. Dowler CC. 1997. Weed survey-southern states: grass crops subsection. *Southern Weed Sci. Soc.* 50:227–246.
14. Environmental Protection Agency (EPA). September 29, 2003. Glufosinate-ammonium; pesticide tolerance. *Fed. Regist.* 68(188):55833–55849.
15. Estorninos LE Jr. 2000. Interference and genetic characterization of red rice (*Oryza sativa* L.) and rice (*O. sativa*). Ph.D. dissertation. Fayetteville, AR: Department of Crop, Soils, and Environmental Sciences, University of Arkansas. 144 pp.

16. Estorninos LE Jr., Gealy DR, Baldwin TL, Baldwin FB, Burgos NR. 2003. Estimates of outcrossing rates between IMI rice lines and red rice based on SSR fingerprinting and phenotypic characteristics. In *BR Wells rice research studies 2002,* Norman RJ, Meullenet JF, Eds., pp. 33–40. Fayetteville: University of Arkansas Agricultural Experiment Station, Series 504. http://www.uark.edu/depts/agripub/publications.
17. Estorninos LE Jr., Gealy DR, Dillon T, Baldwin FB, Burgos NR, Tai TH. 2002. Determination of hybridization between rice and red rice using microsatellite markers. In *BR Wells rice research studies 2001,* Norman RJ, Meullenet JF, Eds., pp. 82–87. Fayetteville, AR: University of Arkansas Agricultural Experiment Station, Series 495. http://www.uark.edu/depts/agripub/publications.
18. Estorninos LE Jr., Gealy DR, Talbert RE. 2002. Growth response of rice (*Oryza sativa*) and red rice (*Oryza sativa*) in a replacement series study. *Weed Tech.* 16:401–406.
19. Estorninos LE Jr., Gealy DR, TeBeest DO, Burgos NR. 2003. Reciprocal outcrossing rates between non-herbicide-resistant rice and red rice. In *BR Wells rice research studies 2002,* Norman RJ, Meullenet JF, Eds., pp. 41–46. Fayetteville, AR: University of Arkansas Agricultural Experiment Station, Series 504. http://www.uark.edu/depts/agripub/publications.
20. Federici MT, Vaughan D, Tomooka N, Kaga A, Wang X, Doi K, Francis M, Zorrilla G, Saldain N. 2001. Analysis of Uruguayan weedy rice genetic diversity using AFLP molecular markers. *Electronic J. Biotechnol.* 4:130–145. http://www.ejbiochem.com.
21. Ferrero A. 2001. Biology and control of red rice (*Oryza sativa* L. var. *sylvatica*) infesting European fields. In *Medoyrzae-interregional cooperative research network on rice in the Mediterranean climate,* Chataigner J, Ed., pp. 2–4.
22. Gealy DR, Black HL. 1998. Activity of glufosinate (Liberty) against red rice biotypes in glufosinate-resistant Gulfmont rice. In *BR Wells rice research studies 1997,* Norman RJ, Johnston TH, Eds., pp. 41–48. Fayetteville, AR: University of Arkansas Agricultural Experiment Station, Series 460. http://www.uark.edu/depts/agripub/publications.
23. Gealy DR, Black HL, Yan W, Rutger JN. 2003. Weed suppression potential in indica rice lines for the southern U.S. *Southern Weed Sci. Soc.* 56:303.
24. Gealy DR, Bryant RJ. 2002. Investigations into seed chemistry characteristics of red rice and red rice-rice crosses. *Rice Tech. Working Group* 29:160–161.
25. Gealy DR, Dilday RH, Baldwin FL, Black HL. 1999. Imazethapyr ('Pursuit') effect on red rice (*Oryza sativa* L.) biotypes. In *BR Wells rice research studies 1998,* Norman RJ, Johnston TH, Eds., pp. 79–89. Fayetteville, AR: University of Arkansas Agricultural Experiment Station, Series 468. http://www.uark.edu/depts/agripub/publications.
26. Gealy DR, Mitten DH, Rutger JN. 2003. Gene flow between red rice (*Oryza sativa*) and herbicide-resistant rice (*O. sativa*): implications for weed management. *Weed Technol.* 17:627–645.
27. Gealy DR, Tai TH, Sneller CH. 2002. Identification of red rice, rice, and hybrid populations using microsatellite markers. *Weed Sci.* 50:333–339.
28. Gealy DR, Yan W, Eizenga G, Moldenhauer K, Redus M. 2004. Insights into the parentage of rice/red rice crosses using SSR analysis of U.S. rice cultivars and red rice populations. *Rice Tech. Working Group* 30. In press.
29. Gealy DR, Yan W, Rutger JN. 2003. Characterization of hybrid populations from rice crossed with awned and awnless red rice. *3rd International Temperate Rice Conference*. 69: abstract 140.
30. Gressel J. 2002. *Molecular biology of weed control.* New York: Taylor & Francis. 520 pp.
31. Gressel J, Segel LA. 1978. The paucity of genetic adaptive resistance of plants to herbicides: possible biological reason and implications. *J. Theor. Biol.* 75:349–371.
32. Groth D, Lee F. 2003. Rice diseases. In *Rice: origin, history, technology, and production-crop production series #6149,* Smith CW, Dilday RH, Eds., pp. 413–436. New York: John Wiley & Sons.
33. Gu XY, Kianian SF, Foley ME. 2004. Multiple loci and epistases control genetic variation for seed dormancy in weedy rice (*Oryza sativa*). *Genetics* 166:1503–1516.
34. Heap IM. 2004. International survey of herbicide resistant weeds. http://www.weedscience.com.
35. Heap IM. 1997. The occurrence of herbicide-resistant weeds worldwide. *Pest. Sci.* 51:235–243.
36. Hu J, Rutger JN. 1991. A streptomycin induced no-pollen male sterile mutant in rice (*Oryza sativa* L.). *J. Genet. Breed.* 45:349–352.

37. Jiang J, Linscombe SD, Wang J, Oard JH. 2000. High efficiency transformation of U.S. rice lines from mature seed-derived calli and segregation of glufosinate resistance under field conditions. *Crop Sci.* 40:1729–1741.
38. Kendig A, Williams B, Smith CW. 2003. Rice weed control. In *Rice: origin, history, technology, and production-crop production series #6149,* Smith CW, Dilday RH, Eds., pp. 457–472. New York: John Wiley & Sons.
39. Langevin SA, Clay K, Grace JB. 1990. The incidence and effect of hybridization between cultivated rice and its related weed red rice (*Oryza sativa* L.). *Evolution* 44:1000–1008.
40. Lee FN, Gealy DR, Dilday RH. 2000. Differential response of United States *Oryza sativa* (red rice) accessions to races of *Pyricularia grisea.* In *BR Wells rice research studies 1999,* Norman RJ, Johnston TH, Eds., pp. 94–101. Fayetteville, AR: University of Arkansas Agricultural Experiment Station, Series 476. http://www.uark.edu/depts/agripub/publications.
41. Livore AB. 1984. Inheritance of shattering and other characters in red rice. *Proc. Rice Tech. Working Group* 20:15.
42. Mackill DJ. 1995. Classifying japonica rice cultivars with RAPD markers. *Crop Sci.* 35:889–894.
43. Mackill DJ, McKenzie KS. 2003. Origin and characteristics of U.S. rice cultivars. In *Rice: origin, history, technology, and production-crop production series #6149,* Smith CW, Dilday RH, Eds., pp. 87–100. New York: John Wiley & Sons.
44. Madsen KH, Valverde BE, Jensen JE. 2002. Risk assessment of herbicide-resistant crops: a Latin American perspective using rice (*Oryza sativa*) as a model. *Weed Technol.* 16:215–223.
45. Magalhães AM Jr., Andres A, Franco DF, Silva MP, Abreu A, Luzzardi R, Coimbra J. 2001. Avalição do fluxo genico entre genótipos de arroz transgenico, cultivado e arroz vermelho. II Congresso Brasileiro De Arroz Irrigado. XXIV Reunieo Da Cultuira Do Arroz Irrigado, ANAIS 768–771.
46. Masson JA, Webster EP, Williams BJ. 2001. Flood depth, application timing, and imazethapyr activity in imidazolinone-tolerant rice *(Oryza sativa). Weed Technol.* 15:315–319.
47. Mese MD, Azzini LE, Rutger JN. 1984. Isolation of genetic male sterile mutants in a semi-dwarf japonica rice cultivar. *Crop Sci.* 24:523–525.
48. Messeguer J, Fogher C, Guiderdoni E, Marfa V, Catala MM, Baldi G, Mele E. 2001. Field assessments of gene flow from transgenic to cultivated rice (*Oryza sativa* L.) using a herbicide resistance gene as tracer marker. *Theor. Appl. Genet.* 103:1151–1159.
49. Messeguer J, Marfa V, Catala MM, Guiderdoni E, Mele E. 2004. A field study of pollen-mediated gene flow from Mediterranean GM rice to conventional rice and red rice weed. *Mol. Breed.* 13:103–112.
50. Mitten DH. 2004. Achieving global registration: status for glufosinate-tolerant rice. *Rice Tech. Working Group* 30. In press.
51. Moldenhauer K, Gibbons J, McKenzie K. 2004. Rice cultivars. In *Rice: chemistry and technology,* Champagn ET, Ed., pp. 49–75. St. Paul, MN: The American Association of Cereal Chemists.
52. Morishima H. 1985. Habitat, genetic structure and dynamics of perennial and annual populations of the Asian wild rice *Oryza perennis. NATO ASI Series* 5:179–190.
53. National Agricultural Statistics Service — United States Department of Agriculture. 2004. http://www.nass.usda.gov.
54. Noldin JA, Chandler JM, Ketchersid ML, McCauley GN. 1999. Red rice (*Oryza sativa*) biology. II. Ecotype sensitivity to herbicides. *Weed Technol.* 13:19–24.
55. Noldin JA, Chandler JM, McCauley GN. 1999. Red rice (*Oryza sativa*) biology. I. Characterization of red rice ecotypes. *Weed Technol.* 13:12–18.
56. Noldin JA, Yokoyama S, Antunes P, Luzzardi R. 2002. Outcrossing potential of glufosinate-resistant rice to red rice. *Planta Daninha* 20:243–251.
57. Oard J, Cohn MA, Linscombe S, Gealy DR, Gravois K. 2000. Field evaluation of seed production, shattering, and dormancy in hybrid populations of transgenic rice (*Oryza sativa*) and the weed, red rice (*Oryza sativa*). *Plant Sci.* 157:13–22.
58. Oard JH, Linscombe SD, Braverman MP, Jodari F, Blouin DC, Leech M, Kohli A, Vain P, Cooley JC, Christou P. 1996. Development, field evaluation, and agronomic performance of transgenic herbicide resistant rice. *Mol. Breed.* 2:359–368.
59. OECD. 1999. Consensus document on the biology of *Oryza sativa* (rice). In *Series on harmonization of regulatory oversight in biotechnology,* ed. Paris: Organization for Economic Cooperation and Development.

60. Olofsdotter M, Valverde BE, Madsen KH. 2000. Herbicide resistant rice: global implications for weedy rice and weed management. *Ann. Appl. Biol.* 137:279–295.
61. Ottis BV, Chandler JM, McCauley GN. 2003. Imazethapyr application methods and sequences for imidazolinone-tolerant rice (*Oryza sativa*). *Weed Technol.* 17:526–533.
62. Ottis BV, Malik MS, Scherder EF, Talbert RE. 2004. Competitiveness of rice cultivars, Wells, CL161, and XL8 at reduced seeding rates. *Southern Weed Sci. Soc.* 57:255–267.
63. Pantone DJ, Baker JB. 1991. Weed-crop competition models and response-surface analysis of red rice competition in cultivated rice: a review. *Crop Sci.* 31:1105–1110.
64. Paterson AH. 2002. What has QTL mapping taught us about plant domestication? *New Phytol.* 154:591–608.
65. Pellerin KJ, Webster EP, Zhang W, Blouin DC. 2003. Herbicide mixtures in water-seeded imidazolinone-resistant rice (*Oryza sativa*). *Weed Technol.* 17:836–841.
66. Rood MA. 2004. Red rice update: southern fields aren't quite clear; new tool coming. *Rice J.* 107(2):17–19.
67. Rood MA. 2004. Switching to dry: California farmers experiment with dry-seeding to fight herbicide resistance. *Rice J.* 107(2):14–16.
68. Sankula S, Braverman MP, Jodari F, Linscombe SD, Oard JH. 1997. Evaluation of glufosinate on rice (*Oryza sativa*) transformed with the *BAR* gene and red rice (*Oryza sativa*). *Weed Technol.* 11:70–75.
69. Sankula S, Braverman MP, Oard JH. 1996. Genetic analysis of glufosinate resistance in controlled crosses between cultivated rice (*Oryza sativa* L.) and red rice (*Oryza sativa* L.). *Southern Weed Sci. Soc.* 50:26.
70. Sankula S, Braverman MP, Oard JH. 1998. Genetic analysis of glufosinate resistance in crosses between transformed rice (*Oryza sativa*) and red rice (*Oryza sativa*). *Weed Technol.* 12:209–214.
71. Scott RC, Boyd JW, Smith KL. 2004. Rice. In *Recommended chemicals for weed and brush control,* pp. 68–82. University of Arkansas Division of Agriculture.
72. Shivrain VK, Burgos NR, Moldenhauer KAK, Gealy DR, Gbur EE. 2004. Phenotypic and genetic diversity of red rice (*O. sativa*) in Arkansas. *Proc. Southern Weed Sci. Soc.* 57:207.
73. Shivrain VK, Burgos NR, Rajguru SN, Sparks OC, Anders MM. 2004. Potential for gene flow between imidazolinone-resistant rice and red rice. *Weed Sci. Soc. Am. Abstracts* 44:225.
74. Shivrain VK, Burgos NR, Scott RC, Sparks OC. 2004. Tolerance of Arkansas red rice biotypes to imazethapyr, glyphosate, and glufosinate. *Rice Tech. Working Group* 30. In press.
75. Smith KL, Scott RC, Kelley MB. 2004. Rice yield and grade as influenced by rice cultivar and seeding rate and red rice density. *Rice Tech. Working Group* 30. In press.
76. Smith RJ Jr. 1989. Economics of weed control in U.S. rice. *Asian-Pacific Weed Sci. Soc.*12:39–51.
77. Song ZP, Lu B-R, Zhu YG, Chen JK. 2003. Gene flow from cultivated rice to the wild species *Oryza rufipogon* under experimental field conditions. *New Phytol.* 157:657–665.
78. Song ZP, Lu B-R, Zhu YG, Chen JK. 2002. Pollen competition between cultivated and wild rice species (*Oryza sativa* and *O. rufipogon*). *New Phytol.* 153:289–296.
79. Steele GL, Chandler JM, McCauley GN. 2002. Control of red rice (*Oryza sativa*) in imidazolinone-tolerant rice (*O. sativa*). *Weed Technol.* 16:627–630.
80. Street JE, Bollich PK. 2003. Rice production. In *Rice: origin, history, technology, and production-crop production series #6149,* Smith CW, Dilday RH, Eds., pp. 271–296. New York: John Wiley & Sons.
81. USDA-FSA. 2003. 2003 crop rice loan rates and grade discounts. ftp://ftp.fsa.usda.gov/public/notices/LP_1919.pdf.
82. USDA-GIPSA-Federal Grain Inspection Service. 2002. United States standards for rice: United States standards for rough rice. http://www.usda.gov/gipsa/reference-library/standards/ricestandards.pdf.
83. Valverde BE, Itoh K. 2001. World rice and herbicide resistance. In *Herbicide resistance and world grains,* Powles SB, Shaner DL, Eds., pp. 195–249. Boca Raton, FL: CRC Press.
84. Valverde BE, Riches CR, Caseley JC. 2002. Prevention and management of herbicide resistant weeds in rice: experiences from Central America with Echinochloa colona. San Jose, Costa Rica: Camara Insumos Agropecuarios. 123 pp.
85. Vandiver VV Jr., Hall DW, Westbrooks RG. 1992. Discovery of *Oryza rufipogon* (Poaceae: Oryzeae), new to the United States, with its implications. SIDA 15:105–109.
86. Vaughan KL, Ottis BV, Bormans CA, Prazak-Havey AM, Sneller C, Chandler JM, Park WD. 2001. Is all red rice found in commercial rice really *Oryza sativa*? *Weed Sci.* 49:468–476.

87. Webster EP, Masson JA. 2001. Acetolactate synthase-inhibiting herbicides on imidazolinone-tolerant rice. *Weed Sci.* 49:652–657.
88. Webster TM. 2000. Weed survey-Southern states: grass crops subsection. *Southern Weed Sci. Soc.* 53:247–274.
89. Wheeler C, Baldwin F, Gealy D, Gravois K. 1997. Weed control in Liberty-tolerant rice. In *BR Wells rice research studies 1996,* Norman RJ, Johnston TH, Eds., pp. 72–74. Fayetteville, AR: University of Arkansas Agricultural Experiment Station, Series 456. http://www.uark.edu/depts/agripub/publications.
90. Wheeler CC, TeBeest DO. 2002. Hybridization of glufosinate-tolerant rice (*Oryza sativa*) and red rice (*Oryza sativa*). In *BR Wells rice research studies 2001,* Norman RJ, Meullenet JF, Eds., pp. 58–64. Fayetteville, AR: University of Arkansas Agricultural Experiment Station, Series 495. http://www.uark.edu/depts/agripub/publications.
91. Willson JH. 1979. *Rice in California.* Butte County Rice Growers Association. pp. 49, 115–116.
92. Xiong LZ, Liu KD, Dai KX, Xu GC, Zhang Q. 1999. Identification of genetic factors controlling domestication-related traits of rice using an F_2 population of a cross between *Oryza sativa* and *O. rufipogon*. *Theor. Appl. Genet.* 98:243–251.
93. Yoshida S. 1981. Growth and development of the rice plant. *Fundamentals of rice crop science,* Los Banos, Philippines: International Rice Research Institute. pp. 55–58.
94. Zhang N, Linscombe S, Oard J. 2003. Out-crossing frequency and genetic analysis of hybrids between transgenic glufosinate herbicide-resistant rice and the weed, red rice. *Euphytica* 130:35–45.
95. Zhang W, Linscombe S, Webster E, Oard J. 2004. Risk assessment and genetic analysis of natural outcrossing in Louisiana commercial fields between Clearfield rice and the weed red rice. *Rice Tech. Working Group* 30. In press.

QUESTIONS AND ANSWERS

Rick Roush: With all the detailed sequencing data available in rice, does anyone know the specific mutations conferring the difference between shattering and non-shattering? If not, why not?

Answer: The QTL analyses of mapping populations derived from crosses between rice and weedy or wild rice species provide valuable insights into the numbers of loci involved, but apparently these results can be easily influenced by environmental interactions. Apparently, several genes are involved and have been described in some detail for *Oryza* sp. Additional details and relevant literature citations are available at http://www.gramene.org. It is my understanding that the same gene or genes controlling shattering may be found in many grass species due to synteny. Thus, if known for oats or rye, etc., this information should be highly applicable to rice and vice versa.

Rick Roush: What is the real difference between wheat and rice when it comes to weediness? Why are there weedy rices, but almost no weedy wheats?

Answer: Aegilops cylindrica (CCDD tetraploid), a weedy relative of wheat (AABBDD hexaploid), is a particularly troublesome weed in wheat, primarily because of its crop mimicry. However, gene flow and introgression of traits between wheat and *A. cylindrica* are much less prevalent than between rice and weedy rice. Presumably this limitation is in part imposed due to the unequal genome composition of wheat and *A. cylindrica*, and their relatively large and complex genomes (hexaploid and tetraploid, respectively) compared to the simple, equal genome composition in rice and weedy rice (both are AA diploid). Additionally, it is my understanding that many of the genes or chromosomal regions conferring weediness (e.g., shattering, dormancy) to weedy rice are also found in wheat and most other cereal crops due to high levels of synteny among these species.

21 Modeling Population Dynamics to Overcome Feral Rice in Rice

Francesco Vidotto and Aldo Ferrero

21.1 SPREAD AND IMPORTANCE OF WEEDY RICE IN EUROPE

Rice is cultivated in Europe on about 420,000 ha, mostly in a continuously flooded, flow-through system, mainly using water from rivers and lakes. Water is usually supplied from the topmost to the bottommost basin and regulated by floodgates. About 80% of the total area is cultivated with *japonica* varieties. At present, most of the rice is mechanically direct-seeded in flooded fields from mid-April to the end of May and harvested from mid-September to the end of October (29).

Weedy rice is undoubtedly the main weed problem in European rice fields. The first available reports of shattering forms of cultivated rice in Europe mainly referred to Italian paddy fields and dated from the first years of 19th century (7). Since then, weedy rice began to spread and is now considered one of the most important constraints to rice cultivation. For several years, the shattering ability and the other traits of this weed were regarded as a syndrome in cultivated rice induced by an unidentified plant pathogen. Since the beginning of the 20th century, weedy rice has been regarded by botanists in Europe as the *varietas "silvatica"* of the species *Oryza sativa* L. (13–15,68).

The adoption of transplanting from 1920 to 1960 helped reduce the spread of weedy rice, mainly because this technique allowed the identification and culling of weeds already at the nursery level (30,32,76). The presence of this weed in rice fields dramatically increased after this period, mainly because of the shift from transplanting to direct seeding and the cultivation of varieties containing grains of weedy plants from seed produced in infested areas. Weedy rice occurs in all main European countries where rice is cultivated (Italy, France, Spain, Portugal, and Greece), particularly in Spain and Italy. The situation further worsened during the 1980s when European Commission policies and the growing demand of European consumers (82) encouraged the cultivation of poorly competitive, *indica*-type semi-dwarf rice varieties. Weedy rice presently infests an estimated 65% of the total European rice growing area and is considered a crucial problem especially in Italy, France, and Spain, where it infests about 70% of the total rice area (12,35). Weedy rice has also been reported to be an important weed in Russian rice fields (A. Ferrero, University of Torino, personal communication, 2004).

Most of the European weedy rice has grains with pigmented pericarp (red rice) (29). As reported in several other areas of the world, weedy rice has exceedingly variable morphological traits. Bocchi et al. (8) isolated seven different clusters based on morphological characteristics in Italian accessions. The most common plants had red-brown pigmented nodes and internodes, erect panicular leaves, pale green blades with purple ribs, pale green ligules and auricles, no awn, and erect panicles.

The origin of European weedy rice and its genetic relationship with cultivated varieties are still far from clear. Using microsatellite markers, Bres-Patry et al. (9) found weedy rice populations in Camargue (France) with a greater genetic diversity than Mediterranean varieties (temperate *japonica*) and the presence of alleles more ascribable to *indica* varieties. They suggested that distant crosses between *japonica* and *indica* varieties could have been the origin of these weedy populations.

Studies including a higher number of weedy rice populations have been carried out in the U.S. (see Chapter 20). The control of weedy rice in rice is much more difficult than other weeds because of the high biological affinity with cultivated varieties and the great morphological variability. With the exception of the transgenic varieties (67), which at present cannot be cultivated in European Union (E.U.) countries, control with postemergence herbicides in the crop cannot be carried out because there are no products that are both effective on weedy rice and safe for the crop. For this reason, weedy rice management necessarily relies upon a combination of preventive, cultural, and mechanical practices.

Even though it is recognized that the first step in controlling weedy rice is prevention of infestation by planting seed free of weedy rice grains, regulations are often not sufficiently restrictive as far as rice seed purity is concerned. According to E.U. regulations, certified seed must have fewer than 3 grains per 400 g (less than 0.03%). At a sowing rate of about 180 g/ha, this would allow planting more than 1300 weedy rice seeds per hectare, leading to severe infestation in 2 to 3 years.

Crop rotation in combination with the use of "clean" seed is considered the best practice to control weedy rice. Most of the European rice is cultivated in extensive, continuous monocropping systems in which rotation is seldom practiced. This is mainly due to organizational reasons at the farm level. Monocropping allows a simplification of cultural practices and specialization of machinery. In some environmental conditions (11), rice monocropping is often the only cultural system that can be applied.

The most widely adopted technique to control weedy rice in Europe is the stale seedbed, also known as false seeding (12). The soil is tilled and the seedbed is prepared as for normal seeding. Usually irrigation is performed to promote germination of weedy rice and other weeds. After emergence (usually at 2- to 3-leaf stage), weedy rice seedlings are destroyed by spraying nonselective herbicides or by mechanical means. The most used herbicides in Europe are glyphosate, dalapon, cycloxydim, and clethodim. Mechanical destruction of weedy rice seedlings (blade or rotary harrowing) is usually less effective (36), but good results are frequently achieved in Spain with puddling (*fangueado* in Spanish), which consists of repeated passes over the flooded field with tractors equipped with cage wheels.

The stale seedbed technique is aimed at reducing the weed infestation in the same season in which it is applied and gradually decreasing its seed bank. It implies a delayed sowing and the use of early maturity, generally lower yielding, cultivars. The delayed planting of rice results in about 6 to 8% yield loss due to the shorter season. Because of the adoption of stale seedbed across Europe, there has been a great demand from rice growers of early maturity variety seeds in the last few years. In some cases, this brought about the production and marketing of seed lots heavily infested by weedy rice seeds.

After seeding of the crop, weedy rice can be controlled when the weed is taller than the crop by using cutting bars that remove weedy rice panicles (4). Similarly, broad-spectrum herbicides (glyphosate, in particular) can be applied using wiping bars. These interventions do not prevent weedy rice from competing with that season's crop, but are mainly aimed at preventing the enrichment of the seed bank and from having weedy rice seed contaminate crop seed.

Breeding of short duration, higher yielding, late-planted varieties is imperative for an integrated management system of weedy rice based on preventive and agronomic means. Still breeding cannot fully compensate for having fewer days to photosynthesize and grow. Conversely, a potential side effect of the widespread adoption of late planting could be the selection of weedy rice populations that mimic the crop cycle.

21.2 WEEDY RICE BIOLOGY IN RELATION TO POPULATION DYNAMICS

Patterns of weedy rice population dynamics are similar to those of other weed species. The negative effects of this plant lie more in its botanical closeness to the crop than in the specific biological features. Like all other weeds, weedy rice population dynamics can be regarded as the combined result of several mechanisms, each one acting during specific steps of the plant life cycle.

The phases that have a major impact on population dynamics can be can be chronologically simplified to only a few important steps: seed, seedling emergence, seed production, and dispersal.

21.2.1 Seed

Contrary to *japonica* varieties, seeds of weedy rice are not usually capable of germinating immediately after their dispersal. Seeds progressively lose primary seed dormancy after ripening, with a rate varying according to the biotype and storage conditions. Significant reduction of primary dormancy can already be observed 2 months after seed production (18). Several studies have aimed at clarifying the physiological mechanisms of secondary dormancy in weedy rice (10,16,17,23,26,38–40), while less information is available on secondary dormancy dynamics under field conditions (55,77). Weedy rice seeds need to be exposed to field conditions for about 150 days to obtain high germination (77). By the following spring, weedy rice seeds are usually capable of germinating if favorable environmental conditions occur. In practical conditions, most seeds can germinate between the end of April and beginning of May. Subsequent flushes can occur during the cropping season, especially after soil disturbance caused, for example, by tractor passes. These later flushes partly negate the effect of stale bedding.

In ordinary field conditions, longevity of the seeds with secondary dormancy, rather than primary dormancy, can have a significant agronomical relevance, as depletion of the seed bank due to mortality can have a strong impact on population dynamics. Crop rotation programs, for example, should necessarily consider this aspect. Several studies pointed out that crop rotation allows for a relatively easy control of weedy rice with selective herbicides (3,69,74). It is likely that crop rotation would select for weedy rice with longer dormancy, but no evidence of this has been reported. One year of rotation with soybean can eliminate more than 80% of the seed bank (33), which is sufficient, in case of low seed bank density, to avoid yield losses in the subsequent year. However, it should be noted that a seed bank capable to give above 5 plants/m^2 of weedy rice can result in over 22% yield loss (53).

Despite its great importance, the sparse information is often conflicting. Appreciable levels of germinability are still reported after burial for 1 (30), to 3 (46), to over 10 years (24). Good weedy rice control has been reported in both U.S. and Europe after a 1- or 2-year rotation with soybean (2,3,5,34,48,51,61,65), supporting in part evidence that in practical conditions seed mortality can be particularly high (sometimes over 90%) during the first year of burial (34). No quantitative information is available on the ratio between transient (longevity <1 yr), short-term persistent (1 to 4 yr), and long-term persistent (>4 yr) seeds in the seed bank, which is considered a useful indicator of the role of seed bank in population dynamics (6).

21.2.2 Seedling Emergence

As with other plants, the percentage of weedy rice seeds that develop into seedlings (rate of emergence) is inversely related to the depth of burial (72). In weedy rice, other factors, such as the soil texture and the presence of a water layer over the soil, can influence the rate of emergence (30,44). In sandy soils, emergence can occur up to about 10 cm of depth, while only more superficial seed can germinate in loam and clay soils (32,75,77). As for most of the other morphological and biological features, there is a great variability among populations. Some populations emerge from a burial depth of 25 cm (62).

Different tillage operations can determine a different distribution of seeds in the soil profile, strongly affecting the rate of emergence. Water management has a similar effect: the presence of a water layer usually results in a reduction of the rate of emergence, regardless the distribution of the seeds in the soil profile.

21.2.3 Seed Production and Dispersal

This step in the cycle of the weed represents the main input to the weedy rice seed bank. The amount of seed per unit area that is produced and reaches the soil can be schematically described as the result of the contribution of single components, which are basically the same as those considered when crop yield is analyzed.

A first component is plant density, which is the result of the initial seedling density (i.e., rate of emergence), the efficacy of control treatments (in the case of stale seedbed), and crop competition.

A second component is the number of panicles per plant. This parameter shows great variability among weedy rice populations, and it is usually higher than in the crop (24,25). The number of panicles per plant is inversely related to weedy rice density (7 and 4.6 panicles/plant with 10 or 40 weedy rice plants/m^2, respectively) and it is influenced by the growth habit of the cultivated variety (27). In other studies, this parameter was less influenced by density (21,78).

The plant density, multiplied by the number of panicles per plant, gives the number of panicles per unit area (e.g., panicles/m^2). As the weedy rice population cannot grow indefinitely, a maximum value of this last parameter should be introduced in a population model. It can be assumed that this value could be between 400 to 600 panicles/m^2, roughly the maximum possible crop panicle density (27,54,56).

The number of grains per panicle is also highly variable, with values from about 40 to over 400 seeds/panicle reported (24,34,35,66,78). Early seed shattering is a specific characteristic of weedy rice. A great proportion of weedy rice seeds typically shatters and falls to the ground by rice harvest. The assessment of the percentage of seeds that shatter before rice harvesting poses several difficulties. It is usually carried out by isolating the panicles with gauze or paper bags or by simulating violent mechanical stresses (such as that caused by wind) in laboratory conditions, with reports ranging between 65 and 80% (27,34,66). More than 50% of the seeds are already able to germinate (after the interruption of primary dormancy) about 20 days after flowering (34).

21.3 MODELING WEEDY RICE INFESTATION DYNAMICS

Several models have been proposed to partially or universally predict the population dynamics in various crop-weed situations (19,28,45,52,63,64), with some specifically for weedy rice (37,78). The annual life cycle of this weed can be summarized in a diagram divided in two main sections: seed bank evolution and weed growth (Figure 21.1). The input variables and the parameters that control weedy rice population growth are related to traits of the weed population itself (seed bank size, seed germination and emergence, etc.), to the agronomic conditions (soil tillage, in particular), or to the practices applied preplanting for weed control. The effect of preplanting control practices (i.e., stale seedbed) is included.

The main input regulating seed bank dynamics is the seed rain from plants that escape control treatments, which reach maturity and are able to produce and disperse seeds. The weedy rice seeds carried with the planted crop are a minor input unless the field had been free of weedy rice. This amount of seed can have a little impact in high infestation conditions, but is crucial in low or uninfested field. The main outputs are the seeds that germinate and become seedlings and those that are lost to aging (mortality) or predation.

The numbers of seeds that germinate and grow to seedlings can be estimated by simply calculating a percentage of the seed bank. A more accurate evaluation also considers the distribution of the seeds into the soil profile, as a consequence of soil tillage.

Several models predict the seed movement through the soil profile due to tillage and the consequent effects on germination and emergence due to modification of burial depth (20,22,41,42,49,50). One of the most common ways to formalize seed movement is to use probability matrices (transition matrices). Several matrix layers of discrete thickness represent the soil profile. For each layer, the probability of a seed moving to another layer after tillage is defined.

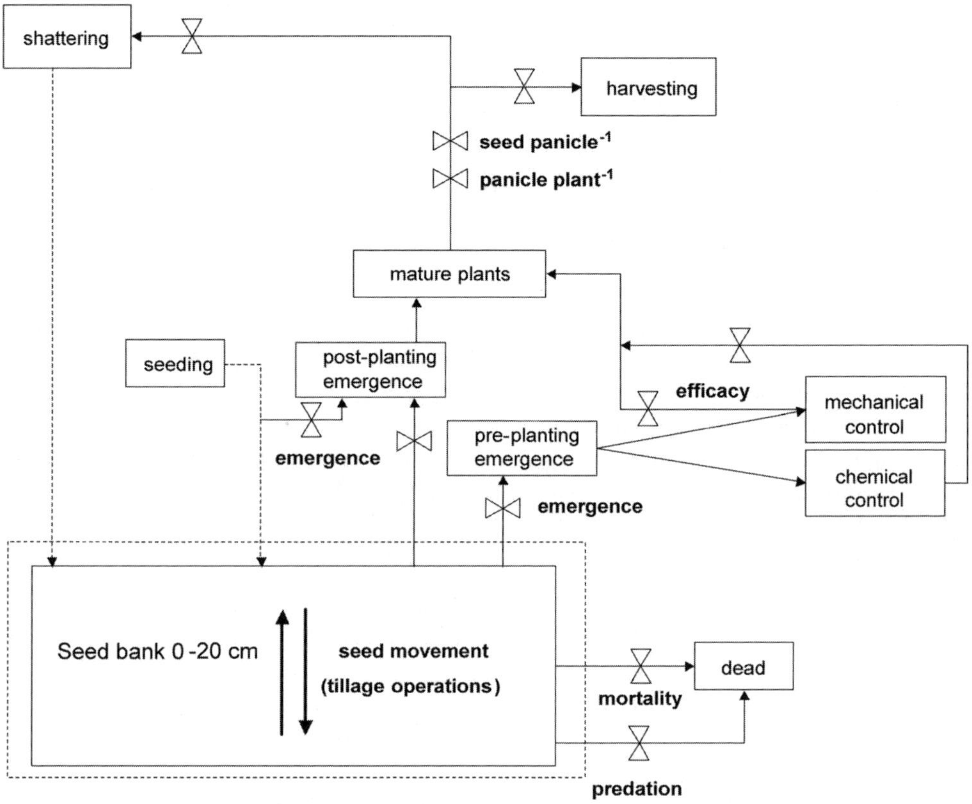

FIGURE 21.1 Interrelated factors influencing weedy rice life cycle. Symbols: □ = state variables; ↑ = fluxes; ⧖ = rates; dashed arrows: seed bank inputs; dotted arrows: seed bank outputs. (Modified from Vidotto et al. (78).)

From: 0-5 cm, 5-10 cm, 10-15 cm, 15-20 cm (columns); To: 0-5 cm, 5-10 cm, 5-10 cm, 15-20 cm (rows)

$$\begin{pmatrix} p_{1,1} & p_{2,1} & p_{3,1} & p_{4,1} \\ p_{1,2} & p_{2,2} & p_{3,2} & p_{4,2} \\ p_{1,3} & p_{2,3} & p_{3,3} & p_{4,3} \\ p_{1,4} & p_{2,4} & p_{3,4} & p_{4,4} \end{pmatrix} \begin{pmatrix} s_{1,t} \\ s_{2,t} \\ s_{3,t} \\ s_{4,t} \end{pmatrix} = \begin{pmatrix} s_{1,t+1} \\ s_{2,t+1} \\ s_{3,t+1} \\ s_{4,t+1} \end{pmatrix}$$

Transition matrix; before tillage, after tillage: Seed distribution

FIGURE 21.2 Estimation of seed distribution after tillage by using a transition matrix on a vector of initial seed distribution.

This approach has been adopted mainly to describe vertical seed movement and only a few attempts have been made to also consider horizontal movement (58,73).

The model proposed by Cousens and Moss (22), for example, refers to four layers of 5 cm each. The number of seeds initially present in each layer at time t can be summarized as a vector. Multiplication of this vector by the transition matrix gives the vector of the number of seeds in each layer present after tillage (at time $t + 1$; Figure 21.2).

In the case of plowing, Cousens and Moss (22) suggest a transition matrix in which a significant seed accumulation is assumed in the deeper layer. The matrix is slightly different in the case of

$$\underset{\textbf{Plowing}}{\begin{pmatrix} 0.02 & 0.21 & 0.37 & 0.29 \\ 0.11 & 0.27 & 0.26 & 0.10 \\ 0.40 & 0.30 & 0.20 & 0.12 \\ 0.46 & 0.21 & 0.18 & 0.48 \end{pmatrix}} \underset{\textbf{Minimum tillage}}{\begin{pmatrix} 0.70 & 0.33 & 0.02 & 0.00 \\ 0.23 & 0.50 & 0.15 & 0.00 \\ 0.06 & 0.15 & 0.68 & 0.16 \\ 0.02 & 0.01 & 0.16 & 0.84 \end{pmatrix}} \underset{\textbf{No tillage}}{\begin{pmatrix} 1 & 0 & 0 & 0 \\ 0 & 1 & 0 & 0 \\ 0 & 0 & 1 & 0 \\ 0 & 0 & 0 & 1 \end{pmatrix}}$$

FIGURE 21.3 Transition matrices for plowing, minimum tillage, and no tillage. (Modified from Cousens and Moss (22) and Vidotto et al. (78).)

minimum tillage, when minor exchanges between layers occur. No seed movement is supposed when no tillage is applied (Figure 21.3), which is probably more applicable in a rice paddy than in a field with earthworms.

An estimate of the seed distribution along the soil profile can be obtained by applying the correct transition matrix. As the rate of emergence is inversely related to the depth of burial, the number of seedlings that can emerge from each soil layer can be estimated. Little information is available about the numerical relationships between depth of burial and rate of emergence of weedy rice. Ferrero and Finassi (30) suggest the following empirical equations:

$$e_{pr} = 0.508 - 0.139d + 0.0093d^2 \qquad \text{For: } 0 < d < 5 \tag{21.1}$$

$$e_{pr} = 0.160 - 0.054d + 0.0043d^2 \qquad \text{For: } 0 < d < 4 \tag{21.2}$$

where e_{pr} is the proportion of the seeds that emerge as seedlings; d is the depth in cm of burial layer. Equation 21.1 and Equation 21.2 refer to moist and flooded (10 cm of water) soils, respectively. The total number of seedlings (E_{pre}) can be then calculated by summing the ($e_{pr}\quad s_{i,t+1}$) values obtained from each layer.

When the stale seedbed technique is applied, the total number of weedy rice seedlings is reduced by a factor that is a function of the efficacy of the mechanical or chemical treatment carried out to eliminate emerged seedlings.

The number of plants that will reach the maturity and eventually produce seeds can be calculated as follows:

$$M = E_{pre}(100 - c) + E_{post} + E_{seed} \tag{21.3}$$

where E_{pre} is the total number of weed seedlings emerged per unit area during stale seed bedding; c is the percentage of efficacy of the chemical or mechanical control treatment; E_{post} is the number of seedlings that emerge from seed bank after the treatment; and E_{seed} is the number of seedlings per unit area recruited from weed seeds carried with cultivar seed. In general, E_{post} can be set at low values, as the main flush of weedy rice germination occurs during the stale seed bedding (32). No information is available on the percentage of seeds carried with cultivar seed that are able to emerge as seedlings. In plot experiments carried out in Italy, this value ranged between 25 and 65% (Ferrero, unpublished). It is assumed that all seedlings reach maturity, both for E_{pre}, E_{post}, and E_{seed}.

From the number of mature plants per unit area it is then possible to estimate the number of seeds produced and shattered (S_{shat}):

$$S_{shat} = M \cdot p_p \cdot s_p \cdot s_h \tag{21.4}$$

where p_p is the number of panicles per plant; s_p is the number of seeds per panicle; and s_h is the proportion of seeds that shatter.

The value S_{shat} represents the number of seeds that replenish the seed bank and can potentially germinate the following season, although considerable depletion of the seed bank could occur during intercropping (from October to April). Apart from mortality due to aging, which can affect even recently shattered seeds, predation from small mammals, birds, and arthropods can play an important role.

The study of seed predation only recently received attention primarily after the first attempts to build models for weed populations suggested that this might be a critical factor. Several techniques have been developed to estimate the causes and extent of post-dispersal losses, but still there is not a commonly adopted method and little information is available for major weeds, in general, and for weedy rice, in particular (45,79,81).

21.4 RUNNING THE MODEL

21.4.1 Validation

The earlier proposed model had been evaluated mainly for its capability to predict seedling density before preplanting weed control (78). Despite a tendency to give overestimations, the model predictions correlated well with experimental data of seedling emergence in seedbeds prepared by plowing. Poor correlations were obtained for minimum or no tillage (78).

A more comprehensive evaluation of the model requires the joint availability of seedling emergence data and assessments of seed banks. Reported weedy rice seedling density values are rarely accompanied by information on tillage before emergence, and only some have seed bank data (32,33) The main reason for this paucity of information is that seed bank assessment in weedy rice is more costly, time-consuming, and messy than for other weeds. Weedy rice seeds are bigger and usually present in the soil at a lower density than other weed seeds. This requires that soil and mud cores should be larger and more numerous than for other weeds. The general lack of combined data sets of seed bank and related emergence is a common Achilles' heel of weed population dynamic models (57,70).

A partial validation of the model has been attempted using the few data available (30–33,35). Under these conditions, the model estimates seed bank dynamics with a proper order of magnitude, particularly in the case of seedbed preparation with plowing (Figure 21.4).

21.4.2 Sensitivity Analysis — Singling Out Crucial Factors

Several parameters have been included in the model. Each of them can have a different importance in influencing weedy rice seed bank evolution. By varying the base value of each parameter and keeping all the others constant, it is possible to single out which parameters should have the strongest influence on weed population dynamics.

The sensitivity analysis has been carried out based on the assumptions starting with values reported in Table 21.1.

The variation of emergence at the end of the stale seedbed period and seed bank after 10 years of simulation with the starting values of main parameters increased or decreased by 10% is shown in Figure 21.5. The results of the simulation suggest that the model is particularly sensitive to parameters related to the reproductive capacity of the weed, in particular the number of seeds per panicle. The largest increase in seed bank size is foreseen with a 10% reduction of stale bed weed control.

There is a highly asymmetrical effect of predation on seed bank variation. According to this evaluation, variations of 10 and +10% in mortality result in less than +10% and more than –20% seeds/m^2, respectively.

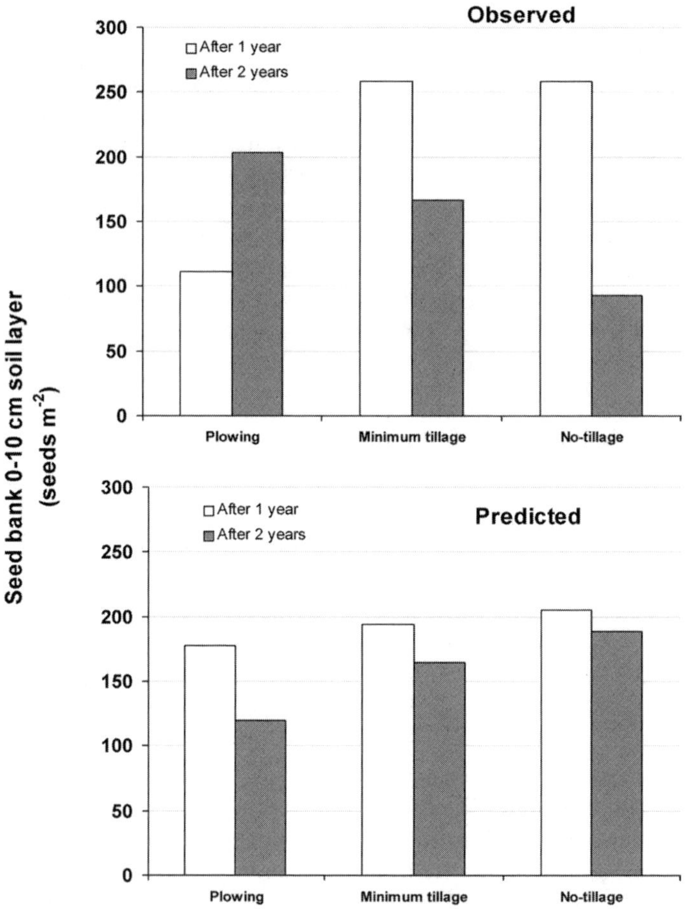

FIGURE 21.4 Observed and predicted weedy rice seed bank dynamics after 1 and 2 years of stale seedbed with plowing, minimum tillage, and no tillage for 1 and 2 years (0 to 10 cm soil layer; starting value: 2200 seeds m^{-2}).

TABLE 21.1
Main Parameters Used for Sensitivity Analysis

Parameter	Value
Starting seed bank	
0 to 10 cm	300 seeds/m^2
10 to 20 cm	100 seeds/m^2
Control technique	Stale seedbed
Tillage	Plowing
Flooding conditions	Flooded (5 cm water)
Control efficacy	85%
Panicles/plant	3
Seeds/panicle	100
Shattering	80% of produced seeds
Predation	29% of shattered seeds
Seed mortality	95% per year of seeds escaped to predation
Postplanting emergence	2.5% of 0 to 5 cm seed bank

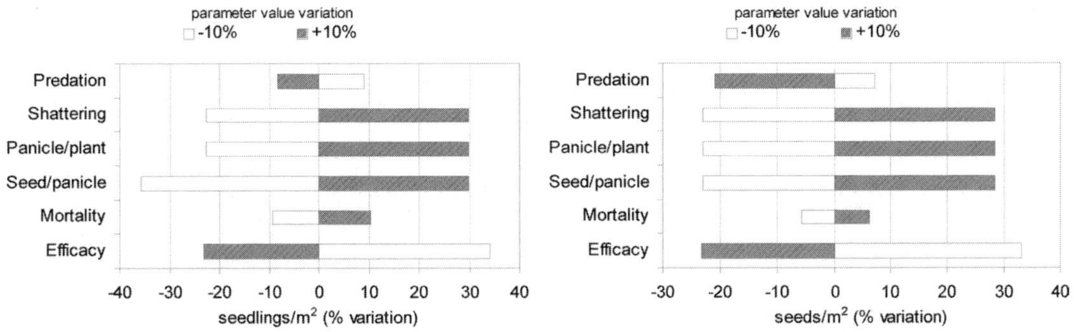

FIGURE 21.5 Sensitivity analysis of the model. The variation of emergence at the end of the stale seedbed period (left) and seed bank (right) after –10 or +10% variation of reported parameters was analyzed. Values are percentages of variation after 10 years of simulation in monoculture.

21.4.3 Simulation of Scenarios

The model can be adopted to predict the medium- and long-term weedy rice population dynamics. Even if the model does not precisely predict what will happen to the seed bank, it can be applied to compare the effect of different management strategies.

A first comparison can be made between the effect of soil tillage when no specific control measures (e.g., stale seedbed technique) are applied to control weedy rice (Figure 21.6A).

In these conditions, regardless of whether plowing or minimum tillage was used to prepare the seedbed, the seed bank roughly grows exponentially starting from the 2nd year of simulation. The growth of the seed bank is particularly evident in minimum tillage and when a high level of seeds is present in the initial seed bank. This simulation highlights a dramatic increase of the weedy rice population if control measures are not adopted. When the seedbed is prepared by plowing, a relatively low population growth rate can be observed at low initial seed bank values. In any case, this is not sufficient to prevent the population from becoming extremely abundant.

Conversely, the introduction of the stale seedbed technique and chemical treatment on emerged seedlings before rice seeding can result in a more moderate variation of the seed bank. In this case, minimum tillage keeps the seed bank within the initial values (Figure 21.6B).

Population dynamics are highly sensitive to the efficacy of the treatment carried out to control seedlings in the preplanting period. Small variations of this parameter (from 95 to 98%) can result in a gradual increase or depletion of the seed bank. This effect is related, in particular, to the soil flooding conditions during the stale seedbed. When the soil is saturated, but not flooded, a greater number of seeds become seedlings that can be chemically controlled and a gradual reduction of seed bank size can be achieved (Figure 21.6C).

The model allows prediction of the seed bank evolution in a non-infested field where the only source of infestation is weedy rice seeds carried with the commercial rice seed (Figure 21.7). In this situation, the weedy rice population remains quite undetectable until the 4th or 5th year. Afterward, the seed bank increases dramatically, even at the threshold of 0.015% weedy rice seeds in commercial rice lots (which is allowed, for example, in Italy).

This simulation points out that the use of commercial rice seed absolutely free from weedy rice seeds is a key to weedy rice control, especially for non-infested fields.

A significant effect on seed bank dynamics can be attributed to the seedlings that emerge after rice planting and that cannot be controlled. By simulating emergence rates ranging from 0.5 to 2% of the seed bank (0 to 20 cm), it is possible to obtain the results shown in Figure 21.8.

Again, the simulation stresses the importance of a step, which is sometimes under estimated. The main flush of weedy rice emergence in European rice conditions occurs between the end of April and the end of May (31), even though subsequent minor flushes can occur.

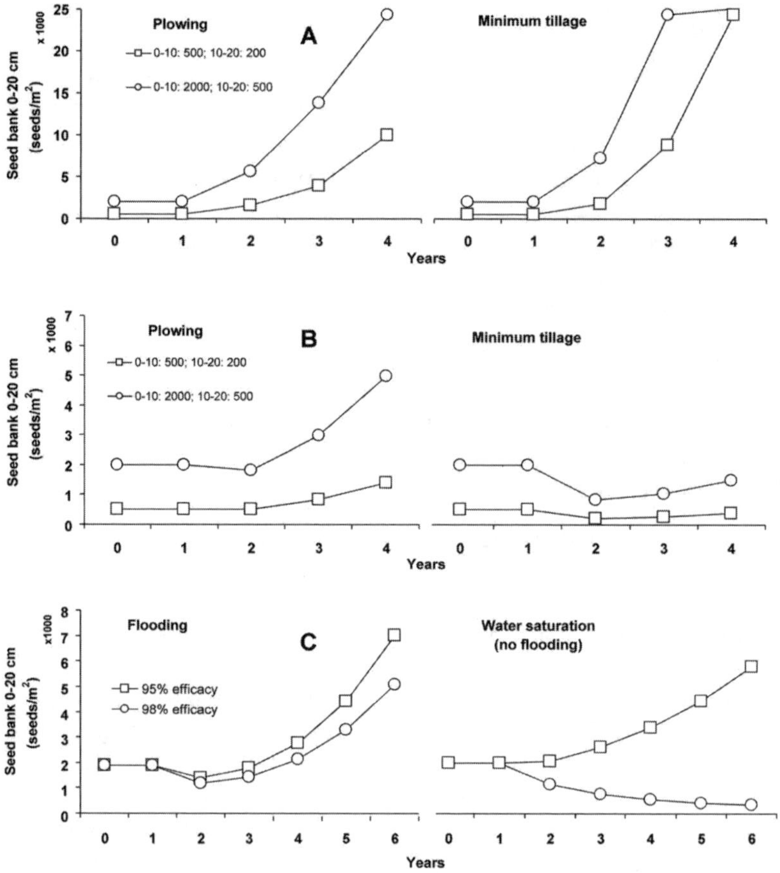

FIGURE 21.6 Simulations of agronomic parameters on weedy rice seed bank enrichment when: (A) no specific control measures are adopted; (B) stale bedding and chemical control are used; and (C) only chemical control is adopted. In A and B the seedbed is prepared by plowing (left) or minimum tillage (right). The starting seed bank is 500 and 200 seeds/m^2 in the layer 0 to 10 and 10 to 20 cm, respectively (–□–), or 2000 and 500 seeds/m^2 in the layer 0 to 10 and 10 to 20 cm, respectively (–○–).

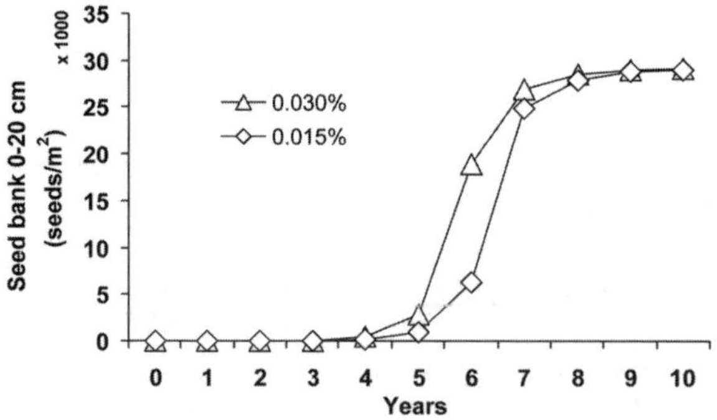

FIGURE 21.7 Simulation of the effect of crop seed contamination with weedy rice seeds on weedy rice seed bank enrichment in non-infested fields when no specific control measures are adopted. Weedy rice seeds carried with rice seed are modeled as the source of infestation. The two values (0.015%: –◇–; 0.030%: –△–) of weedy rice seed content in commercial rice seed allowed in the E.U. are indicated.

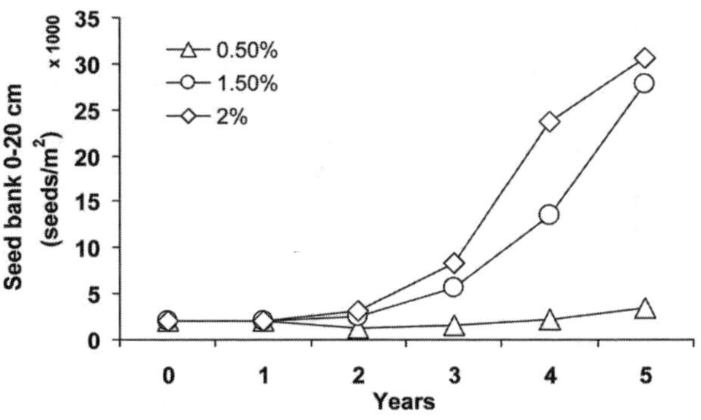

FIGURE 21.8 Simulation of seed bank enrichment under different rates of weedy rice emergence after rice planting. Values corresponding to 0.50% (–△–), 1.5% (–○–), and 2% (–◇–) of weedy rice seed bank in the layer 0 to 20 cm are indicated.

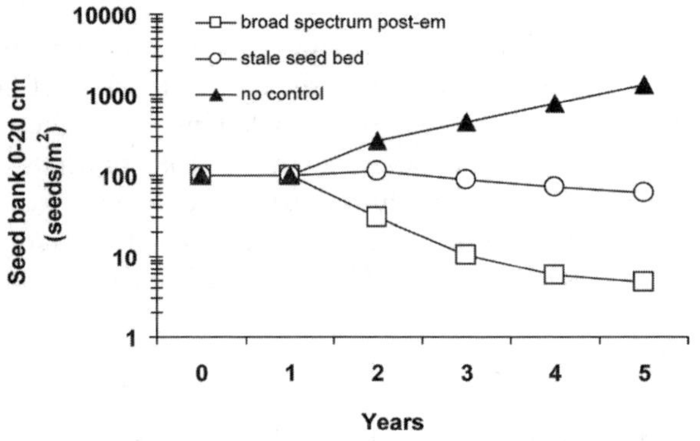

FIGURE 21.9 Simulating weedy rice seed bank dynamics when a broad-spectrum herbicide is used in post-emergence (99% of control) in a herbicide-resistant crop in comparison to stale seedbed and to no control. The starting seed bank is 100 and 50 seeds/m^2 in the layer 0 to 10 and 10 to 20 cm, respectively.

The use of transgenic herbicide-resistant rice could permit effective postemergence control of weedy rice in the rice crop. In this context, repeated treatments with broad-spectrum herbicides would allow control of both the main and secondary flushes, obtaining high levels of control. By assuming that the herbicide resistance gene will not spread throughout the weedy rice population, the seed bank of the weed will decrease faster than in the case of stale seedbed (Figure 21.9). An intermediate behavior can probably be achieved if the stale seedbed and if post-emergence broad-spectrum herbicides are applied in alternate years. A similar intermediate seed bank dynamic can probably occur if imidazolinone-resistant rice varieties are adopted. In this case, values of weedy rice control higher than 97% are reported (80) and can be considered similar or higher to that usually observed with stale seedbed.

One of the major concerns with the use of transgenic-resistant rice is the possibility of gene flow to wild rice. The recent availability of herbicide-resistant rice varieties posed the need to evaluate the risk of transfer of transgenes from crop to weedy rice and several studies have been carried out (59,60). The majority of data indicate that the natural outcrossing rate to weedy rice is usually less than 1% (43). Several techniques have been proposed to overcome the risk of transfer of herbicide resistance, including the maternally inheritance of herbicide resistance (47,71) and transgene mitigation to reduce weed fitness (1).

In the latter case, for example, it is theoretically possible to use transgenic mitigation by linking a dominant transgene that reduces shattering to the herbicide resistance transgene. Assume that both genes will be transferred to the weedy rice population and that a seed bank of 10 and 2 seeds/m^2

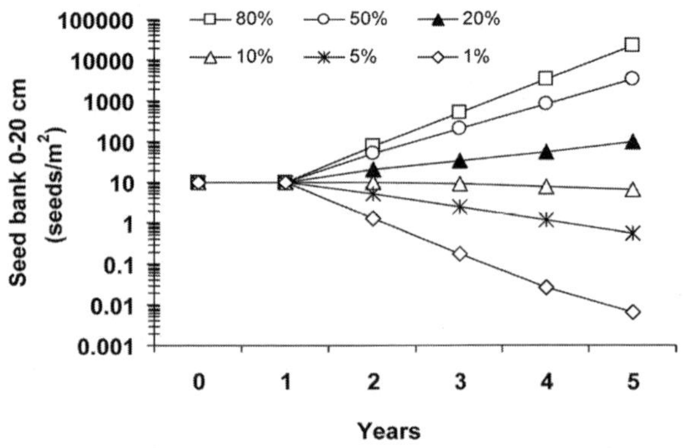

FIGURE 21.10 Simulating weedy rice seed bank dynamics in a population resistant to herbicide in which transgene mitigation determines different rates of shattering (from 1 to 80%). The starting seed bank is 10 and seeds/m^2 in the layer 0 to 10 and 10 to 20 cm, respectively. The level of weed control modeled is 99%.

of herbicide-resistant weeds is present in a field. Simplifying the model, also assume that none of the seedlings from this seed bank is controlled by the herbicide and that no gene exchanges will occur between resistant and susceptible weedy rice populations. Under these assumptions, resistant weedy rice populations will grow only if more than about 10% of seeds are still able to shatter (Figure 21.10). Yet, it should be noted that during harvest, combines can lose up to 5% (and in some conditions more) of harvested grain. Consequently, the higher the starting levels of infestation, the lower the rate of shattering seeds that should be achieved to avoid population growth.

21.5 CONCLUSIONS

The model highlights the importance of knowledge of seed bank structure, size, and distribution in the soil profile, which are the least known aspects for most weeds and of weedy rice, in particular.

According to the model simulation results, the critical aspects of weedy rice population dynamics are soil tillage, water management for stale seedbed application, efficacy of the chemical or mechanical treatment for the weed control after stale seedbed application, presence of weed seeds in commercial rice seed, and the emergence occurring after planting of the crop.

The model also stresses the importance of components of weed seed production, such as the number of seeds per plant.

Another outcome of the model simulations is that the model suggests that long-term effects are strongly influenced by the efficacy of the specific control measures applied. The model underlines the fact that the use of clean rice seed is crucial in the case of fields with low infestations, but can be of little importance when infestations are more severe. In addition to testing the theoretical effects of different strategies, the model can be intended to suggest specific research to better understand some aspects of weedy rice population dynamics.

For instance, post dispersal seed losses due to predation can play an important role among the other traits that can influence weedy rice infestations. It is necessary to carry out investigations on predation, as specific data on this aspect are quite limited. As with some other parameters, predation has been considered to be a simple percentage of the shattered seeds. It is reasonable to think that the rate of predation varies according to the seed density, as predators are more attracted and their activity is more intense when seeds are abundant.

In this simple present form, the current model is mainly suited to account for the effects of agronomic practices on weedy rice populations where pre-planting control techniques are largely adopted. Its extension to other areas is possible by adopting parameter values closer to the specific situations. A further development of this or other models should necessarily account for the large variability of morphological and biological features of different weedy rice populations, as well as the different effects of the control strategies adopted. Other aspects that should be included are related to the variability of weedy rice seed production, in response to inter- and intraspecific competition.

LITERATURE CITED

1. Al-Ahmad H, Galili S, Gressel J. 2004. Tandem constructs to mitigate transgene persistence: tobacco as a model. *Mol. Ecol.* 13:697–710.
2. Askew SD, Shaw DR, Street JE. 2000. Graminicide application timing influences red rice (*Oryza sativa*) control and seed head reduction in soybean (*Glycine max*). *Weed Technol.* 14:176–181.
3. Askew SD, Street JE, Shaw DR. 1998. Herbicide programs for red rice (*Oryza sativa*) control in soybean (*Glycine max*). *Weed Technol.* 12:103–107.
4. Balsari P, Tabacchi M. 1997. Lotta meccanica di soccorso al riso crodo. *Inform. Agrario* 53:56–60.
5. Barrentine WL, Street JE, Kurtz ME. 1984. Post-emergence control of red rice (*O. sativa*). *Weed Sci.* 32:832–934.
6. Bekker RM, Bakker JP, Ozinga WA, Thompson K. 2003. Seed traits: essential for understanding seed longevity. In *Aspects of applied biology*, pp. 1–9. Wellesbourne, U.K.: Association of Applied Biologists.
7. Biroli G. 1807. *Del riso*. Milano, Italy: Tipografia Giovanni Silvestri. 151 pp.
8. Bocchi S, Vandoni G, Maggiore T. 2002. Morphological and physiological characterisation of red rice. *Riv. Agron.* 36:157–162.
9. Bres-Patry C, Bangratz M, Ghesquière A. 2001. Genetic diversity and population dynamics of weedy rice in the Camargue. *Genet. Select. Evol.* 33:S425–S440.
10. Cai HW, Morishima H. 2000. Genomic regions affecting seed shattering and seed dormancy in rice. *Theor. Appl. Genet.* 100:840–846.
11. Catala Forner MdM. 1995. Chemical and cultural practices for red rice control in rice fields in Ebro Delta (Spain). *Crop Protect.* 14:405–408.
12. Catala Forner MdM, Eleftherohorinos IG, Martins L, Vidotto F, Ferrero A, et al. 2002. Red rice control in southern Europe with pre- and post-emergence applied herbicides. In *Proc. 2nd Temperate Rice Conf., 13–17 June 1999*, Hill JE, Hardy B, Eds. pp. 489–492. Los Baños, Philippines: International Rice Research Institute.
13. Chiappelli R. 1929. *Il riso crodo*. Vercelli, Italy: Tip. Gallardi.
14. Chiappelli R. 1948. I risi a cariosside rossa nell'inquinamento delle nostre varietà coltivate (parte 1). *Risicoltura*: 36:128–133.
15. Chiappelli R. 1948. I risi a cariosside rossa nell'inquinamento delle nostre varietà coltivate (parte 2). *Risicoltura*: 36:160–166.
16. Cohn MA. 1996. Chemical mechanisms of breaking seed dormancy. *Seed Sci. Res.* 6:95–99.
17. Cohn MA, Footitt S, Come D, Corbineau F. 1993. *Initial signal transduction steps during the dormancy-breaking process*. Presented at 4th International Workshop on Seeds: Basic and Applied Aspects of Seed Biology, Angers, France.
18. Cohn MA, Hughes JA. 1981. Seed dormancy in red rice (*Oryza sativa*) I. Effect of temperature on dry-after ripening. *Weed Sci.* 29:402–404.
19. Colbach N, Debaeke P. 1998. Integrating crop management and crop rotation effects into models of weed population dynamics: a review. *Weed Sci.* 46:717–728.
20. Colbach N, Forcella F, Johnson G. 2000. Spatial and temporal stability of weed populations over five years. *Weed Sci.* 48:366–377.
21. Coppo B, Sarasso G. 1990. *Il riso crodo*: Istituto Federale di Credito Agrario, Piemonte, Liguria e Valle d'Aosta.
22. Cousens R, Moss SR. 1990. A model of the effects of cultivation on the vertical distribution of weed seeds within the soil. *Weed Res.* 30:61–70.
23. Delatorre CA. 1999. Dormencia em sementes de arroz vermelho. *Cienc. Rural* 29:565–571.
24. Diarra A, Smith RJ, Jr., Talbert RE. 1985. Growth and morphological characteristics of red rice (*Oryza sativa*) biotypes. *Weed Sci.* 33:310–314.
25. Diarra A, Smith RJ, Jr., Talbert RE. 1985. Interference of red rice (*Oryza sativa* L.) with rice. *Weed Sci.* 33:644–649.

26. Doherty LC, Cohn MA. 2000. Seed dormancy in red rice (*Oryza sativa*). XI. Commercial liquid smoke elicits germination. *Seed Sci. Res.* 10:415–421.
27. Eleftherohorinos IG, Dhima KV, Vasilakoglou IB. 2002. Interference of red rice in rice grown in Greece. *Weed Sci.* 50:167–172.
28. Fernandez-Quintantilla C, Gonzales-Andujar J, Appleby A. 1990. Characterization of the germination and emergence response to temperature and soil moisture of *Avena fatua* and *A. sterilis*. *Weed Res.* 30:289–295.
29. Ferrero A. 2002. Rice in Europe and the Mediterranean. In *Rice almanac — source book for the most important economic activity on earth*, Maclean JL, Dawe BH, Hardy B, Hettel GP, Eds., pp. 68–71. Los Baños (Philippines), Bouaké (Côte d'Ivoire), Cali (Colombia), Rome (Italy): IRRI, WARDA, CIAT, FAO.
30. Ferrero A, Finassi A. 1995. Viability and soil distribution of red rice (*Oryza sativa* var. *silvatica*) seeds. *Med. Fac. Landbouw. Toege. Biol. Weten. Univ. Gent.* 60:205–211.
31. Ferrero A, Vidotto F. 1996. Prediction of red rice seedling densities from seed bank. *Med. Fac. Landbouw. Toege. Biol. Weten. Univ. Gent.* 61:1181–1187.
32. Ferrero A, Vidotto F. 1997. Influence of soil tillage on red rice emergence. *Med. Fac. Landbouw. Toege. Biol. Weten. Univ. Gent.* 62:785–789.
33. Ferrero A, Vidotto F. 1997. *Influence of the rotation on seed bank evolution of red rice (Oryza sativa [L.] var. sylvatica)*. Presented at International Symposium on Rice Quality, Notthingham.
34. Ferrero A, Vidotto F. 1998. *Germinability after flowering, shattering ability and longevity of red rice seeds*. Presented at Comptes rendus 6eme symposium Mediterraneen EWRS, Montpellier, France, 13–15 Mai 1998. 1998, 205–211.
35. Ferrero A, Vidotto F. 2002. Biology and control of red rice (*Oryza sativa* L. var. *sylvatica*) infesting italian rice fields. In *Proceedings of the 2nd Temperate Rice Conference, June 13–17, 1999*, Hill JE, Hardy B, Eds., pp. 523–533. Los Baños, Philippines International Rice Research Institute.
36. Ferrero A, Vidotto F, Balsari P, Airoldi G. 1999. Mechanical and chemical control of red rice (*Oryza sativa* L. var. *sylvatica*) in rice (*Oryza sativa* L.) pre-planting. *Crop Protect.* 18:245–251.
37. Fischer AJ, Ramirez A. 1993. Red rice (*Oryza sativa*): competition studies for management decisions. *Int. J. Pest Manage.* 39:133–138.
38. Footitt S, Cohn MA. 1992. Seed dormancy in red rice. VIII. Embryo acidification during dormancy-breaking and subsequent germination. *Plant Physiol.* 100:1196–1202.
39. Footitt S, Cohn MA. 1995. Seed dormancy in red rice (*Oryza sativa*). IX. Embryo fructose-2,6-bisphosphate during dormancy breaking and subsequent germination. *Plant Physiol.* 107:1365–1370.
40. Footitt S, Vargas D, Cohn MA. 1995. Seed dormancy in red rice. X. A 13C-NMR study of the metabolism of dormancy-breaking chemicals. *Physiol. Plant* 94:667–671.
41. Forcella F. 1992. Prediction of weed seedling densities from buried seed reserves. *Weed Res.* 32:29–38.
42. Forcella F. 1993. Seedling emergence model for velvetleaf. *Agron. J.* 85:929–933.
43. Gealy DR, Mitten DH, Rutger JN. 2003. Gene flow between red rice (*Oryza sativa*) and herbicide-resistant rice (*O. sativa*): implications for weed management. *Weed Technol.* 17:627–645.
44. Gealy DR, Saldain NE, Talbert RE. 2000. Emergence of red rice (*Oryza sativa*) ecotypes under dry-seeded rice (*Oryza sativa*) culture. *Weed Technol.* 14:406–412.
45. Gonzales-Andujar J, Fernandez-Quintanilla C. 1991. Modelling the population dynamics of *Avena sterilis* under dry-land cereal cropping systems. *J. Appl. Ecol.* 28:16–27.
46. Goss WL, Brown E. 1939. Buried red rice seed. *J. Am. Soc. Agron.* 31:633–637.
47. Gray AJ, Raybould AF, Daniell H. 1998. Reducing transgene escape routes. *Nature* 392:653–654.
48. Griffin JL, Harger TJ. 1990. Red rice (*Oryza sativa*) control options in soybeans (*Glycine max*). *Weed Technol.* 4:35–38.
49. Grundy A, Mead A, Bond W. 1996. Modelling the effect of weed-seed distribution in the soil profile on seedling emergence. *Weed Res.* 36:375–384.
50. Grundy A, Mead A, Burston S. 1999. Modelling the effect of cultivation on seed movement with application to the prediction of weed seedling emergence. *J. Appl. Ecol.* 36:663–678.
51. Khodayani K, Smith RJ, Jr., Black HL. 1987. Red rice (*Oryza sativa*) control with herbicide treatments in soybeans (*Glycine max*). *Weed Sci.* 35:127–129.
52. Kim DS, Brain P, Marshall E, Caseley JC. 2002. Modelling herbicide dose and weed density effects on crop: weed competition. *Weed Res.* 42:1–13.

53. Kwon SL, Smith RJ, Jr., Talbert RE. 1991. Interference of red rice (*Oryza sativa*) densities in rice (*O. sativa*). *Weed Sci.* 39:169–174.
54. Kwon SL, Smith RJ, Jr., Talbert RE. 1992. Comparative growth and development of red rice (*Oryza sativa*) and rice (*O. sativa*). *Weed Sci.* 40:57–62.
55. Leopold AC, Glenister R, Cohn MA. 1988. Relationship between water content and afterripening in red rice. *Physiol. Plant.* 74:659–662.
56. Luppi G, Finassi A, Cavallero A. 2000. Riso. In *Coltivazioni erbacee: cereali e proteaginose*. Bologna, Italy: Pàtron Editore.
57. Lutman P. 2002. Estimation of seed production by *Stellaria media*, *Sinapis arvensis* and *Tripleurospermum inodorum* in arable crops. *Weed Res.* 42:359–369.
58. Mead A, Grundy AC, Brain P, Marshall EJP. 2003. Development of a model for the joint horizontal and vertical movement of seeds following cultivation. In *Aspects of applied biology*, pp. 179–186. Wellesbourne, U.K.: Association of Applied Biologists.
59. Messeguer J, Fogher C, Guiderdoni E, Marfà V, Català MM, et al. 2001. Field assessments of gene flow from transgenic to cultivated rice (*Oryza sativa* L.) using a herbicide resistance gene as tracer marker. *Theor. Appl. Genet.* 103:1151–1159.
60. Messeguer J, Marfà V, Català MM, Guiderdoni E, Melé E. 2004. A field study of pollen-mediated gene flow from Mediterranean GM rice to conventional rice and the red rice weed. *Mol. Breed.* 13:103–112.
61. Minton BW, Shaw DR, Kurtz ME. 1989. Postemergence grass and broadleaf herbicide interactions for red rice (*Oryza sativa*) control in soybeans (*Glycine max*). *Weed Technol.* 3:329–334.
62. Montealegre SF, Clavijo PJ. 1994. Efectos ambientales y geneticos en la germinacion y dormancia de los arroces rojos. *Arroz* 43:14–18.
63. Munier-Jolain NM, Chauvel B, Gasquez J. 2002. Long-term modelling of weed control strategies: analysis of threshold-based options for weed species with contrasted competitive abilities. *Weed Res.* 42:107–122.
64. Neeser C, Aguero R, Swanton C. 1998. A mechanistic model of purple nutsedge (*Cyperus rotundus*) population dynamics. *Weed Sci.* 46:673–681.
65. Noldin JA, Chandler JM, McCauley GN, Sij JW, Jr. 1998. Red rice (*Oryza sativa*) and *Echinochloa* spp. control in Texas gulf coast soybean (*Glycine max*). *Weed Technol.* 12:677–683.
66. Oard J, Cohn MA, Linscombe S, Gealy D, Gravois K. 2000. Field evaluation of seed production, shattering, and dormancy in hybrid populations of transgenic rice (*Oryza sativa*) and the weed, red rice (*Oryza sativa*). *Plant Sci.* 157:13–22.
67. Olofsdotter M, Valverde BE, Madsen KH. 2000. Herbicide resistant rice (*Oryza sativa* L.): global implications for weedy rice and weed management. *Ann. Appl. Biol.* 137:279–295.
68. Pomini L. 1957. *Saggio di flora della risaia vercellese-novarese*. Vercelli, Italy. 123 pp.
69. Rao SR, Harger TR. 1980. *Field evaluation of postemergence herbicides for red rice control in soybeans*. Presented at Proceedings of the 33rd Annual Meeting of the Southern Weed Sci. Society, 39 pp.
70. Rasmussen IA, Holst N. 2003. Computer model for simulating the long-term dynamics of annual weeds: from seedling to seeds. In *Aspects of applied biology*, pp. 277–284. Wellesbourne, U.K.: Association of Applied Biologists.
71. Raybould A. 2001. Gene flow from genetically modified crops. *Pestic. Outlook* 12:177–180.
72. Roberts HA, Feast PM. 1972. Fate of seeds of some annual weeds in different depths of cultivated and undisturbed soil. *Weed Res.* 12:316–324.
73. Roger-Estrade J, Colbach N, Leterme P, Richard G, Caneill J. 2001. Modelling vertical and lateral weed seed movements during mouldboard ploughing with a skim-coulter. *Soil Til. Res.* 63:35–49.
74. Salzman FP, Smith RJ, Jr., Talbert RE. 1989. Control and seed head suppression of red rice (*Oryza sativa*) in soybeans (*Glycine max*). *Weed Technol.* 3:238–243.
75. Smith RJ, Jr., Fox WT. 1973. Soil water and growth of rice and weeds. *Weed Sci.* 21:61–63.
76. Tarditi N, Vercesi B. 1993. Il riso crodo: un problema sempre più attuale in risicoltura. *Inform. Agrario* 49:91–95.
77. Vidotto F, Ferrero A. 2000. Germination behaviour of red rice (*Oryza sativa* L.) seeds in field and laboratory conditions. *Agronomie* 20:375–382.
78. Vidotto F, Ferrero A, Ducco G. 2001. A mathematical model to predict *Oryza sativa* var. *sylvatica* population dynamics. *Weed Res.* 41:407–420.

79. Watson SJ, Mauchline AL, Brown VK, Froud-Williams RJ. 2003. Post-dispersal losses of *Stellaria media* and *Polygonum aviculare* seeds in spring barley (*Hordeum vulgare*). In *Aspects of applied biology*, pp. 203–208. Wellesbourne, U.K.: Association of Applied Biologists.
80. Webster EP, Masson JA. 2001. Acetolactate synthase-inhibiting herbicides on imidazolinone-tolerant rice. *Weed Sci.* 49:652–657.
81. Westerman PR, van der Werf W, Brussaard L. 2003. Measuring causes of seedbank mortality: a method. In *Aspects of applied biology*, pp. 217–222. Wellesbourne, U.K.: Association of Applied Biologists.
82. Yap CL. 1996. The rice market in the EEC. In *Cahiers Options Méditerranéennes*, Chataigner J, Ed., pp. 160. Montpellier: CIHEAM-IAMM.

22 Molecular Containment and Mitigation of Genes within Crops — Prevention of Gene Establishment in Volunteer Offspring and Feral Strains

Jonathan Gressel and Hani Al-Ahmad

22.1 INTRODUCTION — NEEDS FOR PREVENTING GENE FLOW AND OVERCOMING FERALITY

The previous chapters have indicated where the evolution of *exoferality* (ferality engendered by introgression of genes from relatives) and *endoferality* (back mutations within the crop) can be problems, both with traditionally bred and with transgenic crops. Below, we try to further delineate and accentuate ensuing problems and then describe the status of potential molecular solutions. Molecular solutions to ferality problems for non-transgenic crops may sound oxymoronic in the present climate surrounding transgenics. Still, if the scientifically determined risk of the evolution of ferality in a crop outweighs the risk of having ferality-inhibiting transgenes in the crop, such molecular solutions should be sought.

22.1.1 Molecular Tools to Prevent or Overcome Ferality in Traditionally Bred Crops

There are a variety of reasons to want to use molecular tools to prevent the evolution of feral forms of traditionally bred crops. Despite the popular press hysteria about "superweeds" evolving from transgenics, there have been many needs for preventing or overcoming ferality prior to the advent of transgenic crops. The feral sugar beet situation described in Chapter 4 is a case in point. Crops still not yet fully domesticated, where less domesticated forms can be volunteer *cum* feral weeds are equally important, such as oilseed rape, sorghum, etc., or domesticated crops that have feral forms as a major weed such as rice. The dilution of contemporary oilseed rape varieties by older varieties or wild forms that are still high in erucic acids and glucosinolates suggests that feral material exists. They reduce the utility, quality, and economic value of the crop (16), which is why these two compounds have been bred out of the crop.

There are crops that have been abandoned, as they had to be multiply harvested because of seed shatter and were supplanted by crop species that were less labor intensive. Breeding and selection have eliminated many undesirable traits from crops (e.g., uneven germination, bolting, shattering, bitter compounds, poisons), not always successfully, or the effort was too daunting, or the mission impossible if the genes do not exist in the wild species that will allow domestication. Sometimes the market seems too small to perform the task of removing feral traits by breeding. For example, little of the soybean harvested goes for making infant formulae, but 30% of young infants are allergic to soy-based formula, which is often the only economic choice for a poor mother who cannot nurse. Transgenics can remove the feral allergens (32), which probably had important functions in the wild. When the desired allergen-free varieties are released, gene inflow of the feral trait from non-transgenic varieties would be problematic and require containment of the non-transgenic varieties.

As the world has become overly dependent on a few major crops, there is a danger to food security from a failure of any one major crop. Thus, molecular tools or genes are needed to transform incompletely domesticated crops that have been mostly abandoned into domesticated crops, which could considerably increase the biodiversity of crops available. Examples of such a crop are spelt wheat (*Triticum aestivum spelta*), which has a high price on the European market, but is hard to harvest because of excessive seed shattering; or teff in Africa, a drought-tolerant, but also shattering crop. This mutual problem could be redressed transgenically, both in these crops and in many desirable landraces of wheat that are losing favor with farmers because of undomesticated properties.

Techniques are needed that have considerable power to modulate the feral characters, to prevent feral gene movement into crops, and to domesticate partially domesticated crops. It is posited below that molecular tools can assist in the task. Similar molecular techniques can be used to inhibit crops from becoming volunteer weeds and prevent them from remaining as volunteer weeds that can further dedomesticate to feral forms.

22.1.2 Molecular Tools Are Needed to Prevent or Overcome Ferality with Transgenic Crops

From the proceeding chapters, it is apparent that some transgenic traits can exacerbate the evolution of volunteer weeds to feral weeds. In other cases, there is a need to prevent the transgene from even being in a potentially persistent volunteer that might continually cross-pollinate other varieties of the crop. This is especially the case for transgenic crops bearing genes encoding production of industrial feedstocks, specialty enzymes, or pharmaceuticals.

22.2 METHODS FOR PRECLUDING FERAL TRAITS FROM BECOMING PREDOMINANT IN POPULATIONS

Genes do flow in nature, not only within species, but also among related species that do not readily cross, in a process coined *diagonal gene transfer* (23) to readily distinguish between vertical gene transfer in readily crossing species and horizontal gene transfer between totally unrelated species. For example, a DNA sequence typical of hexaploid wheat, found in modified form in some progenitors of wheat, was not found in 78 accessions of *Aegilops peregrina* (syn. *Ae. variabilis*) but was found in two geographically distinct populations of that species with >99% sequence identity to wheat (61). In agroecosystems, such inadvertent gene flow may be undesirable.

Most discussions so far have dealt with containing gene flow (preventing its movement) from agroecosystems to natural ecosystems (1,18,22–24,29,35,59), with little or no discussion on preventing and mitigating endoferal and exoferal dedomestication of crops as volunteer weeds within the agroecosystem. This chapter is meant to remedy this situation.

When molecular tools are used to overcome ferality in crops, there is a need to prevent the re-establishment of feral traits, whether by endoferal back mutation or exoferal gene flow. There are two general approaches to dealing with this:

1. *Contain* the transgenes in the novel variety so that inflow is precluded.
2. *Mitigate* back-mutational or gene flow effects if there are inevitable leaks in the containment system, which should also prevent volunteer weeds from establishing or reaching maturity so that they cannot evolve into feral problems.

Containment and mitigation are discussed below in the general context of bidirectional containment as well as mitigation.

22.2.1 Containing Gene Flow

Several molecular mechanisms have been suggested for containing genes, especially transgenes within the crop (i.e., to prevent outflow to related species) or to mitigate the effects of transgene flow once it has occurred (11,22,23,59). In the case of exoferality, it is more important to prevent gene flow into the crop to be protected than to prevent outflow, especially when the wild relative is not in the agroecosystem. Even though the hybrids may be the same, the likelihood of a crop-wild hybrid of any of the major crops discussed surviving in the wild is minimal. Models suggesting that crops could cause genetic meltdown of nearby interbreeding wild species (28) were based on many erroneous assumptions. They include ignoring the propensity of pollen dispersal to decrease exponentially with distance, such that the adjacent members of a wild species are most likely to effect cross-pollination and not the transgenic crop. These models also ignore the incredibly strong competition of multitudes of fit non-transgenic offspring, which are likely to out-compete the slightly unfit hybrids. Plants, unlike animals have many offspring competing to replace the parents.

22.2.1.1 Containment by Targeting Genes to a Cytoplasmic Genome

The most widely discussed containment possibility is to integrate the transgene of choice in the plastid or mitochondrial genomes (37,44,45). The opportunity of gene outflow is limited due to the predominantly maternal inheritance of these genomes in many but far from all species. This presently arduous technology, which so far is limited to a few crops, does not preclude the wild species or partially feral volunteer weed from pollinating the crop and then acting as the recurrent pollen parent. The claim of strict maternal inheritance of plastome-encoded traits (6,12,45) was not substantiated. Tobacco (3) and other species (13) often have between a 10^{-3} to 10^{-4} frequency of pollen transfer of plastid inherited traits in the laboratory. Pollen transmission of plastome traits can only be easily detected using both large samples and selectable genetic markers. A large-scale field experiment utilized a *Setaria italica* (foxtail or birdseed millet) with chloroplast-inherited atrazine resistance (bearing a nuclear dominant red leaf-base marker) crossed with five different male-sterile yellow- or green-leafed herbicide susceptible lines. Chloroplast-inherited resistance was pollen transmitted at a 3 10^{-4} frequency in >780,000 hybrid offspring (60). At this transmission frequency, the probability of herbicide resistance movement via plastomic gene flow is orders of magnitude greater than by spontaneous nuclear genome mutations. Thus, chloroplast transformation is probably unacceptable for preventing transgene outflow, unless stacked with additional mechanisms, and as noted above, will not at all impede gene inflow. Maliga (45) discounts the relevance of the findings with tobacco and *Setaria* as being due to an origin of the plastids as due to interspecific (closely related) cytoplasmic substitution, where pollen transmission barriers can break down (38). But as seen in Chapter 6, *Setaria viridis,* the wild progenitor of *Setaria italica,* is basically con-specific with it. There are two problems with this denigration of the relevance of pollen movement of plastome-encoded genes:

1. It is just such interspecific movement that could be a problem between crops and related species.
2. Maliga (45) ignores the discussion in Darmency et al. (13) about cases of intraspecific transmission of plastomic traits by pollen at about the same frequency, within the same species, as reported above between species.

22.2.1.2 Male Sterility Coupled with Transplastomic Traits

A novel additional combination that considerably lowers the risk of plastome gene outflow within a field (but not gene influx from related strains or species) can come from utilizing male sterility with transplastomic traits (60). Introducing plastome-inherited traits into varieties with complete male sterility would vastly reduce the risk of transgene flow, except in the small isolated areas required for line maintenance. Such a double fail-safe containment method might be considered sufficient where there are highly stringent requirements for preventing gene outflow to other varieties (e.g., to organically cultivated ones) or where pharmaceutical or industrial traits are engineered into a species. Plastome-encoded transgenes for non-selectable traits (e.g., for pharmaceutical production) could be transformed into the chloroplasts together with a trait such as tentoxin or atrazine resistance as a selectable plastome marker. With such mechanisms to further reduce outcrossing risk, plastome transformation can possibly meet the initial expectations.

22.2.1.3 Genetic Use Restriction Technologies and Recoverable Block of Function

Other molecular approaches suggested for crop transgene containment include seed sterility, utilizing the genetic use restriction technologies (GURT) ("terminator gene") (9,48) and recoverable block of function (RBF) (41) to prevent transgene flow. Such proposed technologies control both the gene influx of exoferality and endoferal volunteer seed dispersal, but theoretically if the controlling element of the transgene is silenced, expression would occur, rendering a critical defect in principle and practice. The frequency of loss of such controlling elements is yet unclear, as there have been no large-scale field trials to test this.

22.2.1.4 Repressible Seed Lethal Technologies

An impractical technology has been proposed to use a "repressible seed-lethal system" (53). The seed-lethal trait and its repressor must be simultaneously inserted at the same locus on homologous chromosomes in the hybrid the farmer sows to prevent recombination (crossing over), a technology that is not yet workable in plants. The hemizygote transgenic seed lethal parent of the hybrid cannot reproduce by itself, as its seeds are not viable. If the hybrid could be made, half the progeny would not carry the seed lethal trait (or the trait of interest linked to it) and they would have to be culled, which would not be easy without a marker gene. A containment technology should leave no viable volunteers with the transgene, but this complex technology would kill only 25% of the progeny and 50% would be like the hybrid parents and 25% would contain just the repressor. Thus, the repressor can cross from the volunteers to related weeds, and so can the trait of choice linked with the lethal, and viable hybrid weeds could form. The death of a quarter of the seeds in all future weed generations is inconsequential to weeds that copiously produce seed, as long as the transgenic trait provides some selective advantage.

In summary, none of the above containment mechanisms is absolute (Figure 22.1), but the risk could be reduced by stacking a combination of containment mechanisms, compounding the infrequency of gene introgression. Still, even at low frequencies of gene transfer, once gene transfer occurs, the new bearer of the transgene could disperse throughout the population if it has just a small fitness advantage.

22.2.2 Preventing Volunteer Establishment by Transgenic Mitigation (TM)

If a transgene confers even a small fitness disadvantage, the transgenic crop volunteers and their own or hybrid progeny should only be able to exist as a small proportion of the population. Therefore, it should be possible to mitigate volunteer establishment and gene flow by lowering the fitness of transgene recipients below the fitness of competitors, so that the volunteer or hybrid offspring will not reproduce. A concept of transgenic mitigation (TM) was proposed (22), in which mitigator genes are linked or fused to the desired primary transgene. Thus, a transgene with a desired trait is directly linked to a transgene that decreases fitness in volunteers (Figure 22.2). TM could also be used as a stand-alone procedure with non-transgenic crops to reduce the fitness

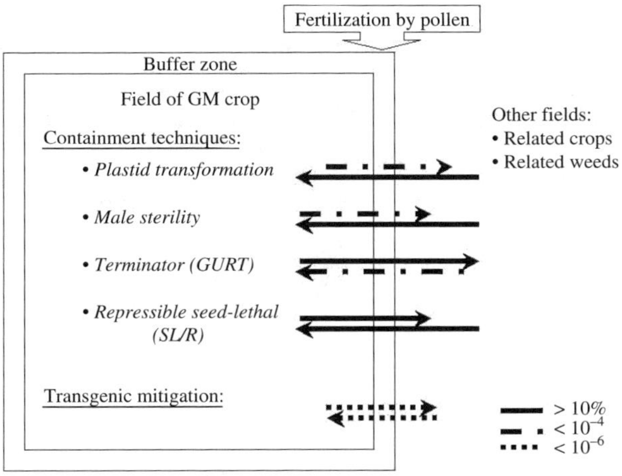

FIGURE 22.1 Containment systems allowing gene flow in one or more directions. GURT technologies will preclude gene flow in both directions in farmers' field, but not in seed production fields, where they only prevent inflow. Such systems indicate a need for stacking containment systems to contain gene flow bidirectionally and to compound the safety factor, as well as to consider mitigation systems that reduce the effects of genes that leaked from containment.

advantage of hybrids and their rare progeny and thus substantially reduce the risk of exoferal hybrid volunteer persistence.

This TM approach is based on the premises that:

- Tandem constructs act as tightly linked genes, and their segregation from each other is exceedingly rare.
- The gain of function dominant or semidominant TM traits chosen are neutral or favorable to crops, but deleterious to volunteer progeny and their hybrids due to a negative selection pressure.
- Individuals bearing even mildly harmful TM traits will be kept at low frequencies in volunteer/hybrid populations because strong competition with their own wild type or with other species should eliminate even marginally unfit individuals and prevent them from persisting in the field (22).

Thus, it was predicted that if the primary gene of agronomic advantage being engineered into a crop will not persist in future generations if it is flanked by TM genes, such as genes encoding dwarfing, strong apical dominance to prevent tillering (in grains) or multiheading (in crops like sunflowers), determinate growth, non-bolting genes, uniform seed ripening, non-shattering, or antisecondary dormancy. When they are in such a tandem construct, the overall effect would be deleterious to the volunteer progeny and to hybrids. Indeed a TM gene such as anti-shattering will lower the number of initial volunteers. (There is typically a small amount of shattering due to imperfect harvesting equipment, which may leave a few seeds behind.) Because the TM genes will reduce the competitive ability of the rare exoferal hybrids, they should not be able to compete and persist in easily measurable or biologically significant frequencies in agroecosystems (22,23).

Once TM genes are isolated, the actual cost of splicing the TM constructs is minimal, compared to the total time and effort in producing a transgenic crop. The cost is even inconsequential in systems where co-transformation allows introducing genes into the same site such that the tandem construct is made by the plant. It is the type of work that can be done by a beginning graduate student, as evidenced in Al-Ahmad et al. (1).

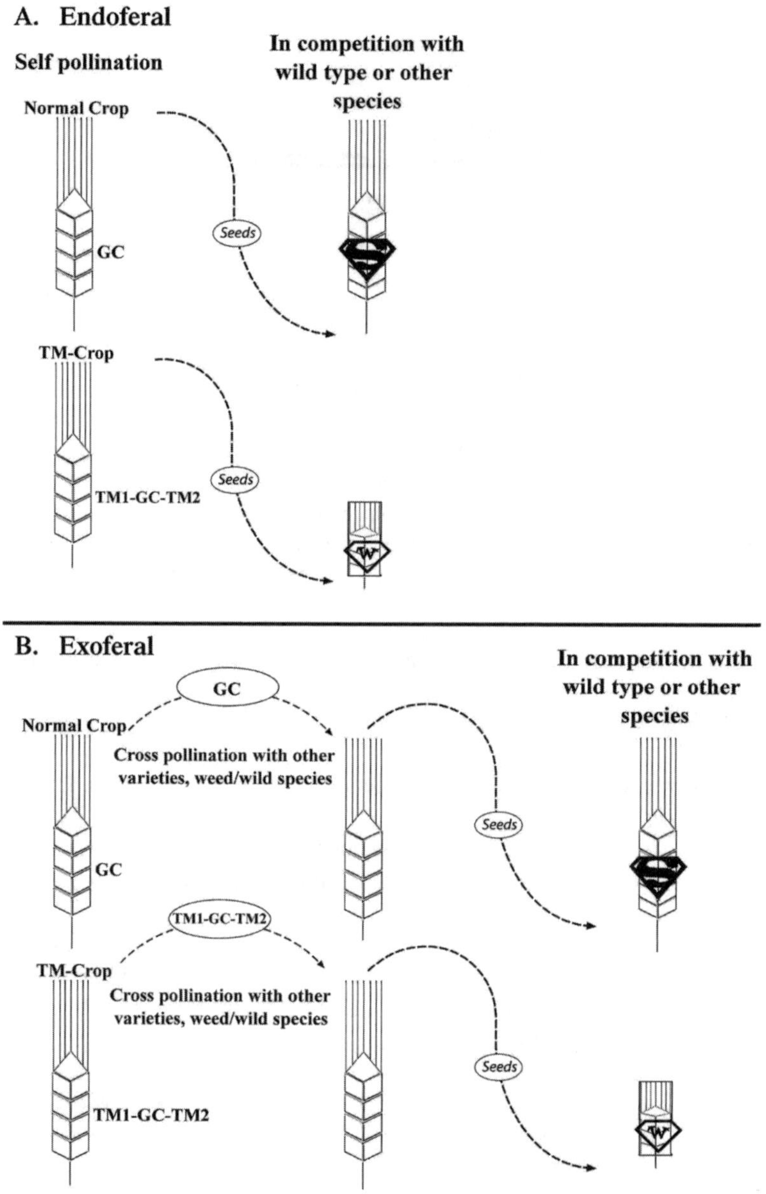

FIGURE 22.2 TM to prevent (A) volunteer endoferal and (B) exoferal hybrid establishment of desirable transgenes, by coupling the transgenes in tandem with genes that are neutral or positive for the crops, but render volunteers or hybrids unfit to compete outside of cultivation.

22.2.2.1 Demonstration of Transgenic Mitigation in Tobacco and Oilseed Rape

We used tobacco (*Nicotiana tabacum*) as a model plant to test the TM concept: a tandem construct was made containing an *ahas*^{R} (acetohydroxy acid synthase) gene for herbicide resistance as the primary desirable gene of choice and the dwarfing Δ*gai* (gibberellic acid-insensitive) mutant gene as a mitigator (1). Dwarfing would be disadvantageous to the rare weeds introgressing the TM construct, as they could no longer compete with other crops or with fellow weeds, but is desirable in many crops, preventing lodging and producing less straw with more yield. The dwarf and

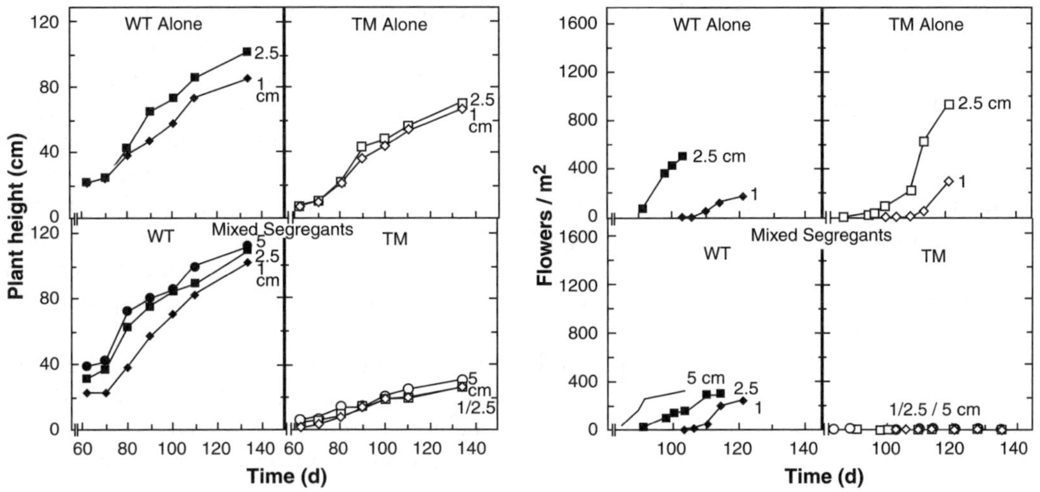

FIGURE 22.3 Suppression of growth and flowering of TM-bearing tobacco plants carrying a dwarfing gene in tandem with a herbicide resistance gene (open symbols) when in competition with the wild type (closed symbols) (right panels), and their normal growth when cultivated separately without herbicide (left panels). The wild type and transgenic hemizygous semidwarf/herbicide-resistant plants were planted at 1, 2.5, and 5 cm from themselves or each other in soil. See (1) for further details.

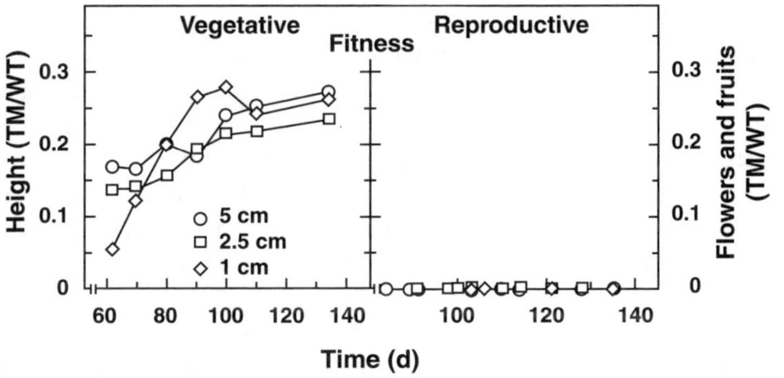

FIGURE 22.4 Suppressed vegetative and reproductive fitness of TM transgenics in competition with wild-type tobacco. The points represent the calculated ratio of data for TM to wild-type plants in Figure 22.3.

herbicide-resistant TM transgenic hybrid tobacco plants (simulating a TM introgressed hybrid) were more productive than the wild type when cultivated alone (without herbicide). They formed many more flowers than the wild type when cultivated by themselves, which is an indication of a higher harvest index (Figure 22.3). Conversely, the TM transgenics were weak competitors and highly unfit when co-cultivated with the wild type in ecological simulation competition experiments (Figure 22.3 and Figure 22.4). The inability to achieve flowering on the TM plants in the competitive situation (Figure 22.4) led to a zero reproductive fitness of the TM plants grown in a 1:1 mixture with the wild type at the spacing used. This is representative of weed spacing in the field (Figure 22.4). The highest vegetative fitness was less than 30% of the wild type.

From the data above it is clear that TM should be advantageous to a crop growing alone, while disadvantageous to a crop-weed hybrid living in a competitive environment. If a rare pollen grain bearing tandem transgenic traits bypasses containment, it must compete with multitudes of wild-type pollen to produce a hybrid. Its rare progeny must then compete with more fit wild-type cohorts during self-thinning and establishment. Even a small degree of unfitness encoded in the TM construct would bring about the elimination of the vast majority of progeny in all future generations, as long as the primary gene provides no selective advantage that counter-balances the unfitness of the linked TM gene.

We have inserted the same construct into oilseed rape and have tested the selfed progeny, as well as hybrids with the weed *Brassica rapa* (syn. *B. campestris*). The results are basically the same as with tobacco: the progeny of the weedy individuals introgressing the transgene in a TM construct were unfit to compete with their weedy cohorts or the crop. Should some seeds of such hybrids fall in an area where there is little competition with weeds, the dwarf TM individuals may further reproduce, but are not a threat to crop production. It is probably wiser to flank the gene of choice with two TM genes, such as dwarfing and non-shattering, so that seeds from the few surviving dwarf plants are harvested and sorted away. The use of two flanking TM genes compounds (the yet unknown) infrequency of mutation to loss of function.

The rare hybrid offspring from escaped pollen bearing TM genes would not pose a dire threat, especially to wild species outside fields, as the amount of pollen reaching the pristine wild environment would only be at a minuscule fraction of the pollen from the wild type. Further large-scale field studies will be needed with crop/weed pairs to continue to evaluate the positive implications of risk mitigation.

22.2.2.2 Risk that Introgression of Transgenic Mitigation Traits Will Affect Wild Relatives of the Crop

Models by Haygood et al. (28) claim to support the premise that demographic swamping by crop genes would cause migrational meltdown of wild species related to the crop, especially if the introgressed genes confer unfitness. This proposition that recurrent gene flow from crops, even TM gene flow, could affect wild relatives deserves some discussion, as it negates the concept of mitigation.

They claim that their model demonstrates that recurrent gene flow from transgenic crops with less fit genes will cause wild populations to shrink. First, conventional crops already belie this possibility. There are few if any major domesticated crops that are fit to live in a wild ecosystem, so their normal genes should confer a modicum of unfitness. Such crop wild hybrids continually form, yet they (28) present no evidence that demographic swamping did occur from recurrent gene flow from the crops, nor could we locate any published data to that effect. Indeed, considerable evidence has been presented in previous chapters that many crops exist near their wild or weedy progenitors, without causing the extinction of the progenitors, despite gene flow (61).

There are other mundane yet fatal flaws in their models that are based on shaky premises and assumptions not borne out by plant or evolutionary biology. Three problematic issues that seem to invalidate the relevance of their model for the vast majority of conceivable crop/wild species systems are discussed below:

1. To get the level of swamping that they (28) discuss, the wild relative and the crop would have to live in the same ecosystem. There is typically geographic separation between agroecosystems and wild ecosystems, with pollen flow decreasing exponentially with distance — usually to a low asymptote due to wind currents or insects not fully following simple physics. There should always be far more wild pollen in the wild ecosystems, so hybridization events in the wild from crop pollen will be rare, even with masses of pollen occurring within the agroecosystem. Thus their basic assumption of crop pollen swamping wild-type pollen in the wild is invalid. Indeed, even when they assume an enormous 10% of hybridizations in the wild each generation coming from crop pollen, according to their model it will take about 20 generations of recurrent pollination for the unfit crop allele to become fixed in half the population and 50 generations for an unfit gene to asymptotically reach 80% of the population (Figure 22.5). As discussed below, their other assumptions leading to these numbers are also off target, so it should actually take much longer.

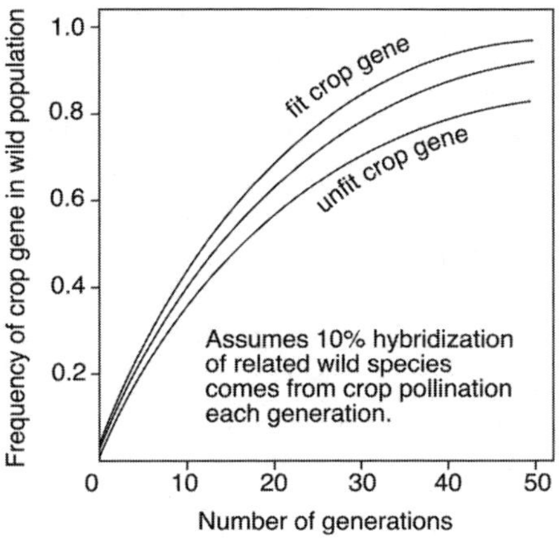

FIGURE 22.5 Modeling of gene introgression under recurrent gene flow from crops to wild relatives, used by Haygood et al. (28) to invalidate the use of TM as a fail-safe mechanism to prevent transgene establishment in volunteers and related weeds in agroecosystems. (Modified from Haygood et al. (28).)

2. They assume synchronous flowering, no self-fertilization, and no genetic or other barriers to cross fertilization; indeed, this negates the definition of speciation. It is exceedingly rare for crop pollen to fertilize another species without any genetic barrier between them. Of the species mentioned in preceding chapters, this might only occur with conspecific wild sunflowers, which might fit this criterion, but even in this case there are genomic deterrents to introgression (reviewed in (59)). Conspecific rice and red (weedy) rice do not fit their assumptions because they are cleistogamous, predominantly self-fertilizing before the flowers open, and the amount of outcrossing possible would be low. Of course, weedy rice is not a wild species (by definition), so it too is not really relevant to their case. There are fertilization barriers of different chromosome numbers, non-homology, etc., which limit fertilization of wild relatives of oilseed rape and wheat, so they are outside the models.
3. Their models assume animal-type replacement rates — a few progeny per mating, where lower fitness can indeed become fixed. Most wild relatives of crops produce copious amounts of seed to replace parents. Hundreds to thousands typically germinate in the area occupied by a parent and the process of self-thinning is ferociously competitive, eliminating less fit individuals (Figure 22.3 and Figure 22.4). The experimental data show that at realistic seed output and seeding rates, unfit individuals are eliminated or remain at a low frequency, just as unfit mutations are maintained in populations at some low frequency (the relative fitness multiplied by the mutation frequency).

Their conclusion that "the most striking implication of this model is the possibility of thresholds and hysteresis, such that a small increase in (unfit gene) immigration can lead to fixation of a disfavored crop allele …" (28) flies in the face of evolutionary evidence and decades of classic and contemporary field data showing that only near-neutral genes exist in pockets of the evolutionary landscape of plants, and blatantly unfit plant genes are not known to exist in such pockets unless all the fit genes are somehow removed. Just as endogenous unfavored gene mutations exist in the wild at a frequency lower than the mutation rate, crop transgenes that have a fitness penalty will exist in the wild at a rate lower than the immigration rate. As discussed above, the immigration rate to the wild is perforce low. Unfit genes are eliminated from populations of plants that produce large numbers of seeds, whereas they could be fixed in populations of animals with few progeny. When a model contradicts reams of data, it is more likely than not that the model is invalid.

They further contend that their model would work if the crop were heterozygous for the unfit gene (and many hybrid crops have the transgene in a single parent and are thus hemizygous). The data in Figure 22.3 and Figure 22.4 clearly show that when even half of the backcross progeny contain a TM construct, they cannot compete with their non-transgenic siblings, let alone the wild type. Part of the problem may be that Haygood et al. (28, p. 1880, column 2) "assume (that) the number of plants surviving to maturity does not vary from one generation to the next," a questionable assumption for unfit phenotypes when they must compete with fit cohorts and other species.

Where might their model have some validity? Even though, despite their claims, the model has limited validity for the wild ecosystems, the model might be valid for a few weeds (not wild species) related to crops. When flowering weeds are at a low density in an agricultural ecosystem (and perhaps close by in ruderal systems), the model might be predictive, especially when the TM genes are introduced in multicopy transposons (see Section 22.3.5) where all their progeny receive the transgene. If it was possible to so debilitate a weed population by "genetic self-biocontrol" instead of using aggressive cultivation or herbicides, would this be so bad? As weeds are man-made domesticated species (of a sort) (see Chapter 2), should not people have the right to eliminate them? The nature of weeds is such that they do not go extinct, as much as the farmer would desire. It is far more likely that such evolutionarily threatened weeds would evolve exclusionary mechanisms that would block evolutionarily threatening gene flow (e.g., they would evolve a shift to predominant self-fertilization that would protect them from crop pollen bearing unfit genes).

In summary, to quote Nobel laureate Manfred Eigen, "A hypothesis has two possibilities, it can be right or wrong; a model has a third alternative, it can be right, but irrelevant." The model of Haygood et al. (28) may be right for certain animal systems, but irrelevant for the vast majority of plant systems, because they fail to mention specific plant systems where their model might be valid. With its peculiar assumptions, it demonstrates that Eigen forgot the fourth alternative applicable to a model: it can be wrong and irrelevant.

22.2.2.3 Following Transgene Flow to Volunteers and Feral Forms

Using the various containment and mitigation strategies, it should be possible to keep transgene leaks below risk thresholds, which have to be specified by science-based regulators on a case-by-case basis (Chapter 24). As the numbers of transgenic species being released increased, and the problems of monitoring for such genes increased geometrically, we suggested that a uniform BioBarCoding™ system be used, where a small piece of non-coding DNA having uniform recognition sites is at the ends (for a single PCR primer pair amplification) with an assigned variable region in between. Thus, PCR-automated sequencing could be used to determine the origin of leaks, contamination, and liability, as well as intellectual property violations (24).

22.3 SPECIAL CASES WHERE TRANSGENIC MITIGATION IS NEEDED — SPECIAL GENES

We have described above some general anti-weediness genes that can be used to engender a modicum of unfitness to volunteer offspring and their hybrids with other cultivars and species. There are some special cases where other genes can be envisaged for use to design an unfitness to volunteers or feral forms coming from the crop. The TM genes are typically still neutral or positive to the crop but give unfit offspring.

22.3.1 Transgenic Mitigation Genes for Crop-Produced Pharmaceuticals and Industrial Products

Pharmaceuticals and industrial proteins, especially enzymes and antibodies, can be produced inexpensively in plants, without the need for animal tissue culture cells grown in a medium of expensive serum albumin that is all too easily contaminated with pathogenic mycoplasms, prions, and viruses. Although there are compelling economic and biosafety considerations propelling the production of pharmaceuticals in crops, there are good reasons to exclude the pharmaceutical and industrial transgenes from introgressing into other varieties of the crop or related species, or to remain in viable volunteers in the field.

The containment of pharmaceutical transgenes has been physical, as evidenced by recent human error that reportedly allowed temporary volunteer escape of "Prodigene" maize containing such genes (2). The biological containment strategies described above may be preferable to depending on physical containment by humans, and the TM strategies should work as well. Maize pharmaceutical transgenes are expressed in embryo tissues, and a potential tandem mitigating gene could be any RNAi-type suppression of genes that affects the endosperm (e.g., the various "shrunken seed" loci), especially those where sugar transformation to starch is inhibited (8). Such shrunken seeds, with their high sugar content, are somewhat harder to store than normal maize, but are extremely unfit in the field, and are unlikely to overwinter and produce volunteers. Hybrids with other varieties would have shrunken seeds and would be culled during seed cleaning. Their volunteers would be unfit and could not overwinter volunteer in the field. Because of the triploid nature of corn endosperm where 67% of alleles are derived from pollen, it is important that expression of pharmaceutical encoding genes be only in the diploid embryo.

22.3.2 Mitigation for Biennial and Other Root Crops

Mitigating genes should easily prevent both the premature and volunteer flowering in sugar beets, carrots, onions, celery, radishes, and in other biennial or two phase crops where the vegetative material is marketed and flowering (bolting) is detrimental. This could easily be effected by preventing gibberellic acid biosynthesis (31), either in a TM construct or by permanent mutation of the kaurene oxidase gene using a chimeraplastic gene conversion system (63), a system that as yet is hard to use in plants. Kaurene oxidase suppression would require the use of gibberellic acid to force flowering for seed production. There should be a concomitant biosafety requirement that seed production areas be far removed from areas where weedy or other feral or wild beets grow to prevent pollen transfer.

Delaying of bolting and flowering by using a different transgene has recently been demonstrated. Curtis et al. (10) engineered a fragment of the *GIGANTEA* gene, the gene encoding a protein that is part of the photoperiod recognition system, into radish using an antisense approach. Bolting was considerably delayed, and thus seed production could come about without reversal mechanisms if seed producers waited long enough.

If despite all isolation distances, a TM construct or a mutant in a seed production area introgresses with a wild species, the progeny will be biennial or be too delayed (i.e., the transgenic hybrid would be uncompetitive with cohorts, which reproduce in a single year and do not need to overwinter).

Other transgenes can be considered for mitigating the risks of introgression with root crops, such as genes promoting partitioning to roots, which would be advantageous to cultivated root crops, but detrimental to feral forms.

22.3.3 Mitigation of Ferality in Species Used for Phytoremediation

Plants have been used to correct human error over the ages. The few species capable of revegetating Roman lead and zinc mine tailing in Wales (56) taught us that there are a limited number of species that can withstand toxicants: some by exclusion and others that can withstand toxic wastes after they have been taken up. Plants with the latter type mechanism are of interest for phytoremediation. Ideally, one might consider that it is best to use the species that naturally take up particular toxic wastes, but these are often slow growing (e.g., mosses, lichens, or the *Thlaspi* species that take up heavy metals) (39) and may have a potential to be weedy. If the desired wild species do not exist locally, there may be a reticence or legal issues about introducing them into the ecosystem, toxic as they may be, due to fear that the plants or their genes may spread to other areas. Two types of multi-cut species are used, with the cut material burnt to extract the heavy metals or to oxidize the organic wastes: herbaceous species such as *Brassica juncea* and *Spartina* spp. (cord grasses), which are most efficient at dealing with surface wastes, and trees such as *Populus* spp., for dealing with deeper wastes (50). Thus, heavy metal tolerance has been brought into *Brassica juncea* (Indian mustard) from *Thlaspi* by protoplast fusion (along with many other genes) (17). *Brassica juncea* wild type had been used commercially, because it grows rapidly, and is easy to cultivate as a crop, but especially because of its inherent ability to take up heavy metals. This ability has been enhanced by mutant selection (in tissue culture) for heavy metal resistance (54), but it was better yet to transgenically transfer genes leading to enhanced glutathione content (4,64) to make the necessary phytochelatins. A single cropping of *B. juncea* does not clean up a toxic site: many growth cycles are required, with multiple harvests and natural reseeding. *B. juncea,* even more than its close relative *B. napus* (oilseed rape) is not fully domesticated, and the multiple cycles of cropping would allow the possibility of selecting for ferality. Thus, mitigation seems necessary to prevent volunteers from becoming feral. The issues with species such as poplars are discussed in detail with issues of forestry (Section 22.3.4). One gene that might specifically fulfill the need is overexpression of a cytokinin oxidase (5), which reduces the levels of isopentenyl- and zeatin-type cytokinins. This in turn led to phenotypes with far reduced shoot systems (unfitness to compete) but faster growing, more extensive root systems (62), all the better for extracting toxic wastes.

22.3.4 Mitigating Endo- and Exoferality by Rendering Crops Obligatively Vegetatively Propagated

Some vegetatively propagated crops such as potato and elite tissue culture propagated forestry material also flower. In forestry, this is especially problematic, as the long-term implications of gene movement are longer than human lifetimes. The introgression of traits from these species to wild populations is discussed in (23,43). Some landscaping trees such as decorative plantings of olives create an urban problem of allergies from tree pollen and a mess from their fruits. Olives can easily become feral as happened in Australia, where feral olives have been replacing native flora (57). Such ornamental trees could be vegetatively propagated if there would be a way to prevent allergy-causing pollen clouds and messy fruits, which would preclude ferality. There is also the possibility that preventing allocation of resources to sex will increase the yield of the vegetative tissue, which could be advantageous in many ornamentals and in forestry, where vegetative propagation is possible. Thus, a TM trait that prevents pollen formation or fruit set could be coupled to herbicide resistance or other primary traits. An ideal gene for doing this is *barnase* under the T29 tapetum-specific promoter (46). The ribonuclease is only produced in the tapetum and prevents pollen formation with no other ill effects. There is a good chance that the shelf life of many flower species (e.g., roses and carnations) could be enhanced as well by preventing pollen production; fertilization starts the process of floral degeneration and fruit set. Additionally, a flower-specific promoter from poplar coupled to a cytotoxin gene caused flower ablation (55), basically preventing exoferality and requiring vegetative propagation of the trees.

If one has an important crop in which transgenics are exceedingly worthwhile, yet the risks of cultivation are too great, one could envisage using a pollen sterility system coupled with flower drop, as described above, and the crop could be propagated by artificial seed, such as artificially encased somatic embryos produced in mechanized tissue culture systems. Transgenic parthenocarpy could be used to get a fruit without seeds or cross-pollination from a non-transgenic crop is another possibility.

22.3.5 Tac-Tics for Eliminating Feral Forms of Pasture Grasses

Lolium spp. (ryegrasses) and *Festuca* spp. are members of closely related, intercrossing genera that are widely cultivated as pasture species. They are barely domesticated, still bearing many feral traits and, especially *Lolium* spp., can be exceedingly pernicious weeds in the non-pasture part of a rotation. Indeed, weedy *Lolium* spp. have evolved resistance to a large number of herbicides used for their control in cropping situations (30). To further domesticate these species into better pasture grasses, it is necessary to remove feral traits such as toxicity to herbivores, which is desirable when aimed at insects, but not when livestock species are affected. To do this, one must provide resistance to the nematode that vectors an alkaloid-producing bacterium (51) or to the bacterium itself. Increased digestibility is being introduced by transgenically modulating lignin content in *Festuca* (7) and the same approaches are beginning with *Lolium* (42). The vegetative phase (with better digestibility) is being extended by transgenically delaying flowering (34). Herbicide resistance would be an advantage during pasture establishment. These approaches to further domesticating pasture grasses generate two problems:

1. Preventing the improved pasture grasses from becoming volunteer weeds in the rotational crops
2. Preventing exoferal gene flow from weeds or other varieties into the improved varieties, lowering their quality and value

22.3.5.1 The Use of Tac-Tics for Insect Control

A mitigation solution, once thought to be in the realm of science fiction, is to transgenically debilitate weedy biotypes in crops (21,23,25). One concept, adopted from entomological biocontrol, is to release organisms with a gene that is normally benign unless triggered by an inducible promoter (14) and have it cross into the wild population. Large numbers of such transgenics would have to be released to obtain a large proportion in the final population bearing the inducibly lethal gene, which may be impractical. The problem of distribution of such a conditionally lethal gene throughout a population was solved by Pfeifer and Grigliatti (27,49). They proposed a means for controlling pests called the Tac-Tic model: Transposons with Armed Cassettes for Targeted Insect Control. They also proposed to use a chemically induced promoter to activate genes that would kill the insect; that is, they postulated that not many transgenic pests would be needed if the chemically assisted-suicide transgenes, termed by one of us (25) as *kev* (Kevorkian) genes, are transmitted in multicopy transposons. These deleterious transposons (DTs) carrying the *kev* genes will quickly disperse in the indigenous population through mating. This is because all the offspring of any cross with one parent carrying the multicopy transposon will also carry the transposon, as will all their future progeny. After DT transmission has spread to most of the population, the use of a chemical that turns on the chemically induced promoter replaces pesticides. Heavy transposons loaded with such a gene will still disperse in laboratory experiments. Even when an initial 0.5% of the population carried the transposon, more than 90% of the population carries the transposon after 5 generations (27).

22.3.5.2 Modification of Tac-Tics for Preventing Ferality of Pasture Grasses

Such an approach could be further modified and work with pasture grasses *cum* pernicious weeds, which are solely or predominantly outcrossing, such as *Lolium* spp. This can be done without the need for *kev* genes, based on mitigator genes alone. A DT/TM construct can be introduced into a transposon cassette and transformed into the pasture grasses to generate the newly domesticated varieties. Transposon proliferation should enable reaching high copy numbers of the DT/TM constructs in the progeny of these sown parents. Basically, the DT/TM constructs will be dominant and in high copy number per cell. Therefore, most if not all the progeny will rapidly contain them. The DT/TM containing new pasture grass variety seed could then be abundantly sown in existing pastures, which would then convert feral pasture grasses into the new variety. The Ac/Ds transposon family, originally found in corn (47), has been shown to be active in all the heterologous plant systems (approximately 15 different species) where it has been introduced (40). Ac is preferentially transposed during DNA replication, and as a result its copy number can increase while it transposes (20). There would be nothing wrong with the pasture phase containing potentially self-lethal *kev* genes, as long as the chemical inducer is not leaky (i.e., cannot get turned on in the pasture phase).

22.3.5.3 TM Genes for Use in Transposons for Pasture Grasses

A plethora of potential TM genes is available off the shelf for use in DT/TM constructs for pasture grasses. Indeed, the primary genes might themselves be expected to lower the weedy potential. Low lignin content, less insect herbivory, delayed flowering would give less fit weeds, but superior pasture, as it would have higher digestibility, be less toxic, and have high nutritional quality for a longer period, respectively. One could go further, inserting transgenes that would abolish flowering altogether, necessitating that a hormone be used to induce flowering in seed production fields, while preventing shed of volunteer seeds due to their absence. Other genes that are neutral or even positive during the pasture phase but that are decidedly deleterious during the rotation to the field crop can be considered without *kev* promoters, as part of DT/TM constructs, including:

- Genes that cause dwarfing by internode shortening or non-recognition of shade would not hurt pastures but would confer unfitness to compete with a field crop.
- Genes suppressing auxin biosynthesis or stimulating cytokinin biosynthesis would suppress the dominance of the shoot apex, stimulating branching (tillering), which is good in pasture but again is deleterious in competition with crops.
- Suppression of long-term (secondary) dormancy would preclude seed remaining in the seed bank for more than one season, ensuring that seeds will not remain overlong in the soil.

All these genes will be beneficial to the pasture species, but will render it less competitive with a wheat crop, even a dwarf wheat crop.

22.3.5.4 Herbicide-Mimic, Lethal *Kev* Genes

Many enzyme systems, when even partially inhibited, cause the accumulation of lethal metabolites in plants. This can be mimicked using RNAi technology by antisensing or over-expressively co-suppressing such genes (33) under a *kev* promoter, as part of a TM construct. Antisensing the same genes that have evolved target site herbicide resistance should be appropriate candidates for TM *kev* constructs under an inducible promoter. For example, two *Lolium* species have evolved resistance to herbicides that inhibit acetolactate synthase (ALS) (52), and acetyl CoA carboxylase (ACCase) (15) by single major gene mutations. Such point mutations should have no effect on the ability of *kev* constructs from suppressing gene and enzyme activity, if the antisense construct is not in an overlapping area of the mutation. Thus, such an antisense *kev* construct would inhibit the susceptible and resistant biotypes.

Additionally, *kev* constructs could be perfectly useful on the pasture biotypes that evolved metabolic herbicide resistance, by RNAi-suppressing a consensus promoter region for a plethora of cytochrome P450s, to suppress the P450 that cause of some multiple herbicide resistances. This could allow control of pasture grass volunteers in the following crop by the previously used herbicides.

A temperature-sensitive *kev* gene could be used for winter annual pasture species to be eliminated before summer rotational crops. Such a genetic system has been proposed in a non-transposon context, but would be appropriate for transposon dissemination. The *barnase* ribonuclease gene was put under the control of a heat shock promoter such that when the leaf temperature rises above 42ϒC, the plants die (58). Such a *kev* system could be used with any primary gene of choice.

22.3.5.5 Chemically Induced Promoters for *Kev* Genes

An array of chemically induced promoters is available for chemically inducing the expression of genes in plants (19,36), and could be used for the Tac-Tic concept in pasture grasses. Some inducers are antibiotics or expensive compounds deemed inappropriate for agronomic use, but copper salts and ethanol are among the inexpensive, simple inducers for known promoters. As noted above, there is the possibility of making DT for *Lolium* that does not need to possess a chemically inducible promoter.

22.3.5.6 Biosafety of Transposing Pasture Grasses with the DTs

TM/DT constructs could force the domesticative evolution of superior pasture grasses, away from being weeds. They could affect wild populations, where such occur. This effect can be given a positive or negative evaluation. In geographic areas where introduced pasture grasses are displacing indigenous species, the effect of rendering the introduced species less fit when in natural habitats, would be considered positive. When *kev* genes are used under a chemical promoter, there should be little danger accruing from the DT transposons entering wild populations of the weed species or its introgressing relatives. The DT transposon should have little negative value if the chemical inducer of the DT gene is not found in the wild.

The above are general theoretical considerations, but it is clearly necessary to perform a risk analysis based on the particular weed, its place in agro- and natural ecosystems, and whether it has readily introgressing relatives. The genes used could only confer a negative value on the weed, and it might attain insignificant levels in fields. Clearly, until one knows how such genes will act, one should test such technologies with introduced weeds, far from their centers of origin. One must consider the potential risks from different *kev* genes under the control of different promoters. One could consider constructing decision trees, similar to those constructed to estimate the risk of transgene introgression from crop to weed to deal with such issues (26). Clearly, despite these paper analyses, experiments should be designed to safely evaluate the risk of environmental hazard (if any) from DT weeds.

When novel control strategies were first tested, such as sterilized male insects, they were tested on islands to limit the possibility of their dissemination, should there be unforeseen problems. *Lolium* is the major weed problem in Australia, an island that actually has some internal deserts as east-west barriers between agricultural regions. Perhaps Australia is the appropriate biosafe place to test the various DT/TM strategies for *Lolium*.

22.4 CONCLUDING REMARKS

Systems exist that can theoretically preclude a crop from becoming or remaining endo- or exoferal, whether by containing gene flow or by preventing the establishment of feral plants in the field by mitigation. Thus, if a risk of ferality is discerned, it should not preclude developing transgenic crops — it should stimulate the imagination to devise and test systems to deal with the potential problems.

ACKNOWLEDGMENTS

The authors thank Norman Ellstrand and Neal Stewart for cogent comments on an earlier version of this manuscript. The research on TM was supported by the Levin Foundation, by INCO–DC, contract no. ERB IC18 CT 98 0391, H. A.-A. by a bequest from Israel and Diana Safer and J.G. by the Gilbert de-Botton chair in plant sciences.

LITERATURE CITED

1. Al-Ahmad H, Galili S, Gressel J. 2004. Tandem constructs to mitigate transgene persistence: tobacco as a model. *Mol. Ecol.* 13:697–710.
2. Anonymous. 2002. What is the compliance history with APHIS' biotechnology regulations? http://www.aphis.usda.gov/brs/compliance9.html.
3. Avni A, Edelman M. 1991. Direct selection for paternal inheritance of chloroplasts in sexual progeny of *Nicotiana. Mol. Gen. Genet.* 225:273–277.
4. Bennett LE, Burkhead JL, Hale KL, Terry N, Pilon M, Pilon-Smits EAH. 2003. Analysis of transgenic Indian mustard plants for phytoremediation of metal-contaminated mine tailings. *J. Environ. Qual.* 32:432–440.
5. Bilyeu KD, Cole JL, Laskey JG, Riekhof WR, Esparza TJ, et al. 2001. Molecular and biochemical characterization of a cytokinin oxidase from maize. *Plant Physiol.* 125:378–386.
6. Bock R. 2001. Transgenic plastids in basic research and plant biotechnology. *J. Mol. Biol.* 312:425–438.
7. Chen L, Auh CK, Dowling P, Bell J, Chen F, et al. 2003. Improved forage digestibility of tall fescue (*Festuca arundinacea*) by transgenic down-regulation of cinnamyl alcohol dehydrogenase. *Plant Biotechnol. J.* 1:437–449.
8. Chourey PS, Taliercio EW, Carlson SJ, Ruan Y-L. 1998. Genetic evidence that the two isozymes of sucrose synthase present in developing maize endosperm are critical, one for cell wall integrity and the other for starch biosynthesis. *Mol. Gen. Genet.* 259:88–96.
9. Crouch ML. 1998. How the terminator terminates: an explanation for the non-scientist of a remarkable patent for killing second generation seeds of crop plants. http://www.biotech-info.net/howto/html. Edmond, WA: The Edmonds Institute.
10. Curtis IS, Nam HG, Yun JY, Seo K-H. 2002. Expression of an antisense GIGANTEA (GI) gene fragment in transgenic radish causes delayed bolting and flowering. *Transgenic Res.* 11:249–256.
11. Daniell H. 2002. Molecular strategies for gene containment in transgenic crops. *Nature Biotechnol.* 20:581–586.
12. Daniell H, Datta R, Varma S, Gray S, Lee SB. 1998. Containment of herbicide resistance through genetic engineering of the chloroplast genome. *Nature Biotechnol.* 16:345–348.
13. Darmency H. 1994. Genetics of herbicide resistance in weeds and crops. In *Herbicide resistance in plants: biology and biochemistry*, Powles SB, Holtum JAM, Eds., pp. 263–298. Boca Raton, FL: Lewis.
14. Davis MS, Solbiati J, Cronan JEJ. 2000. Overproduction of acetyl-CoA carboxylase activity increases the rate of fatty acid biosynthesis in *Escherichia coli. J. Biol.Chem.* 275:28593–28598.
15. Devine MD, Shimabukuro RH. 1994. Resistance to acetyl coenzyme A carboxylase inhibiting herbicides. In *Herbicide resistance in plants*, Powles SB, Holtum JAM, Eds., pp. 141–169. Boca Raton, FL: Lewis.
16. Diepenbrock W, Leon J. 1988. Quantative effects of volunteer plants on glucosinolate content in double-low rapeseed (*Brassica napus* L) — a theoretical approach. *Agronomie* 8:373–377.
17. Dushenkov S, Skarzhinskaya M, Glimelius K, Gleba D, Raskin I. 2002. Bioengineering of a phytoremediation plant by means of somatic hybridization. *Int. J. Phytorem.* 4:117–126.
18. Ellstrand NC. 2003. *Dangerous liaisons — when cultivated plants mate with their wild relatives.* Baltimore, MD: Johns Hopkins University Press. 244 pp.
19. Gatz C, Lenk I. 1998. Promoters that respond to chemical inducers. *Trends Plant Sci.* 3:352–358.
20. Greenblatt IM. 1984. A chromosome replication pattern deduced from pericarp phenotypes resulting from movements of the transposable element, Modulator, in maize. *Genetics* 108:471.
21. Gressel J. 1999. Needed: new paradigms for weed control. Presented at Weed Management in the 21st century: Do we know where we are going? Hobart, IN.
22. Gressel J. 1999. Tandem constructs: preventing the rise of superweeds. *Trends Biotechnol.* 17:361–366.
23. Gressel J. 2002. *Molecular biology of weed control.* London: Taylor and Francis. 520 pp.
24. Gressel J, Ehrlich G. 2002. Universal inheritable barcodes for identifying organisms. *Trends Plant Sci.* 7:542–544.
25. Gressel J, Levy A. 2000. Giving *Striga hermonthica* the DT's. In *Breeding for Striga resistance in cereals*, Haussmann BIG, Hess DE, Koyama ML, Grivet L, Rattunde HFW, Geiger HH, Eds., pp. 207–224. Weikersheim, Germany: Margraf Verlag.

26. Gressel J, Rotteveel T. 2000. Genetic and ecological risks from biotechnologically-derived herbicide resistant crops: decision trees for risk assessment. *Plant Breed. Rev.* 18:251–303.
27. Grigliatti TA, Pfeifer TA, Meister GA. 2001. TAC-TICS: transposon-based insect control systems. In *Enhancing biocontrol agents and handling risks*, Vurro M, Gressel J, Butts T, Harman G, Pilgeram A, et al., Eds., pp. 201–216. Amsterdam: IOS Press.
28. Haygood R, Ives AR, Andow DA. 2003. Consequences of recurrent gene flow from crops to wild relatives. *Proc. R. Soc. London B* 270:1879–1886.
29. Haygood R, Ives AR, Andow DA. 2004. Population genetics of transgene containment. *Ecol. Lett.* 7:213–220.
30. Heap IM. 2004. International survey of herbicide-resistant weeds. http://www.weedscience.org.
31. Hedden P, Kamiya Y. 1997. Gibberellin biosynthesis: enzymes, genes and their regulation. *Annu. Rev. Plant Physiol. Plant Mol. Biol.* 48:431–460.
32. Herman EM, Helm RM, Jung R, Kinney AJ. 2003. Genetic modification removes an immunodominant allergen from soybean. *Plant Physiol.* 132:36–43.
33. Hutvagner G, Zamore PD. 2002. RNAi: nature abhors a double-strand. *Curr. Opin. Genet. Dev.* 12:225–232.
34. Jensen CS, Salchert K, Gao CX, Andersen C, Didion T, Nielsen KK. 2004. Floral inhibition in red fescue (*Festuca rubra* L.) through expression of a heterologous flowering repressor from *Lolium*. *Mol. Breed.* 13:37–48.
35. Jenszewski E, Ronfort J, Chevre AM. 2003. Crop-to-wild gene flow, introgression, and possible fitness effects of transgenes. *Environ. Biosafety Res.* 2:9–24.
36. Jepson I, Martinez A, Sweetman JP. 1998. Chemical-inducible gene expression systems for plants — a review. *Pestic. Sci.* 54:360–367.
37. Khan MS, Maliga P. 1999. Fluorescent antibiotic resistance marker for tracking plastid transformation in higher plants. *Nature Biotechnol.* 17:910–915.
38. Kiang A-S, Connolly V, McConnell DJ, Kavavagh TA. 1994. Paternal inheritance of mitochondria and chloroplasts in *Festuca pratensis-Lolium perenne* intergeneric hybrids. *Theor. Appl. Genet.* 87:681–688.
39. Kramer U, Smith RD, Wenzel WW, Raskin I, Salt DE. 1997. The role of metal transport and tolerance in nickel hyperaccumulation by *Thlaspi goesingense* Halacsy. *Plant Physiol.* 115:1641–1650.
40. Kunze R. 1996. The maize transposable element Activator (Ac), in transposable elements. In *Transposable elements*, Saedler H, Gierl A, Eds., pp. 161–194. Berlin: Springer.
41. Kuvshinov V, Koivu K, Kanerva A, Pehu E. 2001. Molecular control of transgene escape from genetically modified plants. *Plant Sci.*160:517–522.
42. Larsen K. 2004. Cloning and characterization of a ryegrass (*Lolium perenne*) gene encoding cinnamoyl-CoA reductase (CCR). *Plant Sci.* 166:569–581.
43. Llewellyn DJ. 2000. Herbicide tolerant forest trees. In *Molecular biology of woody plants*, Jain SM, Minocha SC, Eds., pp. 439–466. Dordrecht, Netherlands: Kluwer.
44. Maliga P. 2002. Engineering the plastid genome of higher plants. *Curr. Opin. Plant Biol.* 5:164–172.
45. Maliga P. 2004. Plastid transformation in higher plants. *Annu. Rev. Plant Biol.* 55:289–313.
46. Mariani C, Debeuckeleer M, Truettner J, Leemans J, Goldberg RB. 1990. Induction of male sterility in plants by a chimeric ribonuclease gene. *Nature* 347:737–741.
47. McClintock B. 1951. Chromosome organization and genic expression. *Cold Spring Harbor Symp. Quant. Biol.* 16:13–47.
48. Oliver MJ, Quisenberry JE, Trolinder NLG, Keim DL. 1998. Control of plant gene expression. U.S. Patent 5,723,765.
49. Pfeifer TA, Grigliatti TA. 1997. Genetic pest management strategies: a view of targeted pest insect management in the 21st century. *Agro. Food Ind. Hitech* 8:29–35.
50. Pilon-Smits E, Pilon M. 2002. Phytoremediation of metals using transgenic plants. *Crit. Rev. Plant Sci.* 21:439–456.
51. Riley IT. 1995. *Vulpia myuros* and the annual ryegrass toxicity organisms, *Anguina funesta* and *Clavibacter toxicus*. *Fundam. Appl. Nematology* 18:595–598.
52. Saari LL, Cotterman JC, Thill DC. 1994. Resistance to acetolactate-synthase-inhibitor herbicides. In *Herbicide resistance in plants: biology and biochemistry*, Powles SB, Holtum JAM, Eds. pp. 80–139. Boca Raton, FL: Lewis.

53. Schernthaner JP, Fabijanski SF, Arnison PG, Racicot M, Robert LS. 2003. Control of seed germination in transgenic plants based on the segregation of a two-component genetic system. *Proc. Natl. Acad. Sci. USA* 100:6855–6859.
54. Schulman RN, Salt DE, Raskin I. 1999. Isolation and partial characterization of a lead-accumulating *Brassica juncea* mutant. *Theor. Appl. Genet.* 99:398–404.
55. Skinner JS, Meilan R, Ma CP, Strauss SH. 2003. The *Populus PTD* promoter imparts floral-predominant expression and enables high levels of floral-organ ablation in *Populus, Nicotiana* and *Arabidopsis*. *Mol. Breed.* 12:119–132.
56. Smith RAH, Bradshaw AD. 1979. Use of metal tolerant plant populations for the reclamation of metalliferous wastes. *J. Appl. Ecol.* 16:595–603.
57. Spennemann DHR, Allen LR. 2000. Feral olives (*Olea europaea*) as future woody weeds in Australia: a review. *Aust. J. Exptl. Agric.* 40:889–901.
58. Stanislaus MA, Cheng C-L. 2002. Genetically engineered self-destruction: an alternative to herbicides for cover crop systems. *Weed Sci.* 50:794–801.
59. Stewart CN, Halfhill MD, Warwick SI. 2003. Transgene introgression from genetically modified crops to their wild relatives. *Nature Rev. Genet.* 4:806–817.
60. Wang T, Li Y, Shi Y, Reboud X, Darmency H, Gressel J. 2004. Low frequency transmission of a plastid encoded trait in *Setaria italica*. *Theor. Appl. Genet.* 108:315–320.
61. Weissmann S, Feldman M, Gressel J. 2003. Evidence for sporadic introgression of a DNA sequence from polyploid wheat into *Aegilops peregrina* (*Ae. variabilis*). *Proceedings — Tenth International Wheat Genetics Symposium, Paestum (Italy)*: 539–542.
62. Werner T, Motyka V, Laucou V, Smets R, Van Onckelen H, Schmulling T. 2003. Cytokinin-deficient transgenic *Arabidopsis* plants show multiple developmental alterations indicating opposite functions of cytokinins in the regulation of shoot and root meristem activity. *Plant Cell* 15:2532–2550.
63. Zhu T, Mettenburg K, Peterson DJ, Tagliani L, Baszczynski CL. 2000. Engineering herbicide-resistant maize using chimeric RNA/DNA oligonucleotides. *Nature Biotechnol.* 18:555–558.
64. Zhu YL, Pilon-Smits EAH, Tarun AS, Weber SU, Jouanin L, Terry N. 1999. Cadmium tolerance and accumulation in Indian mustard is enhanced by overexpressing gamma-glutamylcysteine synthetase. *Plant Physiol.* 121:1169–1177.

23 Assessing the Environmental Risks of Transgenic Volunteer Weeds

Alan Raybould

23.1 INTRODUCTION

Environmental damage resulting from the unintended creation of plants with increased weediness has long been recognized as a risk of cultivating transgenic crops (17,18). Although much has been written about the potential for increased weediness of transgenic plants (15,34), there is less in the literature about how to assess whether the risks are acceptable. In this chapter, I describe the purpose and structure of risk assessments and discuss how awareness of these components can guide the efficient collection of data to assess the environmental risks posed by transgenic volunteers and feral crops.

23.2 WHAT IS A RISK ASSESSMENT?

Environmental risk assessment is a decision-making tool, not a way of setting an agenda for scientific research (14). A risk assessment is the synthesis of sufficient information to judge whether the risks of a proposed course of action are acceptable; it does not seek to develop theory or acquire data unless they can improve decisions. Although risk assessment does not seek to increase knowledge for its own sake, it is scientific and has things in common with the best scientific research. Excellent research develops and tests theories that make clear predictions about phenomena that are of general interest; theories that predict uninteresting things are poor science, as is the collection of data that does not allow the testing of an interesting theory. Good scientific risk assessment is the same, as it seeks to predict changes in assessment endpoints that are of general concern.

Risk assessment and scientific research differ because assessment endpoints are a restricted subset of the phenomena that are the justifiable interest of research scientists. Furthermore, a more accurate or precise prediction always constitutes a scientific improvement, whereas risk assessment seeks sufficient accuracy and precision to make a decision. If a prediction is adequate to make a decision, attempts to increase accuracy or precision may worsen the risk assessment by confusing decision making and diverting effort from more worthwhile activities. Moreover, if the collection of additional data delays the introduction of a beneficial product, overall environmental risk may be increased rather than reduced (7).

A conceptual framework for risk assessment is vital for clarity in decision making and efficiency of data generation. A generic approach to risk assessment can be described using a small number of terms with precise meanings: endpoint, hazard, exposure, risk, and trigger value. These terms are clear in the environmental risk assessment of pesticides, and therefore, for the purposes of illustration, it is convenient to describe a framework to assess the environmental risks posed by a new pesticide. The use of pesticides as an example is not meant to imply that the risk assessment of pesticides should be the same as that of transgenic plants. There are some common elements; for example, the toxicity of active ingredients in pesticides and transgenic plants can be estimated using similar methods. Unlike pesticides, transgenic plants can replicate and spread (although pesticides can bioaccumulate), which calls for different concepts and experiments to estimate risk. Nevertheless, the terms hazard, exposure, and risk are applicable to any risk assessment and are conveniently illustrated with reference to pesticides.

The elements necessary to develop the framework are:

- A precise definition of the environmental variable to be protected from effects of the pesticide. This is known as the *assessment endpoint* and should comprise an entity (e.g., a population of a particular species in a particular area) and a property of that entity (e.g., the population size). For pesticides, the assessment endpoints are usually the population sizes of non-target organisms.
- An estimate of the *hazard* posed by the pesticide. Hazard assessment usually involves experiments to measure the effect of the pesticide on a test organism. A measurement of the response of the organism in such a test is called the *test endpoint*. For example, a common test endpoint is the LC_{50}, the concentration of a chemical required to kill 50% of the test population.
- An estimate of *exposure* to the pesticide. Exposure estimates are variously referred to as expected, estimated, or predicted environmental concentrations (EEC or PEC). Exposure estimates may be obtained directly by measurement or from laboratory data and mathematical modeling.
- An estimator of *risk*. The measurements of hazard and exposure must be combined to give an estimate of risk. For chemicals, the estimator may be as simple as the ratio of the test endpoint and the value of the predicted environmental concentration (i.e., LC_{50}/PEC, the toxicity exposure ratio [TER]).
- A threshold value of unacceptable risk (*trigger value*). All substances are toxic at some concentration and therefore have a risk that is greater than zero. It is not possible, therefore, to make a decision based on the absence of risk. The determination of unacceptable risk involves a relationship between the value of the risk estimator and an unacceptable change in the assessment endpoint. For some organisms, field and laboratory studies have empirically derived the value of the estimator that gives high confidence that unacceptable effects will not occur. For other organisms, the relationship between the risk estimator and assessment endpoint may be difficult to measure, and mathematical models are necessary to predict population changes from laboratory data. Because of the uncertainty in parameter values and the relationships between parameter, higher safety factors may be employed when validation from field studies is not possible.

23.3 TIERED TESTING AND RISK ASSESSMENT

The above elements are often organized into a tiered approach to risk assessment. A tiered approach begins by assessing risk from measurements of hazard and exposure under worst-case conditions; for pesticides, the hazard is measured under conditions where contact with the test substance is unavoidable and the exposure may be set as the application rate of the chemical. A value of the TER under these Tier 1 conditions is defined, by whatever criteria, as acceptable risk. If the TER from Tier 1 tests is above this value, the risk is deemed acceptable and testing can stop. If the TER is below this value, higher tier tests involving more realistic exposure need to be carried out if acceptable risk is to be demonstrated. Unacceptable risk is called the trigger value because different actions are triggered depending on whether the TER estimate is above or below this number.

Tier 1 tests are simple and consequently are often criticized for being highly unrealistic compared with field conditions. However, these tests are not intended to be realistic; their purpose is to aid efficient decision making by exposing organisms to unrealistically high concentrations of a substance and so prevent unnecessary higher tier testing of substances that present low hazard. If a chemical is safe under worst-case conditions, then sufficient information is available to make a decision and no further information is required for the purposes of risk assessment. If Tier 1 studies indicate an unacceptable risk, further higher tier tests that introduce more realism can be made. A new trigger value that accounts for the greater realism of the test is set, and decisions about the acceptability of risk are made in the same way as at lower tiers.

In theory, with sufficient imagination, a large number of tiers could be designed. However, if a substance repeatedly fails lower tier tests, perhaps exceeding trigger values by a wide margin, a decision that the substance poses unacceptable risks will be made. Small refinements of tests to slightly increase realism are unlikely to yield results that are sufficiently different from the lower tier study to be a convincing demonstration of acceptable risk. Therefore environmental risk assessment schemes tend to be limited to about four tiers.

Tiered testing has been used widely in safety assessments of pesticides. A more-or-less standard procedure is now used and accepted within the European Union (E.U.) (5). Several authors have called for a tiered approach to risk assessment of transgenic crops (9,27,32), although few advocate that risk assessment should seek the minimum necessary information to make a decision and stop testing at Tier I should acceptable risk be demonstrated.

The attitude that transgenic risk assessment research should do more than generate a minimum set of data for decision making and instead should aim to increase our understanding of ecological processes is common (10). Peters (25) has argued that ecology has failed to become a predictive science because it has concentrated too much on the detailed study of small, isolated components of a phenomenon of interest, which seems to be what proponents of greater understanding are advocating, rather than direct study of that phenomenon (in the case of risk assessment, the relevant endpoint). The call for greater understanding acts as a cover for the absence of a predictive theory and allows the uncontrolled collection of data without a view to how they will be used.

Implicit in the call for understanding is that the goal of ecology is to build ever more detailed models of the mechanisms of ecological processes because they come closer to revealing the reality of the way the world works. This view is called *realism.* A different view, called *instrumentalism*, is that the purpose of science is to generate tools (26). Instrumentalism does not imply that science is utilitarian and only concerned with applied problems; a tool is a device for making predictions about nature, regardless of whether the predictions have practical application. In effect, instrumentalism holds that the value of a scientific theory be judged by the quality of the predictions it makes and that prediction is the only criterion that distinguishes science from other systems of thought. Realism invokes other criteria for judging theories, in particular, whether they offer explanatory power. Some authors (see (25)) regard the failure of a theory to predict as insufficient grounds for its rejection because the theory offers satisfying explanations. Risk assessment science must take an instrumentalist view; we are seeking predictions, not explanations.

23.4 GENERAL REQUIREMENTS FOR ASSESSING RISKS FROM VOLUNTEER TRANGENIC CROPS

Risk assessment must begin by defining what concerns us; in other words, we have to define the assessment endpoints. Next we need to analyze existing data to see whether our proposed course of action — the cultivation of a transgenic crop — poses a conceivable risk to the assessment endpoints. If the data indicate a high likelihood that entities comprising the assessment endpoint will not be exposed to the transgenic crop, or that the crop poses no hazard to them, then we decide that there is no detectable risk and no further data are required. If there are insufficient data to indicate lack of hazard or exposure, then new experimental studies are required. The initial studies should be carried out under worst-case conditions (Tier I) to maximize our confidence that our conclusions apply to all conceivable situations in the field.

The identification of assessment endpoints, possible hazards and routes of exposure, and, if necessary, suitable Tier I tests, is the problem characterization phase of the risk assessment; clarity and precision at this stage are vital to focus the collection of data and gain agreement for decisions based on the risk assessment. Experimental studies done to support the risk assessment aim to estimate the expected exposure and hazards, combine these into an estimate of risk, and make decisions based on whether the estimate of risk predicts an acceptable or unacceptable change in the assessment endpoint. Tests should start with unrealistically high exposures and increase realism until acceptable risk is demonstrated. Once acceptable risk is shown, testing should stop; if unacceptable risk is demonstrated, risk management can be used to reduce the risk to an acceptable amount. If the risk cannot be managed, the risks are unacceptable and crop will not gain registration.

23.5 ASSESSMENT ENDPOINTS

23.5.1 Regulatory Obligations

Legislation in the E.U., the U.S., and elsewhere requires that the environmental risks from cultivating transgenic crops are assessed and judged acceptable before permission to sell seed of the crop can be granted. In the E.U., new legislation was introduced to regulate transgenic plants, the most recent revision of which is Directive 2001/18/EC. The scope of the legislation is broad; the directive requires that the risk assessments "identify and evaluate potential adverse effects of the genetically modified organism (GMO), direct [or] indirect, immediate or delayed, on human health and the environment which the deliberate release or placing on the market of GMOs may have." This requirement encompasses all transgenic organisms and every conceivable environmental effect they could have.

In the U.S., existing legislation that dealt with environmental risks of products used in agriculture was adapted to regulate transgenic plants. Hence, U.S. regulations are often regarded as dealing with the product of plant transformation, in contrast to the E.U., which regulates the process of transformation. Although this is true in theory, in practice all transgenic plants are regulated in the U.S. until they are shown to meet the criteria of Non-regulated Articles (see below).

Currently, transgenic plants in the U.S. are regulated under the Federal Plant Protection Act (FPPA) and the Federal Plant Quarantine Act (FPQA), administered by the United States Department of Agriculture Animal and Plant Health Inspection Service (USDA — APHIS). If the transgenic plant produces a pesticide, the U.S. Environmental Protection Agency (EPA) regulates the pesticide and material required for its production (the gene coding for the pesticide, marker genes and their products, promoters, terminators, etc.) under the Federal Insecticide, Fungicide, and Rodenticide Act (FIFRA). At first sight, the FPPA and FPQA have narrower remits than the E.U. Directive, concentrating on the regulation of risks from potential plant pests. Under these acts, APHIS defines transgenic plants as "Regulated Articles" if "the donor organism, recipient organism, vector or vector agent belongs to any genus or taxon … known to have plant pests, and meets the definition of a plant pest, or is an unclassified organism and/or an organism whose classification is unknown, or any product which contains such an organism, or any other organism or product altered or produced through genetic engineering which the Director of BBEP (Biotechnology, Biologics and Environmental Protection division of APHIS), determines is a plant pest or has reason to believe is a plant pest" (1). Plant pests include weeds, bacteria, and viruses, which are often used as transformation vectors and sources of promoters. Therefore, for practical purposes, all transgenic plants are Regulated Articles, although if a plant that was not a weed were transformed with genes from an organism that is not a plant pest, and without the intervention of a plant pest, the resulting transgenic plant would not necessarily be classed as a Regulated Article (22).

Regulated transgenic crops have strict limits on the areas that can be grown and on movement between states. Therefore if someone wishes to sell seed of a transgenic crop, they must petition APHIS to grant the crop Non-regulated Status. To obtain Non-regulated Status, the petitioner must show that the crop does not present a plant pest risk, a plant pest being an organism or substance that "can directly or indirectly injure or cause disease or damage in or to any plants or parts thereof, or any processed, manufactured, or other products of plants" (1). APHIS views this definition broadly; in addition to harm to crops, it includes damage to native wild plants and organisms beneficial to plants, such as pollinators, mycorrhizae, and rhizobia (20).

Although the philosophies of the European and North American regulations are different, their practical effects are the same: the risk of harm to crops, native plants, beneficial organisms, and human health must be assessed. As discussed above, the first step in assessing a risk is to define assessment endpoints. What should we use as assessment endpoints for the effects of volunteer weeds?

23.5.2 Harmful Effects of Volunteer Crops

A common view is that "the transgenic process presents no new categories of risk compared to conventional methods of crop improvement" (22); the validity of this assertion is considered below. If transgenic and non-transgenic crops pose similar risks, assessment endpoints for volunteer transgenic crops can be derived from the harmful effects of non-transgenic volunteers.

Volunteers harm crops directly by way of competition and contamination, and indirectly by acting as "green bridges" for pests and pathogens, and are therefore regarded as plant pests. Competition reduces yield of the affected crop and contamination can lead to lower quality or, in extreme cases, rejection of products intended for human consumption (23).

Competition is a serious problem when a slow growing crop, such as onions, is in the same rotation as a fast growing crop like potatoes. Runham et al. (29) showed that if onions were sown immediately after potatoes, the onion yield was between 0 and 13 tons per hectare if the volunteer potatoes were not controlled; hand weeding of potatoes gave onion yields of 20 to 62 tons per hectare; and the best herbicide regime, 53 tons of onions per hectare.

Loss of quality occurs when seed from the volunteer is mixed with seed from the crop. For example, in the 1980s, the area of linseed grown in the U.K. increased rapidly and volunteer linseed began to contaminate other crops. Vining peas grown for freezing or canning were particularly prone to contamination with linseed because the pea pods and linseed capsules appear similar and were difficult to separate in the factory and some processors refused to buy peas from fields with a history of linseed cultivation (19). Another example is the contamination of low glucosinolate and erucic acid varieties of oilseed rape used for human and animal feed, with volunteers of high glucosinolate or high erucic acid varieties that have industrial uses; contamination can make rapeseed oil unsuitable for consumption by humans and animals (3,21). Similarly, volunteers of wheat for animal feed in wheat for milling can reduce the quality of the grain (11).

Some plant pathogens that use living tissue cannot survive long without living plants to infect. Volunteers that emerge soon after harvest, while viable pathogen inoculum from the previous crop is present and while the temperature is high enough to promote infection, can lead to the build up of inoculum ready to infect newly emerging crops in the following season; in other words, the volunteers act as a green bridge between one crop and the next. Volunteers may also reduce the effectiveness of rotations in controlling root pathogens and maintain seed-borne diseases that can normally be controlled by fungicide applications to crop seeds (35).

Yarham and Gladders (35) reviewed the importance of volunteers as green bridges in U.K. agriculture. Volunteers are crucial to the maintenance of yellow rust (*Puccinia striiformis*) infections, as the fungus has no means of surviving in the absence of living tissue and has no alternative host. Infections are particularly common where a crop is sown early adjacent to a field where cereal stubble is removed late. Other cereal diseases in which green bridges are important are mildew (*Erisyphe graminis*), take-all (*Gaeumannomyces graminis*), and barley yellow dwarf virus. Oilseed rape volunteers are effective at transmitting seed-borne pathogens, such as *Alternaria*, between successive rape crops. Rape volunteers can also act as a source of inoculum for fungus diseases of cereals (e.g., *Sclerotinia*). Volunteer potatoes and plants growing on dumps of discarded tubers are sources of inoculum for potato blight (*Phytophthora infestans*) and viruses. A possible benefit of green bridges is the maintenance of populations of fungi that are susceptible to fungicides.

23.5.3 Do Volunteer Transgenic Crops Pose New Hazards?

Problems caused by volunteers of non-transgenic crops are a normal part of agriculture; they are not newsworthy. The reaction to the appearance of volunteers of transgenic crops has been different. The discoveries of volunteers of oilseed rape with resistance to multiple herbicides that arose after gene flow between two transgenic and one conventional variety (12) and of soybeans contaminated with transgenic maize volunteers containing an experimental vaccine to protect pigs against harmful strains of *Escherichia coli* (21) have been written about in mass circulation newspapers, as well as the specialist literature.

Despite the great interest in the transgenic oilseed rape and maize volunteers, they do not represent new hazards. The herbicide-resistant oilseed rape volunteers do not present a problem in terms of competition with a following crop; the volunteers can be controlled using other herbicides. There is an argument that more harmful herbicides are needed to control the volunteers; however these herbicides are approved by regulators and do not represent a new *class* of hazard compared with herbicides that may have been used in the absence of resistant volunteers. The concern is that transgenic herbicide-resistant seeds will contaminate non-transgenic seed, which is similar to the contamination of low erucic acid varieties of oilseed rape through gene flow from high erucic acid varieties. The difference is that erucic acid is harmful to human and animal health, and therefore raises safety concerns; herbicide-resistant oilseed rape is approved for use as human and animal feed in the U.S. and Canada and therefore contamination of non-transgenic seed is not related to safety, but to regulation. The concern about mixing of transgenic and non-transgenic seed is related to purity (12), which determines whether seed can be used for organic production, or whether it has to be labeled as genetically modified in the E.U. In theory, if the transgenic event has not yet been approved in an export market, human and animal safety could be claimed to be a concern.

The contamination of soybeans with transgenic maize is not a different kind of problem from the presence of linseed in frozen peas. Again, it is debatable whether the contamination of soybeans with small amounts of transgenic maize raised any safety problems (21), whereas eating linseed could be harmful or at least unappetizing.

A general point raised by the stories about transgenic volunteers, and a related case where small amounts of transgenic maize approved for animal feed only (a variety called StarLink) became mixed with maize for human consumption, is that transgenes make extremely low amounts of contamination easy to detect: herbicide-resistant volunteers are revealed after spraying with herbicide; immunological methods can detect small amounts of protein coded by transgenes; and the polymerase chain reaction can detect transgenes themselves at extremely low dilutions. Therefore, while transgenic volunteers may present no new classes of hazard, they may be regulated in a different way. Mere presence of transgenes is seen as a cause for concern, regardless of whether the product of the transgene is present in quantities that could cause harm. In other words, volunteers of transgenic crops may be regulated on the basis of exposure, rather than risk.

23.5.4 Making Assessment Endpoints Operational

The discussion above has narrowed the requirement to assess the risks of transgenic volunteers to plants to a consideration of three subjects: competition, contamination, and green bridges. To develop these concepts as useful assessment endpoints, we need to make them operational, meaning to define "the practical specification of the range of phenomena that [the] concept represents" (25). Although no definition can be sufficiently precise to remove all ambiguity, if a concept is to prove scientifically useful, it must be "sufficiently operational that informed users associate the concept with similar phenomena" (25). In other words, risk assessment will work best if we can agree on how to define and measure the things we need to predict.

Complete operationalization of competition, contamination, and green bridges (i.e., definitions on which everyone agrees) may not be possible; however, as a rule of thumb we should try to define them in terms of simple entities and properties of those entities that can be measured. Illustrations of how these concepts can be made operational, using previously discussed examples, are listed below. Of course, some terms in these definitions may need further definition, but that should be relatively uncontentious.

- *Competition* — negative regression between yield of onions (tons per hectare) and the biomass of potato volunteers
- *Contamination* — the presence of erucic acid (above a tolerance threshold concentration) in rapeseed oil for human consumption
- *Green bridge* — a statistically significant greater frequency of yellow rust infection (measured as coverage of the flag leaf) on cereals next to a field that grew winter wheat the previous season than on cereals where the nearest winter wheat was more than one field away

These definitions give scope for unacceptable risk to be defined and predicted, which is the essence of risk assessment. Unacceptable risk can be defined either in terms of the likelihood of an absolute value (loss of so many tons of a crop, concentration of a hazardous substance, frequency of disease) or in terms of changes in the current situation (greater yield loss, greater likelihood of the presence of a hazardous substance above acceptable limits, higher frequency of disease). Unacceptable risk is discussed further below with respect to trigger values.

23.6 HAZARDS, EXPOSURE, AND RISKS OF VOLUNTEER TRANSGENIC CROPS

The difference between hazard and exposure when considering risks posed by volunteer crops is not as obvious as when dealing with a potentially harmful substance. The *hazard* of the substance is defined in terms of its effects — toxicity, flammability corrosiveness, and so on. *Exposure* is defined as the likelihood of encountering specified amounts of the substance. *Risk* is then calculated as the likelihood of meeting an amount of the substance that will induce an unacceptable harmful effect.

One of the potential detrimental effects of transgenic crops is increased weediness, which for the purpose of this discussion is taken to be synonymous with competition as defined above: a transgenic oilseed rape variety would be regarded as weedier than its non-transgenic progenitor if it was associated with a greater decrease in the yield of a following crop. The increase in weediness could arise through changes that make individual transgenic volunteers worse weeds (e.g., if they grew taller than the non-transgenics) or more abundant (higher density). An increase in the size of individual plants seems conceptually most similar to a substance with higher toxicity, and hence can be regarded as an increase in the hazard of the crop. However, it is a moot point whether increased abundance, without change in individual weediness, should be regarded as increased hazard (equivalent to a *different* substance with greater toxicity) or increased exposure (equivalent to a higher concentration of the *same* substance).

For pragmatic reasons, it is preferable to regard increased abundance as a hazard. Changes to life history parameters that are associated with increased abundance may be the same as those that are associated with increases in the weediness of individual plants; one example could be a increase in the number of seeds produced. Changes in these traits can be regarded as hazards that potentially increase weediness, without having to worry whether they operate at the individual plant or population scale.

If changes to individual plants and abundance are regarded as hazards, exposure can be regarded as the frequency with which the crop at risk (onions in the example above) follows the crop that produces volunteers (potatoes). Transformation might allow a crop to be grown in a new rotation or in a different region, which could expose crops to new volunteers and produce new potential hazards.

Similar debates about whether to regard parameters as hazards or exposure can be had over contamination. Consider the concentration of erucic acid in oilseed rape oil as the assessment endpoint. If a transgenic oilseed rape variety contains higher amounts of erucic acid in its seed than the line from which it was derived, does this constitute greater hazard (of the plant) or greater exposure (to erucic acid)? If a high erucic acid line is transformed and contains the same amount of erucic acid as its non-transgenic equivalent, but it increases in abundance as a volunteer, is this greater hazard or greater exposure? I think arguments can be made for both; however, I prefer to consider both increased erucic acid concentration in the seed and increased abundance without changes to erucic acid concentration as hazards. First, we are assessing the risks of the transgenic plant, rather than the risks of erucic acid poisoning. Second, it is useful to be consistent in describing particular parameters as hazards of volunteer crops for all assessment endpoints.

23.7 HAZARD ASSESSMENTS

Discussions of assessment endpoints and the distinction between hazard and exposure lead to the conclusion that there are two main hazards of volunteer crops — abundance and composition. Abundance relates to competition, contamination, and green bridges and composition, to contamination (and possibly to green bridges if composition changes susceptibility to pathogens). The following sections examine how we might assess these hazards of transgenic crops.

23.7.1 ABUNDANCE

Two methods have been tried for predicting the abundance of volunteers of transgenic crops that would result from commercial release. The first method uses mechanistic systems models of the development of volunteer populations with parameters that describe the life history of the volunteers under various types of management (6). The second method uses a measurement of transgenic volunteer populations in field scale releases of the transgenic crop (13). Kareiva et al. (16) doubt that useful predictions of the abundance of weedy transgenic plants are possible from either method because data and mathematical models will always be insufficient to cope with uncertainty.

Systems models seem least likely to reduce uncertainty. The weak predictive power of complex models has long been recognized; uncertainty in estimates of parameters, the number and complexity of the equations, and the length of time the models run lead to massive propagation of errors (25). Indeed, complex models often give results that are indistinguishable from random variation (24). Despite their inability to predict, systems models remain popular with ecologists because they provide explanations or understanding.

The simple model of measuring volunteer abundance of transgenic and non-transgenic crops under realistic farm management is more likely to reduce uncertainty. However, field scale experiments should not be our first choice of experiment to assess the hazards of abundance of transgenic volunteers:

- Experiments are expensive, time-consuming, and introduce delay.
- It may not be necessary to make precise quantitative predictions about number of volunteers at a given time, only whether the transgenic crop volunteers are likely to be more abundant than non-transgenic volunteers.
- The field scale experiments go against a tiered approach; it may be possible to establish acceptable risk using simpler lower tier methods.

Is it possible to assess the risks from increased abundance of volunteers transgenic crops without recourse to field scale experiments? During development of transgenic crops, their performance is compared with conventional varieties in a series of multi-location agronomic field trials. APHIS (33) suggests that characters can be measured in the trials that can be used to assess whether the transgenic variety is more likely to be become a volunteer weed. Suggestions of traits to measure include:

- Reproduction and survival characters — such as life span, vegetative biomass, overwintering capacity, flowering behavior, seed production, seed dormancy, germination, seedling survival, outcrossing frequency, pollen viability, dispersal ability (panicle shattering, etc.)
- Adaptation to stress
  - Biotic — pathogens, herbivores, other plants
  - Abiotic — atmospheric pollutants, nutrient deficiency, temperature extremes, drought, flood
  - Pesticides
- Nutritional composition (undefined)
- Levels of natural toxicants (undefined)

APHIS concludes, "observed changes may warrant further in-depth studies." The rationale presumably is that if no differences in these traits are observed, the transgenic crop is unlikely to be a worse volunteer weed than the conventional crop. Notice that the requirement is not to predict how serious a weed the transgenic crop is, rather whether it is weedier than the conventional crop that, by implication, poses acceptable risk.

Although the further in-depth studies are not specified, one option is to use the field data to estimate a parameter such as population growth rate for the transgenic and non-transgenic crops; this could be done using matrix models (4). In effect, modeling refines the hazard because not all increases in growth or reproduction lead to increases in population growth rate. For instance, if a transgenic crop produces more seed than its conventional counterpart, this does not necessarily make it weedier; it may be that seed production does not limit the population growth rate of the conventional crop volunteers (2). An alternative would be to carry out more realistic experiments; for example, seeds could be sown into a following crop to test directly the worst-case potential for volunteering.

23.7.2 Composition

There are two general cases where composition will be an important hazard to assess. The first is crops that produce pharmaceutical or industrial compounds. The tolerance of contamination of other crops with volunteers of these transgenic plants may be so low that in effect all volunteers must be eliminated (21), regardless of precise concentration of the compounds. If a tolerance is set, or if the risk of contamination is to be managed by expressing the compounds in particular tissues, rigorous, detailed expression profiles of the pharmaceutical or industrial chemical in multi-location trials will be necessary.

The second case is crops that are known to produce harmful compounds in particular varieties or under certain conditions or that have varieties that are used for different purposes. For example, it would be prudent to check that transformation of oilseed rape has not changed glucosinolate and erucic acid concentrations. Similarly, the quality of the grain of transgenic wheat could be used to assess the impact of volunteers in following crops of different wheat varieties (11). These characters will be measured during normal agronomic evaluation of the crop and events showing significant changes from the non-transgenic isoline are unlikely to be commercialized.

23.8 EXPOSURE ASSESSMENTS

It is possible that transgenic varieties may offer such large improvements over conventional varieties that they can be grown in regions in which the crop has not been cultivated previously. An extension of range of a crop is likely to be due to an intended effect of the transgene and indeed may be the purpose of the transformation. For example, the crop may be designed to tolerate colder springs or hotter summers. Therefore the new areas into which the crop will be grown may be known with some precision and could also be limited by regulatory authorities. Hence the likely rotations could be predicted. Although no new class of hazard is likely to arise from new rotations, risk assessments should consider the hazards of competition, contamination, and green bridges under worst-case exposure (the appearance of transgenic volunteers in the rotation, unless data suggest otherwise). Such data could be demonstrations that seed or vegetative parts of the crop could not persist in the new rotation; for example, seeds could be buried and their viability measured under best-case conditions for survival in the new rotation.

Another way in which exposure to the hazards of weeds could change is hybridization between transgenic crops and wild relatives followed by backcrossing to the crop leading to weedy populations (defined as exoferality in Chapter 1). The likelihood of hybridization between crops and wild relatives can be considered a worst-case estimate of exposure when assessing the risks of exoferality (worst-case because hybridization does not inevitably lead to weedy populations).

Many early papers on transgene risk assessment (8,28,30) were reviews of data on whether certain wild species could hybridize with crops. The information needed for risk assessment is whether crops and wild species will hybridize in the field; the possibility of this occurrence can be studied using a tiered approach as advocated above:

- Tier I
  - Test for hybrid production using laboratory methods (worst-case)
  - No hybrids, no detectable risk, stop testing; hybrids, go to Tier II
- Tier II
  - Test for spontaneous hybrid production (lab/field)
  - No hybrids, no detectable risk, stop testing; hybrids, go to Tier III
- Tier III
  - Search for naturally produced hybrids
  - No hybrids, no detectable risk; hybrids, carry out quantitative risk assessment

Even though there is consensus on the need to identify wild species with sexual compatibility to crops, there is debate about whether precise estimates of rates of past and present introgression are of value for risk assessment. Some authors consider that estimates of the rate of introgression between crops and wild relatives based on neutral marker genes are pointless for the risk assessment of transgenic crops because rates of introgression of transgenes (i.e., gene frequencies) will be determined by the effects (fitness) of the gene. This view is usually summarized as "if it can happen, it will happen", and its advocates recommend that hazard characterization is the most important part of the risk assessment because a gene that increases plant fitness will tend to spread to fixation regardless of its initial frequency.

This must be correct in a population of finite size that persists for an infinite time with no changes to initial conditions. However, transgenic varieties will have limited lives, populations of wild relatives are transient and fragmented, and environmental conditions will change. Therefore, the likelihood of a transgene introgressing into a natural population may be low because hybridization is rare, and when hybrids are formed, the transgenes are lost through genetic drift. Therefore, there may be value in differentiating between wild relatives that have limited sexual compatibility with crops and those that are highly compatible with crops. Indeed, if the hazards associated with a substance or process are well characterized, decision making is in effect based only on exposure estimates.

A recent paper by Wilkinson et al. (31) on hybridization between oilseed rape and *Brassica rapa* indicates that it is possible to define areas within a country were the probability of hybridization between a transgenic crop and a wild relative is negligible, even when the distribution of the wild relative is widespread and the degree of sexual compatibility is high. This offers the possibility of regional risk assessments and possibly risk management based on restricting cultivation of transgenic crops to areas where the probability of hybridization is low. For example, the EPA does not allow the cultivation of transgenic insect-resistant cotton in Arizona, Florida, or Hawaii because of the presence of wild relatives, but does allow its cultivation elsewhere in the U.S.

23.9 MONITORING

E.U. legislation requires that monitoring be carried out following commercial release of all transgenic crops. There are two types: case-specific monitoring for the intended effects of the transgene and general surveillance monitoring for unintended effects. The EPA has also required case-specific monitoring for some transgenic crops expressing pesticidal proteins. There is much debate about the necessity for case-specific monitoring when the risk assessment has identified no unacceptable risks. A properly designed tiered testing approach should make case-specific monitoring unnecessary, because lower tier tests represent situations that are worse than the field.

Case-specific monitoring may be highly relevant when risk management is proposed. For example, a risk assessment may show an unacceptable probability that a transgenic crop producing an industrial chemical will produce volunteers in a food crop. This probability may be reduced to an acceptable amount by the use of management to control volunteers. Monitoring would seek to ensure that the control measures were working.

General surveillance can be seen as an attempt to address the uncertainties inherent in trying to assess risks from unintended effects. General surveillance monitoring usually comprises questionnaires about changes in farmland following introduction of the transgenic crop. One problem with general surveillance is whether we have sufficient data to predict the changes that would have occurred had the transgenic crop not been grown. A related question is how will decisions be triggered when general surveillance data show changes? As with risk assessment, we must avoid the temptation to collect data without knowing how we will use it.

23.10 CONCLUSIONS — ACCEPTABLE RISKS AND TRIGGER VALUES

Volunteers of transgenic crops do not pose new categories of risk. However, the acceptability of risk may be much lower. In the case of pharmaceutical compounds, any contamination of the food chain may be unacceptable, even though the pharmaceutical may be less toxic than contaminants that are permitted (21). On this basis, any data showing that transgenic volunteers could occur in a food crop would indicate unacceptable risk, and the argument that experiments showing zero probability of viable volunteers were worst-case would have to be irrefutable.

In the case of competition, or contamination with a substance for which a tolerance is acceptable, the trigger for demonstration of acceptable risk might be that the transgenic variety is no different from non-transgenic varieties for a range of characters that might be associated with weediness (33). If changes are found, then further higher tier studies can refine the risk assessment. Given uncertainty in predicting weediness, it is unlikely that an acceptable increase in weediness potential could be agreed, and therefore no change may remain the trigger, unless other attributes of the crop diminish the total environmental risk.

Science cannot decide what is an unacceptable risk, although scientists have as much right as anyone to offer an opinion. Where scientists can use their specialist knowledge is in helping to frame unacceptable risk in operational terms (assessment endpoints). For example, society may decide that it is concerned that transgenic crops could reduce farmland biodiversity. It is the job of scientists to translate the concept of reduced farmland biodiversity into something that can be measured. For example, in the U.K., the population sizes of a relatively small number of bird species are taken as indicators of the general state of biodiversity on farmland. Scientists can also put risks into context. For example, *Bt* maize plants tend to have lower amounts of infection by fungi than conventional maize plants, because fungi can enter the plant through injuries caused by insect damage; the theoretical risk of harm from eating *Bt* proteins can be compared with the risk of harm from ingesting highly toxic chemicals (mycotoxins) produced by the fungi.

Scientists also have a responsibility to predict changes in the assessment endpoints in as succinct a fashion as possible consistent with making good decisions about minimizing environmental risks. It is easy to view the collection of data as an end in itself, which at worst is harmless. But as explained above, collection of data is not free of risk because at some point additional data become superfluous (i.e., do not improve the final decision), and if decisions are delayed waiting for superfluous data, environmental risk may be increased if the introduction of a less harmful method of farming is delayed. Rodgers, Graham, and Wiener, quoted in (7), give an interesting perspective:

> The "insatiable pursuit of data facilitates delay; any decision dependent upon extensive data gathering promises to be long in incubation and short on results."
>
> **William Rodgers**

> One must balance "the value of more information to better decisions and the cost, including delay of decisions."
>
> **John Graham and Jonathan Wiener**

In this article, I have tried to emphasize the need to plan the collection of data with a view to how the data will be used. Collection of data that cannot be used to predict changes in relevant assessment endpoints upon which decisions can be made is useless at best and at worst, can increase risk through delaying the introduction of safer methods of producing food.

LITERATURE CITED

1. APHIS (Animal and Plant Health Inspection Service). 1987. 7 CFR Parts 330 and 340, Plant pests; introduction of genetically engineered organisms or products; final rule. *Fed. Regist.* 52:22892–22915.
2. Bergelson J. 1994. Failing to predict invasiveness from changes in fecundity: a model study of transgenic plants. *Ecology* 75:49–252.
3. Bowerman P. 1993. Effects of cultivations on volunteer oilseed rape. *Asp. Appl. Biol.* 35:163–166.
4. Bullock JM. 1999. Using matrix models to target GMO risk assessment. *Asp. Appl. Biol.* 53:205–212.
5. Candolfi MP, Barrett KL, Campbell PJ, Forster R, Grandy N, Huet M-C, Lewis G, Oomen PA, Schmuck R, Vogt H. 2001. Guidance document on regulatory testing and risk assessment procedures for plant protection products with non-target arthropods. Pensacola, FL: SETAC Press. 48 pp.
6. Colbach N, Clermont-Dauphin C., Meynard JM. 2001. GeneSys: a model of the influence of cropping system on gene escape from herbicide tolerant rapeseed crops to rape volunteers I. Temporal evolution of a population of rapeseed volunteers. *Agric. Ecosyst. Environ.* 83:235–253.
7. Cross FB. 1996. Paradoxical perils of the precautionary principle. *Washington Lee Law Rev.* 53:851–925.
8. de Vries FT, van der Meijden R, Brandenburg WA. 1992. Botanical files: a study of the real chances of gene flow from cultivated plants to the wild flora of the Netherlands. *Gorteria, Suppl.* 1:1–100.
9. Dutton A, Romeis J, Bigler F. 2003. Assessing the risks of insect resistant transgenic plants on entomophagous arthropods: Bt-maize expressing *Cry1Ab* as a case study. *BioControl* 48:611–636.
10. Ervin DE, Welsh R, Batie SS, Line Carpentier C. 2003. Towards an ecological systems approach in public research for environmental regulation of transgenic crops. *Agric. Ecosyst. Environ.* 99:1–14.

11. Garstang J. 1993. The effects of volunteers on cereal quality and profitability. *Asp. Appl. Biol.* 35:67–74.
12. Hall L, Topinka K, Huffman J, Davis L, Good A. 2000. Pollen flow between herbicide-resistant *Brassica napus* is the cause of multiple-resistant *B. napus* volunteers. *Weed Sci.* 48:688–694.
13. Heard MS, Hawes C, Champion GT, Clark SJ, Firbank LG, Haughton AJ, Parish AM, Perry JN, Rothery P, Scott RJ, Skellern MP, Squire GR, Hill MO. 2003. Weeds in fields with contrasting conventional and genetically modified herbicide-tolerant crops. I. Effects on abundance and diversity. *Phil. Trans. R. Soc. London Ser. B* 358:1819–1832.
14. Hill RA, Sendashonga C. 2003. General principles for risk assessment of living modified organisms: lessons from chemical risk assessment. *Environ. Biosafety Res.* 2:81–88.
15. Kareiva P. 1996. Developing a predictive ecology for non-indigenous species and ecological invasions. *Ecology* 77:1651–1652.
16. Kareiva P, Parker IM, Pascual M. 1996. Can we use experiments and models in predicting the invasiveness of genetically engineered organisms? *Ecology* 77:1670–1675.
17. Keeler KH. 1989. Can genetically engineered crops become weeds? *Bio/Technology* 7:1134–1139.
18. Kling J. 1996. Could transgenic supercrops one day breed superweeds? *Science* 274:180–181.
19. Knott CM. 1993. Volunteer linseed control in vining peas. *Asp. Appl. Biol.* 35:179–184.
20. McCammon SL. 1998. AgrEvo USA Company Petition 98-329-01p. Determination of nonregulated status for glufosinate tolerant rice transformation events LLRICE06 and LLRICE62. Finding of no significant impact. http://www.essentialbiosafety.info/docroot/decdocs/01-290-077.pdf.
21. Miller HI. 2003. Will we reap what biopharming sows? *Nature Biotechnol.* 21:480–481.
22. National Research Council. 2002. *Environmental effects of transgenic plants: the scope and adequacy of regulation.* Washington, D.C.: National Acad. Press. 320 pp.
23. Orson J. 1993. The penalties of volunteer crops. *Asp. Appl. Biol.* 35:1–8.
24. Oster GF. 1981. Predicting populations. *Am. Zool.* 21:831–844.
25. Peters RH. 1991. *A critique for ecology.* Cambridge, U.K.: Cambridge University Press. 366 pp.
26. Popper KR. *The Logic of scientific discovery.* London: Routledge. 480 pp.
27. Poppy GM 2000. GM crops: environmental risks and non-target effects. *Trends Plant Sci.* 5:4–6.
28. Raybould AF, Gray AJ. 1993. Genetically modified crops and hybridization with wild relatives: a UK perspective. *J. Appl. Ecol.* 30:199–219.
29. Runham SR, Davies JS, Leatherhead MJ. 1993. Weed control strategies for volunteer potatoes in onions. *Asp. Appl. Biol.* 35:113–122.
30. Scheffler JA, Dale PJ. 1994. Opportunities for gene transfer from transgenic oilseed rape (*Brassica napus*) to related species. *Transgenic Res.* 3:263–278.
31. Wilkinson MJ, Elliott LJ, Alainguillaume J, Shaw MW, Norris C, et al. 2003. Hybridization between *Brassica napus* and *B. rapa* on a national scale in the United Kingdom. *Science* 302:457–459.
32. Wilkinson MJ, Sweet J, Poppy GM. 2003. Risk assessment of GM plants: avoiding gridlock? *Trends Plant Sci.* 8:208–212.
33. White JL. 2002. US regulatory oversight for safe development and commercialization of plant biotechnology. In *Ecological and agronomic consequences of gene flow from transgenic crops to wild relatives.* Snow A, Mallory-Smith C, Ellstrand N, Holt J, Quemada H, Eds., Columbus, OH: Ohio State University. http://www.biosci.ohio-state.edu/~asnowlab/Proceedings.pdf.
34. Wolfenbarger LL, Phifer PR. 2001. The ecological risks and benefits of genetically engineered plants. *Science* 292:637–638.
35. Yarham DJ, Gladders P. 1993. Effects of volunteer plants on crop diseases. *Asp. Appl. Biol.* 35:75–82.

24 Regulation Should Be Based on Data, Not Just Models

Richard Roush

"Ignorance ain't what you don't know, it's what you do know that ain't correct."

Will Rogers
American humorist

24.1 INTRODUCTION

As scientists, we prefer to test ideas with experiments before making recommendations or policy decisions. However, there are many questions of interest where the potential interactions are complex, where critical events are rare, and where long time frames are required for meaningful results, particularly for ecological systems that naturally change with time. Such questions are difficult and costly to resolve with convincing experiments. However, even in the absence of experiments that can be generally accepted as informative, decisions on key questions must be often be made, especially by regulators, because indecision and continuation of the *status quo* also has its own risks. Because of this reality, mathematical models gained a critical role in regulatory processes, long before transgenic crops and other controversial issues reached the public consciousness. Still, regulatory decisions should be based on sound data, whenever possible. The aim of this chapter is to suggest a perspective by which data collection, experiments, and modeling might be most usefully linked for examining transgenic crops in general, and ferality, in particular.

Based on my impressions of 5 years service on the Australian government's genetic engineering regulatory committees, I will focus on my understanding of what regulators want and really need. I recognize that the key questions revolve around transgenic crops and that few if any regulators are going to be concerned with ferality alone, because as other chapters in this book show, crop ferality has been with us for many years, but still has not attracted serious regulatory attention. Because I could not find good examples from ferality (except that of Haygood et al. reviewed in Chapter 22), other examples from the literature on pesticide resistance and transgenic crops are used.

By forcing a precise statement of assumptions and the mathematical implications of scientific arguments, models have played and continue to play an essential role in building a conceptual framework for many areas of science, especially ecology and population genetics. In fact, the most fruitful way to think of models might be that they are precisely stated hypotheses. Like any hypothesis, models must be tested against experiments or existing data sets. In many cases, models help clarify what are probably the most critical factors in the behavior of a system, and in some cases encouraged the collection of data that had not previously been thought to be important. For example, models of a male-sterile system for management of a key cotton pest raised the importance of the interval between matings of female moths (22). In considering the effect of within row seed mixtures of transgenic and conventional plants for resistance management, larval dispersal between plants proved to be critical (10,19,28). Models focus on exploration of how various features interact in a system and raise questions about how or why things could work; models cannot really prove that a strategy or recommendation will work. Only carefully designed experiments can do that. Thus, it is critical for the use of models in regulation to remember that the models are hypotheses, some much more reliable than others (perhaps tending toward being recognized as theories). Models are still not fact and are reliable only inasmuch as their formulation is accurate and that key assumptions are supported by data. When predictions of carefully designed models fail to fit the real world, models can themselves raise questions about what we think we know and lead to further investigation. Ideally, the models should be tested in some form by carefully designed and critical experiments. We must respect the advice of American cowboy humorist Will Rogers and be careful that we have not trusted models to convince ourselves of things that are incorrect.

Still, because of their logical rigor, models have greater exploratory power than concepts without such frameworks. For example, much of what is commonly believed about pesticide resistance management is simply not supported by the logical analysis forced by the models (21). This includes such concepts as that high doses (or application rates) are more effective than low doses of pesticides, and pesticide mixtures are more effective than rotations. Although mixtures and high doses can be the more effective strategies in some cases, they certainly are not always so. Indeed, experiments on these tactics have given results that are inconclusive with respect to general rules in managing resistance, and the models have suggested why (17–20). In short, models can help in decision making, but often by challenging ideas that are logically untenable in the face of the facts. Models are no substitute for good experiments, but data alone are not enough either. The key issue for modeling, as with experiments, is a precisely stated and relevant question that can be tested. Although a great deal of observational data can usually be collected, interpretation in the absence of manipulative and controlled experiments is often ambiguous. Experiments are not guaranteed to be successful. Even carefully designed experiments can be inconclusive and collect poor data and adversely influence regulatory policy. A famous example that continues to have broad impacts on public opinion and policy was a widely publicized claim that pollen from transgenic insecticidal corn using a gene derived from the bacterium *Bacillus thuringiensis* (Bt) harmed monarch butterfly larvae (24,25). In contrast, broader considerations suggested that Bt corn was likely to provide overall benefits even to monarchs (12). Eventually, the threat to monarch butterflies was discounted by five papers published together in the *Proceedings of the U.S. National Academy of Sciences* (e.g., 23). Unfortunately, probably most of the general public remains unaware of this because the rebuttals received little media attention compared to the original report.

24.2 TRENDS IN REGULATION — ESTIMATING RISK

Most of the regulatory effort to date on transgenic crops has been qualitative and conservative. In Australia and many other countries, any indication of a potentially significant adverse effect on the environment or human health has been enough to block the release of a transgenic crop. However, there is increasing pressure from risk management experts and activists to develop quantitative risk assessment methods that allow probabilities to be assigned to potential adverse outcomes and to foster cost-benefit analyses. Risk is a function of probability and hazard, but for most transgenic crops released to date, it has been difficult to assign values to either. One continuing problem will be how hazard is perceived. Studies on pollen flow have been interpreted differently depending on the perceptions of the beholder. Pollen flow between oilseed rape fields in Australia seems to be less than 0.2% (14), which currently satisfies all international trading standards. Conversely, organic farming advocates review the same data and argue that because pollen flow could be detected at 3 km with the sensitive seedling testing methods used, produce of organic farms will be at risk if within 3 km. Neither models nor more data will solve this argument; only political decisions can.

For some of the most important concerns about transgenic crops, a strict probability and hazard approach to risk assessment may not even make a lot of sense and needs to be modified. For example, the evolution of resistance in insect pests is arguably the most important environmental concern for Bt transgenic crops, such as cotton and maize, because of the hazard that resistance might mean a return to chemical insecticides. The costs of this hazard are thus serious (and can at least roughly be estimated from the impacts of the pesticides that were replaced), but the interpretation is less clear when combined with a high probability estimate for resistance.

Based on the history of the evolution of resistance in key insect pests, most experts assume that the probability of resistance to Bt crops is nearly 1. Resistance will probably eventually evolve in targeted pests with a history of resistance to insecticides (including the cotton bollworms *Heliothis virescens* in the U.S. and *Helicoverpa armigera* in Australia, China, and India) if the transgenic crops are used widely and long enough. The aim of resistance managers is to delay the inevitable resistance. Given that resistance to Bt crops can be indefinitely delayed by simply not using them, a more precise description of the aim of resistance managers is to optimize the benefits of Bt, for example, reduce pesticide use as much as possible at least until new non-pesticidal control tactics can be developed. As far as I am aware, regulators have not yet looked to this level of detail. Among the few countries where regulators have judged resistance management plans (e.g., the U.S. and Australia), the criteria for accepting the plans seem to be whether the plans are reasonable. Although some claims have been made about predicting how long the strategies will last, these are fraught with potential errors of estimation (as explained further below).

In other cases, risk estimation is easier. For example, even though there may be some chance that oilseed rape can outcross to wild radish (13; but there remains some doubts that any outcrossing occurs at all, S. Warwick, personal communication), there seems to be little concern in Australia about the potential for outcrossing of glufosinate resistance from transgenic oilseed rape to wild radish because the herbicide is not effective against wild radish anyway (although it is more effective against unrelated weeds). In the most trivial of cases, such as with maize and soybeans in the U.S. and Canada, escape of transgenes to weedy relatives cannot happen at all (Chapter 10), so there is no risk, except of gene transfer to other varieties of the crop.

Observational data and experiments alone probably cannot achieve probabilistic estimates of risk and cost-benefit analyses that regulators could use. The most challenging concerns about risk are from rare events, which are difficult to measure in experiments of reasonable cost and safety (one would not want to have an escape only then to discover the risk). In any case, combining data that estimate various sources of hazard and various sources of risk will be difficult without a model of some type, at least a statistical one. Models grounded in the best data possible will probably be needed for any advances in quantitative risk assessment.

24.3 TYPES OF MODELS

For the sake of description, one can classify models into four general categories — verbal or graphic, statistical, simulation, and analytical. The latter three kinds of models are mathematically formulated and are therefore by their nature more explicit in their logic and more readily accessible to analysis. These more explicit models offer potential for more detail and more precise probabilities. As with experiments, disputes about such models can be handled in the literature and peer review, which is a bit more difficult to do with verbal models because the underlying assumptions are often implicit or obscure. Even for mathematical models, models with fatal flaws have been published (see (22) for such a history). As noted in Chapter 23, the mathematical models include explicit and rigorous logic statements with explicit questions.

24.3.1 VERBAL MODELS

Verbal models, which are often illustrated with graphs, diagrams, or cartoons, have a lot of value in their simplicity. They can be presented to general audiences and do not require the time or effort needed to explore other types of models. Conversely, the logic underlying such models is not as explicit as for other types of models. They accept uncertainty without trying to define it explicitly. The models are not at all quantitative, so they will be of little assistance in calculating risk. An example of a verbal model could be given for the use of pesticide mixtures as a resistance management tool, where a common argument would be that individuals that are resistant to one pesticide would be killed by the other. This verbal model is almost too seductive in its simplicity, as will be discussed below.

Similarly, it was once argued that a useful insecticide resistance management strategy could be mosaics, where one pattern of areas is treated with one insecticide and the remaining areas with the other insecticide. By this line of argument, the area not treated with Insecticide A serves as a refuge to generate susceptible insects to dilute resistance in areas treated with Insecticide B. It is even possible to draw convincing diagrams for such a strategy (2). However, more detailed modeling suggested otherwise (15): a refuge only works if it generates a significant density of insects, which it cannot do so if another insecticide is controlling them. Mosaics can allow selection for both insecticides to evolve simultaneously if there is free movement between the patches, as has been shown in experiments with insecticides (16) and transgenic plants (30).

24.3.2 Statistical Models

Statistical models are often simple, but allow one to draw conclusions, with well-established probabilities of error, about datasets from either experiments or observations. Regressions and correlations are often used in statistical models.

Watkinson and colleagues (29) provide an example of how statistical models can be used in their predictions of impacts on biodiversity of herbicide-resistant crops. They considered what could happen if weed populations were reduced to lower densities through extensive use of crops that allowed better weed control. Their index for biodiversity was the effects on the local use of fields by birds that eat weed seeds. Although Watkinson et al. (29) used a simulation model to examine the potential effects on one weed, *Chenopodium album,* they used a statistical regression model based on observations of field use by skylarks to assess the effect on bird populations. In the data set, the relationship between birds and weed seeds is $y = 0.14 + 0.0002x$, where y is the density of birds per hectare and x is the weed seed density on the soil per square meter. Across a range of reasonable seed densities of 100 to 1000 per square meter, this is a change in bird density from 0.16 to 0.34 skylarks per hectare. Not until seed densities reach about 10,000 per square meter do bird aggregations exceed 2 skylarks per hectare, but such weed seed densities would surely be problematic for weed control in the next year.

In a "Perspective" article, Firbank and Forcella (5) wrote that Watkinson et al. (29) provided a "welcome conceptual framework," but that further work will be necessary to determine the effects of herbicide-resistant crops on farmland biodiversity. The model is an interesting framework, but it only measured bird aggregations and still does not tell us what regulators really would like to know. Birds would surely prefer higher weed seed densities, but do birds fail to reproduce normally at low weed seed densities, densities that would be compatible with farming practice, whether herbicide resistant or not? If birds do suffer at weed seed densities of less than 100 seeds per square meter, perhaps nature reserves are needed outside farmed fields to help ensure enough seeds. In sum, Watkinson et al. (29) provided data in a sound statistical model, but what is still needed is an experiment on real effects of better weed control on bird densities, and alternatives for provision of food. This is not an entirely speculative comment; a study of Dutch agrienvironmental schemes found that delays in spring mowing, intended to protect nestlings, showed no positive effects on plant and bird species diversity (8).

24.3.3 Simulation Models

Simulation models essentially make initial assumptions and "run the numbers" to project the outcomes, as Watkinson and colleagues (29) did for *Chenopodium album*. Simulation models are often pervasive in biology, even when we do not think of them as such. For example, Sukopp et al. (Chapter 4) implicitly refer to a simulation when they describe a population increase of 10-fold per generation from 7 weed beet plants to 70,000. This is a model with implicit assumptions, such as that there are no density-dependent effects (such as plant competition or increased herbivory at higher beet densities) that could limit population growth. Although simulation models such as these existed long before computers, computers allowed them to flourish by reducing the tedium of calculations. Even compared to the analytical models described below, simulation allows the modeling of complex systems. Simulation models can include *stochasticity* (random chance or Monte Carlo effects) by combining probability theory and numerical analysis. Data are still critical to make the estimates of variability or chance. Even without stochastic modeling, one can use interval estimates.

Vidotto and Ferrero (Chapter 21) have illustrated the use of simulation models, showed how to explore system behavior, and identified gaps in understanding. This model also showed limits of data collection (more information was needed on seed predation rate) and the need for "sensitivity analyses" when data are limited. *Sensitivity analysis* means that one tries the whole range of plausible values for each parameter, as delimited by existing data and biological realities, in all possible combinations, to elucidate the factors to which the system seems most sensitive. Sensitivity analyses often reveal that small mistakes or variations in the assumptions to the models can vastly change the results. Some of the sharpest examples of this are assumptions about the survival of heterozygotes (dominance) in pesticide resistance, where a small increase in mortality (say from 80 to 95% or 95 to 99%) can make an enormous impact on the time until resistance evolves (11,19).

In addition to the critical role that data play in informing the model construction, and the critical role of experiments in testing the model, it is important to design the model in a way that is consistent with data collection capabilities. It is often possible to design simple ways of modeling processes, but the data to test these assumptions can prove to be extremely difficult to collect in the field. For example, insect dispersal and subsequent egg deposition or pollen flow can be modeled as a process based on individual dispersal patterns of gravid females or pollinators, but these are difficult to measure. It may be more useful to model dispersal via genetic markers, which can be more easily measured.

Returning to the example of insecticide mixtures, a verbal model ("individuals that are resistant to one pesticide would be killed by the other") seems quite attractive, but lacks any detail as to how and when mixtures are likely to work well. Mixtures of insecticides are not supported as a general solution to resistance by experiments (27) or models (4). For a pesticide or toxin mixture strategy to work where both of the toxicants are at significant risk for the evolution of resistance, each of the toxicants must be used at doses that effectively kill completely susceptible individuals twice, or else some resistant individuals will not be controlled (4,7). Models suggest that mixtures are effective only if the mortalities of susceptible insects are high (greater than 95%) when exposed to each individual toxicant, consistent with the failures in experiments (15–21). Mixtures of insecticides are not promising due to incomplete coverage of the treated habitat (i.e., few insecticides provide even 95% control) and residue decay (as pesticides break down, susceptible pests that have dispersed into the habitat or emerged from protected sites will not always be killed). In contrast, transgenic plants that pyramid two Bt genes look promising in models (20) and experiments (30).

Tests of models for management of resistance to Bt transgenic crops provide some hope for optimism on the interaction of simulation models and experiments to generate useful recommendations and regulation for the management of transgene escape into feral plant populations. Many of the predictions of the resistance models, such as the importance of refuges (9) and problems with mosaics, have been tested, as discussed above. Other predictions have also been met, at least in terms of population dynamics. For example, Roush (see Figure 9 in (19)) predicted that resistance in the pink bollworm to Bt cotton in Arizona would take at least 20 generations to evolve, even if the proportion of non-transgenic cotton was only 20%, but that the population of pink bollworm would begin to decline because it would be unable to replace itself. In fact, pink bollworm populations have begun to decline after adoption of about 65% Bt cotton (i.e., when 35% was non-transgenic), and there are still no resistance problems after about 24 generations (6 years) of use (3).

24.3.4 Analytical Models

Analytical models, where they can be devised, are in many ways the most elegant of models, because one can explore the underlying relationship of factors by solving or inspecting the equations used. They more easily allow exploration of the "big picture" and identification of what are likely to be key elements of data (11,22). Perhaps the best-known solution of all time to an analytical model is the equation $E = mc^2$ from which it is easy to see at a glance that the constant (c) (speed of light), because it is squared, has a greater impact on energy production per unit than does mass (m).

Although pesticide resistance can be modeled analytically (11), most models of resistance are based on simulations due to the complexity of factors that are needed for a reasonably realistic model. Still, one simple rule that emerges from the analytical models (and simulation models) is that the extent of survival of toxicant exposure by heterozygotes carrying one resistant and one susceptible allele is critical to the rate at which resistance evolves (11). Because the initial frequency of resistance is lower than can usually be found by sampling (1) and, consequently, heterozygotes cannot be found for further study until after resistance evolves, estimates of these parameters are not reliable and predictions of the number of generations until resistance evolves will continue to be uncertain.

24.4 CONCLUSIONS — WHAT MODELS CAN DO FOR REGULATION AND RESEARCH ON FERALITY

Models have helped considerably to guide research in biology generally and in regulation, perhaps mostly by challenging ideas or proposals that need further study. Models are a poor substitute for experiments, but must often be used when experimental data are unavailable. Models coupled to experimental and observational data can help guide research and provide regulators with some assistance in quantitative risk assessment.

Given the large uncertainties about many aspects of ferality, such as the number of genes involved and evolutionary histories, there are limits to what modeling can promise. Given the existing theory on "selective sweeps" of genes that are under selection (6), modeling could help with the design of experiments to explore potential impact on genetic variation of introgression of transgenes into non-transgenic plant populations. Models could also help to explore the management of crops and gene containment strategies. In time, models may also help to explore the evolution of ferality itself.

LITERATURE CITED

1. Ahmad M. 1999. Initial frequencies of alleles for resistance to *Bacillus thuringiensis* toxins in field populations. PhD thesis. University of Adelaide. 215 pp.
2. Byford, RL, Lockwood JA, Sparks TC. 1987. A novel resistance management strategy for horn flies (Diptera: Muscidae). *J. Econ. Entomol.* 80:291–296.
3. Carriere Y, Ellers-Kirk C, Sisterson M, Antilla L, Whitlow M, Dennehy TJ, Tabashnik BE. 2003. Long-term regional suppression of pink bollworm by *Bacillus thuringiensis* cotton. *Proc. Natl. Acad. Sci. USA* 100:1519–1523.
4. Comins, H. 1986. Tactics for resistance management using multiple pesticides. *Agric. Ecosys. Environ.* 16:129–148.
5. Firbank LG, Forcella F. 2000. Genetically modified crops and farmland biodiversity. *Science* 289:1481–1482.
6. Gepts P, Papa R. 2003. Possible effects of (trans)gene flow from crops on the genetic diversity from landraces and wild relatives. *Environ. Biosafety Res.* 2:89–103.

7. Gould F. 1986. Simulation models for predicting durability of insect-resistant germplasm: a deterministic diploid, two locus model. *Environ. Entomol.* 15:1–10.
8. Kleijn D, Berendse F, Smit R, Gilissen N . 2001 Agri-environment schemes do not effectively protect biodiversity in Dutch agricultural landscapes. *Nature* 413:723–725.
9. Liu YB, Tabashnik BE. 1997. Experimental evidence that refuges delay insect adaptation to *Bacillus thuringiensis* toxin Cry1C in diamondback moth. *Appl. Environ. Microbiol.* 63:2218–2223.
10. Mallet J, Porter P. 1992 Preventing insect adaptation to insect-resistant crops: are seed mixtures or refugia the best strategy? *Proc. R. Soc. London B* 250:165–169.
11. May RM, Dobson AP. 1986. Population dynamics and the rate of evolution of pesticide resistance. In *Pesticide resistance: strategies and tactics for management*, National Research Council, pp. 170–193. Washington, D.C.: National Academy Press.
12. Pimentel DS, Raven PA. 2000. Bt corn pollen impacts on nontarget Lepidoptera: assessment of effects in nature. *Proc. Natl. Acad. Sci. USA* 97:8198–9913.
13. Rieger MA, Potter TD, Preston C, Powles SB. 2001. Hybridisation between *Brassica napus* L. and *Raphanus raphanistrum* L. under agronomic field conditions. *Theor. Appl. Genet.* 103:555–560.
14. Rieger MA, Lamond M, Preston C, Powles SB, Roush RT. 2002. Pollen-mediated movement of herbicide resistance between commercial canola fields. *Science* 296:2386–2388.
15. Roush RT. 1989. Designing resistance management programs: How can you choose? *Pestic. Sci.* 26:423–441.
16. Roush RT. 1993. Occurrence, genetics and management of insecticide resistance. *Parasitol. Today* 9:174–179.
17. Roush RT. 1994. Managing pests and their resistance to *Bacillus thuringiensis*: can transgenic crops be better than sprays? *Biocontrol Sci. Technol.* 4:501–516.
18. Roush RT. 1997. Bt-transgenic crops: just another pretty insecticide or a chance for a new start in resistance management? *Pestic. Sci.* 51:328–334.
19. Roush RT. 1997. Managing resistance to transgenic crops. In *Advances in insect control: the role of transgenic plants*, Carozzi N, Koziel M, Eds., pp. 271–294, London: Taylor and Francis.
20. Roush RT. 1998. Two toxin strategies for management of insecticidal transgenic crops: Can pyramiding succeed where pesticide mixtures have not? *Phil. Trans. R. Soc. London B* 353:1777–1786.
21. Roush RT. 2003. Resistance management strategies: have models helped? CAST Pest Resistance Management Symposium: Management of Pest Resistance: Strategies Using Crop Management, Biotechnology and Pesticides, Indianapolis, IN, April 10 and 11, 2003. www.pestmanagement.info/RMWorkshop.
22. Roush RT, Schneider JC. 1985. An analytical model for genetic control of *Heliothis virescens* incorporating the effects of sterile males. *Theor. Appl. Genet.* 71:472–477.
23. Sears MK, Hellmich RL, Siegfried BD, Pleasants JM, Stanley-Horn DE, Oberhauser KS, Dively GP. 2001. Impact of *Bt* corn pollen on monarch butterfly populations: a risk assessment. *Proc. Natl. Acad. Sci. USA* 98:11937–11942.
24. Shelton AM, Sears MK. 2001. The monarch controversy: scientific interpretations and public relations. *Plant J.* 27:483–488.
25. Shelton AM, Roush RT. 1999. False reports and the ears of men. *Nature Biotechnol.*17:832.
26. Tang JD, Collins HL, Metz TD, Earle ED, Zhao JZ, Roush RT, Shelton AM. 2001. Greenhouse tests on resistance management of Bt transgenic plants using refuge strategies. *J. Econ. Entomol.* 94:240–247.
27. Tabashnik BE. 1989. Managing resistance with multiple pesticide tactics: theory, evidence, and recommendations. *J. Econ. Entomol.* 82:1263–1269.
28. Tabashnik, BE. 1994. Delaying insect adaptation to transgenic plants: seed mixtures and refugia reconsidered. *Proc. R. Soc. London B* 255:7–12.
29. Watkinson AR, Freckleton RP, Robinson RA, Sutherland WJ. 2000. Predictions of biodiversity response to genetically modified herbicide-tolerant crops. *Science* 289:1554–1556.
30. Zhao JZ, Cao J, Li Y, Collins HL, Roush RT, Earle ED, Shelton AM. 2003. Transgenic plants expressing two *Bacillus thuringiensis* toxins delay insect resistance evolution. *Nature Biotechnol.* 21:1493–1497.

25 Epilogue

Ervin Balázs

25.1 GOOD AGRICULTURAL PRACTICE

The growing concerns in some of the societal groups over the transgenic technology in the agricultural business motivated the organization of the workshop on "Crop Ferality and Volunteerism." Volunteerism and ferality are known phenomena in agriculture, and with good agricultural practices, their negative impact on productivity and on the ecosystems could be significantly minimized. Both have attracted more attention today in the dawn of the transgenic era. The fears of gene flow from transgenic plants to wild among some ecologists and mostly among different non-governmental activist organizations supported the idea to answer for these anxieties based on our best scientific knowledge. The main constraint is that until the transgenic era it was hard to follow properly. Due to the strict regulation, it is also almost impossible to obtain permission to study the transgenic gene flow in field conditions, and to obtain statistically significant data for risk assessment and risk management. This is a Catch 22.

Plant biologists and breeders were well prepared for the gene flow from cultivars to wild relatives and vice versa as well as among cultivars and breeding lines. Gene flow can and does happen and can be easily observed to have no significant impact on both natural and agricultural ecosystems. Gene flow is the basis for the regulation of the certified seed production in different countries. Genetic protection distances for seed production of different crops were determined based on the pollination circumstances of the given species (e.g., insect-pollination, wind-pollination, self-pollination, cross-pollination). It is also known that a change in a single gene could alter the plant pollination characteristics. A good example for this is the petunia where the color change by a single gene of the flower will change the species of pollinating insect. As transgene flow can happen mainly via seed and pollen depending on these insects, their importance in volunteerism and ferality should be considered, respectively.

25.2 VOLUNTEERISM

Quite often when people visit agricultural fields in early summer they find volunteer plants in plots. Usually they are spectacular for laymen to see flowering oilseed rape or poppy in wheat field or sunflower among soybeans. These events happen when either the previous harvest was not carefully done, or preemergence herbicide treatment was not properly performed. If good agricultural practice was followed, volunteer crops will not be seen, as such no gene flow occurs via seed dispersal. In the case of transgenic crops, no trace amount of transgene product will contaminate the following harvest. The above-summarized facts could be concluded from the chapters on transgenic oilseed rape and on rice culture from three different regions of the world. Even if the transgenic seed dispersed, they will disappear stepwise as long as those transgenes are not providing selective advantages over wild relatives or in natural habitats.

25.3 FERALITY

Agriculture may be the oldest scientific discipline of mankind. Humans that settled in different regions of the world started the cultivation of crops and always selected for some easily recognized characters such as good taste of the fruit or higher yield. These selection efforts were continuously followed and developed. Plant breeding was developed by gaining experience. This discipline became an independent science after the pioneering work of Darwin and Mendel. The real boost in plant breeding started in the last century, when hybrid plants were developed for large-scale crop production. Maize is a good example: the size of the cobs increased dramatically over 40 years, increasing productivity. Introducing genetic engineering to practical plant breeding in the last decades of the 20th century revolutionized this activity. Precision plant breeding became a powerful tool for crop improvement and accelerated its development. In short, domestication of plant species contrary to the involvement of high technology is still a long and laborious process, yet a cultivar can rapidly evolve to wild/feral form. If breeders abandoned their selection lines or cultivars, a few years later they would not recognize them due dedomestication.

It is also worth mentioning that the fitness of cultivated forms to exist outside cultivation is far less than their wild or weedy relatives, which indicates that the transgenic plants are also less competitive outside agroecosystems. Wild relatives of the cultivated crops are usually resistant against different diseases and pests, which also underline the fact that these new traits introduced into the cultivars will not usually provide a selective advantage to wild plants, if the transgenes escape into the natural ecosystems by cross-pollination.

Both ferality and volunteerism are important in daily agriculture regardless of the technology used — organic farming, conventional agriculture, or transgenic technology. In all three types of agriculture, the impact of their feral and volunteer plants could be significantly eliminated by good agricultural practice.

Index

A

Aaronsohn, 40
Abandoned crops, 5
Abandoned paddies, 295–302
Abati-uaupé, 305
Abscission, 232
Acacia spp., 102
Acceptable risks, 399
Acer negundo, 104
Acer platanoides, 101, 106
Acetolactate synthase (*see also* ALS), 93
Acetyl CoA carboxylase (ACCase), 384
Acetyl CoA carboxylase resistance, 93
2-Acetyl-1-pyrroline, 347
Adventive floristics, 100
Aegilops spp., 23,36
Aegilops cylindrica, 11
Aegilops ovata-Triticum durum complex, 35
Aegilops peregrina, 372
Aegilops tauschii, 168
Aegilops speltoides, 168
Aegilops squarrosa, 35
Aflatoxin, 153
AFLP (amplified fragment length polymorphism), 23, 236, 248, 311
Ageratum conyzoides, 297
Agrestals, 11
Agriophytes, 102
Ailanthus altissima, 108
Albugo candida, 62
Alfalfa, 243
Allelopathy, 129
Allergens, 372
Allohexaploids, 21
Allopolyploidy, 168
Allowance weedy rice, 285
ALS inhibitors, 15, 70
ALS resistance (*see also* imidazolinone resistance), 23
Ameliorating ferality, 24
Amidochlor, 284
Analytical models, 408
Ancient divergence, 189
Ancient gene flow, 93
Anemone nemorosa, 105
Antiquity, *Oryza*, 257–260
Apical dominance, 10, 21
Apiculus, 261
Apis mellifera, 154
Apomixis, 10
Arabidopsis thaliana, 62
Arachis, 18
Archaeobotanical methods, 32–35
Archaeobotanical studies, 38–41
Archaeology, wheat, 33
Archaeophytes, 101
Aroma chemicals, 347
Arrozon, 287
Arthurdendyus triangulatus=Artioposthia triangulata, 103
Arundinaceum, 124
Asian rice, 257–272
Assessment endpoint, 390, 392–395
Atrazine resistance, 373
Auxin biosynthesis, 384
Avena fatua, 72
Avena spp. — *see also* oats, 14, 19, 21,32, 231–234
Awnedness, 261, 285
Awning, 318
Azuki bean, 147, 235

B

Bacillus thuringiensis (*see also* Bt) genes, 152
Bacillus thuringiensis cry1Ac, 23
Backcrossing, rice, 344–345
Barley, 17
Barnase, 382, 384
Beans, 18
Beets, 22, 45–57
Beets, transgenic, 50–53
Beta patellaris X *B. macrocarpa*, 48
Beta vulgaris, 45–57
Bicolor, 124
Biennial crops, 381
Biennial radish, 195
BioBarCoding, 380
Biogeography, *Carthamus*, 245
Biosafety, 144, 236, 371–388
Bird density, 406
Bitter substances, 10
Black cherry, 108
Black hull, 261
Black locust, 97
Blackleg, 62
Blast resistance, 347
BNYVV, 52
Bollworm, cotton, 405, 407
Bolting beet gene "B", 48–50
Bolting, 371
Brachiaria mutica, 306
Brachyantha, 259
Branching (*see also* tillering), 21, 384
Brassica carinata, 243
Brassica juncea, 382
Brassica spp., 15, 18, 21–23,59–79, 378

Brown-black awned, 313
Bt cotton, 407
Bt maize, 153, 400,
Bt protein Cry9C, 161
Bt transgene, 225

C

California, 323
Campanula persicifolia, 105
Cardamine hirsuta, 103
Carpobrotus spp., 110
Carthamus spp., 242–247
Cassava, 19
Catch 22, 411
cDNA microarray, wheat, 168
Center of origin, 225
Ceratocystis ulmi, 103
Certified seed, 71, 272, 283, 285, 319
Chaetosorghum, 124
Chenopodium album, 406
Cherry laurel, 109
Chickpea, 19
Chinese kaoliangs, 125
Chloroplast transformation, 373
Cinnanonum gladuliferum, 109
Cirsium arvense, 72
Clades, rye, 187, 189
Claviceps africana, 128
Clethodim, 356
Cluster analysis, *Oryza* spp., 266, 313–314
Clustering, domestication traits, 264
Clustering, gene, 15
Clustering, olives, 234
Clusters, rye, 187
CMS — *see* cytoplasmic male sterility
CO_2, elevated, 109
Coadaptation, 11
Coconut, 19
Code, seed color traits, 312
Coexistence, 305–317
Coffee, 19
Cold tolerance, 63, 175
Colletotrichum graminicola, 128
Colombian weedy rice, 313–315
Competition, weedy rice, 261, 280–284
Competitive ability, 10
Consequences, unintended, 70–71
Containment, 158
Containment systems, 375
Contamination, genetic, 6
Contamination, rice seed, 285–286, 365
Convallaria majalis, 105
Convergence, 225
Corn — *see* maize
"Corn-grass", 31
Cornus saguinea, 100
Corylus avellana, 100, 101
Costa Rican weedy rice, 312–313
Costa Rican wild rices, 306–311
Cotton bollworm, 405, 407
Cottonseed, 18
Cover crop, rye, 181
Cowpea, 19, 234, 235
cpDNA, 317
Cranberry, 16
Crataegus hybrids, 101, 111
Crop ferality, 15–16
Crop mimicry, 15, 285
Crop seed contamination, 365
Crop to weed gene flow, 129–130
Crops, food, 17–20
Crop-weed hybridization, radish, 201–204
Crop-weed-wild complex, 11, 13
Cry1Ab, 248
Cryptic linkage, 263
Cryptic non-native species, 113
Cultivar abandonment, 287
Cultivar development, U.S. rice, 323–324
Cultivated rye, 180
Cultivated sorghums, 125
Cycloxydim, 284, 356
Cyperus rotundus, 297
Cytokinin biosynthesis, 384
Cytokinin oxidase, 382
Cytoplasmic genome targeting, 373
Cytoplasmic genomes, 262
Cytoplasmic male sterility (CMS), 45, 63, 197, 216, 223, 226

D

Dalapon, 283,356
Damage, weedy rice, 279–289
Dark hull, 318
Dedomestication, 1
Dedomestication, process, 14–15
Definitions, 1–2
Degree of domestication, 16–22
Delayed flowering, 384
Deleterious transposons, 383
Demographic swamping, 130, 221, 378–380
Dendrogram, beets, 47
Detection of plant ferality, 31–44
Determinate growth, 10, 83
Diagonal gene transfer, 372
Diatraea saccharalis, 248
Differential gene expression, 169–171
Dinitroanaline herbicides, 70
Dioscorea deltoidea, 13
Direct seeding, 271
Directionality, outcrossing, 343
Disarticularion, 232
Discontinuous dormancy, 12
Disease resistance, 63
Disequilibrium, linkage, 15, 21
Distribution soybean, 140
Distribution, *Secale,* 179
Diversification, rice, 264–267
Diversity, loss, 21

Diversity, weedy rice, 279–280
DNA diversity, 92
Dockage, 71
Dogs, 1
Domestication, 9–25
Domestication, degree, 16–22
Domestication, genetics, 12–15
Domestication, genetics, rice, 263
Domestication, process, 3
Domestication, radishes, 193–197
Domestication, rice, 260–264
Domestication, rye, 179–187
Domestication, safflower, 245
Domestication, *Setaria,* 82–86
Domestication, soybean, 140
Domestication syndrome, 10
Dominance, apical, 21
Dormancy, secondary, 10, 73,176, 264, 266
Dormancy, rice, 345–346
Downy mildew, 129, 217–218
Drought tolerance, 175
Drought tolerance, rice, 318
Drummondii, 124
Dutch elm disease, 103
Dwarfing gene, 376–378
Dynamics, population, 72

E

Early abscission, 199
Echinocloa spp., 306, 324
Echinops, 111
Ecogeographical distribution, 41
Ecological fitness, 144
Economic impacts, red rice, 325–326
Einkorn, 33, 39
Elaegnus, 109
Elaeis, 19
Eleusine indica, 18, 297
Elevated CO_2, 109
Elodia spp., 104
Elymus condensatus, 185
Emmer wheat, 33, 40, 167
Endoferal rye, 176, 181
Endoferality, defined, 5, 16
Endoferality, examples, 109–110, 287
Endoferality, radish, 198–200
Endpoints, 389, 394–395
Environmental risks, 389–401
Ergot, 129
Erucastrum gallicum, 61
Erucic acid, 62, 371, 394, 396
Escape from cultivation, 99
Escaped sunflower, 212–213
Ethalfluralin, 70
Ethametsulfuron, 70
Ethiopia, 131
Eurasian *Setaria,* 82
European weedy rice, 355
Evolution, weedy rice, 265–267
Evolutionary history, wheat, 168
Evolutionary trends, 267
Exoferal hybrids, 37–38
Exoferal origin, rice, 318
Exoferal rye, 176, 181
Exoferality, defined, 5, 16
Exoferality, examples, 110–112
Exoferality, radish, 200–204
Exposure, 389, 395–397
Expressed sequence tags (ESTs), 169
Extruding stigma, rice, 346

F

Fagus sylvatica, 101
Fallopia spp., 104, 107, 110–111
Farmer attitudes, 286
Fatuoids, 14, 32, 232–233
Fecundity, 66
Federal Plant Protection Act (FPPA), 392
Federal Plant Quarantine Act (FPQA), 392
Feral animals, 1
Feral biodiversity, 5
Feral crop plants, 9
Feral radishes, 193–207
Feral rice, detection, 279–289
Feral rye, 175–192
Feral Tibetan wheat, 171–172
Ferality, 4
Ferality, detection, 31–44
Ferality, safflower, 246
Fertile Crescent, 40, 231
Fertility, 9
Fertility, *Lolium-Festuca*, 238–239
Festuca, 382
Field scale experiments, 396
Fitness, 24
Fitness, weed-crop hybrids, 203–205
Flax, 243
Flea beetles, 62
Flooding conditions, 362
Floret, 176
Floristics, 100
Flower ablation, 382
Flowering duration, *Setaria* spp., 83, 85–86
Food crops, 17–20
Forestry, 382
Forrageiro, 201
Foxes, 1
Foxtail millet, 81–96
Free-threshing grains, 167
Fumonisin, 153

G

Galanthus nivalis, 98,105
Gene clusters, rice, 263
Gene contamination, 6
Gene dispersal, 267–270

Gene establishment, 371–388
Gene flow, *Beta vulgaris*, 46, 51
Gene flow, rates, 311
Gene flow, *Setaria*, 89–90
Gene flow, *Sorghum*, 128
Gene flow, rice, 267, 305–319
Gene linkage, rice, 263
Gene pollution, 6
Gene pool composition, 99
Gene reservoirs, *Helianthus*, 212
Genetic analysis, rye, 186–188
Genetic assimilation, 221
Genetic distances, rice, 327
Genetic distances, rye, 187
Genetic divergence, rye, 188
Genetic diversity, 16
Genetic diversity, *Brassica,* 64
Genetic diversity, radish, 200
Genetic diversity, rice, 305–319
Genetic erosion, 131
Genetic linkage, 22–24
Genetic purity, 6
Genetic self-biocontrol, 380
Genetic similarity, beets, 47
Genetic use restriction, technology (GURT), 374, 375
Genetics of domestication, 12–15, 263
Genome groups, rice, 259
Genomics, wheat, 168
Genotype specific microsatellites, 314
Germination, 10, 84, 299
Giant green foxtail, 87
Giant hogweed, 102, 104, 107, 108
GIGANTEA gene, 197, 381
GISH, 92
Glucosinolates, 62, 371
Glufosinate, 68, 157, 327, 328
Glufosinate resistance, 243, 316, 405
Glufosinate-resistance, rice, 284, 330, 332–335, 340–345
Glume, 36
Glutinous gene, 269
Glutinous rice, 269
Glycine, 17
Glycine max, 137
Glycine spp. (list), 139
Glyphosate, 68, 157, 284, 328, 348, 356
Glyphosate resistance, 149, 152, 153
Goldenrod, 109
Good agricultural practice, 411
Gossypium, 18
Grain fill, rice, 298
Grain handling, 150
Grain marketing, 160–162
Grain production, world, 151
Grain size, 86
Grain sucking insects, 281
Graminicides, 157
Granulata Roschev, 259
Grape, 20
Grassy stunt virus, 272
Green bridges, 393
Greenbug, 129
Groundnut, 18
Growth rate, rice, 300

H

Habitats, natural, 31
Hardiness, winter, 63
Harmful volunteers, 393
Harvest index, 4
Hazard, 389
Hazard assessments, 395–397
Hazel, 99
Heat islands, 108
Helianthus spp., 15, 19, 209–230
Helianthus tuberosus, 104,110, 218–220
Helicoverpa armigera, 405
Heliothis virescens, 405
Hemizygous, 14
Heracleum mantegazzianum, 102, 104, 107, 108
Herbarium collections, 35, 36
Herbicide resistance, 23, 63, 376–378
Herbicide resistance, oilseed rape, 59, 68–70
Herbicide resistance, *Raphanus*, 201–204
Herbicide resistance, red rice, 327
Herbicide resistance, rice, 283–284, 311, 318, 328–330, 332–345, 349
Herbicide-resistant sugar beet, 53
Herbicide-resistant weedy biotypes, 15
Heterosorghum, 124
Hexaploid oats, 232
High oleic acid content (HOAC), 209–211
High stearate *Brassica napus,* 67
Historic records, 31–44
Holocene, 40
Hordeum, 17
Hordeum spontaneum, 40
Horizontal gene transfer, 372
Hull, 261
Hulled, rice, 313
Hulled wheat, 33, 38–41, 167
Hybrid ancestry, rye, 188
Hybrid expansion, 221
Hybrid swarms, 189, 267–270
Hybridization rates, 316
Hybrids, exoferal, 37–38
Hybrids, wheat, 36–37

I

Identity preservation, 157–159
Illegal soybean, 146
Imazethapyr, 327
Imazethapyr/imazamox, 68
IMI — *see* imidazolinone
Imidazolinone herbicides, 288
Imidazolinone resistance, rice, 284, 328–330, 336–337, 349, 365
Imidazolinone-resistant sunflower, 222
Immigration, 301

Impact, ecosystems, 31
Impact, weedy rice, 280–284
Impatiens spp., 104, 107
Incestuous relations, 90
Indehiscent fruits, 195
Indeterminacy, 12
Indica, 261
Indica-japonica evolution, 260, 266
Infant formulae, 371
Instrumentalism vs. realism, 391
Integrated weed management, rice, 319
Interfertility, 238
Intermediate domestication, 16
Internode, 36
Interspecific hybridization, 226, 244
Intertaxon hybridization, 130
Introductions to Europe, 98
Introgression, 73, 93, 311
Introgression, maize, 153–154
Introgression, *Sorghum,* 131–132
Introgression, soybean, 153–154
"Invades-elsewhere" criterion, 113
Invasions by ornamentals, 100–112
Invasive traits, 10
Ipomoea, 19
Isolation distances, 89
Isozyme variation, rice, 260

J

Japan, 270
Japanese honeysuckle, 25
Japanese privet, 25
Japonica, 261
Jerusalem artichoke, 110, 209, 218–220
Johnsongrass, 11, 123

K

Kaoliangs, 125
Karyotype fertility barrier, 239
Kev (Kevorkian) genes, 383
Knotweed, 107, 110–111
Kudzu, 25

L

Labeling transgenics, 158–159
Landraces, rice, 265
Laos, 267
Lapinus polyphyllus, 104
Latin America, 305
Laurophyllisation, 109
Laurus nobilis, 109
Lemma, 36
Leptosphaeria maculans, 62
Leucojum vernum, 105
Lignin, 384
Ligustrum, 25
Linkage disequilibrium, 15, 21
Linkage, genetic, 9–25
Linoleic acid, 209
Linseed, 393, 394
Linum usitatissium, 243
Location, genes, 264
Lodging resistance, 63
Lodging, rice, 261
Lolium spp., 14, 236–242, 382
Lolium-festuca complex, 236–238
Lonicera, 25, 109
Loss of diversity, 21
Louisiana, 323
Lysichiton americanus, 104, 108
Lythrum salicaria, 112

M

Mahonia aquifolium, 109, 110
Maize, 16, 17, 149–165
Maize production, 150–153
Malaysia, 271
Male fertility restoration, 203
Male sterility (*see also* cytoplasmic male sterility), 45, 373
Male sterility, rice, 331, 332, 346
Maleic hydrazide, 283
Management, weedy rice, 281–284
Manihot, 19
Maritime beets, 45–57
Marker-assisted breeding, 197
Masked hybrid, rye, 188
Maternal inheritance, 373
Medicago spp., 243
Methods, archaeobotanical, 32–35
Michels grass, 182, 183, 185, 186
Miconia calvescens, 102
Microarrays, 171
Microsatellite markers, 327
Microsatellite markers, rice, 315, 316
Migrational meltdown, 378
Millet, 18
Millets, little and kodo, 81
Mimicry, 15
Minimum tillage, 362
Mississippi, 323
Mitigating ferality, 376
Mitigation, 371–388
Mitochondrial DNA haplotypes, 194
Mitochondrial genome targeting, 373
Model validation, 361
Modeling gene flow, 221–223
Modeling, feral rice, 355–370
Models, types, 403–408
Molecular containment, 371–388
Molinate, 283
Monarch butterfly, 404
Monitoring, 399
Monoculm 1 mutant, 299
Morphometrics, scattergrams, 37–38
Morphotypes, rice, 312

Mount Hermon, 40
Mountain rye, 177–175, 180–181, 183
mtDNA, 317
Multicopy transposons, 380, 383
Multidimensional scaling (MDS), 327
Multilocus genotype frequencies, 187
Multilocus systems, 272
Multiple correspondence analyses, 315
Multiple origins of radish, 194
Multivariate analysis, 88
Multivariate analysis, rice, 312
Mungbean, 234
Myanmar, 267

N

Narcissus, 111
Natural habitats, 31
Natural selection, 14
Naturalized feral rye, 186
Naturalized populations, rye, 184
Neolithic sites, 35
Neophytes, 101
Net assimilatory rate, 300
New Zealand flatworm, 103
Nicotiana tabacum, 243, 376–377
Non-dormant seeds, 167
Non-regulated status, 392
Non-shattering, 9
Non-splitting fruit, 195
No-tillage, 348, 362

O

Oats, 19, 21, 231–234
Odds of dedomestication, 171
Oebalus insularis, 279
Oil content, 63
Oil palm, 19
Oilseed rape, *see also Brassica*, 21, 59–79, 376, 378, 405
Olea europaea, 233
Olea, 19
Oleaster, 233
Oleic acid, 209
Oleosin, 243
Olive, 4, 19, 233–234, 382
Olive oil, 209
Ophiostoma ulmi, 103
Optical sorting, 325
Origin, wild sunflower, 213
Ornamental snow drop, 98
Ornamental trees, 382
Ornithogalum nutans, 105
Orobanche, 211
Oryza, 15, 17, 21
Oryza eichingeri, 296
Oryza genome, 258
Oryza glumaepatula, 305–307
Oryza grandiglumis, 305, 307–309
Oryza granulate, 296
Oryza latifolia, 281, 306–311
Oryza nivara, 296
Oryza rhozomatis, 296
Oryza rufipogon, 296, 331, 338–340, 342, 349
Oryza sativa complex, 259
Oryza taxonomy, 259
Outcrossing, *Oryza* spp., 330–343
Overcoming ferality, 371–372
Oxadiargyl, 283
Oxadiazon, 283

P

Padi angina, 270, 282
Palea, 36
Palm, 109
Panicle, 300
Panicles, rice, 299
Panicum sumatrense, 81
Parasorghum, 124
Parthenocarpy, 382
Paspalum scrobiculatum, 81
Pasture grasses, 382–385
PCR — *see* polymerase chain reaction
Peas, vining, 393
Pennisetum, 18
Perennial habit, 177
Perennial rice, 258, 306
Pericarp pigmentation, 83, 316
Periwinkle, 105
Persistence, crop alleles, 203–205
Persistence of volunteers, 66–68
Pervenets mutation, 211
Pesticide mixtures, 404
Pesticide risk, 389
Pharmaceuticals, plant made, 242–247, 380–381
Phaseolus, 18
Phenological characterization, 313
Phenology, 176
Phenology, rice, 312
Phenology, rye, 185, 188
Phylogeography *Lolium* vs. *Festuca*, 237–238
Phytolith analysis, 34–35
Phytoremidiation, 381–382
Pieris rapae, 62
Pigs, 1
Pinus spp., 104, 112
Plant hopper, rice, 306
Plant-made pharmaceuticals (PMPS), 242–247, 380–381
Plasmopara halstedii, 217–218
Plasticity, 10, 12
Plastid inherited genes, 373
Plastome-encoded genes, 373
Plowing, 362
Policeman's helmet, 107
Pollen analysis, 34
Pollen flow, 158–160, 240–242, 404

Pollinators, 10
Pollution, genetic, 6
Polploidy, 10
Polygonum convolvulus, 72
Polymerase chain reaction (PCR), 327
Polymerase chain reaction, rice, 317
Peanut, 18
Polymorphic markers, 314
Polymorphisms, 92
Polyphenoloxidases, 84
Polyploid, 11
Polyploid foxtail millet, 90
Poplars, 107
Population dynamics, 72
Population dynamics, weedy rice, 356–358
Population meltdown, 221
Populus spp., 104, 107, 382
Portulaca oleraceae, 297
Potato, 19
Predation, 361–362, 366
Predation, rice seed, 358
Predicted environmental concentration (PEC), 390
Predicting invasions, 112–113
Premature bolting, 46
Preventing gene flow, 371–372
Primary gene pool, rice, 258
Principal coordinate analysis, 313
Prunus spp., 101, 104, 108, 109
Pseudotsuga menziesii, 104
Psylliodes chrysocephala, 62
Puddling, 283, 356
Pueraria, 25
Purity, genetic, 6
Purple loosestrife, 112
Purple marker rice, 340, 344
Pyricularia grisea, 347

Q

QTL (quantitative trait loci), 13, 14, 22, 248, 264
QTL, rice flowering, 314
QTL, rice shattering, 344–345
Quercus rubra, 104

R

Rachis, 36
Radish, 22, 193–207
Random amplified polymorphic DNA (RAPD)92, 172, 209, 218, 231, 248, 311
Rapeseed, 18
Rapeseed oil, 62
Raphanus spp., 61, 193–207
Realism vs. instrumentalism, 391
Recoverable block of function, 374
Recurrent gene flow, 130, 234, 379
Red beet, 45
Red pericarp, 261, 270, 346
Red rice, dilemma/economics, 324–328
Reduced tillage, 326
Regulation, 403–408
Regulatory obligations, 392
Related traits, 264
Replacement rates, 376
Repressible seed lethal technologies, 374
Reproductive capacity, 361
Reproductive fitness, 63, 376–378
Reproductive traits, 346
Resistance, herbicide — *see* herbicide resistance and particular herbicide
Restoration, male fertility, 203
Restriction fragment length polymorphism (RFLP), 195, 248
Rhamnus frangula, 107
Reverse transcription, 169
RHBV (rice joja blanca virus), 306, 313
RHBV-resistance transgenic rice, 317
Rhizomania, 52
Rhododendron spp., 111–112
Rice, 17, 22, 23
Rice AA genome, 258
Rice bean, 235
Rice blast, 306
Rice diversity, 266
Rice domestication, 260–264
Rice evolution, 257–272
Rice production U.S.A., 324
Rice weeds, 22
Rice X red rice, 343–345
Ridleyanae, 259
Risk, 395–397
Risk assessment, 389–401
Risk of crop-allele escape, 223
RNAi, 384
Robinia pseudoacacia, 97, 104
Root crops, 381
Rope-wick applicator, 284
Rosa spp., 104, 111
Rotation crops, 282
Rotations, pesticides, 404
RuBISCO protein, 60
Ruderal areas, 73
Ruderal plants, 11
Rye, 19, 175–192
Rye cultivars, 183
Rye domestication, 179–187
Rye production, 182
Ryegrass complex, 237–242

S

Saccharum, 19, 21, 247–248
Safflower, 242–247
Salinization, 40
Scald, leaf, 306
Scatter plot, 268–269
Schizaphis graminum, 129

Sclerotinia-resistant sunflower, 211, 218
Sea beets — *see* maritime beets
Secale, 19, 175–192
Secondary dormancy, 10, 73, 284, 357, 384
Secondary dormancy, rice, 263, 298
Seed bank, 71, 67, 198, 283, 301
Seed bank assessment, 361–366
Seed bank, *Helianthus*, 210
Seed bank, rice, 358
Seed characteristics, rice, 298
Seed color coding, 312
Seed dispersal, 301
Seed dispersal, rice, 358
Seed distribution after tillage, 359
Seed dormancy, 10, 64, 83
Seed losses, 366
Seed movement, rice, 358
Seed predation, rice, 358
Seed shedding (shattering), 83, 85, 232, 284, 285
Seed viability, 299
Seedling emergence, rice, 357
Segmented capsules, 199
Selective sweeps, 408
Self-compatibility/fertility, 9, 12
Self-incompatibility, *Lolium-Festuca*, 238–239
Self-thinning, 379
Semi-cultivated soybean, 140
Semi-wild soybean, 140
Semi-wild Tibetan wheat, 171–172
Sensitivity analysis, 407
Sensitivity analysis, model, 361–363
Set aside areas, 73
Setaria spp., 81–96, 373
Setaria viridis ssp. *pycnocoma,* 87
Setaria faberii, 91
Setaria italica, 12
Setaria pumila, 92
Setaria verticillata, 90–91
Setaria viridis, 72
Setaria viridis var. major, 87
Setaria viridis X *S. italica*, 83
Sethoxydim, 284
Sexually compatible weeds, 17–20
Sh3 shattering gene, 285
Shade recognition, 384
Shattercanes, 127
Shattering, 63, 176, 264, 266, 272, 282–284, 285, 300, 301, 318, 362, 371
shattering capitulum, 245
Shattering, rice, 257, 262, 345–346
Shattering, rye, 179
Shatterproof genes, 62
Sheath blight, 306
Shrunken seed, 381
Silique morphology, 196
Simple sequence repeat (SSR), 211, 327
Simsia, 212
Simulation models, 363–366, 406
Sinapis spp., 61, 72, 243
Skunk cabbage, 108
Skylarks, 406
Soja, 138
Solanum, 19
Solidago spp., 14, 17, 21, 104, 109
Sorghum almum, 128
Sorghum bicolor, 123
Sorghum bicolor ssp. *drummondii*, 127
Sorghum gene pool, 124
Sorghum halepense, 123
Sorghum introgression, 131–132
Sorghum propinquum, 123, 127
Sorghum sudanensis, 132–128
Sorghum taxa, 124
Sorghum, 123–135
Soybean, 17, 137–147, 149–165
Soybean distribution, 140
Soybean domestication, 140
Soybean gene pool, 143
Soybean production, 150–153
Soybean, transgenic, 143–144
Spartina spp., 107, 382
Species diversity, rice, 258
Spike fragility, 36, 167, 176, 177
Spikelet, 36, 261
Spikelet shedding, 232
Spikelets, rice, 298
Spirea X *billardii*, 101
Spongy seedpods, 195
Spotted grain, 306
Spread, *Sorghum*, 126
Sri Lanka, 295–302
SSR markers, 226
SSR markers, *Helianthus*, 220
SSR, soybeans, 143
Stacking of resistance, 71
Staggered germination, 200
Stale seedbed, 283, 356, 360, 362–365
StarLink, 157, 161
Statistical models, 406
Stearate, high, 67
Stem borer, 129
Stem borer, rice, 306
Sterility, male, *see also* cytoplasmic male sterility, 45
Sticky rice, 265
Stiposorghum, 124
Striga, 129
Subtractive cDNA libraries, 169
Sudan, 131
Sudangrass, 132
Sugar beets, 45
Sugarcane, 19, 21, 247–248
Sunflower, 13,19, 209–230
Sunflower oil, 209
Sunflower volunteers, 210
Sunflower X Jerusalem artichoke, 219
Suppression subtraction hybridization (SSH), 169
Sweet, 19
Swiss chard, 45
Swollen root, 199
Symphoricarpus racemosa, 104
Synchronized flowering, 331
Synchrony, flowering, 10

Syndromes, domestication, weedy, wild, 10
Synteny, 14

T

Tac-Tics (Transposons with armed cassettes for targeted insect control), 382–385
Tandem mitigation, 24
Tapetum, 382
Tapetum-specific promoter, 382
Taxonomy, rye/*Secale*, 176–178
Taxus baccata, 101, 107
Temperate rice, 323–353
Tens rule, 100
Teosinte, 16, 21, 153
Terminator genes, 376
Texas, 323
Thailand, 267, 269
Thermophilic ornamentals, 109
Thiobencarb, 283
Thlaspi arvense, 72, 381–382
Thrips tabaci, 154
Tiered testing, 390–391, 398
Tillage practices, 69
Tillering, 83, 85–86
Tillering, rice, 261, 298
Tithonia, 212
Tobacco — *see Nicotiana tabacum*
Tocopherol, 211
Toxicity exposure ratio (TER), 390
Trachycarpus fortunei, 109
Trait differences, rice, 261
Traits, invasive, 10
Transgene containment, sunflower, 223
Transgene flow, soybean, 142
Transgenes, 22–24
Transgenic beets, 50–53
Transgenic biosafety, 371–388
Transgenic mitigation, 144, 374–382
Transgenic sorghum, 132–133
Transgenic soybean, 143–144
Transgenic wheat, 397
Transition matrix, 359–360
Transplastomic traits, 373
Triazine resistance, 15
Triazine-resistant oilseed rape, 68
Triazine-resistant weeds, 349
Trifluralin, 70
Trifluralin resistance, 93
Trifolium angustifolium, 243
Trigger value, 389, 390
Triploid beet, 50
Triticum, 17
Triticum aestivum, 35
Triticum aestivum spelta, 372
Triticum boeoticum, 39
Triticum turgidum sub-species, 167–168
Tropical America, 305–319
Tulipa sylvestris, 105

U

U.S. Environmental Protection Agency (EPA), 392
U.S. rice industry, 323
Unacceptable risk, 395
Undetected weedy rice, 286–287
Uniform germination, 84–86
Unintended consequences, 70–71
Urban ornamentals, 97–121
Urban-rural gradients, 105–107
USDA-APHIS, 392

V

Vaccinium spp., 16, 104, 109
Varietal diversity, rice, 265
"Varietal" weedy rices, 285
Vegetatively propagated crops, 382
Verbal model, 405, 407
Verbesina, 212
Vernalization, 46–50
Vertical gene transfer, 372
Vietnam, 264–265
Vigna spp., 19, 147, 234–236
Vinca minor, 105
Vining peas, 393
Virus-resistant beet, 52
Vitis, 20
Volunteer contamination, 393
Volunteer density, 69
Volunteer frequency, 68
Volunteer maize, 154–156
Volunteer rye, 184
Volunteer seeding, 271
Volunteer soybean, 154
Volunteer sunflower, 210

W

Water tillage, 326
Water weevil, rice, 306
Weed beet, 46
Weed distribution, 17–20
Weed hybrids, 86–88
Weed importance, 17–20
Weed mimics, 15
Weediness characteristics, 12
Weeds, 9–25
Weedy *Raphanus raphanistrum*, 197–200
Weedy rice, 270–271, 311–319
Weedy rice diversity, 279–280
Weedy rice life cycle, 359
Weedy rice, population dynamics, 356–358
Weedy rice, spread, 284–287
Weedy rye, 180–185
Weedy sorghums, 126–128
Weedy soybeans, 156
Wheat, 17, 167–173
Wheat genomics, 168

Wheat, hulled, 38–41
Wheat hybrids, 36–37
Wheat, morphological distinguishers, 36
Wheat, Tibetan, 4, 171–172
White clover, 243
White mustard, 243
White rust, 62
Wild annual *Helianthus* spp., 214
Wild genome, *Oryza* spp., 258
Wild *Helianthus*, 210
Wild oat — *see also Avena*, 72
Wild plants, 9–25
Wild progenitor, 16
Wild radish, 405
Wild rice evolution, 260
Wild sorghums, 126–128
Wild soybean, 157
Wild sunflower establishment, 213
Wild syndrome, 10
Wild-weed-crop complex, 89
Winter hardiness, 63
Wolves, 1
Woody siliques, 196, 199
World grain production, 151

Y

Yangtze Valley, 262
Yew, 107
Yield drag, 153

Z

Zea mays — *see also* maize, 16, 17